MICROBIAL
PHYSIOLOGY
& METABOLISM

MICROBIAL
PHYSIOLOGY
& METABOLISM

DANIEL R. CALDWELL
University of Wyoming

Wm. C. Brown Publishers
Dubuque, Iowa•Melbourne, Australia•Oxford, England

Book Team

Editor *Elizabeth M. Sievers*
Production Editor *Cathy Ford Smith*
Publishing Services Coordinator *Barbara J. Hodgson*
Permissions Coordinator *Lou Ann Wilson*
Art Processor *Brenda A. Ernzen*

Wm. C. Brown Publishers
A Division of Wm. C. Brown Communications, Inc.

Vice President and General Manager *Beverly Kolz*
Vice President, Publisher *Kevin Kane*
Vice President, Director of Sales and Marketing *Virginia S. Moffat*
Vice President, Director of Production *Colleen A. Yonda*
National Sales Manager *Douglas J. DiNardo*
Marketing Manager *Patrick Reidy*
Advertising Manager *Janelle Keeffer*
Production Editorial Manager *Renée Menne*
Publishing Services Manager *Karen J. Slaght*
Permissions/Records Manager *Connie Allendorf*

Wm. C. Brown Communications, Inc.

President and Chief Executive Officer *G. Franklin Lewis*
Corporate Senior Vice President, President of WCB Manufacturing *Roger Meyer*
Corporate Senior Vice President and Chief Financial Officer *Robert Chesterman*

About the cover
The image in the upper left corner is a photomicrograph of cyanobacteria
(×200). The image in the upper right corner is a false color-enhanced
scanning electron micrograph of *Staphylococcus* (×5000). The image
at the botton is a false color-enhanced scanning electron micrograph
of *Pseudomonas putida* (×3800).

Cover credit—Photos
© Manfred Kage/Peter Arnold, Inc.
© Alfred Pasieka/Peter Arnold, Inc.
© Alfred Pasieka/Peter Arnold, Inc.

Cover and interior design: Lesiak/Crampton Design, Inc.

A Times Mirror Company

Library of Congress Catalog Card Number: 93–73908

ISBN 0–697–17192–2

Printed in the United States of America by Wm. C. Brown Communications, Inc.,
2460 Kerper Boulevard, Dubuque, IA 52001

10 9 8 7 6 5 4 3 2 1

For Wanda

Contents

Preface

This book reflects twenty years of teaching microbial physiology and metabolism to advanced undergraduate and beginning graduate students. During that time, the subjects have been taught, on the one hand, as a combined course and, on the other hand, as two separate courses, physiology in the fall and metabolism in the spring. At the present time, the subjects are taught jointly over a year's time period.

It is difficult to teach physiology and metabolism in meaningful ways, because of the tremendous amount of information available in both subject areas and because it is often difficult to decide where physiology leaves off and metabolism begins. It is also often difficult to decide where physiology and metabolism leave off and the related sciences of nutrition, biochemistry, and genetics begin. These difficulties are confounded by the fact that in the microbial world, a particular structure may simultaneously serve both physical and biochemical functions. The membrane, for example, may serve as a regulator of permeability, a vehicle for attachment of the replicating genome, and as the entity within which both lipid synthesis and energy generation occur. Attempting to explain the manifold functions of the membrane to students in a meaningful way and to indicate the ways in which the activities of the membrane are related to the other activities of the cell are complicated and challenging tasks. Explaining the role of genetic phenomena in microbial physiology and metabolism is equally difficult. While the physiology and metabolism of microbes are not properly understood without substantial consideration of genetic phenomena and of the techniques of molecular biology, microbial physiology and metabolism are subjects whose understanding requires more

than *just* what may be understood from genetic considerations. There are critical cell molecules other than genetic molecules and expression of genetic potential is achieved by the coordinated operation of *all* of the cell's structures and molecules.

This book is written with the hope that it will be useful to a broad audience, including students whose primary focus is the science of microbiology and students for whom microbiology is not a primary focus. The book presumes that the student has taken only beginning courses in microbiology, chemistry, biochemistry, physics, and mathematics. The conscientious and reflective student, by careful supplementary reading, may understand this book without prior exposure to subjects other than microbiology and chemistry. The book reflects the belief that *all microbiology* and all the *critical processes of life* may be understood from the physiological perspective that includes, but is not restricted to, the genetic viewpoint. This book reflects a rhyme that I have found helpful to my students. The rhyme is as follows: "A rake and a hoe and old rusty plow, are *somewhat* alike, but different *somehow!*" The biological implications of the rhyme may be regarded as suggesting that, although they differ in form and function, things that are in some ways different may, at the same time, share common properties. With regard to physiology and metabolism, although they differ in form and function, and in the habitats in which they are found, all life forms share the necessity of accomplishing a common set of tasks so that life may continue. Taken collectively, the microbes, particularly the bacteria, display the greatest degree of physiological and metabolic diversity found in nature. However, in principle, if not in exquisite detail, what occurs in microbes also occurs in animals, flowers, and trees. Thus, through study of physiological and metabolic phenomena in microbes, we may understand the functioning of all cellular life forms.

Critical understanding of the general similarity and, at the same time, significant differences between the ways in which essential processes of cellular life are accomplished in different living forms allows us to exploit those differences. It is precisely our understanding of the overall similarity, and yet subtle differences, between the critical physiological processes of various life forms that allows us to treat bacterially-mediated disease so effectively with antimicrobial chemicals. Conversely, our relative lack of understanding of the subtle interconnections between the processes of viral replication and the formation of host cell components severely limits our present attempts at viral chemotherapy. As our knowledge of microbial physiology and metabolism progresses, it may some day be possible to selectively and effectively inhibit viral disease with chemicals, precisely as a result of a more sophisticated understanding of microbial physiology and metabolism than we currently possess.

No textbook is suitable for all purposes, but it is hoped that this book may serve many users, either as a text for an introductory course or for a more advanced one and that it is written in a sufficiently flexible manner so that instructors may select and emphasize those topics that are pertinent to their particular interests and needs. In addition, the diversity of subjects covered allows the student to appreciate the complexity and implications of microbial physiology and metabolism without the necessity of becoming an expert. Toward this end, some subjects are not covered in the detail that would be the case in other textbooks. The book, however, includes sufficient discussion of methods and problems so that the complexity of both microbial physiology and metabolism and their study is conveyed. The selected references at the end of each chapter place a heavy emphasis on review articles. This is deliberate and reflects the strongly held conviction that a person seeking understanding of a new field may be easily overwhelmed by detail to the extent that the broader picture is obscured. By study and reflection upon a collection of review articles written over an extended period of time, the reader may develop a perspective regarding the nature and development of a discipline while, at the same time, receiving guidance by which to seek out the intimate details. The particular review articles selected are designed to give the student both an understanding of the nature of the structure or process under consideration and a perspective regarding the broader implications of the subject at hand. The student or the instructor has the choice of seeking whatever level of understanding he or she desires. The author welcomes comments and constructive suggestions regarding ways in which this book may be made more useful.

Daniel R. Caldwell
Laramie, Wyoming
May 1994

Acknowledgments

Any major endeavor requires the cooperative participation of many individuals. Such is the case with the present effort. First of all, I thank my many students over the years who challenged me to find ways to make the sciences of microbial physiology and metabolism understandable, meaningful, and useful to them. Efforts to help students understand the complexities and significance of the physiology and metabolism of microbes and how an appreciation for those subjects would enhance their understanding of the nature of critical biological processes were the impetus for this book. The opportunity to bring what began as an adjunct to a local teaching effort to a larger audience was provided by Wm. C. Brown Publishers, at the instigation of Colin Wheatley, to whom I give sincere thanks. Equal thanks to Megan Johnson, Jane Deshaw, Robin Steffeck, Diane Beausoleil, and *especially* to Cathy Smith, all of whom provided valuable guidance through the vagaries of bringing an intellectual effort to a practical reality. Thanks also to Jai Bruno for her conscientious and careful efforts in preparation of the index and the following reviewers: Joan L. M. Foster, Metropolitan State College of Denver; John Ingraham, University of California–Davis; Carl A. Westby, South Dakota State University; Alfred E. Brown, Auburn University; and David W. Smith, University of Delaware. In addition to these individuals, I most heartily thank my colleagues in the Department of Biochemistry at the University of Georgia, Alan Przybyla and Claiborne Glover, for the opportunity to spend uninterrupted summers on this project in a highly supportive environment. Finally, most importantly, I thank my dear, loving, and totally supportive wife, Wanda, who over an extended time period, put up with a husband obsessed with punching the keys of a computer terminal.

CHAPTER 1

The Nature of
Microbial Physiology

What is physiology?

When one begins the study of a new field, it is important to understand precisely what one is about. Such is most certainly the case if the subject is one as complex as microbial physiology. In order to understand the nature of our subject it is necessary to understand two things, *the nature of physiology* and the aspects of *microbial physiology* that make it unique. To begin with, precisely what is the discipline of physiology? If you ask a collection of people who call themselves physiologists, you will find that not all of them agree about the nature of their discipline. Some view physiology simply as the study of the relationship of structure to function and therefore focus on the function of a *particular* structure. Others perceive physiology as the study of all of the structures of an organism. Such people recognize that the brain does not function without the aid of the heart, the lungs, the liver, and the kidney and that, in the microbial world, the genome cannot function without ribosomes

and the remaining structures of the cell. For those who regard physiology as study of the structures of an organism operating together in the accomplishment of the organism's life processes, the meaning and implications of physiology are extremely broad and, encompass all of life. If one adopts such a view, it becomes impossible to understand physiology without considering the subjects of metabolism, biochemistry, genetics, and nutrition. If it is further recognized that organisms, in order to survive and grow, must adapt and respond to changing environmental conditions, physiology becomes even broader. This book views physiology from a very broad perspective. Physiology is regarded as the study of the life processes of an organism, as mediated by its structures operating together to accomplish the common tasks of life. Microbes are regarded as models for understanding all life forms and for understanding of the manner in which through changes in genetic expression, critical life tasks are accomplished under a great variety of conditions.

Where do the microbes fit in the physiological world?

If we understand that physiology is a very encompassing discipline, and that microbes may serve as models for understanding all of life, where do microbes fit in the physiological world? What is it that they share with other life forms, and conversely, how do they differ? The answers to these questions are, in some ways, paradoxical. Microbes must accomplish, *in the space of a few cubic micrometers,* all of the processes required for life that are accomplished by large forms of life. They must take up things from their environment and get rid of waste products. They must respond to environmental change. They must obtain energy and form small molecules. They must make macromolecules. They must make structures. They must synthesize DNA, RNA, and protein. They must regulate all of their activities in a manner compatible with life.

Taken as a group, microbes are highly successful in accomplishing the essential activities required for growth and survival. At the same time, they exhibit substantially greater physiological and metabolic diversity and adaptability than is the case for large life forms. Collectively, microbes display the entire spectrum of ways in which life is possible. By studying physiology at the microbial level, we may understand in principle, and often in great detail, the manner in which the critical life processes of all living forms occur. At the same time, we may understand the subtle differences that may exist in the precise details by which essentially identical processes are accomplished so that we may exploit those differences to our advantage.

The comparative physiology of microbes and large life forms

Although the critical life processes of microbes are the same as those for large life forms, the ways in those processes are accomplished in microbes differ, in some respects, from the ways in which large living forms deal with the same tasks. To a great extent, these differences reflect the fact that, for microbes, the organism *is* a cell, whereas large life forms are *composed of cells.* This difference is critical

for many reasons and has many physiological consequences. To focus attention on the nature of microbial physiology and its relationship to the physiology of other living forms, we will contrast the physiology of a microbe with that of a "model" large form, an elephant.

Size

It is obvious that there is a tremendous difference in the size of an elephant and that of a microbe. The dimensions of an elephant are measured in meters, whereas those of the microbe are measured in *micro*meters, a unit 1,000,000 times smaller. Size, however is not the critical factor from the physiological viewpoint. What is critical are the *physiological consequences* of size. If you are a small organism, your surface-to-volume ratio is large. Consider the surface-to-volume ratio (S/V) of a rod-shaped bacterium whose shape approximates a cylinder. The formula for the surface area of a cylinder is: $S = 2\pi rh$. The formula for the volume of a cylinder is $V = \pi r^2 h$. The surface to volume ratio, S/V is therefore equal to $2/r$. If r is 1, S/V is 2. If r is 0.1, S/V is 20. If r is 0.01, S/V is 200. The smaller r becomes, the larger the S/V. If the organism is spherical in shape, similar thinking applies. The surface area of a sphere is $4\pi r^2$, whereas its volume $4/3\pi r^3$. Division of S by V indicates that it is numerically equal to $3/r$! In this situation, if r is 1, $S/V = 3$. If r is 0.1, $S/V = 30$. If $r = 0.01$, $S/V = 300$. Once again, as r becomes smaller, S/V increases. As r becomes very small, S/V becomes an exceedingly large number. Although the precise number, for a particular r value, is different for a rod than for a coccus, from the physiological viewpoint, this difference is insignificant. What is significant is that whatever shape a microbe might assume, *the ratio of its surface area to its volume is very large* compared to that of an elephant! The particular shape of the organism is of no practical consequence.

The very large S/V ratio of a microbe, regardless of its shape, compared to the S/V of an elephant, has major physiological implications. The large S/V ratio of the microbe allows it to take up things very rapidly from its external environment and to rid itself of waste products at an equally rapid rate. Because of these phenomena, a microbe can carry out metabolism rapidly. This ability is often enhanced by the fact that, in many cases, the enzymes of microbes are substantially more efficient than enzymes that catalyze the similar reactions in large life forms. The ability to metabolize rapidly allows the microbe to reproduce rapidly, as evidenced by the fact that the generation time of a "typical" microbe is between 20 minutes and 3 hours. The "generation time" of an elephant, by contrast, is approximately 3 years, a reflection of its relatively less rapid physiological processes. Because it is a multicellular organism, the elephant cannot communicate, at the biochemical level, with its environment at anything approaching the speed of a microbe.

Structures

The structures of an elephant are distinctly different from those of a microbe. For the elephant, structures are composed of *differentiated* cells, while in the microbe, they are *macromolecular* aggregates. The size of structures also differs markedly between microbe and elephant. For the elephant, the size of an organ is such that one may readily study an individual structure from an individual elephant. One can easily locate, for example, an elephant's liver, and obtain a substantial amount of material for study, *even* if one loses 99% of what one begins with! Such is not the case with microbes. With the latter, study of an individual organelle is normally impossible, not only because we cannot obtain it, but also because, if we did, we could not measure its activity. With microbes, at both the structural and organism level, we must study populations, whereas, with elephants, we may study individual structures or individual elephants. The physiological implications of this difference are profound. With elephants, we may make repeated measures on the same elephant, determine the variability of that elephant, and compare the response of one elephant with that of another. With microbes, none of these things are possible. We cannot determine the activity of an individual microbe or a component of it. We must always observe an *average* response of a population. We assume, in the microbial world, a constancy of response but know that because of mutation, selection, and other factors, such may not be the case. In most circumstances, even without genetic variation, the cells of a microbial population do not behave identically. We assume that what we observe, as a population response, is representative of the whole when, in fact, it may not be.

The necessity of studying populations of microbes is a physiological paradox. On the one hand, we observe a population and assume that its response is representative of all members of the population. Conversely, if we desire a large population, we can readily obtain it. Not so for elephants. Study of genetics is a relatively easy task with microbes, but a difficult task with elephants. Within the space of a day or so, we can obtain billions of microbes in a test tube. Obtaining a similar population of elephants is impossible, in any time frame.

Metabolic versatility and adaptability

Collectively, microbes exhibit unparalleled metabolic diversity and adaptability, allowing them to survive in environments incompatible with large life forms. Microbes produce energy and carry out metabolism from a tremendous variety of organic materials and, in addition, can use radiant energy. Microbes can survive in the hot springs of Yellowstone Park and in Antarctica, in soil and in water, in the bottom of lakes and in compost heaps, and within and on the surfaces of both animals and plants. The ability of microbes to adapt and survive in a diversity of environments exceeds, by orders of magnitude, the survival ability of large life forms. Although, from the morphological viewpoint, microbes are simple, they are *metabolically* and *physiologically* complex. In contrast, collectively, large forms, although structurally complex, are, relatively speaking, physiologically simple. It is precisely because of the physiological and metabolic complexity of microbes that they can survive in such a wider variety of circumstances than can elephants and other large forms.

Nutrition

Diversity is found among microbes in the nature of their nutritional requirements. Some microbes require a great many things for growth while others require only a few. The nutritional requirements of a microbe are, in general, inversely related to its physiological complexity. The more things that an organism makes for itself, the fewer things with which it must be presupplied and, conversely, the more external things it requires, the fewer internal things it can make, at least at a rate compatible with life.

In spite of differences among microbes in their nutritional requirements, most of them require fewer nutrients than do large life forms. Whatever the nutritional requirements of a microbe, however, the function of the nutrient required by a microbe is similar or identical to the function of that nutrient for an elephant. By using microbes as models, we may understand the function of nutrients in all life forms.

Summary

Although microbes and large life forms differ physiologically in many respects, when considered as a group, microbes display at the cellular level all of the physiological possibilities of nature. Although microbes, as organisms, are cells and large forms are composed of cells, study of cellular phenomena in microbial systems is essential for understanding the physiology of large forms. In the microbial world, we must study populations of organisms, which at the same time, are populations of cells. The observed response of the microbes is a population response. In contrast, in the large life forms, collections of differentiated cells constitute an individual organism so it is possible to study the properties of an individual organism in a population.

The widespread use of microbes as models for understanding of general biological phenomena reflects the continuity of nature. Both eucaryotic and procaryotic cells are found in the microbial world, while large forms are composed of eucaryotic cells only. Comparative study of cells in microbial systems allows us to examine the critical differences between the cell types from which all life is made and, at the same time, to understand the processes essential to all life. We may, for example, study photosynthesis in a microbe and understand the process in a tree. Microbes serve as models for the understanding of nutrition, metabolism, and biochemistry; protein and nucleic acid synthesis; energy generation, enzyme regulation, membrane transport, motility, differentiation, cellular communication, and the behavior of populations. Without study of the life processes of microbes we would know very little of the nature of life. With the aid of microbes, it is possible that some day we may understand all of life at the cellular level.

CHAPTER 2

The Subcellular Structures of Microbes

The structures of large and small life forms

Whether one is a microbe or a tree, the essential processes of life are mediated by structures. Although the nature of the structures in microbes differs greatly from the nature of structures in large life forms, the fact remains that structures, of one type or another, are essential for life. When one considers that all of the processes essential to life may be accomplished by a microbe within the space of a few cubic micrometers, one may marvel at the fact that so much is accomplished in such a small space. Microbes compensate for morphological simplicity by physiological and metabolic complexity. Consideration of the nature and functioning of the structures of microbes is instructional for at least three reasons: 1) We may appreciate the commonality of the processes of all life forms, whether large or small, 2) at the same time, we may appreciate the subtle differences between the manner in which those processes are accomplished by procaryotic and eucaryotic cells, and 3) an understanding of the latter differences allows us to exploit them.

The comparative structural properties of procaryotic and eucaryotic cells

Eucaryotic and procaryotic microbes accomplish the same life processes with the aid of structures. However, there are both similarities in the nature of the processes and in the structures that accomplish them and distinct differences in both the nature of and the manner in which organisms accomplish the "same" processes. The essential structural features of procaryotic and eucaryotic cells are shown diagrammatically in Figure 2.1. The diagrams are generalized and schematic and are intended to display the spectrum of possibilities.

Some of the structural and functional differences between procaryotic and eucaryotic microbes are shown in Table 2.1. Consideration of the table reveals that there are a number of significant structural and functional differences as a function of organism and cell type. For example, although all organisms contain genetic material, its nature is not identical in procaryotes and eucaryotes. In addition, although protein synthesis occurs in all

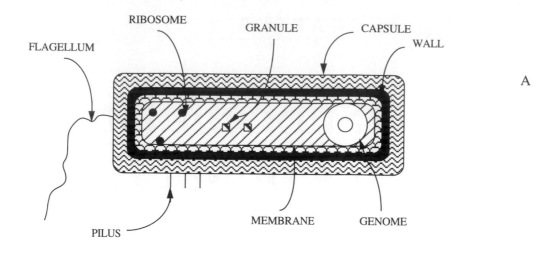

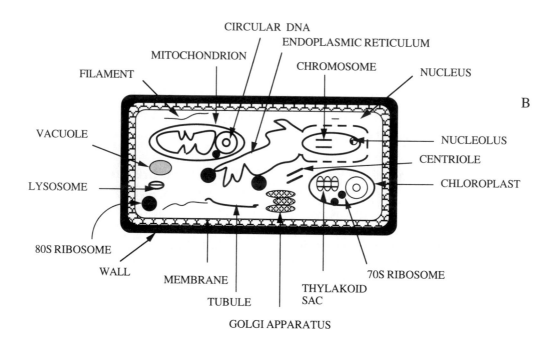

FIGURE 2.1 A schematic representation of the essential elements of a procaryotic (A) and a eucaryotic (B) cell.

life forms, in eucaryotes, it occurs with the aid of both 70S and 80S ribosomes, while in procaryotes it occurs via 70S ribosomes only. In eucaryotic cells, but not in procaryotes, protein synthesis is compartmentalized. It occurs by 80S ribosomes in the cytoplasm, and by 70S ribosomes in mitochondria or chloroplasts. In eucaryotes, DNA replication is also a compartmentalized function, since DNA replication is accomplished in a different manner within the nucleus than it is within the mitochondrion or chloroplast. Because they may replicate both DNA and protein independently of the nucleus, mitochondria and chloroplasts within eucaryotic cells have a certain degree of structural and functional autonomy. Electron transport and selective permeability are both functions of the cytoplasmic membrane of procaryotes, whereas these essential functions are structurally separated

TABLE 2.1 Some Critical Differences in the Structures and Processes of Procaryotic and Eucaryotic Cells

ORGANISM OR CELL TYPE

Structure or Process	Procaryote	Protozoa	Algae	Fungi	Metazoan
Membrane-bound nucleus	Absent	Present	Present	Present	Present
Histones	Absent	Present	Present	Present	Present
Mitochondria	Absent	Present	Present	Present	Present
Chloroplasts	Absent	Absent	Present	Absent	Absent
Phagocytosis	Absent	May occur	Absent	Absent	May occur
Pinocytosis	Absent	May occur	Absent	Absent	May occur
Protoplasmic streaming	Absent	May occur	Absent	Absent	Absent
Endoplasmic reticulum	Absent	Present	Present	Present	Present
Mesosomes	Present	Absent	Absent	Absent	Absent
Electron transport	In cytoplasmic membrane	In mitochondria	In mitochondrion and chloroplast	In mitochondrion	In mitochondrion
Rigid cell wall	Present	Absent	Present	Present	Absent
Ribosomes	Cytoplasmic 70S	Cytoplasmic 80S Mitochondrion 70S	Cytoplasmic 80S Mitochondrion, Chloroplast 70S	Cytoplasmic 80S Mitochondrion 70S	Cytoplasmic 80S Mitochondrion 70S
Golgi apparatus	Absent	Present	Dictyosomes	Present	Present

in eucaryotes. In eucaryotic forms, selective permeability is a function of the membrane, but electron transport occurs in specialized membrane-bound organelles. In addition, pinocytosis, phagocytosis, and protoplasmic streaming require a flexible cell exterior and are impossible in cells that contain rigid cell walls. These examples illustrate the concept that, although both procaryotic and eucaryotic cells accomplish the same essential processes, the manner in which these processes are accomplished differs substantially among different life forms and cell types. In most cases, functional differences reflect structural differences as well.

The structures of microbial cells

It is useful to consider microbial structure in an orderly and comparative way. We will begin our discussion with the cell exterior and proceed inward. In the process, the salient structural differences and their physiological consequences will become apparent.

Gelatinous surface layers

The majority, if not all, cells, whether plant or animal, procaryotic or eucaryotic, have gelatinous material at the cell surface. In ciliate protozoa, the gelatinous layer surrounding the cell is known as a *pellicle* and is often proteinaceous. In amoeboid protozoa this layer is normally a polysaccharide slime. Similarly, in algae and some fungi, a slimy layer, commonly known as a glycocalyx, typically surrounds the cell exterior. In the bacteria, the structure surrounding the cell is commonly known as a slime layer if it is morphologically indiscrete. If the layer is morphologically discrete, it is referred to as a *capsule*. In the vast majority of cases, capsules and slime layers are composed of carbohydrate polymers and similar materials, although in a few cases, most notably *Bacillus anthracis,* they are composed of amino acids.

Proposed functions for capsules

Many functions have been proposed for capsules. It is likely that the proposed functions for capsules may, in large measure, be similar to those for the analogous entities in certain eucaryotic cells. Although capsules are not normally considered a part of the cell proper, their universal presence indicates that they exert an important physiological function or functions. Precisely what these functions may be has been the subject of intense discussion. The capsule is recognized, along with the wall and the flagellum, as one of the major antigenic sites of the cell. In addition, the capsule, or extruded slimy material, may inhibit the process of phagocytosis and, thereby, protect the organism from destruction. The classical example of this phenomenon is the case of *Streptococcus pneumoniae*, which by extensive capsule material extrusion, may avoid phagocytosis and antimicrobial chemotherapy for an extended time period. It has been suggested that, for some organisms, the production of capsular material may serve as a pathogenic mechanism.

In addition to antigenicity, pathogenicity, and protection from phagocytosis, a number of other functions have been proposed for capsules. Some regard production of capsular material as a mechanism for motility, helping organisms devoid of flagella to glide along surfaces. Others have suggested that the capsule may serve as a vehicle for nutrient accumulation in nutritionally sparse environments. Evidence for this suggestion comes from the finding that, in some situations, the concentration of metal ions, particularly Mg^{2+}, is greater per unit of capsule material volume than is the case in the growth medium. It has been suggested that this nutrient sequestering is an effort by the organism to deal with nutritional limitation by altering the immediate cellular environment to one more hospitable for growth. In addition to nutrient accumulation, the capsule may serve as a "cellular garbage dump." Since it is the area immediately adjacent to the cell and has been shown, in some cases, to bind substances that result from microbial metabolism, the capsule may allow the cell to rid itself of potentially toxic materials. Finally, it has been suggested that the capsule may protect the cell from physical injury and dehydration.

Cell walls

A large number of microbes possess rigid exterior layers known either as walls, or when morphologically complex (as with the gram-negative bacteria), envelopes. The possession of a cell wall or envelope allows the organism to maintain its shape and

to resist the effects of changes in osmotic pressure. Organisms devoid of walls or envelopes are extremely sensitive to osmotic pressure changes. The possession of a rigid wall is, however, a mixed physiological blessing since it precludes certain membrane-dependent processes, particularly pinocytosis and phagocytosis.

Contrasts between walls in eucaryotes and procaryotes

When present in eucaryotic microbes, i.e., algae and fungi, walls are of similar chemical composition to the cell walls of higher plants. Typically, they contain cellulose, hemicellulose, lignin, mannoproteins, and, in the case of certain fungi, chitin.

The chemical composition of procaryotic cell walls differs substantially from that of their eucaryotic analogues. Substantial differences are also found both in the chemical composition and in the organization of the components of gram-positive wall and the gram-negative cell envelope. Furthermore, there are substantial differences between the walls and envelopes found in both gram-positive and gram-negative bacteria and those found in the archaebacteria.

The gram-positive procaryotic cell wall

The essential features of the gram-positive cell wall are shown in Figure 2.2. The gram-positive wall is, relatively speaking, both chemically and morphologically less complex than is the envelope of gram-negative organisms. The gram-positive wall contains primarily peptidoglycan (murein), a repeating, β-1,4 linked polymer of N-acetylglucosamine and N-acetyl muramic acid. The latter compound differs from N-acetylglucosamine by possessing a lactic acid residue on carbon 3. Chains of murein are interconnected by amino acid side chains attached to the carboxyl functions of the muramic acid residues. The amino acid side chains are unusual because they contain D-amino acids. In most situations in nature, amino acids are found in the L configuration. The amino acid side chains in murein are also unusual in that chain formation occurs in the absence of ribosomes. In the side chains of murein, peptide bond formation requires the participation of individual enzymes for each bond formed. The number and nature of the peptide-mediated interconnections between

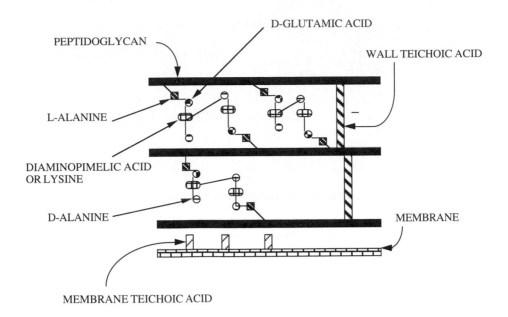

FIGURE 2.2 The essential features of the gram-positive cell wall. The solid black lines are strands of peptidoglycan, a repeating polymer of β-1,4-linked molecules of *N*-acetylglucosamine and *N*-acetylmuramic acid whose structures are shown in Figure 16.4. Chains of peptidoglycan are interconnected by linkages between peptide chains that are attached to the lactic acid residues of *N*-acetylmuramic acid. The diagram shows a typical linkage between the third amino acid of one peptide chain, usually diaminopimelic acid or lysine, and the terminal D-alanine residue of another side chain. Throughout nature, a variety of other linkage mechanisms are known. The heavy striped lines are molecules of "wall" teichoic acids, typically repeating polymers of ribitol and phosphate. The lightly striped lines are molecules of "membrane" teichoic acids, repeating polymers of glycerol and phosphate. The manner of attachment of the components of the wall and many other features of the gram-positive wall are discussed in the text.

murein strands is a major factor that determines both the strength and the three-dimensional structure of the wall. The detailed structure of murein is shown in Figure 16.4.

Teichoic acids

Teichoic acids, polymers of phosphate and either ribitol or glycerol, substituted on their hydroxyl groups with organic materials, are the second major component of the gram-positive wall and membrane. The teichoic acids containing ribitol are commonly referred to as wall teichoic acids and are covalently attached to murein through both muramic acid and the phosphates involved in the connection of ribitol residues. Teichoic acids of the glycerol type are found in the gram-positive cell membrane and, therefore, are often called lipoteichoic acids. The functions of teichoic acids are not entirely defined, although one suggested function is that they allow ion accumulation.

The gram-negative cell envelope

Because of its relative structural and chemical complexity in comparison to the gram-positive rigidity structure, the rigidity layer in gram-negative bacteria is normally referred to as an envelope rather than a wall. The structure of the gram-negative cell envelope is shown in Figure 2.3. Examination of the figure reveals that morphologically and chemically, the gram-negative envelope is a highly complex structure. It contains two layers, the *outer membrane* and the *inner membrane*. The outer membrane is composed chiefly of lipopolysaccharide and lipoprotein, both of which are relatively hydrophobic materials. However, trimeric aggregates of hydrophilic proteins traverse the outer membrane, allowing transport of hydrophilic materials into the cell interior. These protein trimers are known as porins and are, in some cases, highly specific as to the materials that they transport.

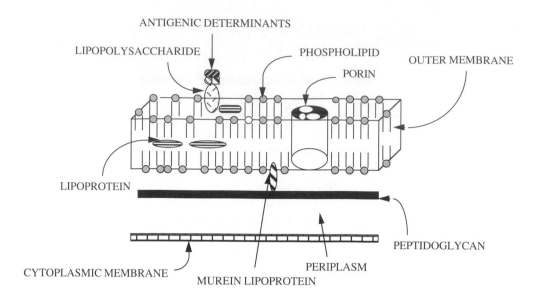

ANTIGENIC DETERMINANTS

LIPOPOLYSACCHARIDE

PHOSPHOLIPID

PORIN

OUTER MEMBRANE

LIPOPROTEIN

CYTOPLASMIC MEMBRANE

MUREIN LIPOPROTEIN

PERIPLASM

PEPTIDOGLYCAN

FIGURE 2.3 The structure of the gram-negative cell envelope. The structure of the envelope is substantially more complex than is the structure of the gram-positive cell wall.

In addition to porins, the gram-negative envelope outer membrane contains a number of other proteins that either participate in maintenance of outer membrane integrity or serve as receptors for particular entities, such as maltose and bacteriophages. In addition to proteins of this type, the outer membrane contains substantial amounts of a lipoprotein known as Braun's, or murein, lipoprotein. It derives the latter name from the fact that it is a major mechanism for maintenance of the physical integrity of the envelope. Murein lipoprotein interconnects the outer membrane with the peptidoglycan layer by covalent bonding.

The periplasm

The region between the outer membrane and the cytoplasmic membrane of gram-negative bacteria contains a unique and rather fluid region, the periplasmic space, within which the peptidoglycan layer resides. Some studies suggest that a functionally equivalent region, a periplasm, also exists in gram-positive organisms, but a true periplasmic space appears to be restricted to gram-negative forms.

Although the periplasmic space is not a rigid component of the gram-negative envelope, it is increasingly evident that the periplasm is involved in a number of critical physiological events. It con-

tains a diversity of proteins, some of which are fluid in the periplasm and some of which, although attached to the cytoplasmic membrane, protrude into the periplasm. The functions of periplasmic proteins are many. Some proteins are hydrolytic enzymes, others participate as receptors for particular substances, and still others mediate transport of substances through the envelope.

Comparisons between the gram-negative cell envelope and the gram-positive cell wall

The gram-negative envelope and the gram-positive cell wall are distinctly different entities, both structurally and functionally. The differences have substantial physiological consequences. To begin with, the gram-positive wall is chemically less complex than is the gram-negative envelope. In addition, the nature and arrangement of the subparts of the two organelles are distinctly different. Murein constitutes the major portion of gram-positive cell walls and is largely exposed to the external environment. As a consequence, lysozyme is generally more destructive to the structure of gram-positive walls than it is for gram-negative envelopes. The substrate for the enzyme is not only more abundant in gram-positive cells than in gram-negative cells, but is also more available. In addition, antibiotics that interfere with murein synthesis are generally more effective with gram-positive cells than with gram-negative cells. Another substantial difference between gram-positive and gram-negative cell walls is the degree to which the two structures participate in transport and selective permeability. Relatively speaking, the gram-negative envelope is much more involved in these processes than is the gram-positive wall. The various receptors in the outer membrane of the gram-negative bacterium facilitate selective accumulation of materials at the cell exterior and, combined with porins, hydrolytic enzymes and receptors in the periplasmic space allow entrance of materials into the cell that might otherwise be excluded.

Finally, structural differences between the gram-positive wall and the gram-negative envelope may account, mechanistically, for their respective Gram stain. Although it is likely that more than one structural explanation accounts for the differential ability of gram-negative and gram-positive cells retain the crystal violet-iodine complex, it is also probable that, in many cases, loss of the blue color by gram-negative bacteria reflects extensive cell envelope damage as a function of alcohol exposure. Conversely, retention of the complex by gram-positive forms is attributable to the less extensive effects of alcohol on the gram-positive wall structure.

Protoplasts and spheroplasts

Many studies have focused on the physiological properties of gram-positive and gram-negative cells from which the wall is removed. Removal of wall material from gram-positive and gram-negative cells and maintenance of the resulting osmotically sensitive structures intact are not easy tasks, but much useful information has been obtained from comparative studies of the properties of cells with and without walls. Osmotically sensitive structures from gram-positive cells are usually referred to as *protoplasts,* while those obtained from gram-negative cells are termed *spheroplasts.* The distinction is made on the basis of the extent to which wall material has been removed. When all wall material is absent, the structure is known as a protoplast, whereas osmotically sensitive spherical structures from which some, but not all, of the wall has been removed are referred to as spheroplasts. It is often difficult to determine which of

the two situations pertains in a particular case. However, in general, it is easier to obtain complete wall removal with gram-positive organisms than with gram-negative cells. The relative ease with which protoplasts are obtained with gram-positive cells reflects both the strategies used for protoplast formation and the structural nature of the gram-positive wall. Protoplasts are typically prepared from gram-positive cells either by lysozyme digestion or by growth of the organism in the presence of cell wall synthesis-inhibiting (murein-inhibiting) antibiotics. Because murein constitutes the major portion of the gram-positive wall, and because it is so readily available, removal of the wall material from a gram-positive organism by lysozyme digestion is relatively easy compared to attempts to accomplish the same process with a gram-negative organism. In addition, because of both substrate availability and quantity, inhibition of wall synthesis with murein-directed agents is more readily achieved with gram-positive than with gram-negative cells.

Complete removal of the wall from gram-negative bacteria is a challenging task, which is seldom completely achieved, although spheroplasts are quite easily obtained. Complete removal of walls from gram-negative cells may require a combination of processes ranging from digestion with enzymes from snails to treatment with detergents. Assessing the completeness of wall removal is difficult and may involve procedures as diverse as chemical analysis, immunological analysis, and electron microscopy.

The comparative properties of intact cells, protoplasts, and spheroplasts

In spite of difficulties in obtaining protoplasts and spheroplasts, a number of researchers have, in fact, obtained both entities from the same organism. Comparative studies of protoplasts and spheroplasts, both with each other and with the cells from which they have been derived, have provided information useful in understanding the physiological functions of the wall and of the cell interior.

Some comparative properties of intact cells, protoplasts, and spheroplasts are shown in Table 2.2. Consideration of the table reveals a number of things. To begin with, it is the wall that determines cell shape and allows osmotic pressure resistance. This conclusion is indicated by the fact that whatever the shape of the cell originally, wall removal

TABLE 2.2 The Comparative Properties of Intact Cells, Spheroplasts, and Protoplasts

PROPERTY	INTACT CELL	SPHEROPLAST	PROTOPLAST
Cell division	Possible	Possible in some cases	Limited or not possible. When possible, normal cells are not formed.
Cell wall regeneration	Possible	Possible in some cases	Impossible
Phage attachment	Possible	Sometimes possible	Impossible
Phage maturation	Possible	Possible	Possible
Spore induction	Possible	Sometimes possible	Impossible
Spore maturation	Possible	Possible	Possible once induced
Protein synthesis	Yes	Yes	Yes
DNA replication	Yes	Yes	Yes
Transport and selective permeability	Yes	Yes	Yes

leads to an osmotically sensitive, and spherically shaped entity. A second major conclusion derived from Table 2.2 is the fact that cell division is possible with intact cells and with some spheroplasts, but is usually impossible, except under rigidly controlled conditions, with protoplasts. When limited division is possible with protoplasts, unusual shapes are obtained because a normal cell cannot be formed from a protoplast. The ability of some spheroplasts to divide into normal cells reflects the fact that, if a sufficient amount of wall material remains during spheroplast formation, wall regeneration is possible, allowing the spheroplast to develop into a new, intact cell. The inability of other spheroplasts to regenerate complete walls and to function as normal cells supports the concept that the cell wall is a cell-regulating entity. Unless a certain amount of cell material is present, a new wall, and therefore, an independent new cell, cannot be formed. To a significant extent, the cell wall regulates the entire course of the cell.

Phage attachment and replication Another interesting fact obtained from Table 2.2 is the finding that, in certain cases, phages attach to intact cells and some spheroplasts, but cannot attach to protoplasts. The implication of this finding is that phage attachment may require the presence of cell wall material that serves as an attachment vehicle because of the presence of specific receptor molecules. When the latter are absent, phage attachment and penetration are impossible. However, phage

synthesis and maturation do not require the presence of the wall since the latter processes involve host cell ribosomes and enzymes, entities interior to the wall.

The wall is required for induction of sporulation, but not for its completion Table 2.2 reveals that the early stages of sporulation require the presence of wall material but that, once sporulation is induced, completion of sporulation is possible in the absence of wall. These findings reflect repeated demonstrations, using cells specifically radiolabelled in their walls, which show that the wall is extensively digested in the early stages of sporulation, but that the later events of sporulation do not require wall material.

Protein synthesis, DNA replication, and transport Table 2.2 shows that protein synthesis, DNA replication, and transport are possible, and may occur with the same kinetics, in intact cells, spheroplasts, and protoplasts. The implication of these findings is that the wall is not required for these processes to occur. In general, this concept is physiologically reasonable, since transport, DNA replication, and protein synthesis require structures and proteins found in or within the membrane, interior to the wall. With regard to transport and selective permeability, recent evidence indicates that substantial differences exist between gram-positive and gram-negative bacteria in the extent to which their walls influence these processes. In many, if not all, cases the receptors and other proteins found in the outer membrane and within the periplasm, as well as the membrane, materially influence both transport and selective permeability in gram-negative organisms. In gram-positive bacteria, both transport and its selectivity, are membrane-associated phenomena.

The comparative properties of cells, spheroplasts, and protoplasts have been studied primarily using procaryotic cells, because it is, in general, easier to obtain these entities from procaryotic cells than from eucaryotic ones. However, the relatively sparse information regarding spheroplasts and protoplasts obtained from eucaryotic cells indicates that the information obtained from study of procaryotic systems is, at least in principle, widely applicable to other cell systems.

Pili

Pili (*hair*; singular, pilus), also known as fimbriae (*fringe*; singular, fimbria) are often found extending from the surface of procaryotic cells. They are of two types, *adhesion* pili and *conjugation* pili. Adhesion pili are involved in the attachment of cells to surfaces and are composed of a single protein, pilin, a 17,000 dalton molecule. They may facilitate the pathogenesis of bacteria by allowing the organisms to colonize mucous membrane surfaces and thereby establish an infection. Conjugation pili, composed of phosphoglycoprotein, may serve as vehicles for transfer of genetic information between microbes. Pili are too small to be seen with the light microscope and can only be seen with the electron microscope. Originally, they were considered by many to be electron microscope artifacts, but it is now known that they are true cellular structures.

Flagella and cilia

A large number of microbes, both procaryotic and eucaryotic, exhibit the phenomenon of motility. In the majority of cases, motility is mediated by flagella or cilia, relatively thin structures that extend from the cell surface, although other motility mechanisms occur in some organisms. It is somewhat unfortunate that the term flagellum was used in the early days of microbiology to describe a motility organelle found in both procaryotic and eucaryotic cells. It is now known that, although there are indeed hairlike motility structures found in bacteria

and in many eucaryotic microbes, both the structural and functional properties of the "flagella" in bacteria, are markedly different from those in eucaryotes. It is equally ironic that, although they are given different names, the flagella and cilia found in eucaryotic microbes are both structurally and functionally similar. The comparative structures of the procaryotic flagellum and the eucaryotic flagellum and cilium are shown in Figure 2.4.

Eucaryotic flagella and cilia

Although they differ somewhat in the precise details of function, the eucaryotic flagellum and cilium are structurally very similar. They are composed of a sheathed doublet of central microtubules and nine peripheral doublets of microtubules. The peripheral microtubule doublets are usually denoted as subtubules A and B, and are connected by microtubules of the protein nexin. Two protein arms composed of dynein, a protein with ATPase activity, extend from subtubule A of each of the peripheral microtubule doublets. In addition, the peripheral microtubule doublets are connected to the central tubules by microtubular spokes. The entire structure of the eucaryotic flagellum or cilium is encased by an extension of the cytoplasmic membrane and attaches to a basal body in the cytoplasm.

The mechanism of motility in eucaryotic cells has been a subject of discussion for a long time. It appears that motility results from the bending and movement of peripheral microtubules under the influence of ATPase activity in the dynein arms that protrude from subtubule A. Somehow, in a manner incompletely understood at present, ATP-mediated contraction of dynein arms allows for sliding and bending of these peripheral microtubules, allowing motility to occur. The type of motility observed as a consequence of peripheral subtubule sliding usually differs between flagella and cilia. In the flagella, a wave may be imparted to the base of the flagellum and transmitted to the tip, pushing the organism along. Alternatively, the wave may originate at the tip and proceed to the base, pulling the organism along. In either case, this type of motility is known as *undulate* motility. In contrast, the cilia's motion is oarlike, and is otherwise known as *tonsate* motility. In many situations, cilia and flagella beat in harmony.

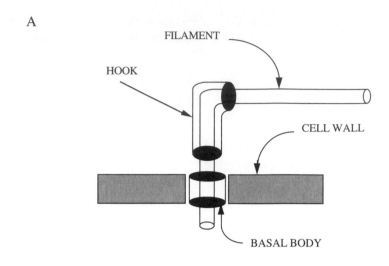

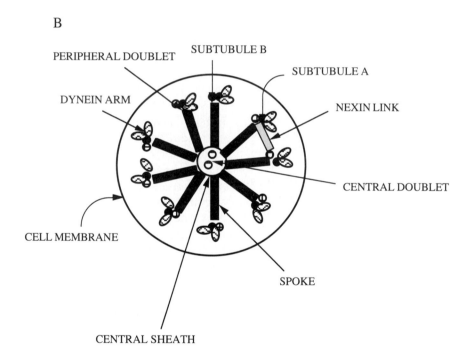

FIGURE 2.4 The comparative structures of the procaryotic flagellum (A) and the eucaryotic flagellum and cilium (B).

Procaryotic flagella

Flagella in the procaryotic world are much simpler in structure than in eucaryotes. Once again, we ironically refer to the typical motility organelles of eucaryotes and procaryotes, collectively, as flagella, as if they were structurally and functionally identical when, in actuality, there are substantial anatomical differences between eucaryotic and procaryotic flagella, and between the flagella of gram-positive and gram-negative bacteria. The comparative structures of gram-negative and gram-positive flagella are shown in Figure 2.5.

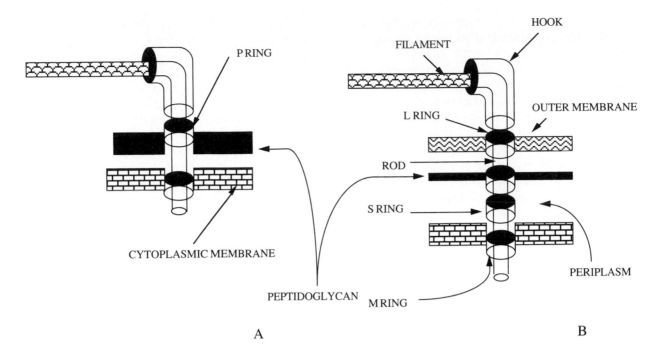

P RING

FILAMENT

HOOK

L RING

OUTER MEMBRANE

ROD

S RING

CYTOPLASMIC MEMBRANE

PERIPLASM

PEPTIDOGLYCAN   M RING

A   B

FIGURE 2.5 The comparative structure of the gram-positive (A) and the gram-negative (B) flagellum.

Study of the figure in comparison with Figure 2.4 reveals that, irrespective of differences between gram-negative and gram-positive forms, the procaryotic flagellum is, comparatively speaking, a substantially simpler organelle than the eucaryotic structure of the same name. The procaryotic structure has no microtubule subunits and, except in a few cases, is not membrane encased. The procaryotic flagellum is composed of three parts: 1) the filament, 2) the hook, and 3) the basal body. The most distal region of the flagellum, the filament, is a hollow tube of a contractile protein called flagellin. The next most proximal region, the hook, is composed of proteins separate and distinct from those that comprise the filament. The hook connects to the basal body, which is found in the cell wall (envelope). Although the basal body constitutes only a small portion of the total flagellar weight and length, it is a complex substructure and is a critical portion of the flagellum because it contains rings, the structures that impart motility to the flagellum. The gram-positive basal body contains two rings, while the gram-negative organelle contains four. Rotation of the rings, under the influence of the proton motive force (PMF), is responsible for procaryotic flagellar motility. The

hook appears to serve in a manner analogous to a universal joint. Under appropriate conditions, counterclockwise rotation of the flagella imparts directed motion to the bacterium.

Flagella synthesis

Flagella are not always formed by potentially flagellated bacteria. In some cases, the nutritional environment determines whether or not flagella are formed. In other circumstances, flagella can be removed without materially affecting the organism, however, flagella that have been removed can usually regenerate themselves. For these reasons, among many others, flagella synthesis has been studied in some detail. The information gleaned from these studies has contributed to our understanding, not only of the nature and function of flagella, but also to understanding of the nature of regeneration.

Much evidence indicates that the flagellum is synthesized from the "inside out," that is, the basal body is synthesized first, followed by the hook, and then the filament. Filament building blocks, synthesized on cytoplasmic ribosomes, apparently migrate through the filament tube and attach to the distal end of the flagellum. Extensive studies have shown that filament protein does not accumulate in the medium.

Procaryotic flagellum synthesis, assembly, and function are complicated processes. Studies with *Salmonella typhimurium* indicate that synthesis begins with the insertion of the M and S rings into the cytoplasmic membrane. This process is followed by attachment of a hollow rod that is subsequently capped. Next, the P and L rings are added, followed by the hook and the filament. Finally, motility proteins are added, allowing the flagellum to be functional. It appears that the synthesis of flagella in all "normal" bacteria occurs in a similar fashion, although certain differences must exist between gram-positive and gram-negative synthetic processes since flagella in the former contain only two rings, while gram-negative flagella contain four.

Flagella mutants   Mutational analysis has shown that all of the processes associated with procaryotic motility are under complex genetic control. At least 40 genes are known to be involved in control of flagella synthesis, assembly, and with motility itself. Hag genes control the formation of filament protein. Hook proteins are formed under the influence of the fla genes. In Salmonella typhimurium three hook-related genes are known: fla W, fla V, and fla U respectively, each of which codes for a separate hook-associated protein (HAP). Formation of the basal body is also under control of a very substantial number of genes.

Flagellar mutants may be classified on the basis of their effect on the motility process. Thus fla- mutants are unable to form flagella, while mutants that result in nonrotating flagella are referred to as mot- mutants.

Chemotaxis

Motility in the bacteria is often a directed process associated with counterclockwise flagellar rotation (clockwise rotation, in contrast, is associated with tumbling). Directed motility involves the movement toward, or away from, particular environmental stimulae. These range from light (*photo*taxis), to magnetic fields (*magneto*taxis), to organic chemicals (*chemo*taxis), and to oxygen (*aero*taxis). Each of these phenomena may be positive or negative, that is, the organism may move either toward or away from the stimulus.

Chemotaxis is, by far, the most studied and most thoroughly understood, of the various tactic phenomena. The mechanism of chemotaxis appears to vary from organism to organism, although there are some common biochemical threads. Chemotaxis involves the recognition of particular materials in the environment by a collection of proteins known as *methyl accepting chemotactic proteins* (MCPs). This recognition is followed by a change in the methylation state of the proteins involved in transmission of a signal to the flagellar motors, causing the latter to rotate in a counterclockwise direction, with its associated directed motility. The precise nature and functioning of MCPs varies from organism to organism. In *E. coli*, four MCPs are recognized: protein I, the product of the *tsr* gene; protein II, the product of the *tar* gene; protein III, coded by the *trg* gene; and protein IV, formed under the influence of the *tap* gene.

The functioning of methyl-accepting chemotactic proteins

The names of the genes for the chemotactic proteins in *E. coli* reflect the functions of their corresponding proteins. Thus, the *tsr* gene product is responsible for chemotaxis in response to serine (s) and repellents (r); the *tar* gene product responds to arginine (a) and repellents (r); the *trg* product responds to ribose (r) and galactose (g), and the *tap* product responds to peptides, particularly dipeptides. Interaction of a MCP with its appropriate tactic molecule causes a change in the methylation state of the MCP that, in turn, transmits a signal of some sort to the flagellar motors. In *E. coli*, interaction of a chemotactic molecule leads to an increase in the methylation state of the MCPs, while interaction with a repellent molecule causes a decrease in the methylation of the molecules. Methylation and demethylation are mediated, respectively, by specific methyl transferases and demethylases.

Although the underlying mechanism for chemotaxis appears similar in all bacteria thus far studied, differences exist in the ways in which MCP-chemotactant interaction affects methylation and in the effects of change in methylation state on motility. Thus, in *Bacillus subtilis*, interaction with an attractant decreases the methylation state of the MCPs, whereas withdrawal of the attractant causes increased methylation, the inverse of *E. coli*. In addition, the specificity of attractants differs between the two organisms and different numbers of genes regulate the organisms' chemotactic response. Only six recognition genes are known for *E. coli*, whereas at least twenty-one are known for *B. subtilis*. In addition to these differences, a number of other differences in the nature and regulation of chemotaxis exist between *E. coli* and *B. subtilis*.

The types of chemotaxis just discussed are collectively known as type I chemotaxis. Type II chemotaxis involves sugars as attractants, as well as the phosphotransferase (PTS) system. Type III chemotaxis requires terminal electron acceptors and oxioreductase enzymes.

Periplasmic flagella

The flexible, spiral-shaped bacteria are made motile by a mechanism different from that of the traditionally flagellated organism. The spirochetes are motile by flagella that are located within the periplasm and are therefore encased in a membranous sheath.

Periplasmic (endo-) flagella are believed to be chemically similar to other flagella and manifest their motility in a manner similar to that of "free" flagella, i.e., by counterclockwise rotation of the organelle around the cylinder of the cell. The structure of the flexible spiral-shaped bacteria is such that their motility allows them to move through high viscosity media.

Other motility modes

Swarming motility   Members of the genus *Proteus* are characterized by extreme motility. This motility is often a practical problem in isolation of pure cultures, because motility is so intense that the entire surface of an agar plate is covered with growth unless the medium contains severely inhibitory agents. Swarming motility may be recognized by concentric rings of growth around a colony. Mechanistically, swarming motility involves periodic morphological changes at approximately 1 to 2 hour intervals. Swarming cells are formed from relatively short and sparsely flagellated cells. Periodically, at the edge of a colony, the latter gives rise to highly flagellated and filamentous cells that have the ability to move rapidly over the surface. After a period of rapid motility, the swarming cells revert to "normal." Cyclical alteration between these two conditions gives rise to the swarming phenomenon.

In *Vibrio parahaemolyticus*, swarming is even more complex than in the *Proteus* species and involves a different morphological change. *Vibrio parahaemolyticus* modifies its flagella in response to medium viscosity. During growth in a liquid medium, the organism forms *polar* flagella, while *lateral* flagella are produced in high-viscosity media. The morphological change is under the control of the *laf* gene.

Nonflagella-mediated motility   In certain microbes, motility may occur in the absence of flagella. In amoebae, motility may occur through cytoplasmic streaming and pseudopod (*false foot*) formation. Precisely how amoeboid motion occurs is not yet clearly understood, but the fact that cell extracts exhibit contraction when ATP is added suggests that motility involves ATPase activity. It appears that, during amoeboid motion, certain areas of the cell may form adhesions, allowing the organism to be pulled over the surface, but the mechanisms by which adhesions may be formed, and released, are speculative.

Gliding motility   Certain organisms move by gliding over surfaces, but the mechanisms by which gliding motility is accomplished are rather obscure. It appears that slime production is a factor in gliding motility, facilitating both the movement of cells and, perhaps, the direction of movement. Surface tension may also be a significant factor. Studies of *Myxococcus xanthus* suggest that such is the case. Cooperative action of cells is known in certain gliding organisms. *Myxococcus xanthus* exhibits both individual and "social" motility, i.e., the movement of individual cells and of cell groups.

The cytoplasmic membrane

The cytoplasmic membrane is a critical structure for all cellular life forms. It is the most exterior cellular entity of eucaryotic microbes, such as the amoeboid protozoa, and of higher animal cells. In many circumstances, the cytoplasmic membrane is surrounded by a gelatinous layer that confers a certain amount of stability to the morphology of the organism or cell compared to a cell devoid of such a covering. However, the morphological integrity of cells that contain a rigid cell wall is much greater than is the case for organisms or cells surrounded by membrane alone.

Although the membrane constitutes a miniscule portion of the cell weight, it is a critical cell structure. The manner in which the membrane participates in the physiology of the cell varies substantially, depending upon whether the cell is procaryotic or eucaryotic. In eucaryotic cells, the cytoplasmic membrane contains *sterols*. The latter are absent in the membranes of all but a few procaryotic forms, e.g., the mycoplasma. In eucaryotes, the membrane serves primarily as an agent of selective permeability and in cell division, but is not extensively involved in a great number of other physiological activities. This situation reflects the fact that the cytoplasm of eucaryotic cells contains a variety of membrane-bound entities that accomplish specialized physiological activities. In procaryotes, however, the cytoplasm, with a few exceptions, is devoid of specialized, intracytoplasmic membrane-bound organelles. As a result, in procaryotes, the cytoplasmic membrane, or invaginations of it, is involved not only in selective permeability, but in a variety of other critical cell processes, such as cell division, sporulation, electron transport, ATP formation, and DNA replication.

The role of the cytoplasmic membrane in selective permeability and transport

Since selective permeability and transport are universal membrane functions for all cells, much attention has been devoted to study of the involvement of the membrane in transport and transport-related processes. These studies have focused on two major physiological problems: 1) How do both hydrophobic and hydrophilic materials traverse an entity that is primarily hydrophobic? and 2) What determines the upper limit of size beyond which materials are incapable of transport? Our increasingly sophisticated understanding of transport and related processes, is coupled with an equally sophisticated understanding of cytoplasmic membrane structure.

To appreciate the extent of progress we have made, both in understanding transport and membrane structure, it is useful to review some of the early theories of membrane structure, and the attendant ideas about the transport process. It was recognized quite early that, in general, charged molecules traversed the cell membrane less readily than did relatively nonpolar materials and that, in general, smaller molecules entered the cell more easily than did larger ones. It was further recognized that molecules above a certain size could not traverse the membrane at all. Attempts to understand membrane transport sought to explain these phenomena.

The sieve and pore theories The sieve and pore theories were early attempts to explain the role of the membrane as a size discrimination agent. According to the sieve theory, the membrane was a homogeneous entity that contained "holes" of a certain size. Transport required that the transported material could "fit" within these holes. The pore theory was similar to the sieve theory, but allowed the "holes" to be of different sizes and shapes, permitting transport of a greater variety of materials. Both the sieve and the pore theories entirely envisioned selective permeability and the upper limit of transported material size as functions of the physical nature of the membrane. Neither theory considered the implications of charge.

The mosaic theory The mosaic theory was, in some respects, an improvement over the sieve and the pore theories. It proposed that the membrane was composed of areas that differed in chemical composition. Some areas were, relatively speaking, hydrophobic, while others were hydrophilic. According to this theory, transport of a material involved physiochemical compatibility between the transported material and appropriate areas within the membrane. It was recognized that the transported materials, as well as the membrane, contained areas of partial charge and that some transported materials were more hydrophobic or hydrophilic than others. The "strategy" of transport was to find compatible membrane regions for the transported material.

The reversible chemical combination theory Early transport theories focussed attention on the properties of the membrane as the determining factor in transport. The reversible chemical combination theory was an improvement over previous theories because it proposed that transport involved an *interaction* between a component or components of the membrane, in a reversible way, in the transport process. The theory did not, however, specify either the nature of the material or the interaction.

The permease theory   The permease theory of transport, a portion of which is still used today, proposed that transport resulted from reversible interaction of the transported material with a stereospecific carrier protein, or *permease* that allowed transport of the material by movement of the complex from one area of the membrane to another. The concept of the involvement of specific proteins in transport processes remains central to our current understanding of the role of the membrane in transport. Our present understanding of the nature and involvement of proteins in transport is substantially more advanced than that of the originally proposed "permease" theory.

The molecular reorientation theory   This theory not only recognized that transport involves interaction of the transported material with specific elements in the membrane, but also proposed that the interaction altered membrane structure to allow transport. Such a concept is an integral part of our current understanding of the role of the membrane in transport. Recent work of G. A. Scarborough and associates suggests that all transport processes may be understood within the framework of membrane reorientation as a consequence of interaction with the transported material.

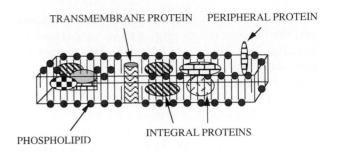

FIGURE 2.6 A schematic diagram of our current concept of cytoplasmic membrane structure.

The fluid mosaic theory of membrane structure and its implications for transport

Our current understanding of transport views the membrane as a fluid in which certain membrane components may move, rather than a stationary mosaic. A diagram of the current concept of membrane structure is shown in Figure 2.6. We regard the membrane to be composed of a lipoprotein bilayer traversed by hydrophilic protein channels that allow transport of hydrophilic substances through a basically hydrophobic layer. The membrane, although composed of hydrophobic and hydrophilic regions, is not a stationary mosaic but rather a fluid one. A structure of this type allows a variety of transport processes, ranging from simple diffusion to active transport.

Intracellular membranous organelles

A number of membranous organelles are found within the cytoplasm of microbial cells. However, the number and nature of these entities differs greatly between procaryotic and eucaryotic organisms. The diversity of such structures in procaryotes is small and the organelles that are found are single-membrane bound. In contrast, a substantial variety of both single-membrane bound and double-membrane bound organelles is found in eucaryotic microbes.

Membranous organelles in procaryotes

There are four major, recognized single-membrane bound organelles in procaryotes: mesosomes, gas vacuoles, photosynthetic vesicles, and carboxysomes. *Mesosomes* are specialized, invaginated areas of the cytoplasmic membrane that participate in processes as diverse as cell septum formation, electron transport, photosynthesis, cell wall formation, DNA replication, and sporulation. Recent evidence suggests that, in some cases, mesosomes are artifacts of electron microscopy, but their repeated demonstration and physical isolation over the years suggests that they are, in many cases, physiological realities. *Gas vacuoles* are found primarily in marine bacteria and are presumed to allow buoyancy. *Photosynthetic vesicles* are found primarily in the genus *Chlorobium*. In most other photosynthetic procaryotes, photosynthesis occurs either within the membrane or in invaginations of it, known as *chromatophores*. Autotrophic bacteria may contain, in addition, single membrane-bound structures known as *carboxysomes* that contain the enzyme ribulose-1,5-bisphosphate carboxylase.

Membranous cytoplasmic organelles in eucaryotes

Eucaryotic microbes contain a variety of membranous intracytoplasmic organelles. Single-membrane organelles include phagosomes, pinosomes, primary lysosomes, secondary lysosomes, and food vacuoles. *Phagosomes* are involved in phagocytosis, the membranous engulfment of particulate material, including food. *Pinosomes,* by contrast, engulf water and things dissolved in it. *Primary lysosomes* contain a diversity of hydrolytic enzymes. Fusion of a primary lysosome with a phagosome results in formation of a phagolysosome while fusion of a primary lysosome with a *food vacuole,* a single-membrane bound entity containing soluble food, results in formation of a *secondary lysosome.*

Double-membrane bound intracytoplasmic eucaryotic organelles

The *mitochondrion* and the *chloroplast* are double-membrane bound organelles found within the cytoplasm of eucaryotes. They are both involved in energy generation, mitochondria by chemotrophic processes and chloroplasts by phototrophic energy generation. Both mitochondria and chloroplasts have been the subject of intense study and speculation for at least two reasons: 1) their possession of double-stranded circular DNA and 2) their possession of 70S ribosomes. These two characteristics have led to the suggestion that they are, in fact, vestigial procaryotic cells that have lost the ability to exist as free-living forms and are thus obligatory inhabitants of eucaryotic cells. It may be further argued that the eucaryotic cell is, in fact, a symbiotic entity.

Mitochondria and chloroplasts are of interest from an additional physiological perspective. Because they contain both DNA and 70S ribosomes, they synthesize certain of their components, particularly those involved in electron transport within the organelle. To a certain extent they possess a degree of physiological autonomy, but on the other hand, most of their components are formed on 80S cytoplasmic ribosomes. Although they are independently able to form a portion of their constituents, they are dependent for their existence on the remainder of the cell.

The Golgi apparatus   The diversity of membrane-bound organelles in eucaryotic cells requires a "membrane factory." The Golgi apparatus is such an organelle and is found in animal-like eucaryotic microbes. A functionally analogous structure, the dictyosome, is found in eucaryotic plantlike microbes. The general functions of the Golgi apparatus and dictyosomes are synthesis of certain substances, packaging of materials within membranes, and modification of proteins so that they may be exported or directed to specific cellular locations. Procaryotic cells are devoid of analogous structures.

The endoplasmic reticulum   Eucaryotic cells contain a membranous network of channels contiguous with the nuclear membrane that are collectively known as the endoplasmic reticulum (ER). No such structure is found in procaryotic cells. To a substantial extent, the endoplasmic reticulum serves to allow transportation of substances within the eucaryotic cell. More recently, the ER has been particularly recognized as a site of ribosome attachment for formation of secreted proteins. Formation of these proteins often involves processing by the ER and still further processing within the Golgi apparatus.

Ribosomes   The cytoplasms of both procaryotic and eucaryotic microbes contain particles composed of ribonucleic acid and protein, whose function is protein synthesis. The number of ribosomes per cell varies with growth rate. In bacteria, the number fluctuates between 10,000 and 20,000 per cell. Procaryotic 70S ribosomes are composed of 30S and 50S subparticles, whereas eucaryotic ribosomes are composed of analogous subparticles that are somewhat larger, i.e., 40S and 60S. In procaryotes, the small particle is composed of a 16S RNA particle and 21 molecules of protein. The small particle of the eucaryotic ribosome may be analogously subdivided into a particle of RNA (18S) and 30 proteins.

The large ribosomal particles of both procaryotic and eucaryotic ribosomes may be also subdivided into particles of RNA and proteins. The large particle of the 70S ribosome consists of a 5S RNA particle, a 23S RNA particle, and 32 proteins. The corresponding particle of the 80S ribosome contains an 18S RNA particle, a 5.8S RNA particle, and a third particle that may range in size from 25S to 28S, plus 50 proteins.

The functions of both eucaryotic and procaryotic ribosomes are at least threefold: 1) attachment to messenger RNA molecules, 2) recognition of tRNA-amino acid molecules, and 3) the joining of amino acids, via peptide bonds, to form proteins. In eucaryotes, formation of secreted proteins requires attachment of ribosomes to membranes of the endoplasmic reticulum, although many proteins are formed by nonmembrane-attached ribosomes. Membrane attachment is apparently nonobligatory for protein synthesis of any type in procaryotes, although it is often the case that the functional procaryotic ribosomes are attached to the cytoplasmic membrane. Protein synthesis, and its control, are discussed in chapters 11 and 13.

Genetic material

All life forms, cellular or otherwise, contain genetic material that dictates the spectrum of proteins that an organism may produce, and thus the limits of its physiological potential, but the nature of the genetic material in various life forms varies substantially among organisms. In cellular life forms, genetic material is composed of DNA. In eucaryotic cells, including microbes, the DNA is encased in a double membrane and the resulting structure is known as a *nucleus*. The eucaryotic nucleus is a highly complicated structure that contains many chromosomes and a sophisticated apparatus for both chromosome replication and for separation of the replicated chromosomes during cell division.

Recognition of the existence of DNA in procaryotes, and the determination that it functioned in a manner functionally identical to the DNA of larger forms, were both long in coming. Difficulties were experienced by early investigators in establishing the existence of DNA in procaryotic cells, primarily because of the lack of proper microscope techniques and because of the inability of stains to distinguish between DNA and RNA. Eventually, by use of sophisticated microscopy and staining procedures on cells treated with selective chemicals or enzymes, it was possible to remove or destroy RNA and to establish that, in fact, procaryotes contained DNA. It was further shown, by elegant electron microscopic techniques in combination with autoradiography, that the DNA of procaryotes is a single, circular, double-stranded, nonmembrane-encased molecule. Confirmation of the circular nature of procaryotic DNA was obtained by genetic experiments that demonstrated that the relationships of procaryotic genes to each other could *only* be understood through the assumption that the procaryotic DNA was circular. This conclusion was reached from studies of the times of conjugation required to allow for the appearance of recombinants in "interrupted mating" experiments.

The recognition that procaryotic organisms contained genetic material that was functionally equivalent to the genetic material of higher forms was a major step forward. However, the substantial structural difference between eucaryotic and procaryotic DNA raised questions regarding the precise terminology that should be applied to procaryotic and eucaryotic genetic material. These questions persist. We normally refer to eucaryotic DNA, and its associated membranes and nuclear apparatus, as a *nucleus*. Various terms such as "procaryotic nucleus," bacterial chromosome, chromatinic body, and bacterial genome have been suggested to describe the procaryotic analogue. Whatever term we adopt to describe it, we should recognize that the genome in procaryotes, although functionally similar or identical to its analogue in eucaryotes, is structurally different.

Plasmids

With the exception of certain yeasts, procaryotes (but not eucaryotes) may contain small, circular, double-stranded, self-replicating, nongenomic DNA molecules called *plasmids*. The latter typically endow the cell with nonessential properties such as drug resistance and the ability to use particular carbon and nitrogen sources. Procaryotic plasmids also participate in the process of conjugation, a major sexual mechanism. In addition, plasmid manipulation offers tremendous promise for genetic engineering and biotechnology because manipulation of plasmid DNA may allow production of nonmicrobial proteins in microbial systems. Plasmids are also useful tools for studying DNA replication. Plasmids and their implications are discussed in detail in chapters 14 and 20.

Spores

Spores of various kinds are found in certain eucaryotic and procaryotic microbes, but the implications of the sporulation process for the two types are different. For eucaryotic microbes, sporulation is a *reproductive* mechanism that allows a single organism to give rise to many new organisms. In the eucaryotic world, spores may be either sexual or asexual. In procaryotes, sporulation has substantially different consequences. In these organisms, spores are asexual only and sporulation is not a reproductive mechanism but rather a *survival* mechanism. For procaryotes, a single spore arises from a single vegetative cell that, upon germination, gives rise to a single vegetative cell. In procaryotes, sporulation is a process that allows the cell to withstand adverse conditions. Relatively speaking, the resistance of bacterial endospores to adverse conditions such as heat, desiccation, and toxic chemicals far exceeds the same ability of the vast majority of spores produced by eucaryotic forms.

Eucaryotic and procaryotic spores differ substantially in the cellular processes that give rise to them. In eucaryotes, sporulation involves cross wall formation and, in some cases, budding. In procaryotes, sporulation involves the surrounding of the spore genome with membranes and synthesis of a variety of carbohydrate and proteinaceous layers that endow the endospore with its resistance to adverse environmental factors. In procaryotes, sporulation is a complex and time-dependent process of differentiation in which certain elements of the spore arise from components of the vegetative cell, while others are synthesized *de novo.* Furthermore, as is the case with mitochondria and chloroplasts, bacterial endospores have a certain amount of functional autonomy. The early stages of bacterial sporulation involve use of vegetative cell components, as well as *de novo* synthesis of components using materials in the extracellular environment. At a later point, however, one may remove sporulating cells from potential spore "building blocks" and sporulation may be completed. This phenomenon, known as *endotrophic sporulation,* allows sporulation to occur in distilled water! The physiological implication of this phenomenon is that at a certain point, the spore becomes independent of its external environment and is an entity unto itself. Because of the complexity of the subprocesses of bacterial endosporulation, and their time dependence, sporulation in bacteria is regarded as a model of differentiation. The details of bacterial endosporulation, and their larger implications are discussed in chapter 18.

Storage granules

Both eucaryotic and procaryotic microbes contain characteristic nonmembrane-encased storage granules. The various classes of algae are distinguished, in part, by the presence of carbohydrates such as laminarin, chrysolaminarin, and paramylon. In contrast, procaryotes but not eucaryotes, may store polybetahydroxybutyrate or volutin. Other storage materials, such as starch, are found in both types of microbes.

Summary

All living things have structures by which they accomplish the essential tasks of life, but substantial differences are found among organisms in the nature of these structures. By understanding the nature and functioning of structures in various organisms, we may understand, simultaneously, the nature of the common tasks of life and the subtle, but significantly different, ways in which various organisms accomplish those tasks. This knowledge allows us to exploit the differences in very practical ways and, at the same time, to appreciate the commonality of life.

Microbes serve as models for understanding all critical life processes, primarily because microbes, taken collectively, exhibit all of the cell types, structures, and functions found in nature. Understanding of the structural and functional distinctions between procaryotic and eucaryotic cells is crucial. Although both cells accomplish the same collection of tasks, the nature of the structures in the two cell types, and the manner of their functioning, are distinctly different. The procaryotic cell compensates for its relatively small size, morphological simplicity, and lack of intracellular compartmentalization by a high degree of metabolic complexity and by the fact that its cytoplasmic membrane accomplishes many of the functions accomplished in eucaryotes by specialized single- and double-membraned organelles.

It is invariably the case that despite the differences between the structures of procaryotic and eucaryotic cells, the life processes accomplished by both cell types are, at least in principle, and often in great detail, identical. In addition, many analogies are found between the two cell types and the nature of their structures. Thus, motility organs, genomes, ribosomes, membranes, cytoplasms, rigidity organelles, and spores, among other structures, are found in both types of cells.

Our ability to exploit differences between microbes and the larger forms they may invade reflects, in large measure, the sophistication of our understanding of the physiological differences between the microbial cells and the cells of larger life forms. Structure has many implications.

Selected References

Cell Walls and Envelopes

Archibald, A. R. 1974. The structure, biosynthesis and function of teichoic acid. *Advances in Microbial Physiology.* **11**: 53–95.

Ballou, C. 1976. Structure and biosyntheis of the mannan component of the yeast cell envelope. *Advances in Microbial Physiology.* **14**: 93–158.

Benz, Roland. 1988. Structure and function of porins from gram-negative bacteria. *Annual Review of Microbiology.* **42**: 359–394.

Brown, M. R. W., and P. Williams. 1985. The influence of environment on envelope properties affecting survival of bacteria in infections. *Annual Review of Microbiology.* **39**: 27–556.

Cabib, E., R. Roberts, and B. Bowers. 1982. Synthesis of the yeast cell wall and its regulation. *Annual Review of Biochemistry.* **51**: 763–794.

Duffus, J. H., C. Levi, and D. J. Manners. 1982. Yeast cell-wall glucans. *Advances in Microbial Physiology.* **23**: 151–181.

Jennings, D. R. 1988. Turnover of cell walls in microorganisms. *Microbiological Reviews.* **52**: 554–567.

Michaelis, S., and J. Beckwith. 1982. Mechanisms of incorporation of cell envelope proteins in *Escherichia coli. Annual Review of Microbiology.* **36**: 435–465.

Neilands, J. B. 1982. Microbial envelope proteins related to iron. *Annual Review of Microbiology.* **36**: 285–310.

Schleifer, K. H., W. P. Hammes, and O. Kandler. 1976. Effect of endogeous and exogenous factors on the primary structures of bacterial peptidoglycan. *Advances in Microbial Physiology.* **13**: 246–292.

Shockman, G. D., and J. Barrett. 1983. Structure, function, and assembly of cell walls of gram-positive bacteria. *Annual Review of Microbiology.* **37**: 501–528.

Thwaites J. J., and N. H. Mendelson. 1991. Mechanical behaviour of bacterial cell walls. *Advances in Microbial Physiology.* **32**: 173–222.

Ward, J. B. 1983. Teichoic and teichuronic acids: Biosynthesis, assembly and location. *Microbiological Reviews.* **45**: 211–243.

Flagella and Motility

Burchard, Robert P. 1981. Gliding motility of prokaryotes: Ultrastructure, physiology, and genetics. *Annual Review of Microbiology.* **35**: 497–530.

Canale-Parola, E. 1978. Motility and chemotaxis of spirochetes. *Annual Review of Microbiology.* **32**: 69–99.

Doetsch, R. N., and R. D. Sjoblad. 1980. Flagellar structure and function in eubacteria. *Annual Review of Microbiology.* **34**: 69–108.

Henrichsen, J. 1983. Twitching motility. *Annual Review of Microbiology.* **37**: 81–94.

Holwill, Michael E. J. 1977. Some biophysical aspects of ciliary and flagellar motility. *Advances in Microbial Physiology.* **16**: 1–48.

Iino, T. Y. Komeda, K. Kutsukake, R. M. Macnab, P. Matsumura, J. S. Parkinson, M. I. Simon, and S. Yamaguchi. 1988. New unified nomenclature for the flagellar genes of *Escherichia coli* and *Salmonella typhimurium. Microbiological Reviews.* **52**: 533–535.

Jones, C. J., and S.-I. Aizawa. 1991. Bacterial flagellum and flagellar motor: Structure, assembly and function. *Advances in Microbial Physiology.* **32**: 110–172.

McNab, R. M., and S.-I. Aizawa. 1984. Bacterial motility and the bacterial flagellar motor. *Annual Review of Biophysics and Bioengineering.* **13**: 51–83.

Manson, Michael D. 1992. Bacterial motility and chemotaxis. *Advances in Mcrobial Physiology.* **33**: 277–346.

Silverman, M., and M. I. Simon. 1977. Bacterial flagella. *Annual Review of Microbiology.* **31**: 397–419.

Pili

Elleman, T. C. 1988. Pilins of *Bacteroides nodosus:* Molecular basis of serotypic variation and relationships to other bacterial pilins. *Microbiological Review.* **52**: 233–247.

Hultgren, S. J., S. Normack, and S. N. Abraham. 1991. Chaperone-assisted assembly and molecular architecture of adhesive pili. *Annual Review of Microbiology.* **45**: 383–416.

Ottow, J. C. G. 1975. Ecology, physiology, and genetics of fimbriae and pili. *Annual Review of Microbiology.* **29**: 79–108.

Paranchych, W., and L. S. Frost. 1988. The physiology and biochemistry of pili. *Advances in Microbial Physiology.* **29**: 53–114.

Plasmids

Assinder, S. J., and P. A. Williams. 1990. The TOL plasmids: determinants of the catabolism of toluene and the xylenes. *Advances in Microbial Physiology.* **31:** 2–69.

Challberg, M. D., and T. J. Kelly. 1982. Eukaryotic DNA replication: Viral and plasmid model systems. *Annual Review of Biochemistry.* **51:** 901–934.

Clewell, Don B. 1981. Plasmids, drug resistance, and gene transfer in the genus *Streptococcus. Microbiological Reviews.* **45:** 409–436.

Couturier, M., F. Bex, P. L. Bergquist, and W. K. Maas. 1988. Identification and classification of bacterial plasmids. *Microbiological Reviews.* **52:** 375–395.

Davies, J., and D. I. Smith. 1978. Plasmid-determined resistance to antimicrobial agents. *Annual Review of Microbiology.* **32:** 469–518.

Foster, T. J. 1983. Plasmid-determined resistance to antimicrobial drugs and toxic metal ions in bacteria. *Microbiological Reviews.* **47:** 361–409.

Gruss, A., and S. Dusko Ehrlich. 1989. The family of highly interrelated single-stranded deoxyribonucleic acid plasmids. *Microbiological Reviews.* **53:** 231–241.

Gunge, N. 1983. Yeast DNA plasmids. *Annual Review of Microbiology.* **37:** 253–276.

Helinski, D. R. 1973. Plasmid determined resistance to antibiotics: molecular properties of R factors. *Annual Review of Microbiology.* **27:** 437–470.

Hiraga, S. 1992. Chromosome and plasmid partition in *Escherichia coli. Annual Review of Biochemistry.* **61:** 283–306.

Nishimura, A., K. Akiyama, Y. Kohara, and K. Horiuchi. 1992. Correlation of a subset of the pLC plasmids to the physical map of *Escherichia coli* K-12. *Microbiological Reviews.* **56:** 137–151.

Novick, R. P. 1989. Staphylococcal plasmids and their replication. *Annual Review of Microbiology.* **43:** 537–566.

Novick, R. P., R. C. Clowes, S. N. Cohen, R. Curtiss III, N. Datta, and S. Falkow. 1976. Uniform nomenclature for bacterial plasmids: a proposal. *Bacteriological Reviews.* **40:** 168–189.

Silver, S., and M. Walderhaug. 1992. Gene regulation of plasmid- and chromosome-determined inorganic ion transport in bacteria. *Microbiological Reviews.* **56:** 195–228.

Silver, S., and T. K. Misra. 1988. Plasmid-mediated heavy metal resistances. *Annual Review of Microbiology.* **42:** 717–744.

Sukupolvi, S., and C. D. O' Connor. 1990. TraT lipoprotein, a plasmid-specified mediator of interactions between gram-negative bacteria and their environment. *Microbiological Reviews.* **54:** 331–341.

Thomas, C. M., and C. A. Smith. 1987. Incompatibility group P plasmids: Genetics, evolution, and use in genetic manipulation. *Annual Review of Microbiology.* **41:** 77–102.

Volkert, F. C., D. W. Wilson, and J. R. Broach. 1989. Deoxyribonucleic acid plasmids in yeasts. *Microbiological Reviews.* **53:** 299–317.

Waters V. L., and J. H. Crosa. 1991. Colicin V virulence plasmids. *Microbiological Reviews.* **55:** 437–450.

Membranes

Andersen, H. C. 1978. Probes of membrane structure. *Annual Review of Biochemistry.* **47:** 359–384.

Cronan, J. E., Jr. 1978. Molecular biology of bacterial membrane lipids. *Annual Review of Biochemistry.* **47:** 163–190.

DiePierre, J. W., and L. Ernster. 1977. Enzyme topology of intracellular membranes. *Annual Review of Biochemistry.* **46:** 210–262.

Downie, J. A., F. Gibson, and G. B. Cox. 1979. Membrane adenosine triphosphatases of prokaryotic cells. *Annual Review of Biochemistry.* **48:** 103–132.

Drews, G., and J. Oelze. 1981. Organization and differentiation of membranes of phototropic bacteria. *Advances in Microbial Physiology.* **22:** 1–92.

Firshein, W. 1989. Role of the DNA/membrane complex in prokaryotic DNA replication. *Annual Review of Microbiology.* **43:** 89–120.

Hancock, R. E. W. 1984. Alterations in outer membrane permeability. *Annual Review of Microbiology.* **38:** 237–264.

Jennings, M. L. 1989. Topography of membrane proteins. *Annual Review of Biochemistry.* **58:** 999–1028.

Karlsson, K. Anders. 1989. Animal glycosphingolipids as membrane attachment sites for bacteria. *Annual Review of Biochemistry.* **58:** 309–350.

Konings, W. N. 1977. Active transport of solutes in bacterial membrane vesicles. *Advances in Microbial Physiology.* **15:** 175–251.

Nikaido, H., and T. Nakae. 1979. The outer membrane of gram-negative bacteria. *Advances in Microbial Physiology.* **20:** 164–250.

Nikaido, H., and M. Vaara. 1985. Molecular basis of bacterial outer membrane permeability. *Microbiological Reviews.* **49:** 1–32.

Osborn, M. J., and H. C. P. Wu. 1980. Proteins of the outer membrane of gram negative bacteria. *Annual Review of Microbiology.* **35:** 369–422.

Rees, D. C., H. Komiya, T. O. Yeates, J. P. Allen, and G. Feher. 1989. The bacterial photosynthetic reaction center as a model for membrane proteins. *Annual Review of Biochemistry.* **58:** 607–634.

Salton, M. R. J., and P. Owen. 1976. Bacterial membrane structure. *Annual Review of Microbiology*. **30:** 451–482.

Wickner, William. 1979. The assembly of proteins into biological membranes: The membrane trigger hypothesis. *Annual Review of Biochemistry*. **48:** 23–46.

Genome

Doolittle, W. F. 1979. The cyanobacterial genome, its expression, and the control of that expression. *Advances in Microbial Physiology*. **20:** 2–102.

Hiraga, S. 1992. Chromosome and plasmid partition in *Escherchia coli. Annual Review of Biochemistry*. **61:** 283–306.

Holloway, B. W., and A. F. Morgan. 1986. Genome organization in *Pseudomonas. Annual Review of Microbiology*. **40:** 79–106.

Krawiec, S., and M. Riley. 1990. Organization of the bacterial chromosome. *Microbiological Reviews*. **54:** 502–539.

Riley, M., and A. Anilionis. 1978. Evolution of the Bacterial genome. *Annual Review of Microbiology*. **32:** 519–560.

Simpson, L. 1987. The mitochondrial genome of kinetoplastid protozoa: genomic organization, transcription, replication, and evolution. *Annual Review of Microbiology*. **41:** 363 382.

Wallace, D. C. 1982. Structure and evolution of organelle genomes. *Microbiological Reviews*. **46:** 208–240.

Ribosomes

Brimacombe, R., F. Stöffler, and H. G. Wittman. 1978. Ribosome structure. *Annual Review of Biochemistry*. **47:** 217–250.

Kurland, C. G. 1972. Structure and function of the bacterial ribosome. *Annual Review of Biochemistry*. **41:** 377–408.

Kurland, C. G. 1977. Structure and function of the bacterial ribosome. *Annual Review of Biochemistry*. **46:** 173–200.

Lake, J. A. 1985. Evolving ribosome structure: Domains in archaebacteria, eubacteria, eocytes and eukaryotes. *Annual Review of Biochemistry*. **54:** 507–530.

Olsen, G. J., D. J. Lane, S. J. Giovannoni, N. R. Pace, and D. A. Stahl. 1986. Microbial ecology and evolution: a ribosomal approach. *Annual Review of Microbiology*. **40:** 337–366.

Pestka, S. 1971. Inhibitors of ribosome functions. *Annual Review of Microbiology* **25:** 487–562.

Schlessinger, D., and D. Apirion. 1969. *Escherichia coli* ribosomes: Recent developments. *Annual Review of Microbiology*. **23:** 391–426.

Warner, Jonathan R. 1989. Synthesis of ribosomes in *Saccharomyces cerevisiae. Microbiological Reviews*. **53:** 256–271

Wittmann, H. G. 1982. Components of bacterial ribosomes. *Annual Review of Biochemistry*. **51:** 155–184.

Wittmann, H. G. 1983. Architecture of prokaryotic ribosomes. *Annual Review of Biochemistry*. **52:** 35–66.

Wool, I. G. 1979. The structure and function of eukaryotic ribosomes. *Annual Review of Biochemistry*. **48:** 719–754.

Mitochondria and Chloroplasts

Kirk, J. T. O. 1971. Chloroplast structure and biogenesis. *Annual Review of Biochemistry*. **40:** 161–196.

Nagley, P., K. S. Sriprakash, and A. W. Linnane. 1977. Structure, synthesis and genetics of yeast mitochondrial DNA. *Advances in Microbial Physiology*. **16:** 158–277.

Pfanner, N., and W. Neupert. 1990. The mitochondrial protein import apparatus. *Annual Review of Biochemistry*. **59:** 331–354.

Wallace, D. C. 1992. Diseases of the mitochondrial DNA. *Annual Review of Biochemistry*. **61:** 1175–1212.

DNA

Krawiec S., and M. Riley. 1990. Organization of the bacterial chromosome. *Microbiological Reviews*. **54:** 502–539.

Matthews, K. S. 1992. DNA looping. *Microbiological Reviews*. **56:** 123–136.

Schleif, R. 1992. DNA looping. *Annual Review of Biochemistry*. **61:** 199–224.

Zimmerman, S. B. 1982. The three-dimensional structure of DNA. *Annual Review of Biochemistry*. **51:** 395–428.

Spores

Dubnau, D. A. (ed.) 1982. The molecular biology of the bacilli. Academic Press Inc., New York.

Hurst, A., and G. W. Gould. 1984. The molecular biology of microbial differentiation. American Society for Microbiology, Washington, D.C.

Losick, R., P. Youngman, and P. J. Piggot. 1986. Genetics of endospore formation in *Bacillus subtilis. Annual Review of Genetics*. **20:** 625–669.

Moir A., and D. A. Smith. 1990. The genetics of bacterial spore germination. *Annual Review of Microbiology*. **44:** 531–553.

Piggot, P. J., and J. G. Coote. 1976. Genetic aspects of bacterial endospore formation. *Bacteriological Reviews*. **40:** 908–962.

Setlow, P. 1988. Small acid-soluble spore proteins of *Bacilus* species: Structure, synthesis, genetics, function, and degradation. *Annual Review of Microbiology.* **42:** 319–338.

Warth, A. D. 1978. Molecular structure of the bacterial spore. *Advances in Microbial Physiology.* **17:** 1–45.

Proteins

Aisen, Philip, and Irving Listowsky. 1980. Iron transport and storage proteins. *Annual Review of Biochemistry.* **49:** 357–394.

Champoux, J. J. 1978. Proteins that affect DNA conformation. *Annual Review of Biochemistry.* **47:** 449–480.

Chase, J. W., and K. R. Williams. 1986. Single-stranded DNA binding proteins required for DNA replication. *Annual Review of Biochemistry.* **55:** 103–136.

Chou, P. Y., and G. Fasman. 1978. Empirical predictions of protein conformation. *Annual Review of Biochemistry.* **47:** 251–276.

Coleman, J. E. 1992. Zinc Proteins: enzymes, storage proteins, transcription factors, and replication proteins. *Annual Review of Biochemistry.* **61:** 897–946.

Cozzone, A. J. 1988. Protein phosphorylation in prokaryotes. *Annual Review of Microbiology.* **42:** 97–126.

Douglas, M. G., M. T. McCammon, and A. Vassarotti. 1986. Targeting proteins into mitochondria. *Microbiological Reviews.* **50:** 166–178.

Fisher, H. W., and R. C. Williams. 1979. Electron microscopic visualization of nucleic acids and of their complexes with proteins. *Annual Review of Biochemistry.* **48:** 649–680.

Friedman, F. K., and S. Beychok. 1979. Probes of subunit assembly and reconstitution pathways in multisubunit proteins. *Annual Review of Biochemistry.* **48:** 217–250.

Gilman, A. G. 1987. G Proteins: transducers of receptor-generated signals. *Annual Review of Biochemistry.* **56:** 615–650.

Glenn, A. R. 1976. Production of extracellular proteins in bacteria. *Annual Review of Microbiology.* **30:** 41–62.

Johnson, P. F., and S. L. McKnight. 1989. Eukaryotic transcriptional regulatory proteins. *Annual Review of Biochemistry.* **58:** 799–840.

Lanyi, J. K. 1974. Salt-dependent properties of proteins from extremely halophilic bacteria. *Bacteriological Reviews.* **38:** 272–290.

Miller, R. V., and T. A. Kokjohn. 1990. General microbiology of rec A: Environmental and evolutionary significance. *Annual Review of Microbiology.* **44:** 365–394.

Oliver, D. 1985. Protein secretion in *Escherichia coli. Annual Review of Microbiology.* **39:** 615–648.

Pfeffer, S. R., and J. E. Rothman. 1987. Biosynthetic protein transport and sorting by the endoplasmic reticulum and golgi. *Annual Review of Biochemistry.* **56:** 829–852.

Pryer, N. K., L. J. Wuestehube, and R. Schekman. 1992. Vesicle-mediated protein sorting. *Annual Review of Biochemistry.* **61:** 471–516.

Quiocho, Florante A. 1986. Carbohydrate-binding proteins: Tertiary structures and protein-sugar interactions. *Annual Review of Biochemistry.* **55:** 287–316.

Randall, L. L., S. J. S. Hardy, and Julia R. Thom. 1987. Export of protein: A biochemical view. *Annual Review of Microbiology.* **41:** 507–542.

Saier, Milton H., Jr. 1989. Protein phosphorylation and allosteric control of inducer exclusion and catabolite repression by the bacterial phosphoenolpyruvate: sugar phosphotransferase system. *Microbiological Reviews.* **53:** 109–120.

Saier, M. H., Jr., P. K. Werner, and M. Müller. 1989. Insertion of proteins into bacterial membranes: mechanism, characteristics, and comparisons with the eucaryotic process. *Microbiological Reviews.* **53:** 333–366.

Salas, M. 1991. Protein-priming of DNA replication. *Annual Review of Biochemistry.* **60:** 39–71

Schatz, G., and T. L. Mason. 1974. The biosynthesis of mitochondrial proteins. *Annual Review of Biochemistry.* **43:** 51–88.

Schmidt, G. W., and M. L. Mishkind. 1986. The transport of proteins into chloroplasts. *Annual Review of Biochemistry.* **55:** 879–912.

Silhavy, T. J., S. A. Benson, and S. D. Emr. 1983. Mechanisms of protein localization. *Microbiological Reviews.* **47:** 313–344.

Singleton, R., Jr., and R. E. Amelunxen. 1973. Proteins from thermophilic microorganisms. *Bacteriological Reviews.* **37:** 320–342.

Sweeney, W. V., and J. C. Rabinowitz. 1980. Proteins containing 4FE-4S clusters: An overview. *Annual Review of Biochemistry.* **49:** 139–162.

Travers, A. A. 1989. DNA conformation and protein binding. *Annual Review of Biochemistry.* **58:** 427–452.

Vallee, R. B., and H. S. Shpetner. 1990. Motor proteins of cytoplasmic microtubules. *Annual Review of Biochemistry.* **59:** 909–932.

Watson, K. 1990. Microbial stress proteins. *Advances in Microbial Physiology.* **31:** 183–223.

CHAPTER 3

Structure Formation

The common properties of structure synthesis in microbes

Microbes are not, as was once thought, "bags of enzymes." Instead, they are highly organized entities with discrete, functioning organelles. Formation of the functioning components of cells involves a collection of complicated processes. Although the details of the processes involved in structure formation differ substantially among structures, it is often the case that analogies exist in the processes involved in structure formation, irrespective of the resulting structure. In other words, the microbial cell must address a common collection of physiological problems so that the small molecules derived from metabolism may be converted into macromolecules that are transported from the point of their synthesis to the nascent structure and, finally, incorporated into it. In this chapter, we will discuss some of the mechanisms by which the components of microbial structures are formed, the mechanisms by which these components are transported to the addition site of structures, and the major factors that influence and control the addition of new material to the structures. Finally, we will consider mechanisms that control all of these processes and the implications of these mechanisms for the entire physiology of the cell. Because understanding of all of these topics is more complete with procaryotic cells than with eucaryotes and because it is often easier to understand underlying principles in relatively simple organisms than in complicated ones, our discussion will focus on processes that occur in procaryotes, particularly the bacteria. In principle, if not detail, much of what we will consider in this chapter applies to eucaryotes as well.

It is *not* the intent of this chapter to delineate in great detail the various processes involved in the formation of each structure. It *is* the intent to discuss example processes, which illustrate the complexity of structure formation and its regulation. The readers may consult a variety of review articles and other sources if they desire detailed information regarding a particular structure not discussed here. The subject of structure formation and function could easily occupy an entire course, or courses, if the intimate details of formation of all of the structures were considered. It is the intent of this chapter to provide a physiological overview, with selected examples. Because they are both complicated processes that involve participation of different areas of the cell and may also serve as general examples of the complexities of structure formation and its regulation, we will focus our attention primarily on formation of the gram-positive and gram-negative cell walls and, to a lesser extent, on the formation of capsules. We have previously discussed the formation of bacterial and eucaryotic flagella in chapter 2.

Critical physiological questions regarding structure synthesis

Irrespective of the nature of the structure whose synthesis we seek to understand, certain physiological questions present themselves. First, we must know precisely where in the cell the structural building blocks are formed. If, as is usually the case, the building blocks are formed at a location other than the site of structure formation, we also must understand how the building blocks are transported to the site of structure formation. Next, we must understand the manner in which structural building blocks are added to the incipient structure or to one of its macromolecular components. In structures composed of more than one major component, we must further understand the manner in which structure subunits are assembled. To fully understand structure synthesis, we must also understand how the processes of building block formation, subunit transport, and structure assembly are regulated and coordinated. Finally, we must understand the interrelationships between synthesis of various macromolecules and structures, since, in certain cases, two or more structures, or macromolecules, may share common intermediates and regulatory mechanisms. Thus, alteration of synthesis of the building blocks for one macromolecule or structure may influence the formation of another such entity. The gram-positive and gram-negative bacterial cell walls, and bacterial exopolysaccharides are excellent examples of the complexity and interrelatedness of microbial cell structure synthesis. Understanding of the physiological problems posed by these entities helps us understand the nature and complexity of the analogous processes in eucaryotic cells.

Synthesis of the gram-positive wall

Since, relatively speaking, it is the least complex of the structures under consideration, we will begin our discussion of structure synthesis with the gram-positive cell wall, starting with the synthesis of peptidoglycan, the unique rigidity polymer in procaryotic organisms, also known as murein. The process of peptidoglycan precursor synthesis occurs in the cytoplasm and is outlined in Figure 3.1. It begins with the conversion of fructose-6-phosphate via glucosamine-6-phosphate and N-acetylglucosamine-6-phosphate to N-acetylglucosamine-1-phosphate, and conversion of the latter, with uridine triphosphate (UTP), to a molecule of N-acetylglucosamine (NAG) bound to a molecule of uridine diphosphate (UDP), UDPNAG. A portion of UDPNAG is converted via the addition of a lactyl residue to its carbon 3, to UDP-bound N-acetylmuramic acid (UDPNAM). This acid is further converted by separate enzyme-mediated peptide bond formation to UDP-bound N-acetylmuramyl pentapeptide (UDPNAMPP). Formation of the pentapeptide moiety of UDP-NAMPP is unusual for two reasons, 1) the absence of ribosome-mediated peptide bond formation and 2) the presence of D-amino acids in the side chains. The amino acids typically found in the peptide side chain are, beginning with the first amino acid attached to the carboxyl group of the lactic acid moiety, and ending with the terminal amino acid, L-alanine, D-glutamic acid, meso-diaminopimelic acid and two D-alanine residues.

Transport of building blocks through the membrane

Transport of building blocks from the cytoplasm to the addition site of the growing polymer is mediated by a 55-carbon isoprenoid lipid carrier, known as either undecaprenol or bactoprenol. Transfer of building blocks from their UDP-activated forms to undecaprenol facilitates their movement to the addition site. Initially, at the cytoplasm-cytoplasmic membrane junction, a molecule of UDPNAMPP reacts with a molecule of undecaprenol monophosphate (UDPRP), giving a molecule of undecaprenol diphosphate (UDPRPP)-bound NAMPP (UDPRPP-NAMPP) and releasing a molecule of uridine monophosphate (UMP). The UDPRPPNAMPP molecule then reacts with a molecule of UDPNAG

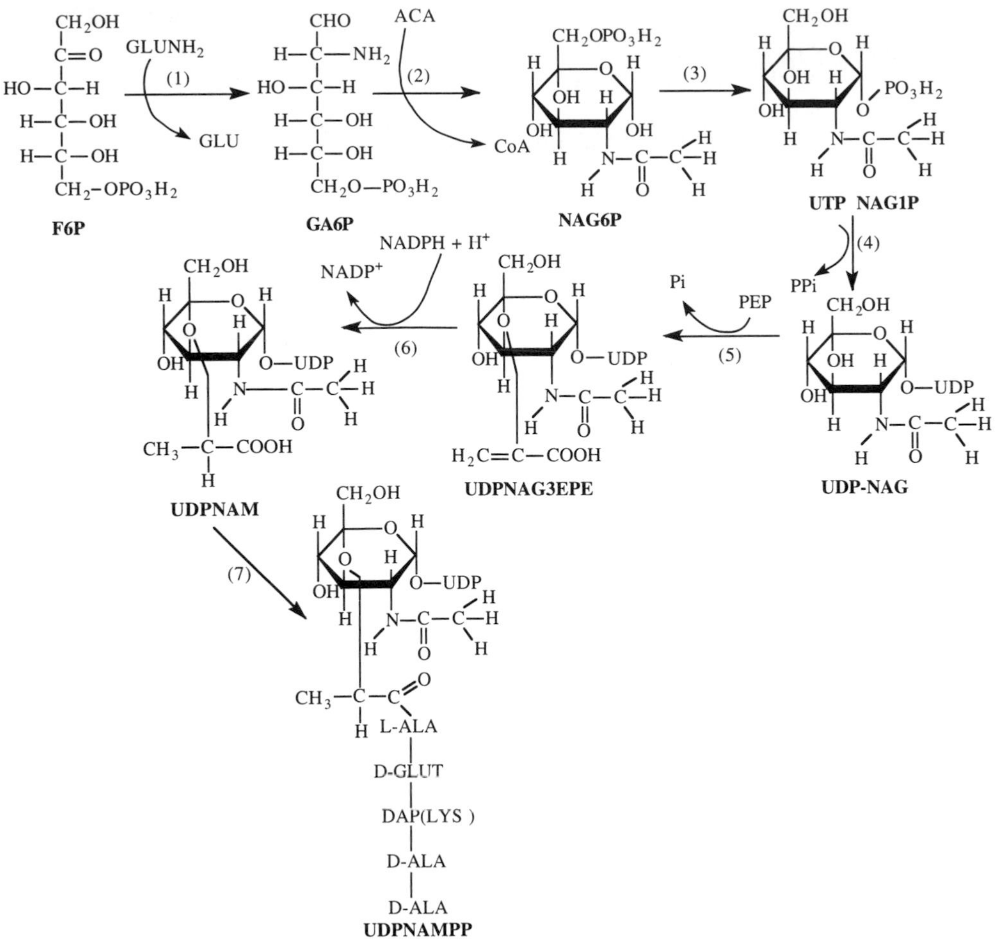

FIGURE 3.1 The synthesis of peptidoglycan precursors in the cytoplasm, F6P, fructose-6-phosphate; GA6P, glucosamine 6-phosphate; NAG6P, N-acetylglucosamine-6-phosphate; NAG1P, N-acetylglucosamine-1-phosphate; UDPNAG, uridine diphosphate-N-acetylglucosamine; UDPNAG3EPE, uridine diphosphate-N-acetylglucosamine-3-enolpyruvylether; UDPNAM, uridine diphosphate-N-acetylmuramic acid; UDPNAMPP, uridine diphosphate-N-acetylmuramyl pentapeptide; GLUNH$_2$, glutamine; GL, glutamic acid; L-ALA, L-alanine; DAP, diaminopimelic acid; LYS, lysine; D-GLUT, D-glutamic acid; D-ALA, D-alanine; UTP, uridine triphosphate; PEP, phosphoenolpyruvate. (1) glutamate: fructose-6-phosphate aminotransferase; (2) glucosamine phosphate transacetylase; (3) N-acetylglucosamine phosphomutase; (4) UDP-N-acetylglucosamine pyrophosphorylase; (5) UDP-N-acetylglucosamine 3-enolpyruvylether synthetase; (6) UDP-N-acetenolpyruvylglucosamine reductase; (7) individual enzymes for the addition of each amino acid to lactyl side chain of UDPNAM.

to form a dimeric molecule composed of joined molecules of NAG and NAMPP, bound to a molecule of UDPRPP, or UDPRPPNAGNAMPP. It is this entity that is transported through the membrane. The formation of UDPRPPNAGNAMPP is accompanied by release of uridine diphosphate (UDP).

During transport of the UDPRPPNAGNAMPP molecule through the membrane of gram-positive or in the periplasm of gram-negative bacteria, the dimeric units of UDPRPP-bound NAGNAMPP are often polymerized to UDPRPP-bound oligomeric units, with the release of a molecule of UDPRPP for each dimeric unit added to the developing UDPRPP-oligomer. Finally, the UDPRPP-bound oligomeric units are added to the growing polymer with the release of yet another molecule of UDPRPP. The UDPRPP is dephosphorylated to UDPRP so that

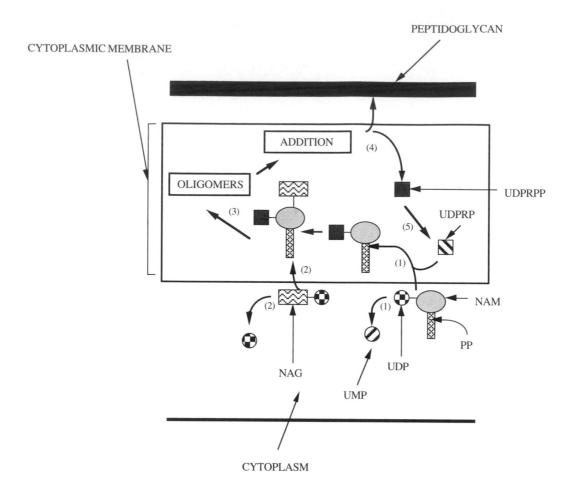

FIGURE 3.2 The formation of peptidoglycan from cytoplasmically synthesized precursors. UMP, uridine monophosphate; UDP, uridine diphosphate; NAM, *N*-acetylmuramic acid; PP, pentapeptide; UDPRPP, undecaprenol diphosphate; UDPRP, undecaprenol monophosphate; NAG, *N*-acetylglucosamine. (1) Transfer of UDP-bound NAMPP to UDPRP to form NAMPP bound to undecaprenol diphosphate (UDPRPP) with UMP release. (2) Transfer of *N*-acetylglucosamine (NAG) bound to UDP to NAMPP bound to UDPRPP to form a dimer of NAMPP and NAG, both bound to UDPRPP. UDP is released during this process. (3) Formation of NAM-NAG oligomers bound to UDPRPP with the release of a UDPRPP molecule with each dimeric addition. (4) Addition of oligomers to the growing peptidoglycan chain, with the release of more UDPRPP. (5) Regeneration of UDPRP from UDPRPP, so that synthesis may continue.

the latter can participate in further peptidoglycan synthesis. The processes involved in peptidoglycan formation from cytoplasmically synthesized precursors are shown in Figure 3.2.

Teichoic and teichuronic acid synthesis

Teichoic acids are the second major component of the gram-positive cell. They are composed of polymers of ribitol or glycerol connected by phosphodiester linkages. Polymers that contain ribitol are associated with the wall through covalent linkages, while glycerol-containing polymers are attached to a glycolipid in the membrane. Under certain conditions of growth, teichuronic acids, polymers containing sugar acids, are formed. The general nature of teichoic and teichuronic acid synthesis is shown in Figure 3.3.

Although the general picture regarding both teichoic and teichuronic acid synthesis is clear, controversy exists regarding the precise details of both processes. It is virtually certain that diversity exists

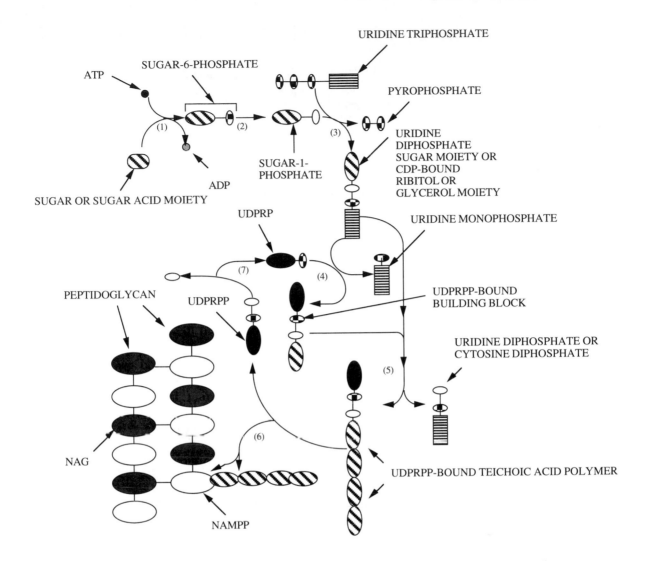

FIGURE 3.3 The general nature of teichoic and teichuronic acid synthesis. (1) Conversion of sugar or sugar acid to sugar or sugar acid-6-phosphate. (2) Conversion of 6-phosphate moiety to a 1-phosphate moiety. (3) Formation of a uridine diphosphate (UDP) derivative of the sugar or sugar acid. (4) Formation of an undecaprenol diphosphate (UDPRPP) derivative of the sugar or sugar acid. (5) Addition of sugar or sugar acid residues to the UDPRPP-bound incipient polymer. (6) Attachment of the teichoic or teichuronic acid polymer to peptidoglycan, with release of the lipid carrier. (7) In some cases, conversion of UDPRPP to undecaprenol monophosphate (UDPRP).

in the precise details of the synthetic mechanisms in various organisms. Particularly perplexing questions are: 1) does synthesis occur on a lipid carrier; 2) in wall-associated teichoic acids, does the polymer synthesis occur on the linkage unit to peptidoglycan or separately, with the polymer added to the linkage units prior to peptidoglycan attachment; and 3) is newly synthesized teichoic acid added to previously synthesized peptidoglycan?

The essential elements of the proposed pathway of wall teichoic acid synthesis in *Staphylococcus aureus* are shown in Figure 3.4. It is likely that many organisms exhibit a similar mode of teichoic acid synthesis. In general, a UDP-bound molecule of an *N*-acetylated sugar is transferred to a molecule of monophosphorylated lipid carrier, in most cases, UDPRP. This process is inhibited by the antibiotic tunicamycin and results in formation of a

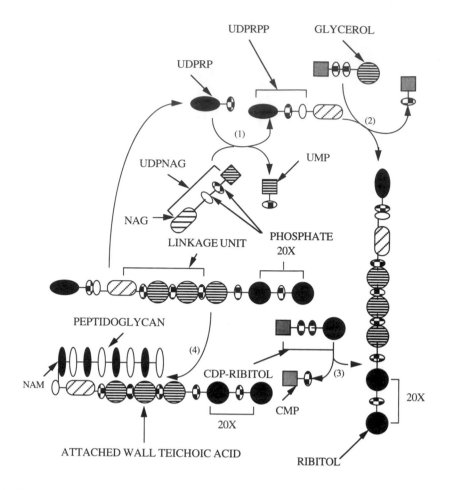

FIGURE 3.4 The apparent mode of synthesis of teichoic acid in *S. aureus* and the manner of its attachment to peptidoglycan. (1) Transfer of *N*-acetylglucosamine (NAG) from uridine diphosphate (UDP) to undecaprenol monophosphate (UDPRP) to form *N*-acetyl-undecaprenol diphosphate (UDPRPP)-bound NAG, with the release of uridine monophosphate (UMP). (2) Addition of glycerol residues to the UDPR-bound NAG. (3) Addition of ribitol residues joined by phosphodiester linkages to the growing polymer, with the release of cytosine monophosphate (CMP) molecules. (4) Attachment of the wall teichoic acid to peptidoglycan, via phosphodiester linkage to *N*-acetylmuramic acid (NAM) residues with the release of UDPRP.

molecule of uridine monophosphate (UMP) and a molecule of *N*-acetylated sugar attached to a molecule of a pyrophosphorylated lipid carrier. This entity then reacts sequentially with three molecules of cytosine diphosphate (CDP)-bound glycerol, producing the linkage unit that at a later time, allows attachment of the teichoic acid to peptidoglycan. The actual formation of teichoic acid occurs by the addition of CDP-bound ribitol residues to the linkage unit, producing a chain of linkage unit-associated ribitol residues bound to each other by phosphodiester linkages. The precise number of residues varies substantially among organisms.

The completed linkage unit-bound teichoic acid is attached to peptidoglycan by the interaction of the phospholipid carrier-bound, linkage unit-associated teichoic acid and a peptidoglycan molecule. This reaction yields a teichoic acid residue, bound via the linkage unit to peptidoglycan. The teichoic acid-peptidoglycan junction requires only one of the phosphate residues associated with the lipid carrier moiety of the linkage unit. The joining

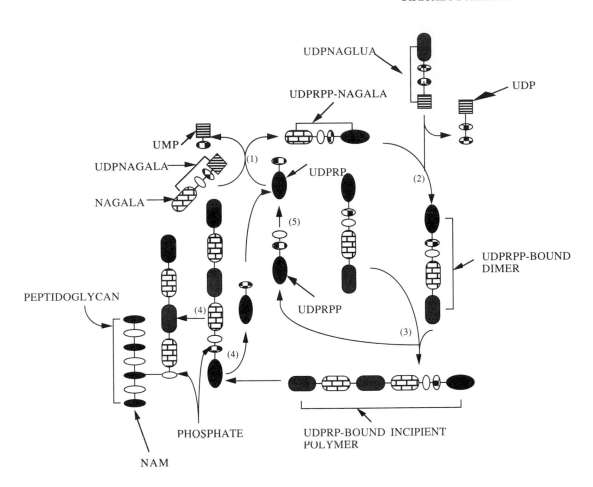

FIGURE 3.5 Teichuronic acid synthesis by *Bacillus licheniformis*. (1) Formation of undecaprenol diphosphate (UDPRPP)-bound *N*-acetylgalactosamine (UDPRPP-NAGALA) from uridine diphosphate (UDP)-bound NAGALA with the release of a molecule of uridine monophosphate (UMP). (2) Addition of *N*-acetylglucuronic acid (NAGLUA) from UDP-bound NAGLUA. (3) Addition of an NAGALA-NAGLUA dimer to the growing polymer, with the release of a molecule of UDPRPP. (4) Attachment of the completed polymer to peptidoglycan by a phosphodiester linkage to a molecule of *N*-acetylmuramic acid (NAM). (5) Regeneration of UDPRP.

of completed teichoic acid to peptidoglycan is thus accompanied by formation of a molecule of monophosphorylated lipid carrier, presumed in most cases to be UDPRP. The bulk of the evidence suggests that teichoic acid is added to newly synthesized peptidoglycan and not to previously existing material, although some studies suggest that teichoic acids may be added to previously synthesized peptidoglycan.

Teichuronic acid appears to be synthesized by mechanisms similar, or identical, to those used for teichoic acid synthesis. Representative examples of teichuronic acid synthesis in *Bacillus licheniformis* and *Micrococcus luteus* are shown in Figures 3.5 and 3.6. The basic synthetic mechanisms in both organisms are similar, but differ in certain details. For both organisms, synthesis is initiated by reaction of a nucleotide diphosphate-activated, *N*-acetylated sugar with a lipid monophosphate (UDPRP) carrier. The mechanism of this reaction leads, in *B. licheniformis*, to the formation of a molecule of *N*-acetylgalactosamine attached to a molecule of undecaprenol diphosphate (UDPRPP) and the concomitant formation of a molecule of UMP. The analogous reaction in *M. luteus* involves the reaction of UDP-bound *N*-acetylglucosamine with UDPRP, but produces a molecule of UDPRP-bound *N*-acetylglucosamine rather than the UDPRPP-bound *N*-acetylgalactosamine formed by *B. licheniformis*. The remaining synthetic events

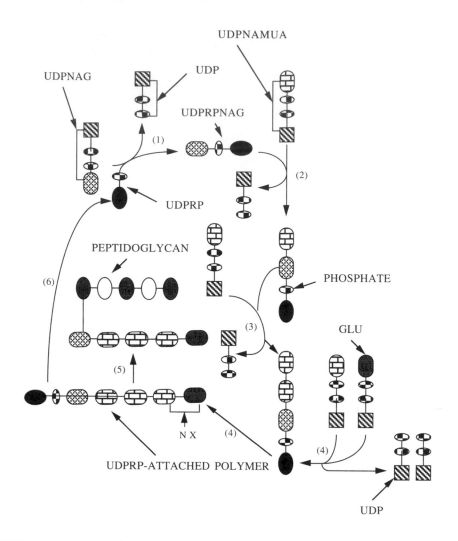

FIGURE 3.6 Teichuronic acid synthesis by *Micrococcus luteus*. (1) Transfer of uridine diphosphate (UDP)-bound *N*-acetylglucosamine (NAG) to undecaprenol monophosphate (UDPRP), with the release of UDP. (2) Addition of a molecule of *N*-acetylmannuronic acid (NAMUA). (3) Addition of a second molecule of NAMUA. (4) Addition of doublets of NAMUA and glucose (GLU). (5) Attachment to peptidoglycan. (6) Regeneration of undecaprenol monophosphate (UDPRP). The precise connection mechanism of teichuronic acid to peptidoglycan is uncertain.

occur on a monophosphorylated lipid carrier for *M. luteus* and on a diphosphorylated carrier in *B. licheniformis.*

For *B. licheniformis,* synthesis of the completed teichuronic acid involves reaction of a molecule of UDP-bound glucuronic acid with the *N*-acetyl-galactosamine-UDPRPP complex, producing a dimeric molecule of glucuronic acid and *N*-acetyl-galactosamine bound to a molecule of UDPRPP. The growing teichoic acid molecule is enlarged by addition of more UDPRPP-bound glucuronic acid-

N-acetylgalactosamine dimers, with the release of a molecule of UDPRPP at each addition event. Finally, the completed teichuronic acid is attached to peptidoglycan by phosphodiester linkage between a muramic acid residue in peptidoglycan and the terminal *N*-acetylgalactosamine of the teichuronic acid. This process is accompanied by release of a molecule of UDPRP.

Teichuronic acid synthesis in *M. luteus* is similar, but not identical, to the process in *B. licheniformis.* The initially formed UDPRP-bound

N-acetylglucosamine molecule reacts with two UDP-activated N-acetylmannuronic acid molecules to yield a UDPRP-bound trimer composed of one N-acetylglucosamine molecule and two molecules of N-acetylmannuronic acid. Elongation of the molecule occurs by the alternate addition of UDP-activated glucose and UDP-activated N-acetylmannuronic acid. Attachment of the completed teichuronic acid to peptidoglycan apparently occurs by phosphodiester linkage between the terminal N-acetylglucosamine residue of the teichoic acid and the terminal muramic acid residue in murein.

Additional features of gram-positive wall synthesis

Although synthesis of molecules of peptidoglycan, teichoic acids, or teichuronic acids and the formation of connections between them are essential elements of gram-positive cell wall synthesis, other processes are also important to wall formation. Most, if not all, of the teichoic acids are modified by the addition of amino acids, most notably D-alanine, and sugars to hydroxyl groups of the ribitol or glycerol residues. Although modification of teichoic acids is frequent, great variation is found in the nature and extent to which the modification occurs.

Formation of peptide linkages between peptidoglycan chains is the final major feature of gram-positive wall synthesis. During this process, the terminal D-alanine residue of the NAMPP component of murein is lost so that completed peptidoglycan involves side chains that contain only four amino acids in the peptide side chain. The precise linkages between peptidoglycan chains vary from organism to organism, but generally are one of four types: 1) direct connection between the carboxyl group of the terminal D-alanine of one tetrapeptide with the amino group of the third amino acid in the tetrapeptide side chain of another murein chain; 2) interconnection by a pentameric bridge of glycine or another L- or D-amino acid; 3) interconnection by a bridge composed of repeating peptide moieties identical to the tetrameric side chain of the muramic acid component of the peptidoglycan chain; and 4) interconnection between the carboxyl groups of the D-alanine and D-glutamic acid residues on one chain with two amino groups on the amino acids of another tetrapeptide side chain. It is the interconnections between murein chains that give the wall

its strength. Because of its greater peptidoglycan content and the larger number of peptide interconnections between murein strands, the gram-positive wall is inherently stronger than the gram-negative wall.

Gram-negative cell envelope formation

The gram-negative cell envelope is an inherently more complex structure than is the gram-positive cell wall. It is because of this complexity that we designate the rigidity structure of the gram-negative organism as an envelope, in recognition of the fact that it contains both an inner and an outer membrane. The relative morphological complexity of the gram-negative envelope is accompanied by more complicated biosynthetic and assembly mechanisms than those required for gram-positive wall synthesis. Despite differences in the nature and complexity of their walls, however, formation of both walls requires the accomplishment of similar tasks.

Synthesis of the gram-negative envelope requires all of the processes necessary for gram-positive organisms and quite a few more. Both organisms must synthesize murein but, in addition, the gram-negative organism must form lipopolysaccharide. Both types of organisms synthesize precursors of polymers in the cytoplasm and have mechanisms to transport polymer building blocks to their addition sites. The gram-positive microbe must transport murein precursors, while the gram-negative microbe must transport lipopolysaccharide precursors as well.

To a great extent, transport of cytoplasmic polymer precursors occurs in the same general way for both gram-positive and gram-negative organisms, that is, via lipoidal carrier molecules, undecaprenol for murein and lipid A for lipopolysaccharide. Although both gram-positive and gram-negative organisms transport cytoplasmically synthesized precursors outside the cytoplasmic membrane, the gram-positive organism has one major transport system, while the gram-negative organism has both lipid-carrier mediated transport and Bayer's junctions which are cytoplasmic membrane-outer membrane adhesions. For both gram-positive and gram-negative bacteria, addition of carrier-associated building blocks to extracytoplasmic membrane-associated polymers results in the release of carrier molecules. The regeneration of carrier molecules in an active state is a common requirement for cell wall biosynthesis in both gram-positive and gram-negative organisms.

Lipopolysaccharide synthesis

We have shown the similarity and dissimilarity of cell wall synthesis in gram-positive and gram-negative bacteria and have emphasized the greater complexity of the process in the latter. The presence of lipopolysaccharide in gram-negative microbes and the absence of lipopolysaccharide in gram-positive forms is a major factor in this greater complexity. The general nature of lipopolysaccharide is shown in Figure 3.7 and its synthesis is shown in Figure 3.8. The process may be divided into three major subprocesses: 1) lipid A synthesis; 2) synthesis of core polysaccharide; and 3) synthesis of the O antigen. Lipid A is synthesized at the interior edge of, or within, the cytoplasmic membrane with the aid of thymidine diphosphate (TDP)-bound glucosamine, cytosine diphosphate (CDP)-bound ethanolamine, and acyl carrier protein (ACP)-bound fatty acids. The core polysaccharide is synthesized on lipid A, which serves both as a primer and as a vehicle for transport of the combined lipid A-core polysaccharide moiety to its joining site with the O antigen portion of the incipient lipopolysaccharide molecule. This joining forms the complete lipopolysaccharide molecule at a location within the periplasm.

The building blocks for core polysaccharide are cytosine monophosphate (CMP)-bound ketodeoxyoctulonic acid, ATP-bound heptose, UDP-bound glucose, and CDP-bound ethanolamine. The O antigen portion of the lipopolysaccharide molecule is constructed at the inner surface of the membrane with UDP-carbohydrate precursors and undecaprenol as the vehicle for transport of the O antigen to the point at which it is added to form the complete lipopolysaccharide molecule. The proton motive force is generally accepted as the energy provider for transport of lipopolysaccharide precursors across the membrane.

Phospholipids and outer membrane proteins

The outer membrane of the gram-negative bacterium contains an abundance of both phospholipids and proteins. These materials are synthesized at either the cytoplasmic membrane or within the cy-

toplasm. Phospholipids are transported from their site of synthesis to the outer membrane with the aid of the proton motive force. Outer membrane proteins are apparently transported to the cell perimeter by passage through Bayer's junctions. Evidence for this suggestion comes from studies of the cellular fate of gold-labelled antibody proteins. Such proteins have been shown to be associated with Bayer's junctions through the use of sophisticated electron microscopy.

Assembly of the gram-negative envelope

The intrinsic chemical complexity of the gram-negative cell envelope dictates that its assembly is a more complicated process than that of the gram-positive wall. The multilayered nature of the gram-negative envelope contributes to this complexity since not only must the components of the outer membrane be assembled, but the outer membrane must be attached to the remainder of the envelope, in particular to peptidoglycan. The precursors of lipopolysaccharide, e.g., the O antigen portion and the lipid A-core polysaccharide moiety, are assembled within the periplasm by a specific translocase enzyme. It appears that completed lipopolysaccharide precursors are primarily incorporated into the outer membrane under the influence of hydrophobic forces as the result of the interaction of newly synthesized lipopolysaccharide precursors with existing lipopolysaccharide molecules. It also appears that direct interaction with existing lipopolysaccharide molecules is a major factor influencing the ultimate location of exported proteins within the outer membrane. Phospholipids seem to be primarily incorporated by hydrophobic interactions. The incorporation of outer membrane proteins into the outer membrane layer is apparently also a function of hydrophobic interactions.

Murein lipoprotein

The integrity of the gram-negative envelope is enhanced by interconnection between the outer membrane and murein. This interconnection is mediated by a lipoprotein known as murein lipoprotein, or Braun's lipoprotein (in honor of its discoverer). Murein lipoprotein is synthesized within the cell and contains a leader sequence of about 20 amino acids at its amino-terminal end. After transport through the cytoplasmic membrane, the leader sequence is removed, and the

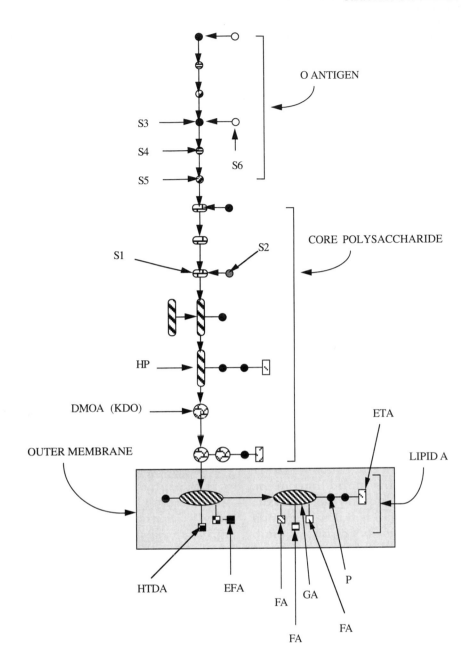

FIGURE 3.7 The general nature of lipopolysaccharide. The lipid A region is within the outer membrane. The core polysaccharide region is just exterior to the outer membrane. The O antigen region is most exterior and most hydrophilic. The abbreviations are the following: glucosamine (GA), ethanolamine (ETA), phosphate (P), 3-hydroxytetradecanoic acid (HTDA), fatty acid (FA) esterified to a glucosamine molecule, fatty acid (EFA) esterified to a glucose residue and also esterified to another fatty acid, deoxymannooctulosonic acid DMOA (KDO), heptose (HP), sugar residues (S). The various numbers associated with S symbols are different sugars. The lipid A and core polysaccharide regions are chemically similar or identical for taxonomically similar organisms, but the sugars in the O antigen region vary substantially within isolates of the same organism. The O antigen sugars occur as repeating tri, tetra, or penta sugar residues, allowing extensive antigenic variation within a common fundamental molecule. The O antigen residues often contain deoxy sugars.

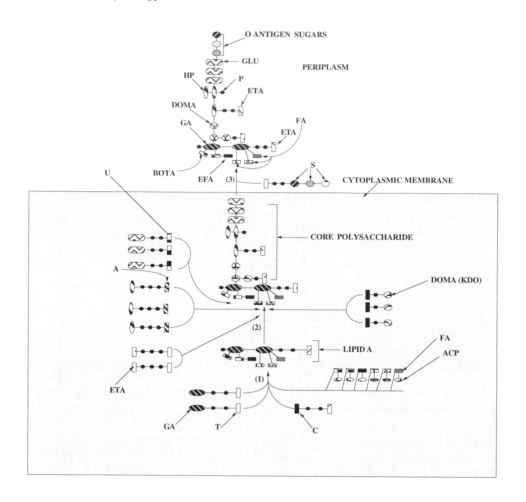

FIGURE 3.8 Lipopolysaccharide synthesis. (1) Formation of lipid A from thymidine (T) diphosphate-bound glucosamine (GA) and cytosine (C) diphosphate-bound-ethanolamine (ETA), plus acyl carrier protein (ACP)-bound fatty acids (FA). (2) Formation of core polysaccharide on lipid A from adenosine (A) triphosphate-bound heptose (HP), cytosine monophosphate-bound 3-hydroxy-mannoctulosonate (DOMA), thymidine triphosphate-bound ethanolamine, and uridine (U) diphosphate-bound glucose (GLU). (3) Addition of uridine diphosphate-bound sugar oligomers to the lipid A-core polysaccharide moiety to form the complete lipopolysaccharide molecule. The reactions involved in lipid A and core polysaccharide formation occur either within the cytoplasm or in the cytoplasmic membrane, but addition of the O antigen sugars occurs in the periplasm. Additional abbreviations: phosphate (P), sugar residues (S), Beta-hydroxytetradecanoic acid (BOTA).

terminal cysteine residue is modified by the addition of a glycerol molecule as a thioether. The remaining glycerol hydroxyl groups are esterified with fatty acids. These modifications collectively impart a hydrophobic character to the amino-terminal end of the lipoprotein and facilitate its integration into the outer membrane. The carboxy-terminal end of murein lipoprotein, which is oriented toward the cell interior, contains a lysine residue. A portion of murein lipoprotein molecules become attached to murein via peptide bond for-

mation between the amino group of a diamino-pimelic acid residue in the side chain of a muramic acid residue in the murein and the carboxyl group of the lysine residue in the murein lipoprotein. At the present time, it is not known why only a portion of the lipoprotein molecules are interconnected to murein or what determines which molecules are involved. What *is* known is that this interaction adds structural integrity to the wall. The general processes involved in gram-negative cell wall assembly are shown in Figure 3.9.

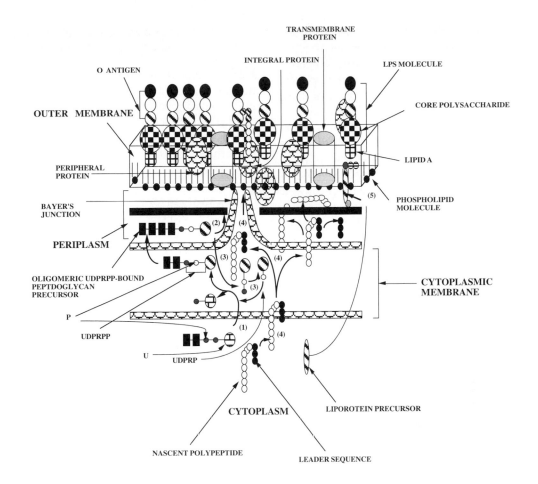

FIGURE 3.9 Critical processes involved in gram-negative envelope assembly. (1) Transfer of peptidoglycan precursor from uridine (U) diphosphate (P) to undecaprenol monophosphate (UDPRP) for transport to the periplasm. (2) Addition of periplasmically formed peptidoglycan oligomers to growing peptidoglycan polymer. (3) Regeneration of UDPRP from undecaprenol diphosphate (UDPRPP). (4) Transport and processing of nascent polypeptide. Removal of the leader sequence may be followed by many consequences. Membrane proteins incorporate into the membrane because of their relatively hydrophobic character, while periplasmic proteins reside in the periplasm, at least in part because of their relative hydrophilicity. Incorporation of outer membrane proteins involves interaction of their amino-terminal ends with lipopolysaccharide (LPS) molecules. (5) Modification of lipoprotein precursor molecule and its attachment both to the outer membrane and to peptidoglycan. See the text for additional commentary.

Exopolysaccharide synthesis

Exopolysaccharides are produced by the great majority of bacteria and by a variety of eucaryotes as well. When morphologically discrete, they are referred to as capsules, but when they are extruded into the extracellular environment and are not cell-associated, they are considered slime layers. Most researchers do not consider exopolysaccharides as part of the cell proper, however, this in no way diminishes their physiological importance.

Whether or not one regards exopolysaccharides as part of the cell or as associated molecules, exopolysaccharide synthesis is a process that involves both the external and internal environment in the majority of cases. In all but one case, exopolysaccharide synthesis involves uptake of materials from the environment, intracellular polymer synthesis, and return of the completed polymer to the cell surface or to the external environment. Exopolysaccharide synthesis may thus be used as a model for understanding transport of both small and large molecules. In addition, both the physiological state of the organism and the nature of the external conditions influence exopolysaccharide formation. These facts allow exopolysaccharide to be used as a model of how the cell responds to environmental change. Many of the building blocks for exopolysaccharides, and the manner of their utilization, are similar to the sequences involved in synthesis of components for both the gram-positive and gram-negative cell walls. Study of the relationship of exopolysaccharide synthesis to cell structure synthesis allows understanding of both the similarities and differences between these processes and the ways in which synthesis of cellular entities that share common processes and intermediates are coordinated and regulated.

Physiological questions posed by exopolysaccharide synthesis

Understanding of exopolysaccharide synthesis requires consideration of a number of questions. In most cases, the answers to the questions are incomplete, but reflecting upon them provides a framework for understanding exopolysaccharide synthesis and its relationship to similar processes. Initially, we must understand what spectrum of substrates can be used for synthesis. Why does an organism use one, or a few, materials for synthesis? If, as is usually the case, synthesis is intracellular, what are the factors that regulate substrate uptake? It is equally important to understand the manner in which the completed intracellularly synthesized large molecule is returned to the cell exterior and the mechanisms that determine whether, and in what manner, the polymer is attached to the cell surface. One must also understand precisely how the polymer is synthesized. This necessity requires that we understand the precise nature and form of the building blocks and what, if any, auxiliary molecules may be required for synthesis and transport. Finally, exopolysaccharide synthesis shares much in common with synthesis of the gram-positive and gram-negative cell envelopes as well as the synthesis of intracellular polysaccharide storage granules. Full understanding of the implications of exopolysaccharide synthesis requires understanding of the details of polymer synthesis, but it is equally important to understand how exopolysaccharide synthesis relates to other similar, but not identical, cell synthetic processes.

Physiological factors that influence exopolysaccharide formation

Exopolysaccharide formation is influenced by a number of physiological factors. Taken collectively, microbes form exopolysaccharides from a diversity of starting materials, including carbohydrates, amino acids, and hydrocarbons. It is interesting to note that exopolysaccharide formation may occur in organisms that do not use carbohydrates efficiently, e.g., *Myxococcus xanthus.*

Exopolysaccharide synthesis is often influenced by whether or not growth is limited. In many cases, exopolysaccharide formation is more extensive during growth limitation than when growth is abundant, but the type of limitation is normally not of great consequence. In *E. coli* and *E. aerogenes,* exopolysaccharide formation has been shown to be affected by limitation of nitrogen, phosphorus sulfur, or potassium, with little or no effect upon the chemical nature of the polymer. Metal ions, particularly Ca^{2+}, Mg^{2+}, K^+, and Fe^{2+} may affect exopolysaccharide formation. Both stimulatory and inhibitory effects have been shown. Ions that are required for uptake of substrates or for their use in polymer synthesis are stimulatory. Sometimes, change in the ion concentration can also influence polymer composition, via modification of enzyme activity.

Modes of exopolysaccharide synthesis

Exopolysaccharide synthesis may be characterized as occurring by one of five modes. The various modes of exopolysaccharide synthesis are outlined in Figure 3.10. In four of the modes, synthesis occurs intracellularly—using externally supplied precursors or their metabolites. In the fifth mode, synthesis occurs at the cell surface, negating the necessity for transport of either precursors to the intracellular environment or of the completed polymer to the cell surface. This mode of synthesis is exhibited by members of the genera *Streptococcus* and *Leuconostoc*. These organisms require a specific carbohydrate, sucrose, and produce exopolysaccharide with the aid of two enzymes, dextran sucrase or levan sucrase. With dextran sucrase, the sucrose molecule is cleaved to a dextran (glucose) polymer and free fructose is liberated. With levan sucrase, the inverse process occurs, that is, a poly-

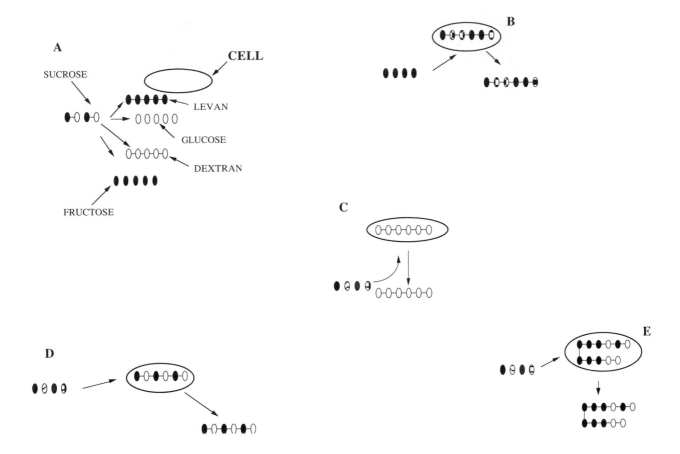

FIGURE 3.10 The various modes of exopolysaccharide formation. (A) Formation of a levan or a dextran at the cell surface from sucrose with the release of either glucose (levan) or fructose (dextran). (B) Intracellular formation, and subsequent excretion, of a heterogeneous polymer from a specific homogeneous extracellular substrate. (C) Formation, and subsequent excretion, of a homogeneous intracellular polymer from a diversity of heterogeneous extracellular substrates. (D) Formation of a heterogeneous intracellular polymer with a repeating unit from a diversity of extracellular substrates. (E) Intracellular formation, and subsequent excretion, of a heterogeneous polymer without a repeating unit from a diversity of extracellular substrates.

mer of fructose (levulose) is formed and free glucose is released. This mode of exopolysaccharide synthesis does not require either nucleotide-activated sugars or lipid carrier molecules.

The remaining modes of exopolysaccharide synthesis involve intracellular synthesis of the polymer and usually require nucleotide-activated sugar precursors. Some organisms produce intracellularly synthesized heteropolysaccharide from a single specific extracellular substrate. Others produce homogeneous intracellularly formed polymers from a variety of extracellular substrates. A third mode of nucleotide-sugar-mediated exopolysaccharide formation produces an intracellular heteropolymer with a repeating unit from a variety of extracellular precursors. The final mode of nucleotide-sugar-mediated exopolysaccharide formation produces intracellular polymers composed of two monomeric units, but with no repeating unit.

The importance of dinucleotide precursors

Except for levans and dextrans, exopolysaccharide production requires formation of polysaccharide monomers in a nucleotide-activated form. In most cases, the monomers are activated to the dinucleotide form with the aid of nucleotide pyrophosphatase enzymes. In the pyrophosphatase reaction, a monomeric sugar-1-phosphate reacts with a nucleotide triphosphate to form a monomeric sugar-nucleotide diphosphate and inorganic pyrophosphate. Although various nucleotide diphosphates may participate in carbohydrate polymer formation, UDP and GDP derivatives are most common. In procaryotes, UDP-glucose is centrally important because it is a precursor to intermediates of both exopolysaccharide and lipopolysaccharide. Glucose-1-phosphate, from which UDP-glucose is formed, is most important as a procaryotic carbohydrate polymer precursor because it may be acted upon by two pyrophosphatase enzymes, UDP-glucose pyrophosphorylase and ADP-glucose pyrophosphorylase that produce UDP-glucose and ADP-glucose, respectively. UDP-glucose gives rise to exopolysaccharide and lipopolysaccharide, while ADP-glucose leads to carbohydrate storage, e.g., glycogen formation. UDP-glucose also serves as a precursor of other substances such as D-galactose and D-glucuronic acid. Although UDP-glucose is important in procaryotic polysaccharide synthesis, guanosine diphosphate (GDP) is also important. In most cases, bacterial exopolysaccharide formation requires the dinucleotide form of the building block, but in a few cases, such as sialic acid (a polymer of N-acetylneuraminic acid), a monomer mononucleotide phosphate precursor (cytidine monophosphate) is required.

The involvement of lipid-bound precursors

Numerous studies have shown that, for many organisms, exopolysaccharide synthesis requires participation of lipid carrier molecules as intermediates. Studies with both *Enterobacter aerogenes* and *Acetobacter xylinum* implicate undecaprenol diphosphate as the most probable lipoidal substance involved. In some systems, e.g., sialic acid formation in *E. coli,* a lipid monophosphate is involved. In contrast, it appears that colanic acid formation by the organism requires both an isoprenoid lipid diphosphate and an undecaprenol monophosphorylated form of glucose. Thus it is possible that more than one type of lipid may be required for exopolysaccharide synthesis. However, in a number of systems, no evidence for lipid involvement in exopolysaccharide formation has been found.

Polymer formation

By whatever means the building blocks for polymers are formed, with or without the involvement of lipid intermediates, the polymeric molecule must be formed. Two general mechanisms are known by which this process might occur: 1) addition of the building block to the nonreducing end of the growing polymer and 2) addition of the building block to the reducing end of the existing, growing polymer. Both mechanisms are known to exist, the first exemplified in the formation of glycogen by *E. coli* and the second in the formation of heteropolysaccharide side chains in *Salmonella* species.

Excretion of the completed polymer and attachment to the cell surface

The precise mechanisms by which intracellularly synthesized exopolysaccharides are excreted to the cell surface or to the external environment remain, in large measure, elusive. It is likely that Bayer's junctions are substantially involved in gram-negative bacteria, but other mechanisms must exist in gram-positive species. Perhaps lipid carrier-mediated and proton motive force-facilitated processes occur in certain organisms, but at this point, the precise ways in which exopolysaccharide is returned to the cell surface or the cell exterior in gram-positive bacteria are not well understood. The factors that determine whether or not the exported polymeric materials are attached to the surface or are excreted to the external environment are equally obscure. Questions also remain regarding the attachment mechanisms for those polymers that adhere to the cell surface. What are the mechanisms by which attachment occurs? Three reasonable locations for polymer attachment seem to be: 1) attachment to the cell wall surface; 2) attachment within the cell wall; and 3) attachment interior to the cell wall. By what mechanism does attachment occur, what determines precisely where it occurs and, when does a cell "decide" that it contains "enough" cell-associated material? Evidence suggests that the answer may revolve around the presence of receptor sites at the cell surface, although it is unlikely that a consistent type of acceptor applies in all situations. In *E. coli,* some studies suggest that the attachment materials are outer membrane pro-

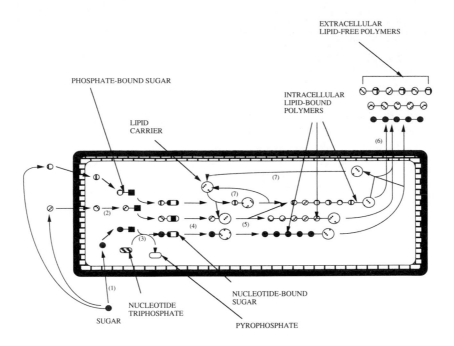

FIGURE 3.11 Processes involved in intracellular synthesis of exopolysaccharide. (1) Transport of sugars into the cell. (2) Conversion of the intracellular sugar to the sugar 1-phosphate. (3) Formation of activated nucleotide phosphate-bound sugar building blocks. In different circumstances, the latter may be either diphosphated or monophosphated. In most cases, the process is accomplished by pyrophosphatase enzymes, with release of a molecule of pyrophosphate. (4) Transfer of the activated sugar to a lipid carrier, which serves to allow both polymerization (5) and transport to the cell surface for export (6). (7) Regeneration of the lipid carrier.

teins specifically associated with capsule formation. Whatever the attachment sites, it is reasonable to suggest that there is an upper limit of them on a cellular basis, and that when the available sites are occupied, the additional exopolysaccharide is released into the extracellular environment as slime. The various processes involved in exopolysaccharide synthesis and cell surface attachment are shown in Figures 3.11 and 3.12.

Regulation of exopolysaccharide formation and function

Many factors regulate exopolysaccharide synthesis and function. Any factor that alters the uptake of substrates for intracellularly synthesized polymers may alter exopolysaccharide synthesis. Regulation of formation and function of the enzymes responsible for building block synthesis and activation, regulation of polymerizing enzymes, and, in the case of lipid-mediated polymer synthesis, of carrier synthesis are all potential means of regulating the formation of exopolysaccharide. With regard to

lipid-mediated synthesis, not only the amount of the material, but also its form must be considered, since it is well documented that certain processes require the diphosphorylated form of the carrier and others require the monophosphated form. Finally, with regard to exopolysaccharide, regulation of surface receptor site formation may regulate polymer function, since it is likely that such regulation determines the extent to which potential exopolysaccharide is either surface-associated or released to the external environment.

Relationships between exopolysaccharide synthesis and cell wall synthesis

---

Exopolysaccharide formation and cell wall formation are complicated processes. However, in spite of their complexity, and the mechanistic differences between them, they share some commonalities. In all cases, building blocks must be formed

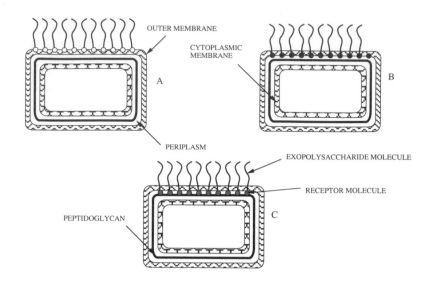

FIGURE 3.12 Possible modes of attachment of exopolysaccharide to the cell surface. (A) Attachment to the cell wall exterior with the aid of cell surface protein receptor molecules. (B) Attachment to receptors within the cell wall. (C) Attachment within the cell at a location other than the wall. Some researchers suggest a periplasmic site. The diagrams depict an idealized gram-negative cell, although analogous locations are possible with gram-positive cells (except for the periplasm).

and combined into polymeric molecules. In most cases, building block precursors must be transported from the cell exterior to their site of metabolism within the cell. In the gram-positive and gram-negative bacteria, it is often the case that precursors of polymeric molecules are formed within the cell and added to a polymeric molecule outside the cell membrane. In exopolysaccharide formation, the completed polymer is usually formed within the cell and transported to the cell periphery or to the external environment. However, the mechanisms of transport for polymers and their building blocks are not identical in the three situations. In gram-positive bacteria, lipoidal substances are involved, to a great extent, as carrier molecules for transport of polymer building blocks from their synthesis site to their addition site, as well as vehicles upon which synthetic activities can occur. Similarly, in gram-negative bacteria, lipoidal materials are intimately involved in membrane transport-associated processes and are also used as vehicles for synthesis, but the nature of the lipoidal materials involved in the various processes of wall synthesis by gram-positive and gram-negative bacteria are, in many cases, different. Similar processes, e.g.,

polymer synthesis, must be accomplished in all three situations but subtle differences separate them.

The extent to which gram-positive and gram-negative cell wall synthesis and exopolysaccharide synthesis are interrelated reflects many factors but is particularly a function of the extent to which the potentially related processes share common intermediates. Many processes accomplished in all three situations require activated nucleotides and are thus reflective of the general energy state of the cell. However, processes may differ in the particular nucleotides involved, allowing separation of processes in the midst of similarity. Although many of the processes accomplished during cell wall synthesis of any kind and during exopolysaccharide synthesis involve lipid materials, regulation of these processes is, in many cases, accomplished by regulation of the relative amounts of required lipids. Regulation and coordination of cell wall and exopolysaccharide synthesis may also be achieved by regulation of the amounts and forms of the sugars that are available for nucleotide activation. In addition to these considerations, full appreciation of the implications of cell wall and exopolysaccharide synthesis cannot be obtained

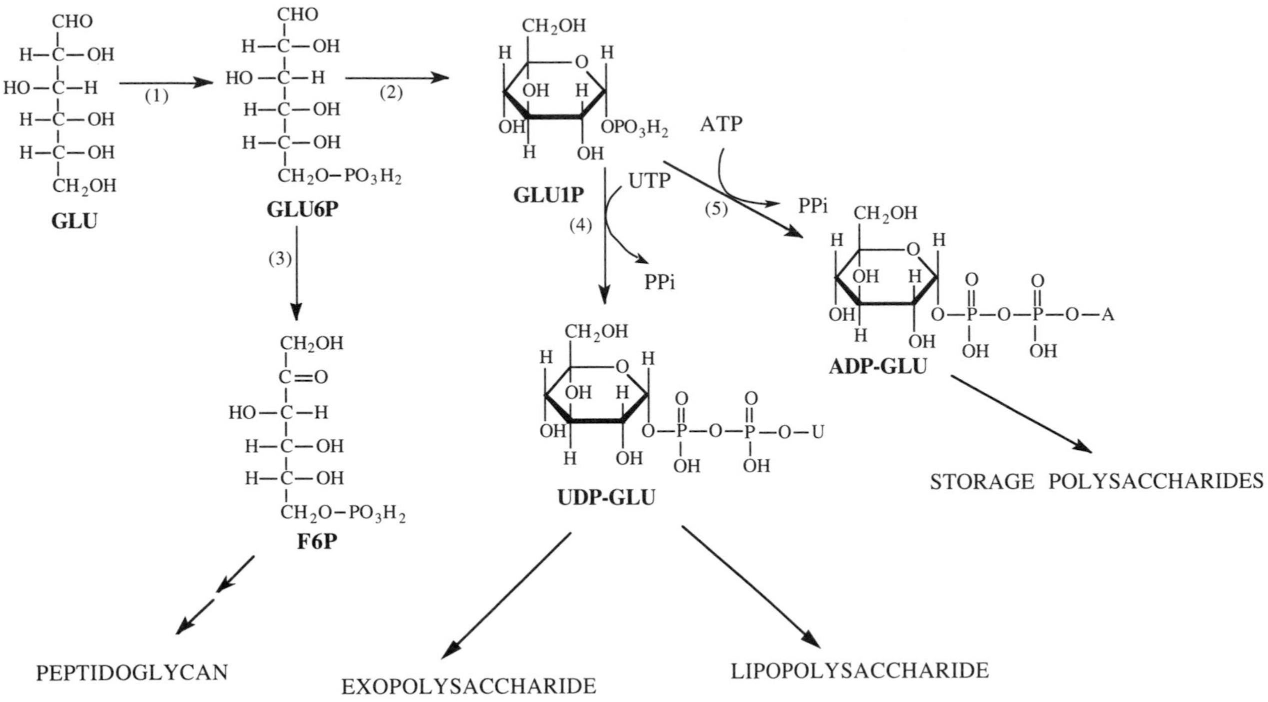

FIGURE 3.13 Metabolic interrelationships between synthesis of intracellular and extracellular carbohydrate polymeric compounds. The central role of glucose-6-phosphate is apparent. (1) The conversion of glucose (GLU) to glucose-6-phosphate (GLU6P), by hexokinase. (2) Isomerization of GLU6P to glucose-1-phosphate (GLU1P) by phosphoglucomutase. (3) Conversion of GLU6P to fructose-6-phosphate (F6P) by phosphohexose isomerase. (4) Formation of uridine diphosphate glucose (UDP-GLU) by uridine diphosphate (UDP)-glucose pyrophosphorylase. This process leads to the formation of exopolysaccharide and lipopolysaccharide. (5) Conversion of GLU1P to adenosine diphosphate glucose (ADP-GLU) by adenosine diphosphate (ADP)-glucose pyrophosphorylase. This process leads to formation of polysaccharide storage granules, particularly glycogen, in procaryotic organisms.

without consideration of the manner in which these processes are interrelated to the formation of nonstructural materials that also arise from common intermediates, for example carbohydrate polymer storage compounds. Understanding of the full implications of structure synthesis is not an easy task. Some relationships between exopolysaccharide formation and other cell processes are shown in Figure 3.13.

bial physiology, it is not possible to fully understand the consequences of a particular cellular process without considering, at the same time, its influence on other cell processes and the effect of the environment upon the total physiology of the cell.

Summary

This chapter has shown, through selected examples, the diversity and complexity of microbial structure formation and, at the same time, the extent to which apparently different processes are, at least in principle, similar. The interrelatedness of structure synthesis is also apparent. In all of micro-

Selected References

Archibald, A. R. 1974. The structure, biosynthesis and function of teichoic acid. *Advances in Microbial Physiology*. **11:** 53–95.

Cabib, E., R. Roberts, and B. Bowers. 1982. Synthesis of the yeast cell wall and its regulation. *Annual Review of Biochemistry*. **51:** 763–794.

Costerton, J. W., J. M. Ingram, and K. J. Cheng. 1974. Structure and function of the cell envelope of gram-negative bacteria. *Bacteriological Reviews*. **38:** 87–110.

Michaelis, S., and J. Beckwith. 1982. Mechanisms of incorporation of cell envelope proteins in *Escherichia coli*. *Annual Review of Microbiology*. **36:** 435–465.

Rogers, H. J. 1979. Biogenesis of the wall in bacterial morphogenesis. *Advances in Microbial Physiology*. **19:** 1–62.

Rothfield, L., and D. Romeo. 1971. Role of lipids in the biosynthesis of the bacterial cell envelope. *Bacteriological Reviews*. **35:** 14–38.

Shockman, G. D., and J. F. Barrett. 1983. Structure, function, and assembly of cell walls of gram-positive bacteria. *Annual Review of Microbiology*. **37:** 501–528.

Sutherland, J. W. 1982. Biosynthesis of microbial exopolysaccharide. *Advances in Microbial Physiology*. **23:** 79–180.

Ward, J. B. 1983. Teichoic and teichuronic acids: Biosynthesis, assembly and location. *Microbiological Reviews*. **45:** 211–243

CHAPTER 4

The Physiological Implications of Nutrition

Understanding of nutrition is critical to understanding the physiology of microbes. We use the spectrum of nutrients required by an organism as a means of classifying it. In addition, the study of an organism requires that we grow it and we cannot grow an organism without an understanding of the nutritional and physiological conditions that it requires for growth. Some of the substances required for growth of microbes are shown in Tables 4.1, 4.2, and 4.3. Consideration of these tables reveals that certain general substances are required by all organisms while variation occurs in other requirements. In addition, materials required by particular organisms may be either inorganic or organic and may be required in either macro or micro amounts.

Nutrition as an indication of physiological complexity

It is usually the case that the nutrients required by an organism directly reflect its physiological capacities. In general, the fewer things the organism requires, the more extensive its physiological complexity. Conversely, the more nutrients an organism requires, the less complex its physiological machinery. For example, many members of the genus *Lactobacillus* may require a number of amino acids, B vitamins, and other nutrients before growth is possible. In contrast, autotrophic microbes may require only sunlight, carbon dioxide,

TABLE 4.1 Some Functions of Various Macroelements Required for the Growth of Microbes

ELEMENT	MAJOR FUNCTION(S)
Carbon	Major component of all components of protoplasm
Nitrogen	Major component of amino acids in all cell proteins; Major component of purine and pyrimidine rings of nucleic acids; Substantial component of murein and chitin; Substantial component of complex lipids; As nitrate, alternate acceptor of electrons in electron transport
Oxygen	A substantial part of all of the organic components of protoplasm; As oxygen gas, acceptor of electrons in aerobic electron transport
Phosphorus	In the form of phosphate, a major component of nucleic acids, teichoic and teichuronic acids, and various membrane phospholipids; As pyrophosphate, inorganic energy storage compound for some microbes
Sulfur	Significant component of the amino acids cysteine, methionine, homocysteine, and cystathione; A significant part of lipoic acid; Various reduced forms may serve as energy sources for chemotrophs or as sources of reducing power for phototrophs; As sulfate, may serve as an electron acceptor; By sulfate reduction, may be assimilated

TABLE 4.2 Some Functions of Various Inorganic Ions Required for the Growth of Microbes

ION	FUNCTION
K^+	Maintenance of osmotic balance; Cofactor for some enzymes
Na^+	Required for use of certain organic compounds, e.g., citrate, by certain microbes; Obligately required as enzyme activator and for transport in certain halophiles
Fe^{2+}	Component of cytochromes, heme-containing enzymes, and nonheme iron electron transport compounds; By oxidation to Fe^{3+}, energy source for members of the genus *Ferrobacillus*
NH_4^+	Preferred nitrogen source for many bacteria; Energy source for members of the genus *Nitrosomonas* and physiologically similar organisms
Mg^{2+}	Enzyme activator, particularly for kinase reactions; Component of chlorophyll
Co^{2+}	Component of vitamin B_{12} and its coenzyme derivatives
Ca^{2+}	Activator of enzymes, particularly protein kinases; Chelated by picolinate in the bacterial endospore
NO_2^-	Energy source for *Nitrobacter*
NO_3^-	Acceptor for electron transport for certain bacteria

and nitrogen gas. The complexity of their physiological machinery is such, that from these three things only (plus a few minerals), they may synthesize all that is required for growth. Most organisms fall between these two extremes. Study of the physiological differences between organisms that require different spectra of nutrients allows us to understand the differences in both their physiological properties and in the ways in which they respond to environmental change. At the same time, through the study of nutrient utilization we may obtain enhanced understanding of the ways in which various organisms accomplish the common tasks of life.

The commonality of growth requirements

Whatever is specifically required to allow the growth of an organism, all cellular organisms must be supplied with certain things (Table 4.1). This fact reflects the commonality of protoplasm and the necessity of accomplishing certain synthetic tasks. Although great variation is found in the specific requirements for growth, in general terms, the nature and functions of growth substances are common for all organisms. All microbes require a carbon source because carbon is a component of all the substances that constitute protoplasm. In addition, all organisms require energy and it is frequently the case that carbonaceous compounds serve as energy sources, in addition to their role in biosynthesis. Nitrogen is universally required because it is a component of proteins, nucleic acids,

TABLE 4.3 Some Functions of Various Organic
Growth Factors Required for the Growth
of Certain Microbes

SUBSTANCE	FUNCTION
Heme and related tetrapyrroles	Component of cytochromes
Vitamin K-like compounds	Electron transport carrier
Methionine	Methyl-donating agent
Coenzyme M	Electron carrier for methanogenic bacteria
Purines and pyrimidines	Components of nucleic acids
Amino acids	Components of proteins
Nicotinic acid	Electron carriers in the form of NAD and NADP; As NADPH source of reducing power for biosynthesis; Dehydrogenation reactions
Riboflavin	Electron transport as FAD or FMN; Dehydrogenations
Biotin	Carboxyl transfer reactions; Carbon dioxide fixation
Thiamine	Decarboxylations
Folic acid	One carbon transfers, Methyl donation
Cobalamin	Rearrangements
Pyridoxine and similar substances	Transamination and deamination
Pantothenic acid	Coenzyme A; Activation, energy metabolism; Fatty acid metabolism

phospholipids, and both chitin and murein, among other substances. Sulfur is obligatorily required because it is an essential part of both methionine and cysteine and is also a component of coenzyme A and lipoic acid, as well as iron-sulfur proteins. In addition, some organisms use inorganic sulfur compounds in energy metabolism. Phosphorus is uniformly required because it is a component of essential cell components such as nucleic acids, teichoic and teichuronic acids, and phospholipids, and in addition, is essential for energy generation in all living things. Among other reasons, water is required because it constitutes the majority of the cell's weight and because of its essentiality to life processes. Minerals are required primarily to facilitate critical cell activities such as enzyme reactions

or transport processes. Whether or not an organism requires additional materials reflects its physiological complexity. Those organisms with substantial physiological complexity may be capable of synthesizing all of their essential substances and accomplishing all their life processes from the materials listed in Tables 4.1 and 4.2. Others may require additional materials because of unique features of their physiology. When organic materials are required (Table 4.3), it is often the case that the requirement results from a deficiency in synthesis of the required substance. An organism may be unable to synthesize the required material at all or unable to synthesize it fast enough to allow growth under a particular set of conditions.

The bacteria as nutritional models With regard to their nutrient requirements, the bacteria are the most diversified organisms in nature. Within the bacteria, examples can be found that represent the entire spectrum of nutritional types. This diversity reflects, in large part, the tremendous versatility of bacteria in their ability to use various carbon and nitrogen sources and also their remarkable metabolic capacity. Since the bacteria and other microbes are a part of the continuum of cellular life, they may be used as models for understanding the nature and functions of the nutrients required by all cells. Normally, if we understand the nature and function of a nutrient in a microbial system, we may understand its nature and function in all cellular forms.

Nutrition implies much more than what an organism needs to grow We frequently think of nutrition as a description of the collection of materials required to support the growth and reproduction of an organism. In this context, we may distinguish *macronutrients*, substances required in large amounts, and *micronutrients*, materials required in only small amounts. In addition, we may describe nutrients in terms of whether they are organic or inorganic, and classify organisms accordingly. We classify organisms as heterotrophs if they use organic carbon and as autotrophs if carbon dioxide is their carbon source. Along with consideration of the nature of an organism's energy source and its source of reducing power, we may use an organism's nutritional requirements to distinguish it from other organisms for taxonomic as well as for physiological purposes.

Critical physiological implications
of a nutrient requirement

Although we classify organisms on the basis of the things they require for growth, full understanding of the significance of a nutrient requirement depends on understanding the functions of the nutrients. To understand the function of a nutrient, we must understand the answers to a number of questions. We must know how much of the nutrient is required, the specificity of the requirement, and the form in which the nutrient is used. In addition, we must understand the mechanism, or mechanisms by which the nutrient is taken up from the external environment and the factors that influence uptake. Comparison of the intracellular form of the nutrient with its extracellular form is also important because it indicates whether or not the material is modified during transport. In addition, study of the precise intracellular location of the nutrient is useful for at least two reasons. Determination of the precise cellular location of a nutrient may reveal whether or not the nutrient serves one or several functions, which provides guidelines regarding the most fruitful approaches to the study of the details of nutrient utilization. Secondly, it is useful to study the relationship between uptake of a nutrient and its intracellular use. Such information provides insight into the relative rates of uptake and utilization and allows estimation of the extent to which an intracellular pool of the material may accumulate. Information regarding all of the aspects of nutrient use is required in order to understand, precisely, the function(s) that the nutrient serves for the organism. Only when all of these aspects are studied can the full implications of the nutrient requirements for an organism be understood.

The amount of a nutrient required does
not determine its importance

The quantitative requirements for a nutrient are often indicative of its general function. Unless it is used for energy, we may suspect that a quantitatively large nutritional requirement reflects its use for synthesis of cell macromolecules or structures, while a quantitatively small requirement is likely to reflect its use as a cofactor or a coenzyme. However, the quantitative requirement for a nutrient does not indicate its importance to the organism.

While it is certainly true that one cannot synthesize proteins without amino acids that are required in relatively large amounts, the proper functioning of those proteins often requires cofactors and carrier molecules. Although these molecules are required in much smaller amounts than amino acids, they are no less important to the microbe than are substances required in much larger quantities. To properly assess the importance of a particular nutrient we must relate it to function.

Specificity and form

The specificity of a nutrient requirement for an organism reflects the extent to which the material itself is essential to the organism or whether the substance can be replaced by other substances. When the latter situation exists, it is a function of the metabolic versatility of the organism.

The implications of metabolic versatility are many. Many organisms could be used as examples of this statement, but consideration of the use of nutrients by *Prevotella* (*Bacteroides*) *ruminicola subsp. ruminicola*, a predominant gastrointestinal anaerobe, is particularly instructional. This organism was originally thought to require protoheme for growth. Spectral and biochemical studies showed that the protoheme requirement reflected the presence of a functional, cytochrome b-containing electron transport system in the organism. The protoheme requirement resulted from the organism's inability to synthesize the tetrapyrrole nucleus. Taxonomically similar organisms (e.g., *Prevotella ruminantium subsp. brevis*) that did not require protoheme synthesized the tetrapyrrole nucleus by classical pathways.

Although it was shown that *Prevotella ruminantium subsp. ruminicola* required protoheme, further study showed that the organism grew when protoheme was replaced with a wide variety of metalloporphyrin compounds, substances similar to protoheme but with minor chemical modifications of the tetrapyrrole nucleus. In addition, the organism used both iron-containing and iron-free tetrapyrroles, as well as nonferrous metal-containing tetrapyrroles, as growth factors. Eventually it was shown that the organism can remove nonferrous metals from a wide variety of tetrapyrroles, replace them with iron, and form a functional cytochrome with tetrapyrroles other than protoheme. The requirement for protoheme is, in fact, a requirement for tetrapyrroles, a variety of which may serve a physiological function similar to that of protoheme. The organism uses a variety of metallotetrapyrrole compounds because it contains metalloporphyrin cheletase enzymes. The organism is specific in its requirement for tetrapyrroles, but because of its metabolic versatility, can use a variety of tetrapyrrolic compounds and their various metal chelates to achieve a common functional purpose.

Prevotella ruminicola displays another nutritional principle. It uses a wide variety of materials as nitrogen sources, but the diversity of the sources is physiologically unusual. The organism uses the ammonium ion, the cyanide anion, peptides such as oxytocin and vasopressin, and oligopeptides containing five to ten amino acids, but under many growth conditions, fails to use amino acids efficiently, if at all. The ability of the organism to use a wide variety of nitrogen sources and its relative inability to use amino acids reflects both the organism's metabolic diversity and its difficulty in amino acid transport. It can transport ammonia, cyanide, and oligopeptides into the cell and use them to make proteins and other nitrogenous cell components, but under most conditions, cannot transport amino acids and high molecular weight, organic nitrogen compounds. The ability of the organism to use relatively high-molecular weight organic nitrogenous materials is a function of extracellular peptidases that degrade these substances to transportable forms of nitrogen. A wide variety of utilizable nitrogen sources are converted to a collection of similar or identical intracellular nitrogen compounds. The ability of the organism to use particular nitrogen sources reflects both its enzymatic complement and its transport systems.

Uptake

If a nutrient is to be used, it must be taken into the cell. The exact details of this nutrient uptake are discussed in chapter 6. Factors such as the form of the material, the precise uptake mechanism(s), and the nature of the physiological environment to which the organism has been previously exposed influence nutrient uptake. In many cases, these factors are interrelated. Study of factors that influence nutrient uptake and use has been extremely useful in understanding phenomena as diverse as membrane transport, structure synthesis and function, the control of protein formation, and the manner in which microbes respond to environmental change.

Nutritional mutants as physiological probes

Much information about cellular processes has been obtained from study of nutritional mutants. Early work with mutants of *Neurospora crassa* led to an understanding of the nature of a gene, a concept central to all of biology. The relationship between genes and enzymes was also recognized. The discovery that nutritional mutants could grow when a required substance was supplied led to the recognition that mutation prevented synthesis of the required material through inactivation or inhibition of a synthetic enzyme(s). This understanding led to the realization that in normal metabolism, materials are synthesized by the coordinated action of a collection of genes, each of which is responsible for a distinct protein. Such interconnected enzymatic sequences are known as *metabolic pathways*. Further study revealed that the genes required for synthesis of many substances in procaryotes are often physically adjacent on the chromosome, a less frequent occurrence in eucaryotic cells.

It was largely through study of nutritional mutants that it was recognized that development of a nutritional requirement sometimes reflects the loss of a transport system. In addition, it was recognized that transport as well as metabolism, is often mediated by proteins and that the kinetics and activity of transport proteins are often similar to, and may be described by, the equations that describe enzyme activity. Our understanding of all of these concepts was derived, in large measure, from study of nutritional mutants.

Structure formation and function

Nutritional mutants are useful for understanding of structure formation and function. For example, flagellar mutants have been obtained that fail to form flagella, form structurally altered flagella, or form nonmotile flagella. It is often the case that mutants of this nature result from disruptions or deficiencies in amino acid synthesis. Similarly, mutants that require cell wall building blocks have been used to study cell wall formation and function and the consequences of disrupting normal cell processes. Mutants deficient in synthesis of membrane components have been used to study the structure and function of membranes. Although studies of this nature have not always

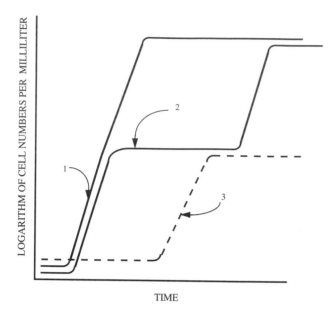

FIGURE 4.1 Diauxic growth of *Escherichia coli*. (1) Lactose-grown cells transferred into a medium containing a mixture of glucose and lactose. (2) Glucose-grown cells transferred into a medium containing glucose and lactose. (3) Glucose-grown cells transferred into a medium containing only lactose as a carbon source.

involved direct use of nutritional mutants, the conceptual basis for understanding the nature and consequences of mutation has been derived, to a substantial extent, from nutritional studies.

Response to environmental change

Nutritional mutants have been used extensively as agents for understanding the manner in which microbes respond to environmental change. The classical studies of this nature concerned the ability of *Escherichia coli* to use glucose and lactose. The essentials of these experiments are shown in Figure 4.1. The studies revealed that the ability of *E. coli* to use glucose was independent of the previous nutritional history of the culture, although the organism's ability to use lactose was not. These findings were accompanied by the recognition that formation of certain critical proteins is constitutive, i.e., occurs at all times, while other proteins, e.g., those required for lactose use, are inducible, and are formed only in the presence of lactose. When the organism was grown in the presence of glucose and subsequently transferred to a glucose-containing medium, growth was immediate and rapid. Similarly, when lactose-grown cells were transferred into a lactose- or glucose-containing medium, rapid growth was obtained. However, when glucose-grown cells were transferred to a medium containing lactose, a delay was observed before growth occurred. Finally it was shown that when presented with both glucose and lactose, *E. coli* used the glucose first and that lactose was used only when glucose was no longer available. This result was manifested as diauxic growth. During the first growth period, glucose was used and lactose utilization occurred only after glucose exhaustion. It was eventually shown that lactose exposure was required for lactose utilization. Study of this phenomenon led to understanding of induction and repression. In addition, it was recognized that glucose metabolism controlled lactose metabolism through alteration of the cellular concentration of adenosine triphosphate (ATP), adenosine diphosphate (ADP), and, particularly, cyclic adenosine monophosphate (cAMP). Further study of lactose utilization by *E. coli* led to understanding the importance of cAMP as a regulatory substance, the phenomenon of catabolite repression (activation), the natures of induction, repression, and diauxic growth, and the metabolic control of the use of alternate energy sources. In addition, study of the effects of lactose upon its utilization by *E.coli* allowed understanding of the nature and function of operons and the manner in which regulation of gene function is controlled. Concepts originally derived from studies of glucose and lactose use are now known to apply to phenomena as diverse as structure formation and function, penicillin resistance, heat shock protein formation, hydrocarbon utilization, nutritional deficiency diseases, the control of metabolic sequences, comparative biochemistry, and genetic engineering. We owe much to those who carefully studied the consequences of microbial nutrition.

Summary

Nutrition has many implications beyond a simple determination of the substances required for growth of an organism. Although determination of the growth requirements of an organism is useful for many practical purposes, study of the functions of nutrients is critical for understanding the physiological properties of an organism. In general, the complexity of the nutrient requirements of an organism is reflective of its physiological complexity, in an inverse fashion. Careful evaluation of a particular nutrient's significance is required, but critical studies of nutrients' functions in a microbial system are extremely useful in assessing their possible role in nonmicrobial systems. Critical study of the mechanisms of nutrient use allows understanding of a diversity of physiological phenomena in microbes and other life forms.

Selected References

Guirard, B. M., and E. E. Snell. 1962. Nutritional requirements of microorganisms. *In* I. C. Gunsalus and R. Y. Stanier (eds.), *The Bacteria,* vol. 4, pp. 33–95. Academic Press, Inc., New York.

Harder, W., and L. Dijkhuzien. 1983. Physiological responses to nutrient limitation. *Annual Review of Microbiology.* **37:** 1–23.

Hunter, S. H. 1972. Inorganic nutrition. *Annual Review of Microbiology.* **26:** 313–346.

Matin, A. 1978. Organic nutrition of chemolithotrophic bacteria. *Annual Review of Microbiology.* **32:** 433–446.

CHAPTER 5

Growth

The general implications of microbial growth

Growth is the ultimate expression of the physiology of an organism, but for microbes, it is also essential to being alive. The ability to reproduce is the major criterion by which we determine whether or not a microbe is alive. Microbial growth reflects the operation of all of the structures of the organism in a coordinated manner, so that life is possible. Since microbes, other than the viruses, are cells, study of microbial growth processes allows us to understand the critical factors involved in the growth and metabolism of all cells and the spectrum of responses that cells may exhibit in the face of environmental change.

The nature of microbial growth

Microbial growth implies an increase in the amount of protoplasm, the formation of new structures, and eventually, the formation of new cells. When all cell components increase in proportion to each other, growth is *balanced*. When such is not the case, and components of new cells increase in a nonconstant relationship to each other, growth is *unbalanced*.

The synthetic activities of microbes that ultimately result in cells, may be thought of from two perspectives, how much growth has occurred, and how fast it has been accomplished. *Yield* is a measure of the amount of growth obtained and *rate* is a measure of its speed. Both measures are important to understanding of the factors that are critical to the growth process.

Cell division is obligatory for microbes

Division is an obligatory requirement for growth in the microbial world. For microbes, life and growth are impossible without cell division. For large life forms, this is not the case, since they are *composed* of cells and substantial cell proliferation may occur without a whole new organism. In contrast, the continuance of life is not possible for microbes without the division event. Because cell division is critical for the microbe's life, much attention has been devoted to precisely why, and how a microbe divides. These studies are of interest not only because they are essential to understanding microbial life processes, but also because the control of cell growth is a subject of general concern. For example, the cancer phenomenon may be viewed as a situation in which cell growth is out of control. Understanding cell growth at the microbial level will aid in understanding general factors that influence the control of growth of all kinds of cells.

Why does a microbial cell divide?

Because cell division is so critical to understanding microbial life, many studies have sought to discern precisely why a microbial cell divides. Definitive answers to this seemingly simple question are not easy to obtain. In order to get a meaningful answer, we must understand the answers to a number of subquestions. We must particularly understand:

1. the relationship of cell size to the division event;

2. the manner in which the components of the cell are replicated;

3. the ways in which the replication of critical cell components are interrelated and coordinated;

4. the manner in which DNA replication is related to other cellular processes; and

5. the minimum components required for a functional cell.

Complete answers regarding these questions are not entirely apparent at present. However, partial answers and reasonable suggestions are available for all of them. Some of these questions will be considered in this chapter and consideration of the others is left for other chapters.

In regard to cell size, the importance of surface-to-volume ratio appears critical. When an organism decreases in size, its surface-to-volume ratio increases, and conversely, when cell size increases, the surface-to-volume ratio decreases. When the cell is small, its metabolic processes occur rapidly and metabolic end products are easily removed. As cell size increases, and the surface-to-volume ratio declines, the rates at which metabolism and waste product removal occur also decrease. Many theorists suggest that the obligatory requirement for cell division in microbes is a reflection of the fact that beyond a certain size, cellular life is impossible because the rate of cellular metabolism is insufficient for life and the lessened rate of waste product produces a toxic intracellular environment. The cell avoids these difficulties by dividing, thereby restoring a size and surface-to-volume ratio compatible with normal cell processes.

A second possible suggestion for the obligate requirement for division is the concept of *genetic control*. According to this idea, a certain amount of genetic material is required to exercise control over a given amount of protoplasm. Support for such a concept is found in the general observation that the amount of DNA generally parallels the complexity of an organism. For example, the amount of DNA in bacteria is greater than that in viruses and the DNA in eucaryotic cells is substantially more abundant than that in procaryotic cells. In general, as the complexity of life increases, the amount of DNA required increases. Those who propose genetic control as a means of cell division control in microbes suggest that as cell size increases, a critical point is reached, beyond which the DNA per cell is insufficient to allow for control of the processes associated with life. When such a point is reached, the cell responds by dividing, restoring an appropriate relationship between genetic material and the protoplasm over which the latter exerts control.

A third suggestion for the reason a cell divides reflects a consideration of the minimum requirements for the existence of individual functioning cells. For a particular cell type, a certain collection of entities must exist in order for a new cell to exist as a discrete entity. Each particular cell must contain a genome, a sufficient collection of ribosomes, a functioning cytoplasmic membrane, a certain number of proteins, and other entities that allow it to exist as a new cell. According to this concept, division occurs when sufficient synthetic activities have occurred in growing cells to allow a new cell to exist as a separate entity.

The events of cell division in a population

Regardless of the factors that determine precisely when a microbial cell divides, division eventually occurs. In many types of microbes, particularly bacteria, growth is logarithmic, so during the period of most rapid growth, growth is normally a multiplicative function of time. Because one cell gives rise to two cells, each of which in turn, gives rise to two cells, etc., during the most rapid period of growth a plot of the logarithm of numbers as a function of time is linear with a positive slope. Eventually, because of many factors, the rate of viable cell production becomes equal to the rate of death. At this point, no net increase in viable cells occurs and, mathematically, a plot of the logarithm

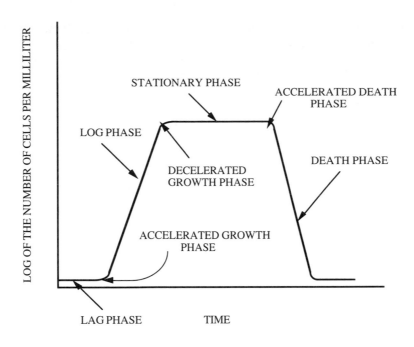

FIGURE 5.1 The general nature of batch culture.

of numbers versus time is a straight line with zero slope. After a further period of time, fewer viable cells are produced than die. As a result, the net number of viable cells decreases. With further time, the relative rates of viable and nonviable cells formed become constant with respect to each other. As a result, net cell death occurs at a constant and logarithmic rate and therefore, a constant fraction of cells dies per unit of time.

Phases of batch culture

The events just described are those pertaining to a typical batch culture, in which a vessel containing broth is inoculated with a relatively small number of organisms from a previous culture. Batch culture is shown diagrammatically in Figure 5.1. It is usually the case that in batch culture, a time interval occurs between inoculation and the beginning of rapid growth. This period is known as the *lag phase* and is a period of adjustment. The end of the lag phase is rapidly followed by the *logarithmic (log) phase,* during which the logarithm of viable cells is a constant function of time. Because not all cells reach the end of lag phase at precisely the same time, a short time is observed between the end of the lag phase and the beginning of the log phase.

This time is known as the *accelerated growth phase* because a net increase in cells occurs, but is not a constant function of time. The log phase then occurs, followed by a short period, the *decelerated growth phase,* in which the relationship between cell death and new cell production is changing, but the rate of life and death are not yet equal. The decelerated growth phase is followed by the *stationary phase,* during which the rates of cell production and death are equal. Following the stationary phase, cell death begins to exceed new viable cell production and net cell death begins but is not yet logarithmic. This time period is known as the *accelerated death phase.* The latter is followed rapidly by the *logarithmic death phase.* The logarithmic death phase, in practice, ceases after a period of time.

Many factors influence the phases of batch culture. The lag phase, in particular, is influenced in many ways. It is generally a period of adjustment and although no net increase in cells occurs, it is a period of intense metabolic activity during which the cells adjust to their new environment. The duration of the lag phase depends on many things. In general, its length reflects the extent to which the conditions for growth at the present time compare to those previously found. If present conditions closely approximate previous conditions, the lag phase may be virtually unnoticed. However, if substantial differences exist between present and previous growth conditions, the lag phase may be extended. Exactly what happens depends on the differences between previous and present conditions.

Shift up and shift down changes   If the present environment is more nutritionally abundant than the previous one, the change is known as a *shift up* change. Conversely, if the present environment is nutritionally less auspicious than the previous one, the change is referred to as a *shift down* change. Each type of change elicits a different response by the microbe. If the organism was previously supplied with abundant nutrients and is subsequently exposed to nutritionally sparse conditions, it must synthesize a great number of substances with which it was previously supplied. It may require new transport systems so that materials not previously used may be utilized in the present circumstance or transport systems previously formed may be discontinued or their operation modified. Because of nutritional limitation, the organism cannot grow as rapidly as it had under nutritionally abundant conditions. It adjusts by forming fewer ribosomes per cell and by reducing the percentage of them that are actively synthesizing protein. Although it produces less total protein, the diversity of the proteins an organism forms during nutritional limitation is substantially greater since the organism must make a variety of things that were previously supplied. It is even possible that under conditions of nutritional deprivation, the growth rate of the organism during its most rapid period may cease to be logarithmic.

In contrast to the events described above, transfer from a condition of nutritional sparsity to one of nutritional abundance evokes a different set of physiological changes. Because the cells are now supplied with nutrients in abundance, many substances synthesized in the nutritionally deprived environment are no longer formed. As a reflection of this condition, the diversity of proteins formed also diminishes, although the organism may now grow rapidly because of nutritional abundance. The number of ribosomes per cell and the rate of protein synthesis both increase markedly, allowing the organism to form more total protein, and to form it more rapidly than before.

The period of adjustment required by microbes upon introduction into a new culture system is in-

fluenced by many factors other than nutrition. Changes in temperature, pH, and gas phase are often influential, since all of these changes require adjustments on the part of the organism. The physiological age of the cells is also important. Logarithmic-phase cells, for example, are intrinsically capable of more rapid metabolism than are stationary-phase cells. However, because of this very substantial metabolic capacity, log-phase cells are usually more affected by environmental change than are stationary-phase cells.

Inoculum size   The size of the inoculum substantially influences the lag phase. In general, if other factors are equal, inoculum size and lag time are inversely related. However, it is seldom the case that all factors are strictly comparable and the length of the lag phase reflects the simultaneous operation of many factors. In general, the lag phase is shortest when the organism is faced with the fewest possible adjustments and is longer in proportion to the nature and extent of its new environment.

The logarithmic phase

Logarithmic growth, which rapidly follows the lag phase, is dependent upon lack of growth restriction by environmental conditions. During logarithmic growth, it is only the organism itself that influences its growth. The organism is supplied with an abundance of nutrients and the accumulation of inhibitory substances is of no physiological importance. The organism can grow as rapidly as possible, limited only by its own genetic potential. The metabolic systems of the organism are operating at their maximum rate and efficiency. The maximal growth rate of a microbe is an intrinsic property of it, and can be used to characterize it. Because of their intense metabolic capacity, logarithmic-phase cells are normally more sensitive to physiological change than cells from other growth phases.

The stationary phase

At a particular point, logarithmic growth ceases and batch culture cells enter the stationary phase. The precise reasons for entrance into stationary phase are many and are not entirely understood. In the logarithmic phase, the supply of nutrients is abundant and little toxic end product accumulation has occurred, but by the time of the stationary phase, one or more critical nutrients is diminished or exhausted and toxic products, chiefly acids and alcohols, have accumulated. The combination of these factors is such that growth of the organism is no longer solely limited by its own genetic potential. Logarithmic growth, therefore ceases, and the population shifts to a survival mode. Theoretically, survival would require the number of new organisms to be equal to those that die and in many circumstances, this happens. However, in other situations, it does not and the slope of the line logarithmically relating growth to time during the stationary phase is not zero. Rather, it displays a hump that reflects the growth of organisms on the end, or lysis, products formed by dead relatives. During the stationary phase, the organisms adjust their activities and their chemical composition significantly.

The death phase

At some point, determined by a combination of factors, the stationary phase ends and net death begins. Death is measured by determining the proportion of total cells that are alive at particular times. In certain situations, death is logarithmic. Speculation regarding this phenomenon proposes the target theory, which states that because of inactivation of a particular target, a constant fraction of cells dies in a given time period. Although the target theory is widely accepted, debate continues regarding the precise nature of the target. DNA is a likely candidate because of its delicate nature and its essentiality for life, but it is possible that other critical cell molecules may also serve as targets. The target situation is analogous to randomly shooting at 1,000 targets, and by chance hitting 90% of them in a particular time interval, thereby reducing the population to 100. If 90% of the remaining targets are hit, the population is reduced from 100 to 10 in the second time interval, making death logarithmic.

The logarithmic nature of death is independent of the actual number of cells and proposes only that a constant fraction of the population is killed in a given time. Logarithmic death is illustrated in Figure 5.2. Because of changing environmental conditions, logarithmic death eventually ceases and the culture becomes sterile or maintains a small number of viable organisms that may initiate growth if conditions change again.

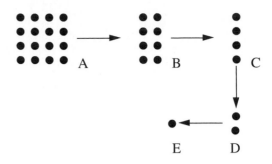

FIGURE 5.2 The kinetics of logarithmic death. At each of the various time intervals (A–E), the number of viable cells is one half the number of the preceding time interval. Thus, in the present example, 50% of the population is killed in each time interval.

Continuous culture

During batch culture, the environment to which the cells are exposed is constantly changing. In the early growth stages, nutrients are abundant and toxic product accumulation is minimal. As growth proceeds, and cell numbers increase, the nutrient supply decreases and accumulation of end products becomes significant. Since the cell population is larger, the rate of nutrient depletion and toxic product formation both increase until the organism is, in effect, both starving and "drowning in its own garbage."

As a function of the continually changing environment of batch culture, the various components of the growth process frequently are not in a constant relationship to each other, i.e., growth is unbalanced. It is often useful to study the growth of microbes in a way such that a constant environment is maintained. This may be achieved by using a continuous culture. Continuous culture is achieved by simultaneously removing waste products and adding new nutrients to the system in a controlled manner. Continuous culture may be regarded as a means of maintaining microbes continually in the logarithmic phase. By growing cells in such a fashion, three things are achieved: 1) a constant microbial population is present at all times; 2) the environment is constant; and 3) the rate of growth is both constant and controlled.

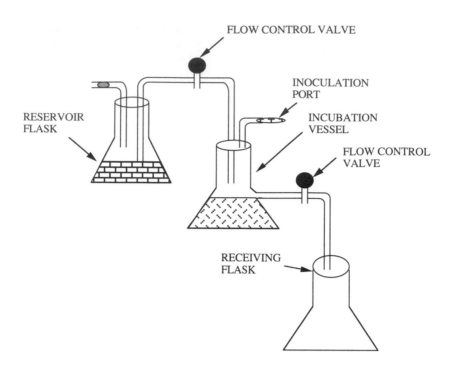

FIGURE 5.3 The general nature of a continuous culture apparatus.

Mechanisms of continuous culture control

In continuous culture, regulation of the cell population, waste product removal, and nutrient addition may be controlled in two ways, by *turbidostatic* control or by *chemostatic* control. In the former case, the turbidity, or cloudiness of the culture itself is used to control the system through the aid of a photocell. Because of this, the system is said to be controlled internally since growth of the culture is the critical regulating factor. Turbidostats are generally used when the cell population is dense and the effects of large population changes are of interest.

Chemostatic continuous culture

In chemostatic control of continuous culture, the population is controlled by a *limiting nutrient*, which determines the growth rate of the culture, since all other required nutrients are supplied in excess. The rates of cell and end product removal are adjusted to equal the rate of growth achievable at a given nutrient concentration. Since nutrient concentration is controlled by the investigator, outside the culture vessel, chemostatic control of continuous culture is regarded as external control. Chemostatic continuous culture is widely used to study the effects of small environmental changes. In addition, the method allows the study of small populations and the effects of extremely slow growth rates on physiological phenomena.

During chemostatic control of continuous culture, a delicate balance exists between culture growth and removal of "spent" medium, that is, the medium in the culture vessel that has already been acted upon by the microbes. The nature of this balance is influenced by several factors. To study very slowly growing organisms, the rate of medium and organism removal must be proportionally slow, so that the culture will not be washed out. It is also possible to study rapidly growing organisms in small populations, but the dilution rate must be substantially more rapid. In the first case, growth rate is limited by the presence of a very small concentration of nutrient and, in the second, growth is substantially more rapid and abundant and the dilution rate must be increased proportionately. The relationships between the existing cell population, the dilution rate, growth rate, and population change are shown in the following equation:

$$dX/dt = \mu X - DX \qquad (1)$$

This equation says that the change in cell population or mass (dX), as a function of change in time (dt), is equal to the difference between the product of the growth rate (μ) and the existing population (X) and the product of the dilution rate and the existing population (DX). The equation emphasizes the concept that both dilution and growth depend on the existing population or mass.

The equation may be modified to factor X on the right-hand side of the equation:

$$dX/dt = X(\mu - D) \qquad (2)$$

Consideration of equation (2) indicates that the critical matter in regard to population change is not the absolute value of the growth rate or the dilution rate, but whether or not they are equal. If D exceeds μ, the difference is negative and the population decreases. Conversely, if the dilution rate is less than the growth rate, the population will increase until such time as it is the maximum population obtainable in the chemostatic vessel. It is only when the growth rate and the dilution rate are equal that the population remains constant. However, the magnitude of the constant population is not specified and may be large or small. Its nature depends only upon the relationship between the growth rate and the dilution rate, not the absolute value of either.

The change in concentration of a limiting nutrient in a chemostatic vessel is influenced by several factors: 1) the concentration of substrate at a given time in the effluent (S), 2) the concentration of substrate added to the culture vessel from the nutrient reservoir (S_o), 3) the dilution rate (D) in terms of culture volumes per time, and 4) the rate of utilization of substrate within the culture. The utilization rate is equal to the product of the growth rate (μ), the mass or number of organisms in the culture vessel (X), and the amount of material consumed in the production of a given amount of cells ($1/Y$), where Y is the yield of cells per unit of energy con-

sumed. Mathematically, the relationship between all of these factors is depicted in the following equation:

$$dS/dt = D(S_o - S) - \mu X/Y \qquad (3)$$

The growth rate of a microbe (μ), whether in a chemostat or not, is a function of the available substrate (S) in relation to the total substrate amount. Up to the organism's genetic potential, its growth rate is a function of the amount of critical nutrient available to it, which determines both the maximal rate of critical nutrient uptake from the environment and the maximal rate of the enzyme reactions associated with the growth-limiting process. These two factors, in turn, determine the growth rate of the organism. The equation that describes growth rate as a function of substrate concentration is:

$$1/\mu = K_s/\mu_{max} \times 1/[S] + 1/\mu_{max} \qquad (4)$$

In the above equation, μ is the rate of growth as a function of limited substrate concentration (S), K_s is the substrate concentration associated with half-maximal growth rate and μ_{max} is the maximum growth rate of the organism.

Since the maximal potential growth rate of the organism is dictated by the maximum rate of the growth-limiting, enzyme-catalyzed process associated with growth, the relationship between enzyme reaction rates (V) at a particular molar substrate concentration (S) and maximal enzyme velocity (V_{max}) is described by an equation identical in form to that which describes the relationship between substrate (that is, nutrient) concentration and growth. For enzymes, the relationship between substrate concentration and initial enzyme reaction velocity is:

$$1/V = K_m/V_{max} \times 1/[S] + 1/V_{max} \qquad (5)$$

In equation (5), V is the initial enzyme reaction velocity, K_m is the substrate concentration of half-maximal reaction speed, S is the molar substrate concentration, and V_{max} is the maximum enzyme reaction velocity obtainable. All of the parameters that describe enzyme reaction velocity as a function of substrate concentration were originally formulated by Michaelis and Menten and the reciprocal equation relating velocity and substrate concentration was developed by Lineweaver and Burke.

Uptake of nutrients from the environment determines both the rate of critical enzyme reactions

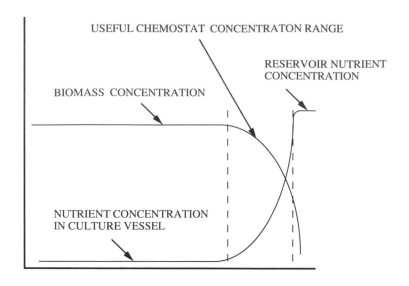

USEFUL CHEMOSTAT CONCENTRATON RANGE

RESERVOIR NUTRIENT
CONCENTRATION

BIOMASS CONCENTRATION

NUTRIENT CONCENTRATION
IN CULTURE VESSEL

DILUTION RATE

FIGURE 5.4 The effects of dilution rate on the biomass of cells and the nutrient concentration in a chemostatic continuous culture system.

and growth. It is interesting that the equation that describes nutrient uptake as a function of external nutrient concentration uses the same form as those that describe the relationships between nutrient concentration, enzyme activity, and the growth:

$$1/V_{uptake} = K_{up}/V_{upmax}1/[C_{ex}] + 1/V_{upmax} \quad (6)$$

The above equation shows the relationships between the reciprocals of the uptake rate of critical materials from the external environment ($1/V_{up}$), the rate of half-maximal uptake (K_{up}), the maximum uptake rate (V_{upmax}), and the external substrate concentration (C_{ex}). A similar equation describes the rate of removal of toxic material from the intracellular environment.

When a chemostat is operating at equilibrium, the dilution rate is equal to the growth rate. For this reason, the equation relating the dilution rate (D), the maximal growth rate of the culture (K_{max}), half-maximal growth rate substrate concentration (K_s), and the reciprocal of the nutrient concentration in the culture vessel is in the same form as the growth rate:

$$1/D = K_s/K_{max} 1/[S] + 1/K_{max} \quad (7)$$

The similarity of all of these equations reflects the fact that all of the processes described are protein-mediated phenomena and interdependent. The similarity of the dilution rate equation (7) to the other equations results from the fact that when cells are growing at a particular rate, the summation of all of the processes associated with growth must be balanced by the rate of cell and end product removal in order for the system to maintain equilibrium.

Dilution rate and growth rate effects

At a particular point, the ability of an organism grown in a chemostat to maintain a constant population is lost. This situation occurs when the dilution rate of the culture exceeds the organism's maximum possible growth rate. Further increase in dilution rate at this point causes the population to diminish, and it is eventually eliminated. Conversely, complete nutrient utilization is no longer possible because of the decreasing population. The concentration of nutrient in the chemostat, as a function of further dilution rate, thus increases until it is the same as that of the nutrient reservoir, as shown in Figure 5.4. In a given situation, the maximal concentration of cells obtainable in the chemostat is determined by the intrinsic properties of the organism and the concentration of the limiting nutrient.

Growth yields and maintenance energy

The yield of cells obtained during a microbe's growth reflects both the amount of available nutrient and the proportion of that nutrient used for growth, as opposed to other essential nongrowth processes, collectively known as *maintenance.* Maintenance processes are transport, motility, and sustaining structural integrity, among others. The amount of energy required for maintenance is a reflection of both the amount of existing mass and the rate of mass increase. At very small substrate concentrations that fail to allow growth, all of the available substrate is used simply for maintenance. As substrate concentration increases and growth begins, a smaller portion of the total available substrate is required for maintenance. Eventually, the proportion of the total energy required for maintenance becomes a very small portion of the total available, although the overall use of substrate for *both* growth and maintenance increase. The mathematical relationships between total change in substrate concentration (dC/dt) and the proportions used for growth (dC/dt_g) and maintenance (dC/dt_m) are shown below:

$$dC/dt = dC/dt_g + dC/dt_m \qquad (8)$$

The yield of microbes obtainable from a given amount of substrate is influenced by the proportion of substrate required for maintenance. The greater proportion of substrate available for synthetic processes, the greater the yield obtainable within the genetic limits of the organism. The relationships between yield (X), yield per unit of substrate consumed (Y), change in time (dt), nutrient concentration (C), and the specific maintenance rate constant (a), are shown by the equation:

$$dX/dt = YdC/dt - ax \qquad (9)$$

Synchronous culture

Continuous culture allows maintenance of a constant population of cells at all times. Because of growth rate control and the balancing of growth rate with end product removal, the environment to which the cells are exposed is constant, as opposed to a batch culture, in which it is continually changing. In the continuous culture, the constancy of growth conditions is accompanied by constancy in cell composition because the coordinated processes of growth are affected in a constant way, and therefore, maintain a constant relationship to each other. On the average, the behavior of a cell population in continuous culture is constant, although the behavior of individual cells is not uniform. For certain purposes, particularly for study of the relationship of events in the cell cycle to each other, it is useful, if not essential, that all of the cells in a microbial population behave in the same way. *Synchronous culture* is the mechanism by which this objective is obtained. The usefulness of such a procedure is exemplified by the knowledge that DNA replication is semiconservative. If it were not for the ability to achieve synchronous culture, or to closely approximate it, the demonstration of the semiconservative nature of DNA replication would not have been possible.

Methods for achieving synchronous culture

A number of methods have been devised to approximate synchronous culture. These include: 1) nutrient limitation, 2) temperature oscillation, and 3) cell selection. In nutrient limitation, the strategy is to starve the cells to the point that growth is no longer possible and, at a particular time, to resupply the organism with the required nutrient. Achievement of synchrony in this manner is sometimes difficult because of the necessity of extreme depletion of intracellular reserves of the limiting nutrient. The strategy for attempting synchronous growth by nutrient limitation is depicted in Figure 5.5.

Temperature oscillation is a second method for achieving cell synchrony and is attempted by repeatedly altering the temperature to which the organism is exposed, both above and below the optimum growth temperature. Such a procedure results in rapid changes in the rates of growth subprocesses and in the manner in which cellular regulatory events occur. In response to this situation, the cells cease to divide until such time as the temperature is adjusted to the organism's optimum temperature. The cells then divide synchronously. Synchronous growth as achieved by temperature oscillation is shown in Figure 5.6.

A number of strategies are available for attempting synchronous growth through cell selection. In one strategy, synchrony is achieved by obtaining a homogeneous cell population through

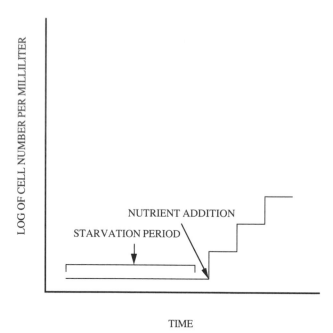

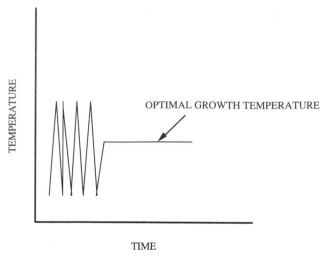

FIGURE 5.6 Achievement of synchronous growth by temperature oscillation. The temperature is repeatedly varied, both above and below the optimum growth temperature of the organism. After a period of time, the temperature is adjusted to the optimal growth temperature, at which point the cells divide synchronously.

FIGURE 5.5 Achievement of synchronous growth by nutrient limitation. The organism is starved for a period of time sufficient to deplete endogenous reserves of a critical nutrient. During this time, little or no growth occurs. The nutrient is then added to starved nongrowing cells and synchronous growth occurs for a short time period.

differential centrifugation or filtration. Such a strategy attempts to obtain a population of cells that are uniform in size. It presumes that under a given set of physiological conditions, cells of the same size are in a similar or identical physiological state.

The method of Helmstetter and Cooper is a method of cell selection that is highly useful and perhaps least subject to criticism, since it involves a minimum of stress during the selection process. In this procedure, a bacterial culture is filtered through a cellulose nitrate filter. The cells adhere to its surface. The filter is inverted and the cells are washed off it with fresh medium. After removal of loosely-attached bacteria from the filter, subsequent cells released from the filter are those that have arisen from division of the adhered cells. Since the latter have adhered to the filter as a result of a similar or identical physiological state, their immediate progeny are presumed to be in an identical physiological state and, in fact, at the same stage of the cell cycle, thus achieving synchrony. The strategy for achievement of synchrony by cell-surface attachment is shown in Figure 5.7.

Growth measurement

By whatever means we grow microbes, it is necessary to determine the extent and rate of growth. We may do this in a variety of ways. If is often the case that we measure growth indirectly. In such a situation we measure a property of the cells that result from growth. With large cell populations ($\geq 10^7$ cells per milliliter), we may measure growth by turbidity since the cloudiness of the culture increases as a result of the growth process. This phenomenon results from interaction of light rays with bacterial particles, scattering the incident light and resulting in apparent absorption. When conditions are constant, the numbers of cells per milliliter are a linear function of the logarithm of turbidity, commonly known as optical density (*OD*):

$$N = K \times OD \qquad (10)$$

When growth is measured by turbidity, the logarithm of the difference between the transmitted and incident light is assessed. Growth is determined as apparent absorption (OD).

When the microbial population is sparse, we may sometimes measure microbes by direct measurement of the light scattered by them using the

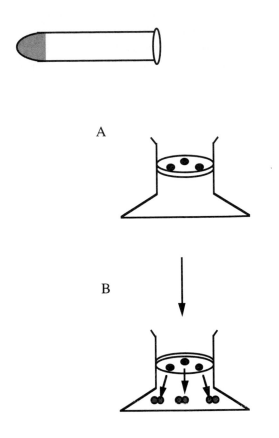

FIGURE 5.7 The Helmstetter-Cooper method for achieving cell synchrony. Material from the test tube is used to attach cells to a filter so that isolated colonies are obtained (A). The filter is then inverted and the progeny from attached cells are washed off the filter (B). Progeny of adhered cells are presumed to be in the same physiological state and, thereby, to be synchronous.

technique of *nephelometry*. This procedure allows detection and quantitation of smaller populations than are detectable by turbidity. With either procedure, one must assure that the optical measurements obtained do not reflect absorption of light by cell components.

Direct measures of growth

Because of their speed and simplicity, optical methods are often used to assess growth, these are indirect procedures since they measure light scattering rather than the cells or their components. For critical studies, indirect measurements must be standardized against direct measures of growth, many of which are available. As a general rule, more than one procedure should be used.

Determination of cell numbers is often used to assess growth. This measure may be accomplished in several ways, depending upon the objectives desired. It is often useful to determine cell numbers by a *direct count*, a procedure that assesses both living and dead cells. This procedure is accomplished by quantitative dilution of a cell suspension in a way that does not injure the cells. The number of cells in a portion of the diluted material is determined and the number of cells in the original is calculated by considering the portion of the original sample evaluated and the extent of dilution. We may, for example, adhere a known fluid volume (typically 0.01 ml) to a known area of a slide, stain the resultant and count the average number of cells in a microscope field at a particular magnification. We may then relate the area of a microscope field to the volume of diluted fluid to which it corresponds and determine the cell population in the undiluted material. Alternatively, we may place a known volume of diluted material in a chamber of known volume, count the cells in the chamber, and calculate the original number of cells.

The Coulter counter is a third way commonly used to determine total cells. This method detects the passage of a microbial cell through an orifice of a particular size that results in an electrical signal and assumes that a proportional relationship exists between the number of cells and the extent of electrical disturbance. With appropriate standardization, it is possible to determine the number of cells in the population by the magnitude of the signals that they generate. The instrument is equipped with a set of orifices of various sizes. By careful use of these orifices, it is possible to assess not only the total cells in a population, but the distribution of their sizes.

Viable cell measurement

Instead of determining total cells, it is sometimes desirable to measure only those that are alive. In other situations, particularly when we are studying organisms from natural environments, it is helpful to determine both measures and to compare them. Such a comparison is useful for purposes as diverse as determination of the extent of viability of a pure culture or the adequacy of a habitat-simulating medium for evaluating the habitat under study. In general, if we would determine viable organisms in a population, we dilute the material in a quantitative way and distribute a portion of the resultant either within, or on, the surface of an agar medium that is subsequently incubated at an appropriate temperature. After a period of time, the number of resulting colonies is evaluated and the number of viable cells in the original culture is calculated.

There are a number of commonly held assumptions regarding viable counts that are not absolutely true. It is assumed that a one-to-one relationship exists between colonies and cells, that is, that each colony is a clone. Although such is often the case, it is not always true. In critical situations, it is necessary to experimentally determine the relationship between colonies and cells. This may be determined by comparing the number of cells observed in total counts with those that are culturable. To critically determine precisely what is occurring, it is necessary to determine the proportion of viable cells. This measure may be studied with vital stains and by study of the kinetics of culture growth in relationship to the total cells observed.

Additional factors may substantially influence viable counts. Particular attention must be paid to: 1) the adequacy and representative nature of the sample, 2) the extent to which mixing is adequate, 3) the extent to which organisms may be attached to particulate material in the sample, 4) cell death or inactivation during dilution, 5) the possibility of growth during dilution, and 6) the adequacy of the physiological and nutritional conditions. When all of these factors are carefully considered and attended to, biologically and physiologically reliable results may be obtained.

Measures of mass

It is frequently the case that in addition to measuring cells, we may measure protoplasm without regard to the cellular packages it is found in. During growth, the synthetic activities of the cells result in major increases in cell mass. If done with care, we may use mass increase as a measure of growth. The most frequently used measures of mass are dry weight, protein, and DNA.

Dry weight

Determination of cell dry weight is not an easy task. However, under appropriate conditions, it is a useful measure of growth. The most frequent way in which dry weight is determined is to sepa-

rate cells from their growth medium by centrifugation, wash the separated cells, dry the resultants at a temperature sufficient to dry the cells (normally about 70° C), and weigh the dry cells until a constant weight is obtained which gives the total dry weight. The temperature often used to remove water, 70° C, is sufficient to remove water without causing loss of organic material. It is possible to use a slightly higher temperature, but care must be exerted to ensure that the temperature used to remove water is not high enough to damage cellular material. Treatment of the dry cells at substantially higher temperatures (400 to 600° C) causes combustion of the organic cell components, leaving inorganic ash. Subtraction of the ash weight from the total dry weight yields the total organic dry weight, which is a reasonable measure of growth.

Dry weight measurements must be made with caution and are quite subject to error. In particular, washing with fluids that are not isotonic with cells can lead to weight loss through lysis. In contrast, isotonic washing can lead to erroneously high weights as a function of the incomplete removal of washing solution. In addition to these concerns, nonspecific precipitation of inorganic materials during centrifugation or filtration may occur. If this material is confused with protoplasm, erroneously high dry weight values may again be obtained. Finally, there are occasions when extensive accumulation of intracellular storage materials occurs, leading to dry weights that do not represent true protoplasm. All critical dry weight measurements should be accompanied by a study of the chemical nature of the dry weight so that its true significance can be assessed.

Protein measurement methods

Protein is a reasonable measure of growth since it normally constitutes the majority (50–70%) of the organic cellular dry weight. The method by which we assess cell protein depends upon our purpose and the amount of material involved. For microbial physiology, three methods of protein measurement are commonly used: 1) the biuret procedure, 2) the Folin reagent, and 3) the Coomassie blue dye-binding assay. Under different conditions, each procedure is useful.

The biuret procedure depends upon the interaction of the cupric ammonium ion with the peptide bond. The procedure is considered by many to be the most unbiased protein measurement, since it depends only on the peptide bond and is independent of the amino acid composition. Reaction of copper with the peptide bond yields a complex that is blue-colored and absorbs broadly over a range between 500 and 650 nm. At an appropriate wavelength that may be selected, within a reasonable range, by the investigator, a linear relationship is found between the copper-protein complex and protein concentration. Relatively large amounts of protein are required to obtain substantial absorbance readings if selected wavelengths are in the visible wavelength range.

In addition to absorbance in the visible range, the copper-protein complex absorbs quite intensely between 280 and 320 nm. Making absorbance readings within this range, increases the sensitivity of the biuret procedure, but the procedure is complicated by the fact that nonproteinaceous cellular materials will also absorb in the near ultraviolet.

The Folin reaction   The relative insensitivity of the biuret procedure limits its usefulness in microbial physiology. Accordingly, more sensitive methods are often used, such as the Folin reaction. This method depends upon interaction of protein with the cupric ion (Cu^{2+}), followed by oxidation of the complex by the Folin reagent. The oxidation of the complex is coupled with the reduction of a mixture of phosphotungstic and phosphomolybdic acids in the reagent, producing a blue color that is proportional to the amount of copper-protein complex.

Coomassie blue   An alternate sensitive procedure for protein measurement takes advantage of the fact that, in highly acidic solution, interaction of protein with Coomassie blue dye leads to a change in the dye's absorption from 465 nm to 595 nm that is proportional to the concentration of the protein-dye complex. When carefully standardized, the procedure is highly sensitive, rapid, and less subject to interferences than the Folin procedure.

DNA

DNA is often used as an index of growth. Even when it is not used for this purpose, DNA is a frequently measured parameter. Typically, DNA is measured in one of three ways: 1) by the absorption of intact material at 260 nm, 2) by deoxyribose measurements, and 3) by phosphate measurements.

When 260 nm absorption is used to measure DNA, the DNA must be purified since other cell components may absorb at 260 nm. The ratio of absorption at 280 and 260 nm is often used as an index of purity in preparations that contain protein. With highly purified DNA, 5 μg of DNA per milliliter gives an absorption of 0.10 at 260 nm with a 1 centimeter light path. If DNA is not measured by 260 nm absorption, it may be assessed by colorimetric deoxyribose or phosphate measurements and the calculation of DNA content from the proportion that deoxyribose or phosphate constitutes of the total weight of a DNA nucleotide building block.

The relationship of cell mass measures to other parameters

Under carefully controlled conditions, it is possible, and useful, to relate mass measures to each other and to other growth measures. If properly done, it is possible to relate any measure of mass to OD or cell numbers. By so doing, we can study the concentration of a particular cell component per OD unit or cell and compare the relative amounts of these materials in cells at different times and under different growth conditions. By use of a combination of methods, it is possible to study not only the synthetic events that attend growth, but to assess their relationship to each other and the influence of various physiological factors upon them.

Growth yield and rate calculation

Any of the above described measures may be used to study yield, the amount of growth obtained, or its rate. If rate is determined, it is also possible to determine whether or not growth is logarithmic. To determine the rate of microbial growth during the most rapid growth period, one must assess the following: 1) the amount of cells or material at two different times and 2) the number of times in which the material or cells, have doubled in the time period observed. From the above calculations, we may determine the rate at which cells and their components double, that is, the exponential growth rate (R) or the time required for doubling (G).

The ability to assess doublings is critical to rate measurements. The principles involved in the calculation of doublings are the same whether we are concerned with cells or their components. When dealing with cells, it is customary to refer to the period of time for doubling as the *generation time* while for components, the same time period is simply referred to as the *doubling time.* For illustrative purposes, we will consider cells (*N*), but any measure of mass might be considered.

To calculate the number of doublings we may proceed in one of two ways. We may use logarithms (logs) to the base 10 or logarithms to the base 2. From the mathematical point of view, logs to the base 10 are easier but from the physiological viewpoint, logs to the base 2 are preferable since they emphasize that we are, in fact, dealing with doublings. If we use logarithms to the base 10, we:

1. determine the logarithm of the number at the end of the time period;
2. determine the logarithm of the number at the beginning of the time period;
3. subtract (2) from (1); and
4. divide the result by 0.301, the logarithm to the base 10, of the number two.

The result of these processes is the number of doublings that have occurred in the time period under consideration, commonly denoted by the letter "*n.*" Once we have determined *n*, the remaining calculations are easy. To determine *G*, the generation time, we divide *n* into the time (*t*) that has elapsed. To determine *R*, the exponential growth rate, we take the reciprocal of *G*. Mathematically stated:

$$G = t/n \qquad (11)$$
$$R = n/t \qquad (12)$$

In either of the above calculations, the time unit may be selected by the investigator, so we may express the generation time in minutes, hours, or seconds, although it is customary to use either hours or minutes.

It is often useful to calculate doublings by conversion of logarithms to the base 10 to their equivalents to the base 2. We convert logarithms to the base 10 to their equivalents to the base 2 by multiplying logarithms to the base 10 by the number 3.32. If we do so, subtraction of the logarithm of the number at the beginning of a time period from that at the end gives the number of doublings directly, without the necessity of dividing a logarithm difference by the number 0.301. The validity of such a procedure reflects the application of a generalized

equation for the interconversion of logarithms from one base to another. The equation is as follows:

$$Log_b N = Log_a N / Log_a b \qquad (13)$$

The above equation states that the logarithm of any number to a second base is equal to the logarithm of that number to a first base divided by the logarithm to the first base of the second base. The equation is generally applicable to conversion of logarithms from one base to another. If we let *a* be 10 and *b* be 2, the following is the case:

$$Log_2 10 = log_{10} 10 / log_{10} 2 \quad or \qquad (14)$$
$$Log_2 10 = 1/0.301 = 3.32 \qquad (15)$$

As a result of equation (15) it is possible to convert the logarithm of any number from base 10 to base 2 by multiplying that number by 3.32. Let's consider an example of these ideas in order to determine the generation time and the exponential growth rate by conversion of cell numbers to their base 2 logarithms. Assume that an organism increases from 10 cells per milliliter to 80 cells per milliliter in 1 hour. To calculate our parameters using direct logarithms to the base 2, we proceed in the following way:

a. We calculate the logarithm to the base 2 of 10, $1 \times 3.32 = 3.32$.
b. We calculate the logarithm to be base 2 of 80, using the logarithm to the base 10 of 80, and the number 3.32. The logarithm of 80 to the base 10 is 1.90. Thus the base 2 logarithm of 80 is $1.90 \times 3.32 = 6.32$.
c. Subtraction of 3.32 from 6.32 tells us that the number of doublings is 3.0, precisely what we obtain if we consider the consequences of an organism increasing from 10 to 80 cells per milliliter in 1 hour. The first doubling gives 20 cells; the second doubling gives 40; and the third doubling gives 80 cells per milliliter. In our example, the generation time is:

$$G = 1/3 = 0.33 \text{ hour or } 20 \text{ minutes} \qquad (16)$$

The exponential growth rate is:

$$R = 1/G = 3/1 = 3.0/\text{hour} \qquad (17)$$

Although our sample calculations were rather simplistic, it is seldom, if ever the case that one can determine *G* and *R* from inspection. Therefore, it is useful to be able to calculate them in systematic ways.

Calculating the lag time

Normally, rate calculations are made on the assumption that growth is logarithmic during the entire time over which growth is observed. This is often not the case because of the necessity of adjustment when cells are introduced from an old culture system into a new one. If a measure of an organism's maximal growth rate is known from previous study of its growth under the present conditions, it is possible to determine the lag period of the organism. The argument is the following:

1. If an organism of known growth rate initiates growth immediately upon inoculation, we can predict, knowing the initial number of cells, what the population will be at a particular time in the future.

2. If at some time in the future, the observed population is less than it would have been if the organism had initiated growth immediately, the fact that we observe a less abundant population than expected reflects the fact that in actuality, growth did not begin immediately upon inoculation. Instead, a lag period occurred before growth began. Since a smaller population was observed than was expected, the total time over which the culture was observed did not allow growth. The difference between the total time that elapsed and the time required for the observed population change is the lag time, the time between inoculation and the initiation of rapid growth. To illustrate, consider the following:
   a. An organism with an exponential growth rate of 3 per hour increases from 10 cells per milliliter to 80 cells per milliliter in 66 minutes. Calculate the lag time.
   b. Since the organism has an exponential growth rate of 3 per hour, it would undergo three doublings in an hour.
   c. In the present example, the cells would thus increase from 10 cells to 80 cells per milliliter in 1 hour or 60 minutes. However, the total elapsed time is 66 minutes or 1.1 hours. Therefore, the lag time is:

$$[\text{Total time (66 minutes)} - \\ \text{Time required for growth (60 minutes)} \\ = 6 \text{ minutes or } 0.1 \text{ hours.}] \qquad (18)$$

The instantaneous growth rate

During the most rapid period of growth, it is normally assumed that growth is logarithmic and thus, a multiplicative function of 2. It is on such assumptions that G and R, both functions of time, are calculated. However, it is also the case that growth may be calculated as a function of the existing number of cells, or cell substance, at a particular time. Rather than asking the question, "How long will it take for doubling to occur?" we may ask, "What amount of increase will occur in the next unit of time?" From this perspective, the rate of increase in the next time interval is a function of the amount of cell substance existing at the present time. Mathematically, for cells, we have the following:

$$dN/dt = \mu N \qquad (19)$$

In the above equation, N is the number of cells, dN and dt are changes in number and time, and μ is the instantaneous growth rate.

Dividing by N and multiplying by dt, we have:

$$dN/N = \mu dt \qquad (20)$$

Integrating equation (20) we obtain:

$$[\text{Log}_e N = \mu t, \text{ where, } e \text{ is the base of the natural} \\ \text{logarithm and is, in this case, numerically equal} \\ \text{to } 2.7128.] \qquad (21)$$

The exponential form of equation (21) is:

$$[N = N_0 e^{\mu t}, \text{ where } N \text{ is the number of cells at a} \\ \text{particular time, } N_0 \text{ is the number of cells at an} \\ \text{initial time (time zero), } e \text{ is the base for the} \\ \text{natural logarithm, } \mu \text{ is the instantaneous growth} \\ \text{rate, and } t \text{ is time.}] \qquad (22)$$

If we divide both sides of equation (22) by N_0 and take the natural logarithm of the resultant, we obtain:

$$\log_e N/N_0 = \mu t \qquad (23)$$

By dividing equation (23) by t, we discern that the instantaneous growth rate constant, μ, is equal to the natural logarithm of the ratio of N to N_0 divided by time. The instantaneous growth rate constant (μ) and the exponential growth rate (R) are similar in that they are both expressions of the logarithm of the ratio of cells at a given time to the cells at the original time (t_0). However, with the instantaneous growth rate (μ), the logarithm base is *e*, the natural logarithm, while for the exponential growth rate (*R*), logarithm base is 2!

The relationships between the various commonly used measures of growth rate (for example, *G*, *R*, and μ), are apparent when we consider the situation when *N* is the number at the generation time, *G*. In this circumstance, $N/N_0 = 2$ and:

$$Log_e N/N_0 = \mu G = Log_e 2 = \mu G \qquad (24)$$

Since $G = 1/R$, equation (24) becomes:

$$[0.693 = \mu \times 1/R, \text{ where } 0.693 \text{ is the } log_e \text{ of } 2.] \qquad (25)$$

Thus:

$$\mu = 0.693\, R \text{ and:} \qquad (26)$$

$$\mu = K/0.693 \qquad (27)$$

Finally, since $G = 1/R$:

$$G = 0.693/\mu \qquad (28)$$

Arithmetic growth

As stated earlier, growth during the most rapid period of growth is usually logarithmic. In this situation, the logarithm of cells is a linear function of time. However, in certain situations, the number itself, rather than its logarithm is a function of time. In this circumstance:

$$N = \mu t \qquad (29)$$

It is often possible to distinguish logarithmic from arithmetic growth by plotting growth data in two ways, on semilogarithmic graph paper and on linear graph paper. If growth is logarithmic, the relationship between growth and time during the most rapid growth period, when plotted semilogarithmically should be linear. If growth is arithmetic, the plot of growth data on *nonlogarithmic* paper should be linear. In logarithmic growth, growth is limited only by the amount and genetics of the organism. In arithmetic growth, by contrast, growth is a function of some factor other than the organism, frequently the presence of a limitedly soluble nutrient.

Microbial growth in the laboratory and in nature

Understanding the nature of microbial growth and the multitude of factors that affect it, is a formidable and challenging task. For a variety of reasons, we often study the growth of pure laboratory cultures. However, such studies may not be directly applicable to understanding of microbial growth in nature. Whenever possible or practicable, it is highly desirable to study the relationship between the growth of microbes in laboratory culture and in the environment from which the organism came. Such studies may reveal subtle ways in which the microbes interact in an ecosystem that would not be discernable otherwise.

Summary

Microbial growth is a complicated process that involves a great variety of subprocesses. Growth is a function of many factors that include, among other things: 1) the number of organisms present; 2) the nature and concentrations of nutrients; 3) the genetics of the organism; 4) the current physiological conditions; 5) the previous history of the organism; and 6) the conditions of culture. All of these factors influence growth substantially.

We may measure growth in a variety of ways including *yield,* the total amount of growth that occurs, and *rate,* its speed. Within these possibilities, a variety of methodologies are available. We may measure cells directly or indirectly, we may measure their dry weight, or we may measure their protoplasmic components. The precise methods that we use primarily depend upon the nature of the microbial population and our purposes.

In the laboratory, microbes are cultivated typically in *batch culture,* in which organisms from a previous culture are dispensed into a stationary culture vessel; *continuous culture*, in which the cell population is maintained in a constant environment and at a constant density by simultaneous removal of cells and waste products and the addition of nutrients; or by *synchronous culture*, in which all of the cells in the population are induced to divide at the same instant in time.

We must be particularly careful in assessing the extent to which the laboratory culture of microbes approximates the manner of microbial growth in nature and, whenever possible, determine as critically as practicable, the extent to which laboratory studies describe the microbial response to the conditions that exist in nature.

The implications of microbial growth study are many. Because cell division is an obligatory requirement for growth in the microbial world, and because cell division is a universal biological phenomenon, we may learn much regarding the general nature of cell division from study of microbial systems. By study of microbial cell division, we may learn about the interconnections among its subprocesses and may also discern factors that control the division event. This information may be useful to understanding cell growth control in nonmicrobial systems. The implications of microbial growth transcend the microbial world.

Selected References

Bremer, H., and G. Churchward. 1991. Control of cyclic chromosome replication in *Escherichia coli. Microbiological Reviews.* **55:** 459–475.

Button, D. K. 1985. Kinetics of nutrient-limited transport and microbial growth. *Microbiological Reviews.* **49:** 270–297.

Cooper, S. C. 1991. Bacterial growth and division. Academic Press, Inc., New York.

Ingraham, J. L., O. Maaloe, and F. C. Neidhardt. 1983. Growth of the bacterial cell. Sinauer Associates, Inc., Sunderland, Massachusetts.

Jenen, K. F., and S. Pedersen. 1990. Metabolic growth rate control in *Escherichia coli* may be a consequence of subsaturation of the macromolecular biosynthetic apparatus with substrates and catalytic components. *Microbiological Reviews.* **54:** 89–100.

Koch, Arthur L. 1977. Does the initiation of chromosome replication regulate cell division? *Advances in Microbial Physiology.* **16:** 50–98.

Marr, A. G. 1991. Growth rate in *Escherichia coli. Microbiological Reviews.* **55:** 316–333.

Mendelson, Neil H. 1982. Bacterial growth and division: Genes, structures, forces, and clocks. *Microbiological Reviews.* **46:** 341–375.

Meyer, Hans-Peter, Othmar Käppeli, and Armin Fiechter. 1985. Growth control in microbial cultures. *Annual Review of Microbiology.* **39:** 299–320.

Neidhardt, F. C., J. L. Ingraham, and M. Schaechter. 1990. Physiology of the bacterial cell. Sinauer Associates, Inc., Sunderland, Massachusetts.

Newton, A., and N. Ohta. 1990. Regulation of the cell division cycle and differentiation in bacteria. *Annual Review of Microbiology.* **44:** 689–719.

Nierlich, Donald P. 1978. Regulation of bacterial growth, RNA, and protein synthesis. *Annual Review of Microbiology.* **32:** 394–432.

Norbury, C., and P. Nurse. 1992. Animal cell cycles and their control. *Annual Review of Biochemistry.* **61:** 441–470.

Payne, W. J. 1970. Energy Yields and Growth of Heterotrophs. *Annual Review of Microbiology.* **24:** 17–52.

Payne, W. J., and W. J. Wiebe. 1978. Growth yield and efficiency in chemosynthetic microorganisms. *Annual Review of Microbiology.* **32:** 155–183.

van Uden, N. 1969. Kinetics of Nutrient-Limited Growth. *Annual Review of Microbiology.* **23:** 473–486.

CHAPTER 6

Transport

The critical importance of transport

Transport is a critical activity for microbes. It allows the organism to take up nutrients from the external environment and is, therefore, often studied from such a perspective. However, transport is equally important as a mechanism for removal of toxic materials from the internal environment, allowing the cell to maintain an intracellular environment compatible with life. It is transport that, to a substantial extent, is responsible for the fact that certain microbes can exist in environments, which would initially seem incompatible with life.

In general, we may distinguish two types of transport, *nutrient transport* and *electron transport*. These two types of transport are often intimately intertwined. However, nutrient transport concerns uptake and removal of materials associated with the formation and function of cells, and electron transport concerns the generation of energy. This chapter will focus on the transport of materials. Electron transport and its relationship to energy generation will be considered at a later time (Chapter 9).

Types of transport

Substance transport may be viewed from a number of perspectives. With the exception of simple diffusion, transport always involves proteins, but the manner in which proteins facilitate the transport process varies from system to system. It is important to know the nature of the system under study, so that we may appreciate its physiological consequences. Initially, we may consider the direction of transport. If protein-mediated transport occurs in a single direction, it is called *uniport*. Often, a protein may mediate the transport of more than one material. When this is the case and the transport of two materials occurs in the same direction, it is referred to as *symport*. If the same protein mediates, simultaneously, the transport of two materials, but the transport of one material is in the net direction of inward, while the net direction of the second material is outward, the process is referred to as *antiport*.

Simple diffusion

Mechanistically, transport can occur in a number of ways. In simple diffusion, transport requires no biologically generated energy and is not protein-mediated. In this type of transport, the direction of transport is dictated solely by the relative extracellular and intracellular concentrations of the material. If the intracellular concentration of the transported material exceeds that of the extracellular environment, movement proceeds in the net outward direction. Conversely, if the external substance concentration exceeds its internal concentration, net transport is inward. Irrespective of which is the case, in simple diffusion, accumulation against the concentration is impossible.

Facilitated diffusion

Facilitated diffusion is similar to, but different from, simple diffusion. It is concentration-dependent and does not require biologically derived energy, but it is protein-mediated. Because of this, it exhibits saturation kinetics. Its velocity is, within limits, a function of the difference in concentration between the cell exterior and interior. This is because, at low material concentrations, only a portion of the potential reactive sites on carrier molecules are occupied at a given moment in time. As the concentration of transported material is increased, the proportion of total transport sites occupied increases, with a corresponding increase in transport rate. Eventually, the concentration is such that all of the available sites for transport are occupied. At this point, further increase in the concentration of the transported material does not result in an increase in transport rate. Although facilitated diffusion differs from simple diffusion by requiring a protein, the net effects of facilitated and simple diffusion are the same. In neither case is accumulation against the concentration gradient possible. In both cases net movement is from a region of higher concentration of the transported substance to a region of lower concentration and the net transport is a function of the difference in concentration.

Active transport

Active transport is the most frequently encountered mode of transport in microbes. In most systems, it is intimately related to electron transport, which usually provides the force by which active transport is accomplished. Although electron transport usually generates the energy for active transport, some anaerobes generate transport energy in the absence of electron transport. Such organisms obtain energy for transport either by ATP hydrolysis or by the symportic transport of fermentation end products and protons to the cell exterior.

The details of electron transport are discussed elsewhere (Chapter 9), but with regard to the transport of substances, electron transport produces a proton motive force (PMF). PMF is composed of a proton gradient and a membrane potential. In many, if not most, systems, the bulk of the total PMF is attributable to the proton gradient or a secondary gradient generated from it.

Active transport differs substantially from both simple and facilitated diffusion. Although active transport and facilitated diffusion are both protein-mediated and, therefore, exhibit saturation kinetics, active transport allows accumulation against the concentration gradient while facilitated diffusion does not. It is thus possible for organisms that possess active transport to exist in nutrient-sparse environments. Active transport obligately requires biologically generated energy, which is not required for either simple or facilitated diffusion. The elements of diffusion, facilitated diffusion, and active transport are shown in Figure 6.1.

Chemical modification transport

In active transport, simple diffusion, and facilitated diffusion, the form of the transported material is unaltered during the transport process. In chemical modification transport, by contrast, the form of the transported molecule is changed by phosphorylation. Also, active transport and diffusion of any kind are primarily membrane-mediated phenomena. In chemical modification transport, commonly known as the phosphotransferase system (PTS), both the membrane and the cytoplasm are intimately involved in the transport process. Furthermore, chemical modification transport is a function of a multicomponent system, some of whose components are *inducible*, that is, formed only under certain conditions and some of which are *constitutive*, formed continuously without regard to environmental conditions. Chemical modification transport is found exclusively in procaryotes and has been studied most thoroughly with sugars and sugar alcohols.

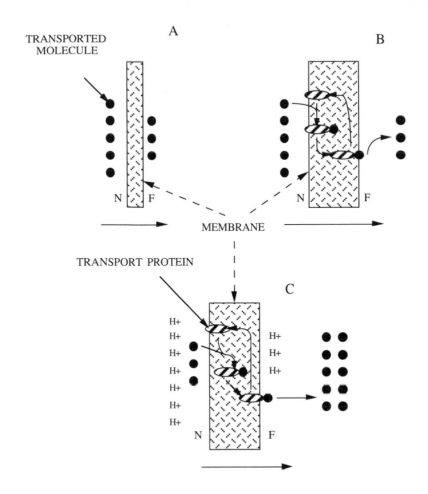

FIGURE 6.1 The elements of simple diffusion (A), facilitated diffusion (B), and active transport (C). The letters N and F indicate the near and far sides of the membrane, respectively. The horizontal arrows indicate the direction of net transport.

The essential elements of chemical modification transport are shown in Figure 6.2. Initially, phosphoenol pyruvate reacts with a cytoplasmic, constitutive protein known as enzyme I (EI) to produce a phosphorylated form of enzyme I (EIP). EIP then reacts with a second constitutive cytoplasmic protein, histidine protein (Hpr), yielding histidine protein phosphate (HprP). HprP reacts with a membrane-associated protein, enzyme II (EII), producing enzyme II phosphate (EIIP). Finally, EIIP reacts within the membrane with the transported material, converting it to a phosphorylated form. In some systems, yet another protein is involved in the transport process, enzyme III (EIII). When such is the case, HprP transfers its phosphate to EIII, producing enzyme III phosphate (EIIIP). EIIIP then transfers its phosphate to produce EIIP, which, in turn, phosphorylates the transported substance.

Regulation of PTS is extremely complex. The state of cellular energy metabolism exerts a substantial effect on the process since the rate and extent of formation of PEP dictates the phosphorylation state of the remaining components of the system. The nature of the external environment also regulates the system. EII and EIII are specifically induced by particular materials, and are typically distinguished by subscripts that denote the substance that leads to their induction. Thus EII_{gluc} is induced by the presence of glucose and EII_{gal} is formed in response to galactose. Because EII and EIII proteins are involved, another level of regulation is possible. Although both EII and EIII proteins are inducible, the specificity of their interaction with transported molecules is not absolute. The EII_{gluc} protein, for example, may react, with a lesser affinity, with a molecule other than

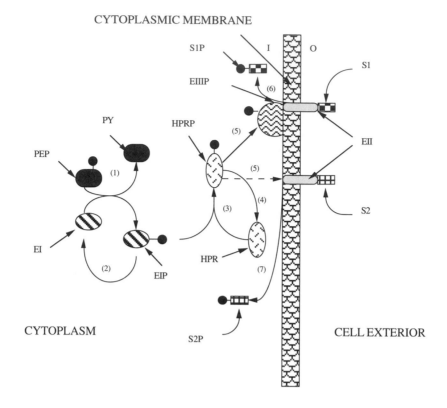

FIGURE 6.2 The essential elements of chemical modification transport. (1) The conversion of phosphoenol pyruvate (PEP) to pyruvate (PY) with the concomitant conversion of enzyme I (EI) to enzyme I phosphate (EIP). (2) Regeneration of EI. (3) Formation of phosphorylated histidine protein (HPRP) by transfer of a phosphate group to histidine protein (HPR). (5) Transfer of phosphate to enzyme II (EII), either directly or by the intermediate formation of phosphorylated enzyme III (EIIIP). (4) Regeneration of nonphosphorylated HPR. (6) Phosphorylation of a substrate (S1) to form substrate phosphate (S1P) during transport of the substrate from the outer side (O) to the inner side (I) of the membrane with the participation of EIIP and EII. (7) Phosphorylation of a second substrate (S2) to form S2P, with the participation of a second, and different, EII molecule only.

glucose, perhaps galactose. Regulation of the operation of the PTS system may reflect not only the presence of particular substrates, but when multiple transport substrates are present, their relative concentrations as well. Competition for HprP by various EII and EIII proteins is yet another level at which control may be exerted. Finally, the intracellular concentration of phosphorylated transported materials exerts a diversity of regulatory effects on the system.

Transport in gram-negative bacteria

For many years it was believed that selectivity of transport and the transport process itself were entirely membrane-mediated phenomena. Study of the PTS system has caused a rethinking of that concept, and has revealed that transport may involve substantial participation by other areas of the cell,

e.g., the cytoplasm. For a long time, little attention was devoted to the cell wall (envelope) as an agent in selective permeability. However, studies of transport in the gram-negative bacteria have shown that in these organisms, the envelope and the periplasm are intimately involved in the transport process.

The structure of the gram-negative cell envelope is shown diagrammatically in Figure 6.3. In contrast to the gram-positive bacterium, in which the cytoplasmic membrane is bounded by layers of murein, the gram-negative cytoplasmic membrane is bounded by a second, fundamentally hydrophobic layer, the outer membrane. A material that would enter or leave the gram-negative bacterium must traverse, not only the cytoplasmic membrane, but also the outer membrane. The fundamentally hydrophobic nature of the gram-negative envelope has raised substantial questions regarding the

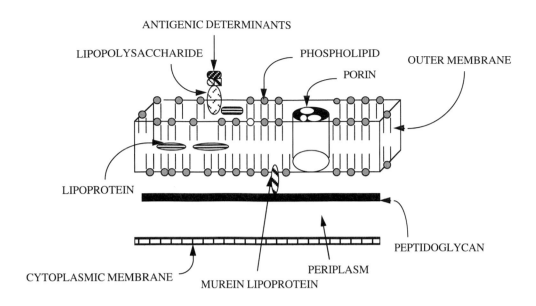

FIGURE 6.3 The general nature of the gram-negative cell envelope.

manner in which hydrophilic molecules accomplish transport. It is now recognized that both the cytoplasmic and outer membrane of gram-negative organisms contain proteins that serve as channels for transport of hydrophilic substances. In particular, a number of outer membrane-associated proteins are known that facilitate the transport of hydrophilic substances across the outer membrane. These proteins are denoted as outer membrane proteins and are abbreviated by the letters "Omp," followed by a capital letter (e.g., OmpA).

The functions of outer membrane proteins are increasingly understood. It is now recognized that many of them are multifunctional and that some of them participate not only in transport but also in other processes. Thus OmpA serves as a receptor for phage Tul a, participates in an interaction with the lipopolysaccharide of the outer membrane, and helps to stabilize mating pairs in F factor-mediated conjugation. OmpB serves as a channel for the diffusion of maltose and other substances. OmpC serves as a diffusion channel for a number of small molecules and also as a receptor for both phage Tul b and T4. OmpF serves as a general channel for small molecules and, in addition as a receptor for phages Tul a and T2.

Certain outer membrane proteins have letter designations other than Omp and are substantially restricted in their action. The LamB protein, for example, mediates the specific transport of maltose and maltodextrin and serves, incidentally, as a receptor for phage lambda. The outer membrane transport proteins are typically trimeric, but the components of the various trimers exist in different physical relationships to each other within the membrane. Trimeric outer membrane transport proteins are collectively referred to as porins.

Ion transport

In addition to organic compound transport and other activities, outer membrane proteins participate in transport of inorganic ions, particularly Fe^{2+} and PO_4^{3-}. Phosphate transport is mediated both by outer membrane and by periplasmic proteins. Thus, the PhoE outer membrane protein facilitates anion transport when phosphate is limited. In contrast, the specific transport of phosphate in *E. coli* involves a periplasmic binding protein (PhoS). That PhoS is periplasmic was revealed by the findings that spheroplasts do not carry out phosphate-specific transport (PST) and that phosphate specific transport (PST) can be restored by the addition of purified PhoS protein to spheroplasts that do not contain it. In addition to PhoS-mediated PO_4^{3-} in *E. coli.*, the organism also contains a phosphate transport system (PIT system) that does not require the PhoS protein and operates in normal spheroplasts.

Transport systems involving periplasmic components can often be distinguished from those that do not by subjecting the suspected organism to osmotic shock, which releases periplasmic components. If transport is abolished or altered by

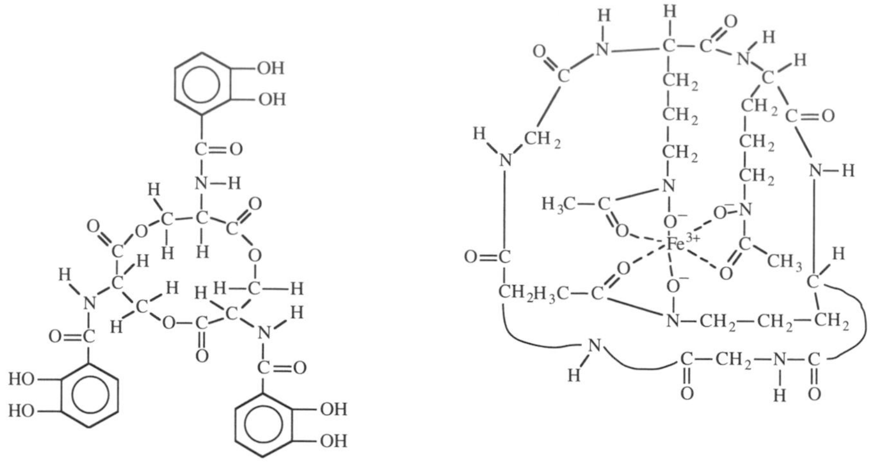

FIGURE 6.4 The general nature of (A) a catechol (enterochelin) and (B) a hydroxamate siderophore (ferrichrome).

osmotic shock treatment, one may suspect that periplasmic components are involved in the transport process. If osmotic shock treatment does not alter transport, it is likely that the periplasm is not involved in the system under study.

Iron transport

Iron is a critical substance for all life, particularly because of its role in energy metabolism. Competition for iron is a mode of pathogenesis for some organisms. Many organisms secrete specific iron transport proteins commonly known as siderophores. Although a great number of siderophores have been studied in a variety of organisms, the majority are one of two types, hydroxamates or catechol compounds. The general structure of these materials is shown in Figure 6.4. Both of these materials bind the Fe^{3+} ion by chelation, but the precise mechanisms by which the bound iron enters the cell varies among organisms. Certain organisms incorporate iron by cell-associated reduction of organically-bound iron without entrance of the siderophore into the cell. In other situations, for example, *Bacillus megaterium,* the intact ligand-bound iron complex may enter the cell by facilitated diffusion. Finally, in some organisms, more than one mechanism for iron uptake is known.

Methods for study of transport

However it occurs and whatever cellular structures are involved, substance transport may be studied from a variety of viewpoints and with a diversity of methods. Before beginning transport studies, it is necessary to carefully consider the most appropriate system for study. Precisely how we proceed is determined by our objectives and the extent to which removal or alteration of cell components alters the transport process. Careful reflection is required with regard to whether it is advisable to study transport in intact cells, spheroplasts, or liposomes. Liposomes are spherical, osmotically sensitive structures that may be derived from treatment of purified cytoplasmic membranes under specific conditions.

In addition to consideration of the type of preparation most appropriate for transport study, careful attention must be given to the methods used to assess the rate and extent of transport. The nature of the methods used will reflect the nature of the substance whose transport is studied. If we are measuring ion transport, changes in electrical conductivity or osmotic phenomena may suffice. However, it is usually advisable to confirm the validity of such measurements by specific chemical or physical determinations of the ions in question.

When measuring the transport of organic materials, specific methods for assessment of the compound under study are required. The latter may be chemical or radiochemical. Frequently, both methods are used. When radioactivity is used, care must be taken to study the relative specific activities of the transported material and the material supplied so that the effects of radioactivity dilution during the transport process may be assessed. In addition, the use of uniformly labelled materials is advisable so that, irrespective of the manner of uptake, the starting material may be accounted for. In some cases, position-labelled materials may also be used for transport studies. Use of these materials may reveal whether or not the transported material is cleaved during the transport process. Finally, it is helpful to use "marker" substances, materials whose general mode of utilization in the physiology of the organism is known. Whenever possible, it is advisable to study both the disappearance of the material and its appearance, either within or without the cell, rather than measuring either parameter separately.

Study of protein-mediated transport

Protein-mediated transport processes are of special interest both because they are widely distributed and because they may be used not only to study transport but also as models for understanding control of protein formation and function. In addition, study of protein-mediated transport facilitates understanding the diversity of ways in which microbial cells respond to environmental change.

Study of protein-mediated transport poses a number of physiological questions. To begin with, we must understand the nature of the system. This necessity requires that we characterize both the nature of the transport system and the manner in which it functions. Furthermore, we must know the manner in which the system responds to changes in the external or internal environment. All of these elements can be addressed, to a considerable extent, by study of the kinetics of transport. Study of transport kinetics, combined with isolation and study of transport proteins, provides much useful information regarding both the nature of the transport process and its physiological consequences.

Transport kinetics

Transport, growth, and enzyme activity are all processes that involve or reflect molecular interactions with proteins. For this reason, although the processes are physiologically distinct, they may be described by a common set of equations, the equations of Michaelis and Menten (see Chapter 5). The relationship between the concentration of the transported material and the rate of transport of a substance whose transport is not affected by competition with other materials is shown below:

$$1/Vt = Ks/V_{max} \times 1/[S] + 1/V_{max} \qquad (1)$$

In this equation, Vt is the initial velocity of transport, Ks is the molar substrate concentration at half maximal transport velocity, S is the initially supplied molar substrate concentration and V_{max} is the maximum rate of transport when all of the reactive sites on the transport molecules are occupied. The operation of this equation reflects the fact that the transport process is a protein-mediated process and is, therefore, subject to saturation kinetics. Various transport proteins have differing affinities for particular transport substrates and thus differ in their Ks values. Determination of the Ks value of a transport protein is useful as a means of characterizing it. Like the growth rates of microbes, intrinsic differences in transport proteins characterize the maximal transport rates obtainable in a given system. In addition to Ks values, determination of the maximal transport rate possible is highly useful in characterization of transport systems.

If a transport protein mediates transport of more than one material, transport of a particular material is influenced by the presence of other materials. The extent to which other materials influence a particular material's transport reflects both the relative affinity of the transported entities for sites on the transport protein and the relative concentrations of the materials. For both of these reasons, transport under these conditions may be described by kinetics analogous to those for competitive enzyme inhibition:

$$1/Vt = (1 + I_t/K_{I_t}) \times Ks/V_{max} \times 1/[S] + 1/V_{max} \qquad (2)$$

In the above equation, Vt is the initial velocity of transport, It is the molar concentration of the substance that competes with transport of another substance, KIt is the equilibrium constant that describes the relationship between the free and protein-bound inhibitory substance, S is the initial molar concentration of the normal transport substance, and V_{max} is the maximum velocity of transport. Use of equations (1) and (2) is helpful in understanding the nature of particular transport systems and the interrelationships among them.

Because of distinctive properties of different transport proteins, the properties of a particular transport system characterize it. In a particular situation, transport of given material will occur in a reproduceable way. Careful study of the operation of various transport systems allows determination of whether they are constitutive or inducible, their relative affinities for particular transport substrates, the optimal conditions for their action and, when inducible, the conditions that facilitate or inhibit their induction. Understanding all of these parameters, as well as characterization of particular transport proteins is most useful to understanding the relationship of transport to the growth of an organism and in assessing the manner in which various organisms may respond to changes in their external environments.

The physiological consequences of transport

A number of factors determine the physiological consequences of transport. Although it is critical for cells, transport in either direction cannot be divorced from the other activities of the cell. For optimal growth to occur, the rates of transport of critical cell materials must be balanced by the rate of their utilization. In addition, the rate of removal of potentially harmful substances must balance their production. The complexity of physiological processes and their relationship to transport is such that we must carefully consider the relationship between optimal conditions for transport and optimal conditions for growth. It is well recognized that changes in growth rate are associated with changes in cell size and that differences in cell size are found both between organisms and between the various substages of the growth cycle of any particular organism. Size differences are also reflected in differences in surface-to-volume ratios. It is important to carefully consider the relationship between size, transport, and growth.

It is instructional to consider precisely what determines the size of an organism under a given set of conditions. In many cases, it is likely that the maximum size of an organism is a compromise between the rate of nutrient transport and the rate of toxic end product removal. According to this concept, each organism, under optimal growth conditions, has a "critical radius" that corresponds to a particular cell size. Cell size increase beyond this limit is impossible because of the inability of nutrient transport to occur at a rate compatible with optimal growth and/or the accumulation of toxic materials to an extent incompatible with life.

Comparisons of transport rates

How may we compare relative transport rates among organisms of differing sizes and shapes in a physiologically useful manner? The answer to this question is not easy and is one that requires reflection. Many possible comparisons can be made but some may be more helpful than others. If we choose, we may measure the transport process in relationship to the physiological process in which it is involved. Thus, we may ask how many molecules of an amino acid are transported per unit of protein formed. Or, if we choose, we may relate both measures to time. Alternatively, we may relate a transport process to a general measure of mass. Another approach is to consider the number of molecules transported per cell and to relate this measure to time. Since cells differ substantially in size and surface-to-volume ratio, it may be necessary or useful to relate transport measurements to surface area. Such comparisons require accurate determinations of cell size and reasonable assumptions about cell shapes so that cell volumes may be calculated accurately. Precisely how comparative measures of microbial cell transport are made is a function of the purposes of the study.

Relative internal and external concentrations of substances

Particularly with protein-mediated transport processes, it is necessary or advisable to determine the relative concentrations of materials in the external and internal cell environment. Such measurements allow determination of the extent to which the transported material is concentrated. Accurate assessment of the intracellular concentration of a transported material is not an easy task, but with careful attention to methodology, reliable measures of the intracellular concentration of a material may be obtained and compared to the extracellular concentration.

In order to accurately assess the intracellular concentration of a material, it is necessary to determine the *volume* occupied by cells that contain the material, in comparison to the *space* that surrounds the cells. The necessity for such a measure is based on the fact that the volume occupied by individual cells is far too small to allow accurate determination of the amount of material within them, so determinations of the cellular concentrations of materials employ a population of cells. Typically, we expose a cell population to the material to be transported and, at intervals, separate a portion of the cells by filtration or centrifugation from the fluid to which they have been exposed. The concentration of the cell-associated material is then determined. The difficulty with such a procedure is that the measured material is found both within *and* between the cells. Failure to correct measurements for space between the cells leads to an inaccurate estimation of the intracellular concentration of the transported material.

Correction of transport measurements for intracellular space may be accomplished by mixing a known volume of cells, obtained by centrifugation of a cell population in a calibrated test tube, with a known concentration of a radiolabelled material, such as inulin, that is impermeable to the cells and which the cells cannot metabolize. After a period of time, the cells are removed from the fluid and the extracellular concentration of radioactive material is assessed. The assumption in these measurements is that the nonmetabolized material will be diluted by the volume of fluid in the cell pellet that is outside the cells. The *expected* concentration of the diluted material that would result if all of the space were available to it can be calculated from the relative volumes of the cell pellet and the diluted material. The *observed* extracellular concentration of the material will be greater than the calculated value, precisely because a portion of the cell pellet is unavailable to the diluted material. Dilution of the nonmetabolized, impermeable material is accomplished only with the space outside of the cells in the pellet. The amount of that space can be calculated by consideration of the total additional space required to produce the observed concentration of the diluted material from the volume originally supplied.

To clarify the ideas just described, consider the following example. A 50 ml volume of 2.0 M inulin is mixed with 50 ml of a packed cell pellet to produce a final extracellular inulin concentration of 1.02 M. What are the volumes of the extracellular space and the space occupied by the cells? Reflection on this question reveals that if all of the cell pellet space were available to the diluted material, the expected inulin concentration would be 1.0 M, since mixing of 50 ml with an additional 50 ml would produce a twofold dilution. The fact that the concentration of extracellular inulin is 1.02 M rather than 1.0 M reflects the fact that not all of the pellet volume is available for dilution. What portion of the cell pellet is available? We may calculate that value by using the following equation:

$$D = \frac{A}{A + B} \qquad (3)$$

In this equation, D is the dilution, A is the original volume, B is the added volume, and A + B is the total volume. Applied to our model measurements, we recognize that dilution is numerically equal to 2.0 divided by 1.02, or 1.9758. Multiplication of the original volume, 50 ml by the number 1.9785, gives the total volume required to dilute 50 ml of 2.0 M inulin to produce a concentration of 1.02 M, or 98.75 ml of fluid. Since 50 ml of that total came from the original solution, the *remaining* 48.75 ml is attributable to the space in the cell pellet that is outside the cells. Since the total cell pellet is 50 ml, the volume attributable to the cells themselves is, in this instance, 50 – 48.75, or 1.25 ml.

Transport measures are inherently fraught with problems and must be done with great care. A problem of frequent occurrence is an incorrect assessment of the amount of material that is *actually* within the cells. Not only must one correct for intercellular space, but care must be taken to avoid loss of material that is truly transported. In some cases, it is necessary to determine whether the transported material is, in fact, intracellular or simply cell-associated. The accuracy and physiological significance of transport measurements are only as good as the adequacy of the methods used and the skill and care with which they are employed.

Summary

Transport is a critical process for microbes. It occurs by a variety of mechanisms in both directions and affects the totality of microbial physiological functions. The process often reflects cellular metabolism because of the direct participation of PEP or because the process is driven by the PMF generated during electron transport. In many cases, the transport process involves the participation of cell structures other than the cytoplasmic membrane. Substantial care and reflection are required to obtain physiologically meaningful transport measurements. Particular attention must be paid to the appropriateness of the method, and the extent and manner in which the transport process reflects the coordinated actions of various cell structures and regions. Equal attention is required with regard to the manner in which transport may affect and be interconnected with other critical physiological processes. When carefully and skillfully done, transport measurements provide an abundance of information regarding critical aspects of the physiology of microbes and the manner in which they respond to, and thrive in, a diversity of environments.

Selected References

Ames-Luzzi, G. F. 1986. Bacterial periplasmic transport systems: Structure, mechanism, and evolution. *Annual Review of Biochemistry*. **55:** 397–426.

Antonucci, T. K., and Dale L. Oxender. 1986. The molecular biology of amino acid transport in bacteria. *Advances in Microbial Physiology*. **28:** 146–180.

Chopra, I., and P. Ball. 1982. Transport of antibiotics into bacteria. *Advances in Microbial Physiology*. **23:** 184–240.

Dills, S. S., A. Apperson, M. R. Schmidt, and M. H. Saier, Jr. 1980. Carbohydrate transport in bacteria. *Microbiological Reviews*. **44:** 385–418.

Eddy, A. A. Mechanisms of solute transport in selected eukaryotic micro-organisms. 1982. *Advances in Microbial Physiology*. **23:** 2–78.

Gibson, Jane. 1984. Nutrient transport by anoxygenic and oxygenic photosynthetic bacteria. *Annual Review of Microbiology*. **38:** 135–139.

Glynn, I. M., and S. J. D. Karlish. 1990. Occluded cations in active transport. *Annual Review of Biochemistry*. **59:** 171–205.

Henderson, P. J. F. 1971. Ion transport by energy-conserving biological membranes. *Annual Review of Microbiology*. **25:** 393–428.

Kaback, H. R. 1970. Transport. *Annual Review of Microbiology*. **39:** 561–598.

Konings, Wil N. 1977. Active transport of solutes in bacterial membrane vesicles. *Advances in Microbial Physiology*. **15:** 175–251.

Maloney, P. C., S. V. Ambudkar, V. Anantharam, L. A. Sonna, and A. I. Varadhachary. 1990. Anion-exchange mechanisms in bacteria. *Microbiological Reviews*. **54:** 1–17.

Meadow, N. D., D. K. Fox, and S. Roseman. 1990. The bacterial phosphoenolpyruvate: glucose phosphotransferase system. *Annual Review of Biochemistry*. **59:** 497–542.

Nikaido, H., and M. Vaara. 1985. Molecular basis of outer membrane permeability. *Microbiological Reviews*. **49:** 1–32.

Poolman, B., A. J. M. Driessen, and W. N. Konings. 1987. Regulation of solute transport in streptococci by external and internal pH values. *Microbiological Reviews*. **51:** 498–508.

Postma, P. W., and S. Roseman. 1976. The bacterial phosphoenolpyruvate: sugar phosphotransferase system. *Biochemica Biophysica Acta*. **457:** 213–257.

Rosen, B. P. 1986. Recent advances in bacterial ion transport. *Annual Review of Microbiology*. **40:** 263–286.

Scarborough, G. A. 1985. Binding energy, conformational change, and the mechanism of transmembrane solute movements. *Microbiological Reviews*. **49:** 214–231.

Silverman, M. 1991. Structure and function of hexose transporters. *Annual Review of Biochemistry*. **60:** 757–794.

Simoni, R. D., and P. W. Postma. 1975. The energetics of bacterial active transport. *Annual Review of Biochemistry*. **44:** 523–554.

Tanford, C. 1983. Mechanism of free energy coupling in active transport. *Annual Review of Biochemistry*. **52:** 379–410.

Vaara, M. 1992. Agents that increase the permeability of the outer membrane. *Microbiological Reviews*. **56:** 395–411.

Wickner, W., A. J. M. Driessen, and F. U. Hartl. 1991. The enzymology of protein translocation across the *Escherichia coli* plasma membrane. *Annual Review of Biochemistry*. **60:** 101–124.

Wilson, David B. 1978. Cellular transport mechanisms. *Annual Review of Biochemistry*. **47:** 933–966.

Wirtz, K. W. A. 1991. Phospholipid transfer proteins. *Annual Review of Biochemistry*. **60:** 73–99.

CHAPTER 7

Catabolic Metabolism

The nature of metabolism

Metabolism is the summation of the interconnected biochemical reactions of an organism. These reactions are connected in such a way that, for the most part, metabolism leads to the formation of large molecules from smaller ones. This type of metabolism is commonly known as *biosynthesis* or *anabolism* and is required so that a microbe may make the materials of protoplasm and organize them into the functioning structures of the cell. For synthetic reactions to occur, however, energy must be provided. The energy may be derived from the trapping of radiant energy, such as light, or from chemical oxidation. Many organisms derive energy from the oxidation of organic compounds. In such a situation, *catabolism*, the inverse of synthesis occurs. In catabolism, large molecules are degraded to smaller ones, and a portion of the energy released is trapped in a biologically useful form, usually adenosine triphosphate (ATP). ATP is the universal energy compound found throughout nature and is used to accomplish cellular life pro-

cesses. From a particular perspective, all metabolism may be thought of as the coupling of energy production and energy use.

The interconnectedness of biochemical reactions

Certain chemical reactions, biochemical and otherwise, occur spontaneously because their occurrence leads to the net release of energy. Other reactions do not occur spontaneously, and must be "forced" to occur, either by energy input, as in synthetic processes, or by the coupling of energetically unfavorable reactions that do not occur spontaneously, with other reactions whose occurrence leads to energy release, so that the *net reaction sequence* yields energy. Catabolic metabolism operates in this manner, allowing energy to be extracted from materials which are not, by themselves, oxidizable, by coupling an energy-unfavorable reaction (or reactions) with other reactions, some of which are energetically favorable, thereby allowing the entire sequence to proceed. We refer to such a sequence of interconnected biochemical reactions as a *catabolic reaction sequence* or *pathway*. By contrast, synthetic, or anabolic, pathways are interconnected reaction sequences that require energy input.

FIGURE 7.1 The principle of LeChâtelier as applied to microbial metabolism. Substance A would not spontaneously be converted to substance E because the reactions catalyzed by enzymes E1, E2, and E3 are all thermodynamically unfavorable, as shown by the relative lengths of the arrows in the forward and reverse directions. However, the continual conversion of metabolite D to substance E by enzyme E4 forces the remaining interconnected reactions to proceed in the net direction of substance E.

Ways in which reactions are interconnected

We may regard the reactions of metabolism as being interconnected in either of two ways. A reaction may be connected with another so that the product of the first reaction is the starting point for the next one, a situation referred to as *chemical interconnection*. Alternatively, a reaction may be connected with another because the energy released in the second reaction is sufficient to allow the first one to occur when it would not do so by itself. Metabolism allows many reactions to be interconnected as long as the terminal reaction in the sequence releases at least enough energy so that the remaining reactions in the sequence may occur. This situation is referred to as *thermodynamic interconnection*.

Catabolic metabolism may be thought of as a biological example of the principle of LeChâtelier, which states that disruption of a system in equilibrium leads to modification of the system in an effort to restore the original equilibrium. By continuous removal of the product of a reaction as a consequence of its conversion to yet *another* product, the original reaction is forced to generate its product. The process is shown in Figure 7.1. It is somewhat analogous to a train, in which the cars are connected to each other, but cannot move by themselves. However, the engine that is connected to one end of them causes the entire train to move.

The bacteria are the most metabolically diverse life forms The bacteria are, without a doubt, the most versatile of all life forms in their ability to extract energy from chemical oxidation or from phototrophic (light-requiring) processes and to use that energy for the accomplishment of their essential life processes. We will therefore use the bacteria as models for understanding both catabolic and anabolic metabolism. The bacteria, taken collectively, exhibit all of the metabolic sequences found in other life forms and, in addition, display sequences that are uniquely their own. By study of metabolism in the bacteria, we may understand, not only the nature of the metabolic processes in higher life forms, but also, precisely how the bacteria are unique.

Central pathways of metabolism

Although the bacteria are highly diverse in their metabolism, they do not contain separate metabolic pathways for each of the substances they use. In most cases, they channel a diversity of materials into a relatively few number of *central pathways*, pathways that are essential to life and are found in all cellular forms, not just the bacteria. To a great extent, the ability of bacteria to use a wide diversity of carbon and nitrogen sources for growth reflects their ability to channel unusual substances, or their metabolites, into central pathways. In this way, the organisms generate energy from a variety of materials through the use of only a few metabolic sequences. Furthermore, the intermediates of catabolism are usually the beginning points for synthesis. It is thus possible for all life forms, through the use of a common collection of central pathways, to both generate energy and synthesize the essential materials of life.

In a very real sense, synthesis begins where catabolism leaves off. Because of this, it is sometimes hard to decide whether a particular metabolic sequence is catabolic or anabolic. Because of this fact, we often regard metabolic pathways as *amphibolic*. In an amphibolic sequence, the same central pathway participates *simultaneously*, in both energy-yielding and energy-requiring processes. Through study of catabolic sequences, we may understand not only the mechanisms by which energy is obtained but also how energy generation and energy utilization are connected.

Common biosynthetic intermediates

In order for an organism to grow, it must synthesize its essential constituents, for example, DNA, RNA, proteins, lipids, and carbohydrates. Although the cell contains a diversity of materials, they may all be formed from metabolism of twelve compounds, the common biosynthetic intermediates. All life forms, irrespective of the way in which they obtain energy or their nutritional requirements, must synthesize these twelve critical metabolites. The metabolites and their major biosynthetic functions are shown in Table 7.1.

In chemoheterotrophic organisms, organisms that degrade organic materials and use organic carbon to produce the building blocks for synthesis, it is the task of the organism to provide both energy and the critical twelve metabolites. No single metabolic sequence allows the formation of all metabolites but, somehow, they must be formed.

The Embden-Meyerhof-Parnas pathway

The Embden-Meyerhof-Parnas pathway, also known simply as the Embden-Meyerhof (EM) pathway, or as glycolysis, is a central metabolic sequence that is found throughout nature. It is a major pathway for the conversion of hexose sugars to pyruvate and is shown diagrammatically in Figure 7.2. The operation of the pathway results in the formation of two molecules of reduced nicotinamide adenine dinucleotide (NADH) and two molecules of ATP per molecule of glucose degraded to pyruvate. The pathway results in the formation of six of the critical biosynthetic intermediates, glucose-6-phosphate, fructose-6-phosphate, triose phosphate, 3-phosphoglycerate, phosphoenol pyruvate (PEP), and pyruvate. The pathway is symmetrical since cleavage of the glucose molecule between carbons 3 and 4 converts a six-carbon molecule into two three-carbon molecules in such a way that each carbon from the top half of the glucose molecule has a *metabolically equivalent* carbon in the bottom half of the molecule. Thus, the phosphorylated carbons in triose molecules arise from glucose carbons 1 and 6, the carbons adjacent to the phosphates of the trioses arise from glucose carbons 2 and 5, and the carbons distal from the phosphated carbons of trioses arise from glucose carbons 3 and 4. Although the pathway generates NADH, it does not generate NADPH so it cannot, by itself, provide reducing power for biosynthesis.

The pathway begins by the conversion of glucose to glucose-6-phosphate. This process is accomplished in various systems in different ways. In many organisms, the enzyme hexokinase, with expenditure of a high-energy phosphate from ATP, yields glucose-6-phosphate, but in organisms that have the phosphotransferase system (PTS) of transport, glucose may be converted to glucose-6-phosphate with the expenditure of a PEP molecule. In the next step of the sequence, glucose-6-phosphate is converted to fructose-6-phosphate by the operation of phosphohexose isomerase. This step is followed by the phosphorylation of fructose-6-phosphate with the enzyme phosphofructokinase and ATP expenditure, yielding fructose-1,6-bisphosphate. Fructose 1,6-bisphosphate is cleaved by aldolase into two three carbon molecules yielding dihydroxyacetone phosphate from the top half of the fructose molecule and glyceraldehyde-3-phosphate (triose phosphate) from its bottom half. Triose phosphate and dihydroxyacetone phosphate are in equilibrium with each other through the enzyme-triose-phosphate isomerase, but the equilibrium is such that dihydroxyacetone would predominate in the absence of other factors. This situation is overcome because of the action of other enzymes in the sequence. The majority of the carbon is metabolized as triose phosphate, which is converted with the aid of glyceraldehyde-3-phosphate dehydrogenase and inorganic phosphate to 1,3-bisphosphoglycerate. Through 3-phosphoglycerate kinase, 1,3-bisphosphoglycerate transfers its acid phosphate to adenosine diphosphate (ADP), yielding ATP and 3-phosphoglycerate. By the combined action of glyceraldehyde 3-phosphate dehydrogenase and 3-phosphoglycerate kinase, triose phosphate is converted to 3-phosphoglyceric acid, ATP and a molecule of NADH are formed, using the electrons generated from conversion of the aldehyde of triose phosphate to the acid group of 3-phosphoglycerate. The remaining steps of the

TABLE 7.1 The Twelve Critical Biosynthetic Metabolites

NAME	STRUCTURE	MAJOR BIOSYNTHETIC FUNCTION
1. Glucose-6-phosphate	$$\begin{array}{c} CHO \\ H-C-OH \\ HO-C-H \\ H-C-OH \\ H-C-OH \\ CH_2O-PO_3H_2 \end{array}$$	Central intermediate in many catabolic pathways; carbohydrate polymers
2. Fructose-6-phosphate	$$\begin{array}{c} CH_2OH \\ C=O \\ HO-C-H \\ H-C-OH \\ H-C-OH \\ CH_2O-PO_3H_2 \end{array}$$	*N*-acetylglucosamine, *N*-acetylmuramic acid murein
3. Ribose-5-phosphate	$$\begin{array}{c} H \quad O \\ \diagdown\!\!\diagup \\ C \\ H-C-OH \\ H-C-OH \\ H-C-OH \\ CH_2OPO_3H_2 \end{array}$$	Nucleic acids, histidine
4. Erythrose-4-phosphate	$$\begin{array}{c} H \quad O \\ \diagdown\!\!\diagup \\ C \\ H-C-OH \\ H-C-OH \\ CH_2OPO_3H_2 \end{array}$$	Aromatic amino acids
5. Triose phosphate	$$\begin{array}{c} H \quad O \\ \diagdown\!\!\diagup \\ C \\ H-C-OH \\ CH_2O-PO_3H_2 \end{array}$$	Precursor to dihydroxyacetone; lipids
6. 3-phosphoglycerate	$$\begin{array}{c} O \\ \diagup\!\!\diagup \\ C-OH \\ H-C-OH \\ CH_2O-PO_3H_2 \end{array}$$	Beginning point for the serine family of amino acids
7. Phosphoenol pyruvic acid	$$\begin{array}{c} O \\ \diagup\!\!\diagup \\ C-OH \\ C-O-PO_3H_2 \\ C \\ H \quad H \end{array}$$	Key intermediate in several carbohydrate degradation sequences; aromatic amino acid precursor

---

TABLE 7.1—*Continued*

NAME	STRUCTURE	MAJOR BIOSYNTHETIC FUNCTION
8. Pyruvic acid	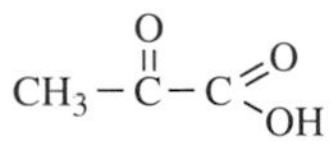 $CH_3-\overset{\overset{O}{\|\|}}{C}-C\overset{\nearrow O}{\diagdown OH}$	Key intermediate in degradative metabolism; "gateway compound" to the TCA cycle; precursor to the pyruvate family of amino acids
9. Acetyl-CoA	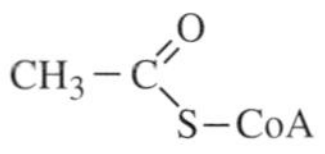 $CH_3-C\overset{\nearrow O}{\diagdown S-CoA}$	Key energy compound; beginning point for lipid synthesis; participant in many synthetic processes
10. Alpha-ketoglutaric acid	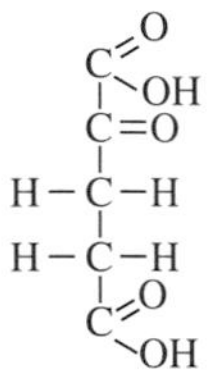 $C\overset{\nearrow O}{\diagdown OH}$ $C=O$ $H-C-H$ $H-C-H$ $C\overset{\nearrow O}{\diagdown OH}$	TCA cycle intermediate; precursor to the glutamate family of amino acids
11. Oxaloacetic acid	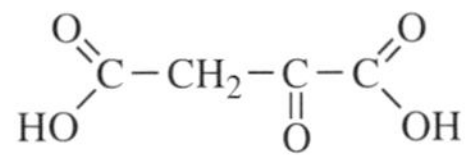 $\overset{O}{\overset{\diagup\diagdown}{HO}}C-CH_2-\overset{\overset{O}{\|\|}}{C}-C\overset{\nearrow O}{\diagdown OH}$	TCA cycle intermediate; precursor to the aspartic family of amino acids
12. Succinyl-CoA precursor to heme	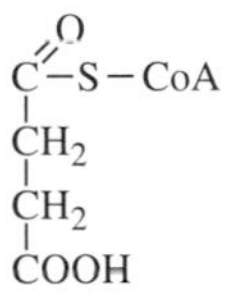 $C\overset{\nearrow O}{\diagdown S-CoA}$ CH_2 CH_2 $COOH$	Key intermediate in the TCA cycle; major energy intermediate

---

Embden-Meyerhof sequence involve the conversion of 3-phosphoglycerate to 2-phosphoglycerate by phosphoglyceromutase, the dehydration of 2-phosphoglycerate to PEP by enolase and conversion of PEP to pyruvate with pyruvate kinase.

Although many of the reactions of the EM sequence would not occur spontaneously, certain reactions of the sequence are essentially irreversible and allow the entire sequence to occur. Three reactions in the EM pathway have critical regulatory functions: 1) the phosphorylation of glucose to glucose 6-phosphate, 2) the conversion of fructose-6-phosphate to fructose-1,6 bisphosphate, and 3) the conversion of PEP to pyruvate. Collectively, these enzymes regulate the flow of carbon and formation of energy via the EM pathway.

Because the EM pathway is an amphibolic pathway and is involved both in energy generation and in the formation of critical biosynthetic intermediates, it is not surprising that both energy compounds, such as AMP and ATP, and biosynthetic intermediates exert regulatory effects upon the pathway. Both phosphofructokinase and pyruvate kinase are regulated by adenine nucleotides. It is often the case that enzymes involved in ATP formation are inhibited by ATP and stimulated by AMP and that reactions involved in ATP expenditure are stimulated by ATP and inhibited by AMP. Such is the case with the EM pathway, among many others.

One theory of metabolic regulation suggests that all metabolism is regulated by the "energy charge" state of the cell. According to this concept, originated by Daniel Atkinson and colleagues, the energy charge is numerically described as:

$$EC = \frac{[ATP] + 1/2\,[ADP]}{[AMP] + [ATP] + [ADP]} \tag{1}$$

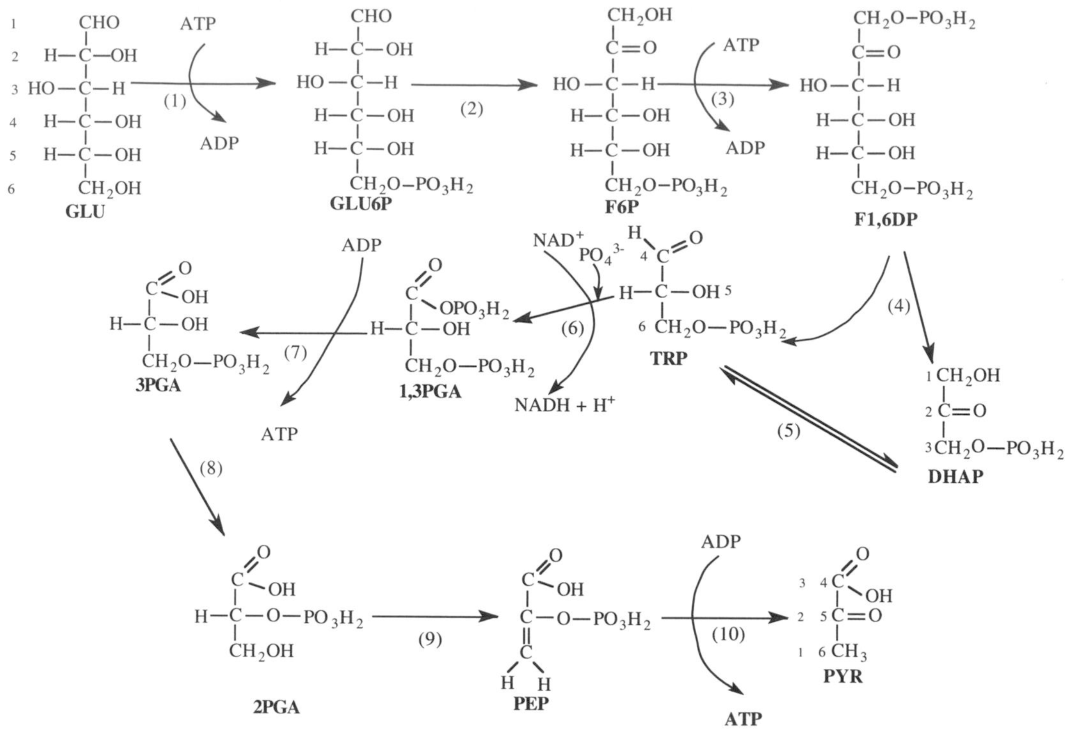

FIGURE 7.2 The Embden-Meyerhof-Parnas pathway. (1) Conversion of glucose (GLU) to glucose-6-phosphate (GLU6P) by hexokinase or glucokinase. (2, 3) Formation of fructose-1,6-di(bis)phosphate (F1,6DP) with the intermediate formation of fructose-6-phosphate (F6P) by the actions of phosphohexose isomerase (2) and phosphofructokinase (3). (4) Cleavage of F1,6DP to dihydroxyacetone phosphate (DHAP) and triose phosphate (TRP) by F1,6DP aldolase. (5) Interconversion of DHAP and TRP by triose phosphate isomerase. (6, 7) Conversion of TRP to 3-phosphoglyceric acid (3PGA) with the intermediate formation of 1,3-di(bis)phosphoglyceric acid (1,3DPGA) by TRP dehydrogenase (6) and 3PGA kinase (7). In the process of these reactions, both ATP and NADH are formed. (8, 9, 10) Conversion of 3PGA to pyruvate (PYR) with the intermediate formation of 2-phosphoglyceric acid (2PGA) and phosphoenol pyruvate (PEP) by phosphoglyceromutase (8), enolase (9), and pyruvate kinase (10). The small numbers adjacent to carbons denote the carbons of the original glucose molecule that appear in intermediates and products.

It is not a single reaction that determines the manner in which metabolism occurs, but the balance between reactions that produce and those that utilize energy. Application of the energy charge concept to the EM pathway explains the manner in which phosphofructokinase, hexokinase, and pyruvate kinase exert their regulatory effects on the EM pathway. ATP inhibits both phosphofructokinase and pyruvate kinase, since both enzymes are involved in energy generation. The manner of inhibition for phosphofructokinase by ATP is interference with the binding of fructose-6-phosphate to the enzyme. The mechanism of ATP inhibition of pyruvate kinase is quite different, apparently by allosteric inhibition of an isoenzyme. Irrespective of differences in mechanism, the net effect is the same, diminishing energy formation when energy is in abundant supply, and with the aid of AMP, stimulating ATP formation when its concentration falls as a result of synthetic activities. The inhibition of phosphofructokinase leads to an accumulation of fructose-6-phosphate, which in turn, leads to a buildup of glucose-phosphate. High glucose-6-phosphate levels diminish the activity of hexokinase. These results make physiological sense, because glucose-6-phosphate conversion is not required when the product is present in abundance.

The amphibolic nature of the EM pathway is further illustrated by the fact that biosynthetic products also regulate its action. Thus, alanine, an immediate synthesis product from pyruvate, regulates pyruvate kinase, and citrate, an indicator of the concentration of other biosynthetic intermediates, also regulates the sequence.

Hexose monophosphate pathways

The EM pathway is the major pathway for glucose conversion to pyruvate in many microbes. However, although it supplies energy and six of the critical biosynthetic intermediates, it does not supply all of the materials needed for synthetic purposes. In particular, the EM pathway fails to provide five carbon sugars required for DNA and RNA formation. It also fails to provide erythrose-4-phosphate, a critical intermediate for aromatic amino acid synthesis. Finally, the EM pathway fails to provide NADPH, the major source of reducing power for biosynthesis. Hexose monophosphate (HM) pathways provide all of these materials. Slight differences are found in the precise details of HM pathways, but all organisms must have the materials that HM pathways provide, irrespective of differences in particulars.

The hexose monophosphate shunt pathway

The hexose monophosphate shunt (HMS) pathway, also known as the oxidative pentose (OP) pathway, and variations of it are widely found in microbes. The classical pathway is shown in Figure 7.3. The pathway begins with the conversion of glucose-6-phosphate, formed from the action of hexokinase on glucose, to 6-phosphogluconogamma lactone, a process mediated by glucose-6-phosphate dehydrogenase that also allows conversion of NADP to NADPH. The lactone is hydrolyzed with lactonase, yielding 6-phosphogluconic acid, which is decarboxylated with 6-phosphogluconic acid dehydrogenase in the presence of NADP, yielding NADPH, carbon dioxide (CO_2), and, via ribulose-5-phosphate, a mixture of ribose-5-phosphate and xylulose-5-phosphate. In many microbes, the ratio of xylulose-5-phosphate to ribose-5-phosphate is 2:1, because of the relative actions of ribulose-5-phosphate isomerase and ribulose-5-phosphate epimerase. When this ratio occurs, pentose molecules are converted back to hexoses through the action of the enzymes transaldolase and transketolase. These two enzymes transfer 3-carbon and 2-carbon moieties, respectively, from ketose sugars to aldose sugars. The enzymes have, comparatively speaking, broad specificities that allow them to mediate a wide variety of ketose to aldose transfer reactions.

When the "complete" OP pathway occurs, a molecule of xylulose-5-phosphate transfers two carbons in a transketolase-mediated reaction, to yield the seven-carbon molecule, sedoheptulose-7-phosphate, which arises from the top two carbons of a xylulose-5-phosphate and the five carbons of a ribose-5-phosphate molecule. Glyceraldehyde-3-phosphate is also formed from the bottom three carbons of the xylulose-5-phosphate molecule. Next, sedoheptulose-7-phosphate interacts with glyceraldehyde-3-phosphate, transferring its top three carbons to the carbons of glyceraldehyde-3-phosphate, by a transaldolase reaction. The result is the formation of a molecule of fructose-6-phosphate and a molecule of erythrose-4-phosphate, which arises from the bottom four carbons of the sedoheptulose molecule. At this point, the erythrose-4-phosphate reacts with a second molecule of xylulose-5-phosphate via a transketolase reaction to yield a second molecule of fructose-6-phosphate, in which the bottom four carbons are derived from erythrose-4-phosphate and the top two carbons originate from xylulose-5-phosphate. In addition, a molecule of glyceraldehyde-3-phosphate is formed from the bottom three carbons of xylulose-5-phosphate. In the complete OP pathway, yet a third molecule of fructose-6-phosphate is formed through the condensation of two glyceraldehyde-3-phosphate molecules by a reversal of aldolase and subsequent conversion of fructose-1,6-bisphosphate via a phosphorylase, to fructose-6-phosphate. Isomerization of the fructose-6-phosphate to glucose-6-phosphate allows reinitiation of the cycle.

When the complete OP pathway operates, the products are CO_2, NADPH, and intermediates for biosynthesis, such as erythrose-4-phosphate and ribose-5-phosphate. If the sequence operates completely, six hexose molecules are converted to six pentoses and six CO_2 molecules and the six pentose molecules are reconverted to hexoses. No ATP formation is possible from its operation, unless it results from conversion of NADPH to NADH by a transhydrogenase reaction and the use of NADH in

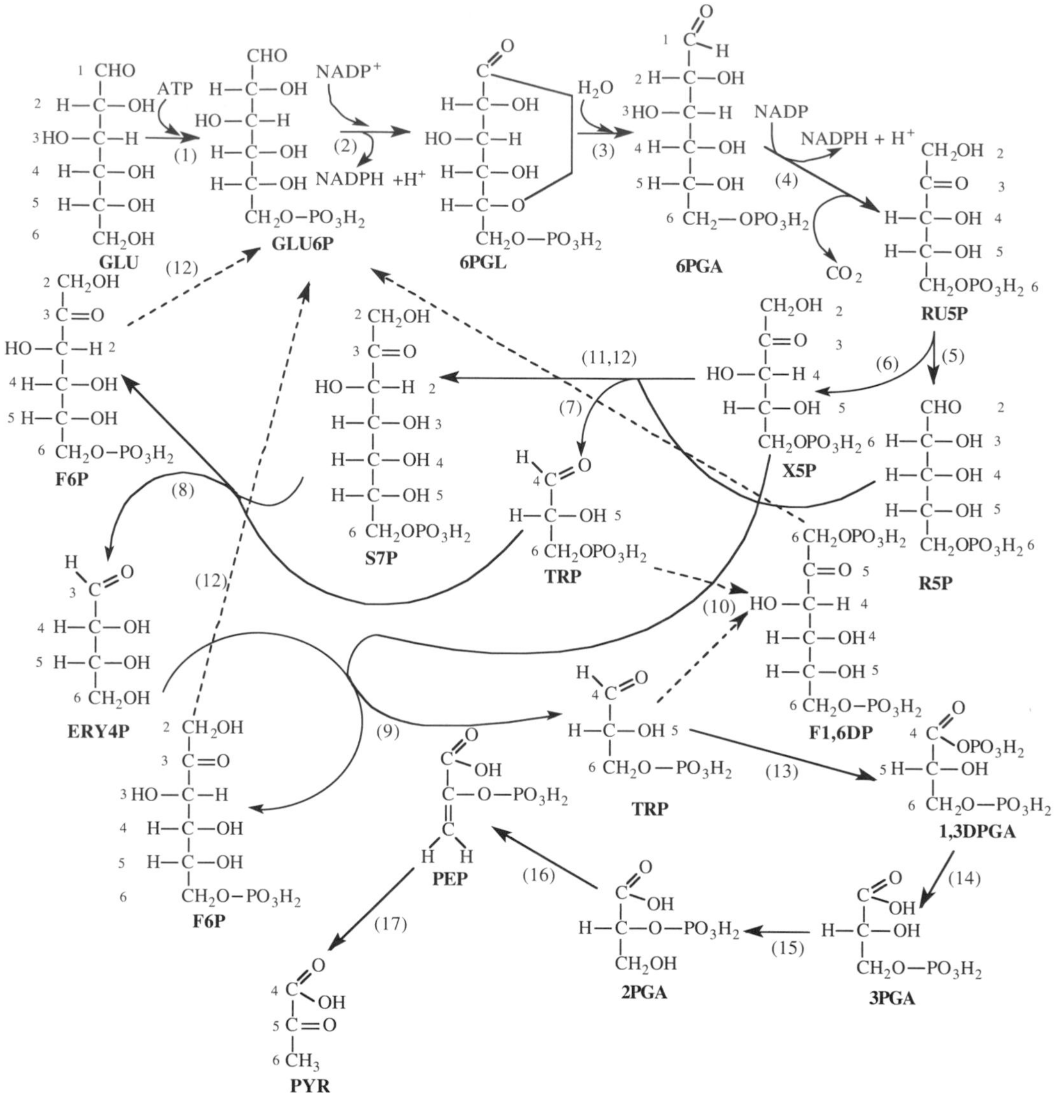

FIGURE 7.3 The classical hexose monophosphate pathway. (1) Conversion of glucose (GLU) to glucose-6-phosphate (GLU6P) by hexokinase. (2) Conversion of GLU6P to 6-phosphogluconolactone (6PGL) by glucose-6-phosphate dehydrogenase. (3) Conversion of 6PGL to 6-phosphogluconic acid (6PGA) by lactonase. (4) Formation of ribulose-5-phosphate (RU5P) and carbon dioxide by 6-phosphogluconic acid dehydrogenase. (5,6) Isomerization (5) and epimerization (6) of RU5P to ribose-5-phosphate (R5P) and xylulose-5-phosphate (X5P). (7) Formation of sedoheptulose-7-phosphate (S7P) and triose phosphate (TRP) by the action of transketolase. (8) Formation of fructose-6-phosphate (F6P) and erythrose-4-phosphate (ERY4P) by the action of transaldolase. (9) Formation of F6P and TRP from X5P and ERY4P by the action of transketolase. (10) Conversion of TRP to fructose 1,6-diphosphate by reversal of glycolysis. Regeneration of G6P by phosphatase (11) and isomerase (12) activity. (13–17) Conversion of TRP to pyruvate. The small numbers adjacent to the carbons of intermediates and products indicate the numbers of the carbons of the original glucose molecule that are found in them.

electron transport phosphorylation. It is often the case, however, that a substantial amount of the glyceraldehyde-3-phosphate formed is metabolized via EM reactions to pyruvate, which makes ATP formation through substrate-level processes possible.

It is impossible for the OP pathway to serve as the sole carbohydrate pathway in an organism since it fails to provide many of the required intermediates for growth. The pathway must be accompanied by other sequences so that these materials are provided. Frequently, the OP pathway operates jointly with the EM pathway. When this happens, a number of factors regulate the relative extent to which the two pathways are operative. The relative concentrations of NADP and NADPH regulate the activity of glucose-6-phosphate dehydrogenase, since NADPH inhibits the enzyme and NADP stimulates it. The proportion of total carbon metabolized in tandem by the two sequences reflects not only the nature of the factors that directly control the operation of phosphofructokinase and glucose-6-phosphate dehydrogenase, but also the particular physiological requirements of the organism under a given set of conditions, such as its requirements for reducing power and biosynthetic intermediates.

The Entner-Doudoroff pathway

The Entner-Doudoroff (ED) pathway may be considered an alternate hexose monophosphate (HM) pathway to the oxidative pentose pathway. The sequence provides a minimum of five of the critical biosynthetic intermediates—glucose-6-phosphate, triose phosphate, 3-phosphoglycerate, phosphoenol pyruvate, and pyruvate. The sequence is shown in Figure 7.4. It begins with glucose, which is converted via hexokinase, to glucose-6-phosphate. This intermediate is further metabolized by glucose-6-phosphate dehydrogenase and lactonase to 6-phosphogluconic acid. At this point, the pathway departs from the OP pathway. In the ED sequence, 6-phosphogluconic acid, rather than being converted to pentoses and carbon dioxide, is dehydrated through 6-phosphogluconic acid *dehydratase* to yield 2-keto, 3 dehydro, 6-phospho-

gluconic acid. Under the influence of 2-keto, 3-dehydro, 6-phosphogluconic acid aldolase, the acid gives rise to pyruvate *directly* from the top three hexose carbons and glyceraldehyde-3-phosphate from the bottom three glucose carbons. The glyceraldehyde-3-phosphate is further converted via EM pathway reactions to pyruvate.

Contrasts between the Embden-Meyerhof and Entner-Doudoroff pathways

Although the ED and EM pathways both convert a glucose molecule to two molecules of pyruvate, and accomplish this by cleavage between carbons 3 and 4 of glucose, the consequences of operation of the two pathways are markedly different. In the EM pathway, both molecules of pyruvate arise by the intermediate formation of glyceraldehyde-3-phosphate, while in the ED pathway, pyruvate arises directly from the top half of the glucose molecule. In both sequences, pyruvate formed from the bottom half of glucose is formed from metabolism of glyceraldehyde-3-phosphate. This difference has critical physiological consequences because both molecules of pyruvate arise from glyceraldehyde-3-phosphate (triose phosphate) in the EM pathway, while only one of the pyruvates arises from glyceraldehyde in the ED pathway, the energy yield by substrate-level phosphorylation per molecule of hexose degraded by the EM pathway is twice that obtained by the ED pathway. This occurs whether or not glucose 6-phosphate rises from glucose by a hexokinase mechanism or by the phosphotransferase system (PTS). If the PTS operates with the EM pathway, only 1 ATP molecule is required to activate glucose. Although only 3 ATP molecules are formed in this situation, only 1 is required by phosphofructokinase for activation, so that the *net* ATP yield per hexose molecule is 2. The net result is the same when 4 ATP moles are produced and through the actions of both hexokinase and phosphofructokinase, 2 molecules of ATP are required to activate a hexose.

With the ED pathway, a net of only 1 ATP is formed per glucose mole degraded, whether or not the PTS operates. If the PTS is operative, only one high-energy phosphate molecule is required per glucose molecule degraded. Such is also the

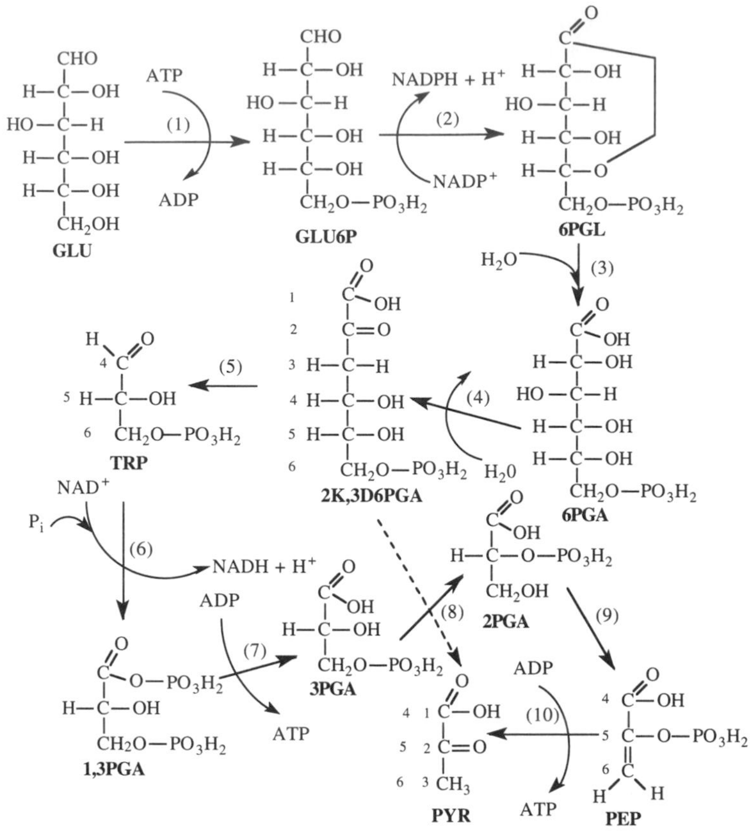

CHO
H—C—OH
HO—C—H
H—C—OH
H—C—OH
CH₂OH
GLU

ATP
(1)
ADP

CHO
H—C—OH
HO—C—H
H—C—OH
H—C—OH
CH₂O—PO₃H₂
GLU6P

NADPH + H⁺
(2)
NADP⁺

H—C—OH
HO—C—H
H—C—OH
C—O
CH₂O—PO₃H₂
6PGL

H₂O
(3)

C—OH
H—C—OH
HO—C—H
H—C—OH
H—C—OH
CH₂O—PO₃H₂
6PGA

1 C—OH
2 C=O
3 H—C—H
4 H—C—OH
5 H—C—OH
6 CH₂O—PO₃H₂
2K,3D6PGA

(4)
H₂O

4 C—O (H)
5 H—C—OH
6 CH₂O—PO₃H₂
TRP

(5)

NAD⁺
Pᵢ
(6)
NADH + H⁺

C—O—PO₃H₂
H—C—OH
CH₂O—PO₃H₂
1,3PGA

ADP
(7)
ATP

C—OH
H—C—OH
CH₂O—PO₃H₂
3PGA

(8)

C—OH
H—C—O—PO₃H₂
CH₂OH
2PGA

4 1 C—OH
5 2 C=O
6 3 CH₃
PYR

ADP
(10)
ATP

(9)

4 C—OH
5 C—O—PO₃H₂
6 C (H)(H)
PEP

FIGURE 7.4 The Entner Doudoroff pathway. (1–3) The conversion of glucose (GLU) to 6-phosphogluconic acid (6PGA) with intermediate formation of glucose-6-phosphate (GLU6P) and 6-phosphogluconolactone (6PGL) by the actions of hexokinase (1), glucose-6-phosphate dehydrogenase (2), and lactonase (3). (4,5) Cleavage of 6PGA to pyruvate (PYR) and triose phosphate (TRP) by the actions of 6PGA dehydratase (4) and 2-keto,3-dehydro, 6-phosphogluconate (2K,3D6PGA) aldolase (5). (6–10) Conversion of TRP to PYR by Embden-Meyerhof (EM) reactions. For the enzymes and intermediate names for the EM reactions, see Figure 7.2. The small numbers adjacent to carbons of intermediates and products are the carbons that correspond to the carbons of the original glucose molecule.

case if transport occurs without PEP expenditure. Whatever the transport mechanism, only 1 high-energy bond is expended per glucose mole degraded via the ED pathway. Since, in the ED pathway, ATP formation by substrate-level phosphorylation occurs only from the triose phosphate formed from the bottom half of the glucose molecule, a net of only 1 ATP is formed per glucose molecule catabolized.

In addition to ATP yields per hexose molecule, the labelling of fermentation end products produced from degradation of position-labelled glucose by the ED and EM pathways is different.

Products formed from triose phosphate will not be labelled from position-labelled carbohydrates labelled in the top half of the glucose molecule if the ED pathway is operative, but will be labelled from the top half of the glucose molecule if the EM pathway is present. In addition, although lactate may be formed from the top three carbons of glucose metabolized by either the EM or the ED pathway, the label position in the lactate formed from the two pathways is different. In the EM pathway, 1-labelled glucose gives rise to methyl-labelled lactate and 3-labelled glucose gives rise to carboxyl-labelled lactate. In the ED pathway, the inverse occurs, that is, 1-labelled glucose gives rise to carboxyl-labelled lactate and 3-labelled glucose gives rise to carboxyl-labelled lactate. The labelling pattern in end products formed from the bottom half of hexoses is identical in both sequences. The ED pathway yields NADPH and, by metabolism of the bottom half of glucose molecules through triose phosphate, NADH as well, while the EM pathway produces only NADH. Neither pathway produces ribose, sedoheptulose, or erythrose, critical intermediates in nucleic acid and aromatic amino acid formation, although both pathways, by conversion of triose phosphate to dihydroxyacetone phosphate (DHAP) may provide the alpha glycerol phosphate required for lipid synthesis.

The xylulose-5-phosphate phosphoketolase pathway

Certain microbes contain what is sometimes called a heterofermentative pathway for glucose catabolism. The critical enzyme of this sequence is xylulose-5-phosphate phosphoketolase. For this reason, the pathway is often referred to as the phosphoketolase (PK) pathway, even though it is now recognized that other pathways involving phosphoketolase enzymes are also found. It thus seems proper to designate which phosphoketolase is operative in the heterofermentative pathway and to refer to it as the *xylulose-5-phosphate phosphoketolase* pathway, shown in Figure 7.5. It begins with glucose, which it converts to glucose-6-phosphate by hexokinase or glucokinase. The glucose-6-phosphate is then converted to 6-phosphogluconic acid by the action of glucose-6-phosphate dehydrogenase with the concomitant formation of NADPH and lactose. The 6-phosphogluconate is decarboxylated by 6-phosphogluconic acid dehydrogenase, yielding additional NADPH, CO_2, and ribulose 5-phosphate. These reactions are similar or identical to normal hexose monophosphate pathways. However, in contrast to the OP pathway in which ribulose-5-phosphate is converted to a mixture of xylulose-5-phosphate and ribose-5-phosphate, in the xylulose-5-phosphate phosphoketolase pathway, the ribulose-5-phosphate is converted entirely to xylulose-5-phosphate, which is cleaved to yield a two-carbon compound, from the top two carbons of the xylulose-5-phosphate molecule. The remain-

ing three carbons of xylulose-5-phosphate become glyceraldehyde-3-phosphate that is converted to pyruvate via EM pathway reactions.

Consequences of the xylulose-5-phosphate phosphoketolase pathway

The operation of the PK pathway has number of consequences different from operation of the EM, ED, and OP pathways. In the PK pathway, no three-carbon compound can result from metabolism of the top three carbons of hexose since, in the conversion of hexose to pentose, carbon 1 of the hexose molecule is lost as carbon dioxide. Cleavage of the resultant pentose allows the formation of only two-carbon products from the top portion of the xylulose molecule. Metabolism of the bottom three xylulose-5-phosphate carbons, which may result from the bottom three carbons of hexose, cannot be distinguished from metabolism of the bottom carbons of hexoses metabolized by the EM and the ED pathways since glyceraldehyde-3-phosphate is the substrate produced from the bottom carbons of hexose in all of these sequences. The PK pathway can be distinguished, however, from the EM and ED pathways by study of the metabolism of 2-labelled hexose. In both the EM and ED sequences, 2-labelled hexose metabolism gives rise to 1-labelled, two-carbon products, for instance, carboxyl-labelled acetate, whereas in the PK pathway, 2-labelled hexose will produce methyl-labelled acetate. The significance of this distinction will become more apparent during the discussion of fermentation in Chapter 8.

The fate of the two-carbon moiety produced from xylulose-5-phosphate cleavage

In the PK pathway, the top two carbons of the xylulose-5-phosphate molecule become, initially, acetyl phosphate. Further metabolism of the latter varies, depending upon the circumstances. Typically, when hexoses are used by this sequence, the initially formed acetyl phosphate is converted, by phosphotransacetylase, to acetyl coenzyme A (CoA), which, via aldehyde dehydrogenase, is converted to acetaldehyde. The acetaldehyde, through alcohol dehydrogenase, is then converted to ethanol. Disposal of reducing power produced in the conversion of hexose to pentose is accomplished, in part, by use of the two-carbon moiety as an "electron sink" compound and the potential

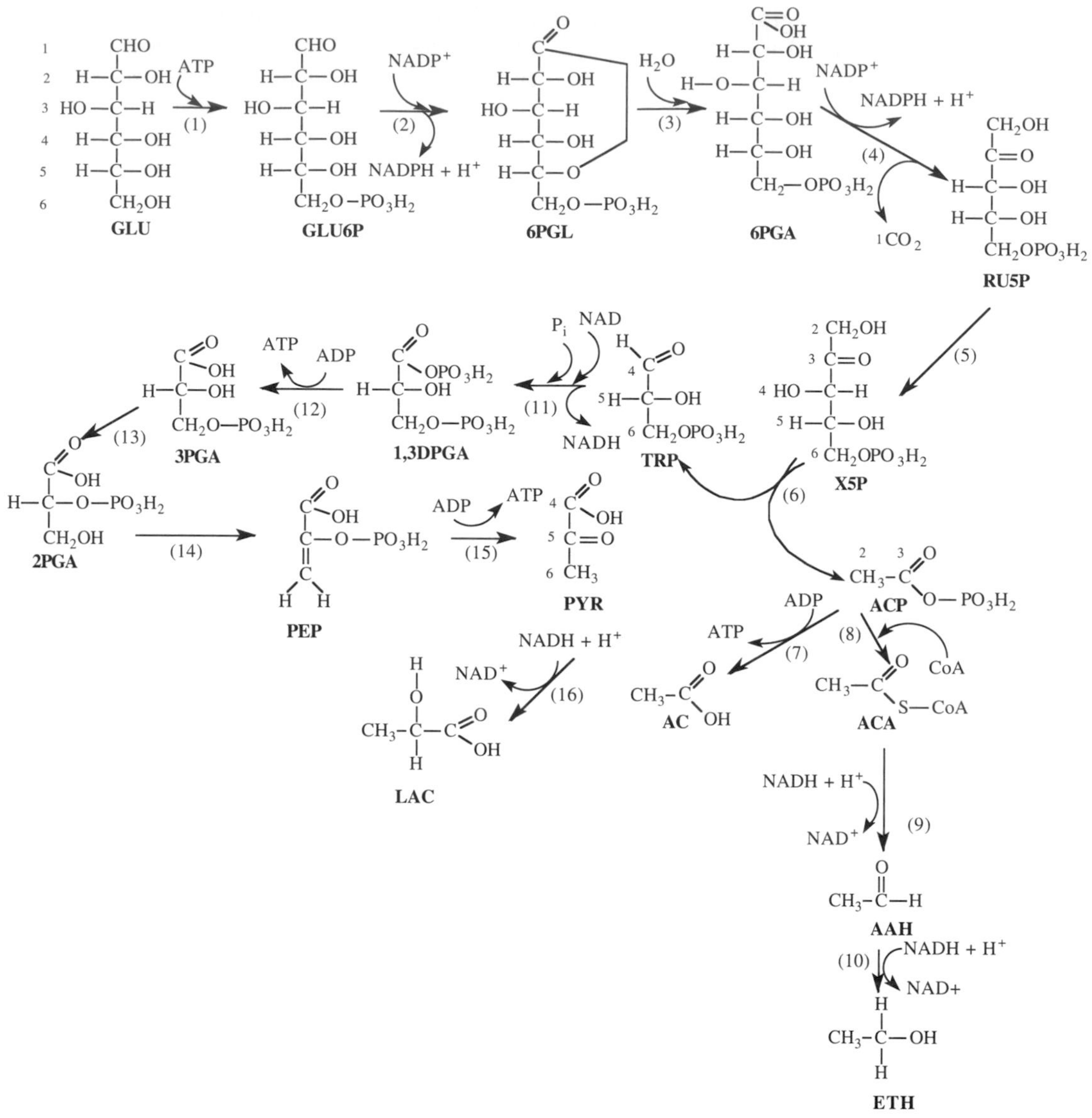

FIGURE 7.5 The xylulose 5-phosphate phosphoketolase pathway. (1) The conversion of glucose (GLU) to glucose-6-phosphate (GLU6P) by hexokinase or glucokinase. (2) The formation of 6-phosphogluconolactone (6PGL) from GLU6P by glucose 6-phosphate dehydrogenase. (3) Formation of 6-phosphogluconic acid (6PGA) from 6PGL by lactonase. (4,5) Formation of carbon dioxide and ribulose-5-phosphate (RU5P) from 6PGA by 6PGA dehydrogenase (4) and conversion of RU5P to xylulose-5-phosphate (X5P) by RU5P epimerase (5). (6) Formation of acetyl phosphate (ACP) and triose phosphate (TRP) from X5P by xylulose-5-phosphate phosphoketolase. (8–10) Formation of acetyl CoA from ACP by phosphotransacetylase (8) and reduction of ACA to ethanol (ETH) by the actions of actacetaldehyde (AAH) dehydrogenase (9) and alcohol dehydrogenase (10). (7) Conversion of ACP to free acetate and ATP by acetyl kinase. (11–15) Conversion of TRP to pyruvate (PYR) by Embden-Meyerhof reactions (Figure 7.2). (16) Reduction of PYR to lactate (LAC) by lactic dehydrogenase (16). The small numbers adjacent to the carbons of selected intermediates and products represent the carbons of the original hexose molecule.

conservation of the phosphate bond energy as ATP, through the operation of acetyl kinase, does not occur.

In contrast, when pentoses are metabolized directly by the PK pathway and do not arise from hexose, a different result may occur. In this situation, reducing power, as NADH, is formed only from metabolism of glyceraldehyde and may be disposed of by reduction of pyruvate to lactate. The acetyl phosphate produced from xylulose-5-phosphate cleavage, may thus be used as an energy compound rather than for reducing power disposal. By the action of acetyl kinase, free acetic acid and ATP are formed. The energy yield obtained from the PK sequence by substrate-level processes is, therefore, variable. When hexoses are used, the ATP yield per carbohydrate molecule attacked is 1 since glucose activation requires 1 ATP molecule and the potential ATP from acetyl phosphate is lost. Since ATP formation occurs only from glyceraldehyde-3-phosphate, which yields 2 ATP molecules per molecule of glyceraldehyde catabolized to pyruvate, the net ATP yield is 1. When pentoses are used directly, and acetyl phosphate is converted to ATP, the ATP yield per carbohydrate molecule is 2.

The PK pathway may produce six of the critical biosynthetic intermediates, glucose-6-phosphate, acetyl CoA, triose phosphate (glyceraldehyde-3-phosphate), 3-phosphoglyceric acid, PEP, and pyruvate. However, as with the ED and EM pathways, the PK pathway fails to produce ribose, erythrose, and sedoheptulose.

The double phosphoketolase pathway

Bifidobacterium bifidum and certain *Acetobacter* species possess a carbohydrate degradation scheme that contains two phosphoketolase enzymes, one of which operates on xylulose-5-phosphate and a second enzyme that attacks fructose-6-phosphate. The joint operation of these enzymes results in a pathway that combines aspects of the PK and OP pathways. The sequence found in *Bifidobacterium* is shown in Figure 7.6. It begins with the conversion of glucose to fructose-6-phosphate via the actions of hexokinase and phosphohexose isomerase. The fructose-6-phosphate is then cleaved to yield erythrose-4-phosphate, from the bottom four carbons of fructose and acetyl phosphate from the top two carbons of fructose. By the action of acetyl

kinase, the acetyl phosphate is converted to acetic acid and ATP. The erythrose-4-phosphate formed from the action of the first phosphoketolase enzyme reacts with a second molecule of fructose-6-phosphate in a transaldolase reaction, generating sedoheptulose-7-phosphate and glyceraldehyde-3-phosphate. The sedoheptulose is of such a nature that its top three carbons come from the top three carbons of fructose and its remaining four carbons come from fructose carbons 3,4,5, and 6. The glyceraldehyde-3-phosphate is formed from carbons 4, 5, and 6 of fructose. The sedoheptulose-7-phosphate and the glyceraldehyde-3-phosphate now participate in a transketolase reaction, transferring the top two carbons of sedoheptulose to glyceraldehyde and yielding a ribose-5-phosphate molecule labelled, in terms of the original hexose molecule, 1,2,4,5,6. In this reaction, the bottom five carbons of sedoheptulose-7-phosphate become a molecule of xylulose-5-phosphate labelled, in terms of the original hexose, 3,3,4,5,6. At this point, the second phosphoketolase enzyme intervenes, yielding acetyl phosphate and triose phosphate from the molecule of xylulose-5-phosphate formed from the bottom carbons of sedoheptulose and a second xylulose-5-phosphate molecule that arises from metabolism of the ribose formed by the transketolase reaction. The acetyl phosphate molecules are labelled 1,2 or 3,3 in terms of the original fructose, depending upon precisely how the various xylulose-5-phosphate molecules arise. Irrespective of differences in their labelling patterns, both of the acetyl phosphate molecules, by the action of acetyl kinase, yield free acetic acid and ATP. The molecules of glyceraldehyde, all of which arise from carbons 4, 5 and 6 of hexose, are converted by EM pathway reactions to lactate.

Critical features of the double phosphoketolase pathway The double phosphoketolase may produce many of the critical biosynthetic intermediates—glucose-6-phosphate, fructose-6-phosphate, ribose-5-phosphate, erythrose-4-phosphate, sedoheptulose-7-phosphate, triose phosphate, 3-phosphoglycerate, phosphoenol pyruvate, and pyruvate. The pathway provides relatively large amounts of energy by conversion of acetyl-phosphate to acetate and ATP, and by ATP generation from the 1,3-diphosphoglyceric acid and PEP during conversion of glyceraldehyde 3-phosphate to pyruvate. Under anaerobic conditions, it appears

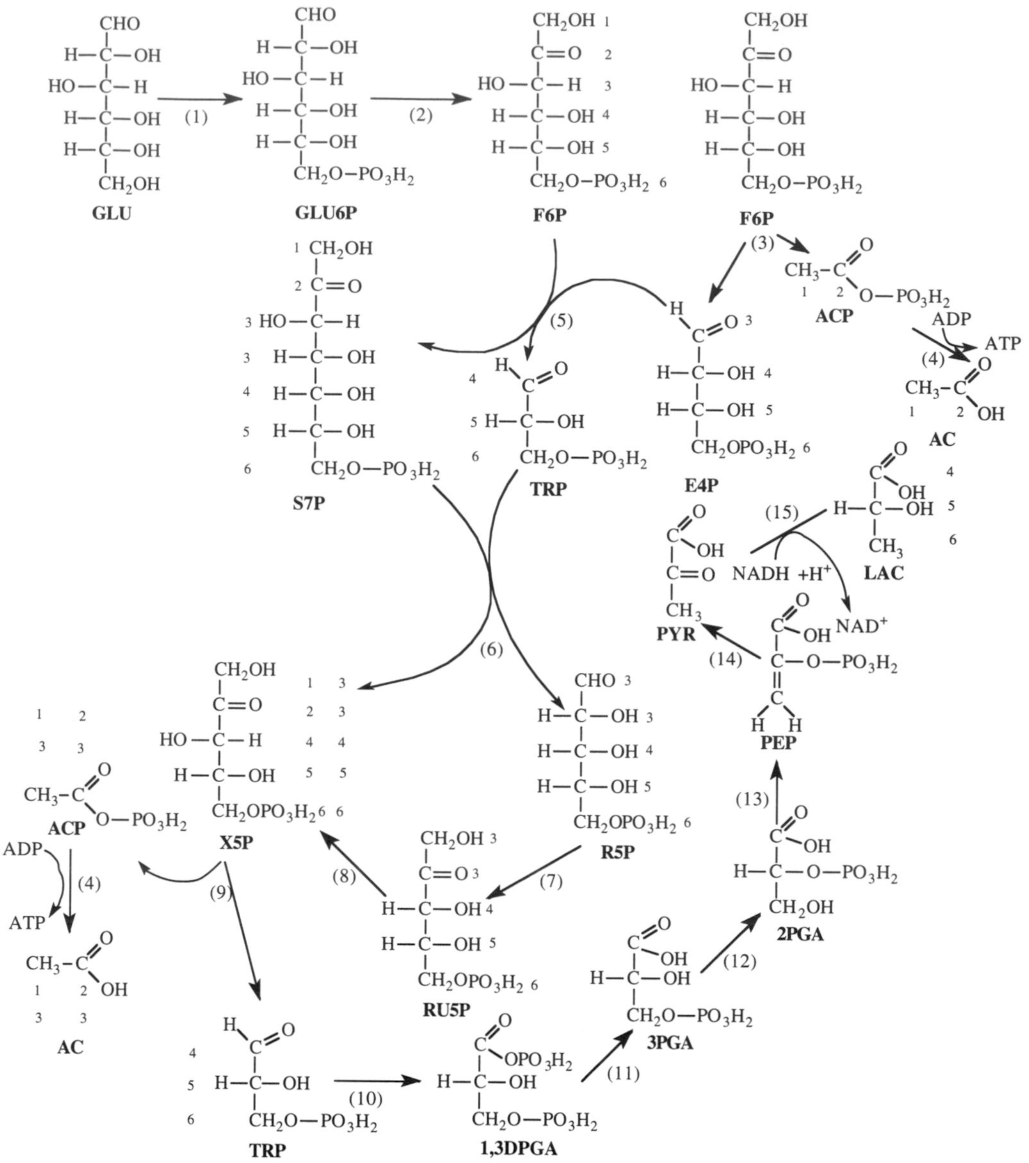

FIGURE 7.6 The double phosphoketolase pathway. (1,2) Conversion of glucose (GLU) to fructose-6-phosphate (F6P) via glucose-6-phosphate (GLU6P) with hexokinase (1) and phosphohexose isomerase (2). (3) Cleavage of F6P to erythrose-4-phosphate (E4P) and acetyl phosphate (ACP) by F6P phosphoketolase. (4) Formation of free acetic acid (AC) and ATP from ACP with acetyl kinase. (5) Transaldolase reaction with F6P and E4P to yield sedoheptulose-7-phosphate (S7P) and triose phosphate (TRP). (6) Transketolase reaction with S7P and TRP to yield xyulose-5-phosphate (X5P) and ribose-5-phosphate (R5P). (7,8) Conversion of R5P to X5P via ribulose-5-phosphate (RU5P) by the actions of R5P isomerase (7) and RU5P epimerase (8). (9) Formation of ACP and TRP by the action of X5P phosphoketolase. (10–14) Conversion of TRP to pyruvate (PYR) by Embden-Meyerhof reactions See Figure 7.2 for reactions and the names of abbreviated intermediates. (15) Reduction of pyruvate to lactate with lactate dehydrogenase. The small numbers adjacent to the carbons are the carbons in intermediates and products equivalent to carbons of the original hexose.

that the pathway is interconnected to glycolysis by conversion of 2 molecules of glyceraldehyde-3-phosphate to fructose-1,6-diphosphate. It also appears that, in certain circumstances, the double phosphoketolase pathway may be cyclic, by conversion of fructose-1,6-diphosphate to fructose-6-phosphate with phosphatase.

Distinguishing features of the double phospho-ketolase sequence The double phosphoketolase (DPK) pathway shares certain features with other mechanisms of carbohydrate degradation while, at the same time, differing substantially from the other common sequences. It is a hexose *monophos-phate* pathway, but uses fructose-6-phosphate as well as glucose-6-phosphate. It is a phosphoketolase pathway, but uses two phosphoketolases, while the PK pathway uses only one. The double phosphoketolase pathway generates all of the critical biosynthetic intermediates, except those that result from the citric acid cycle. It produces glyceraldehyde-3-phosphate from the same hexose carbons as do all the other major sequences. However, the way in which it produces acetyl phosphate differs from the way in which the same material is produced in the xylulose-5-phosphate phosphoketolase pathway. In the latter sequence, 2-labelled hexose yields methyl-labelled acetyl phosphate and, thus, methyl-labelled acetate. By contrast, in the DPK pathway, 2-labelled hexose gives rise to carboxyl-labelled acetate. In addition, a portion of the acetyl phosphate is formed in the DPK pathway so that both the carboxyl and methyl carbons will contain label if the DPK-containing organism is grown in the presence of 3-labelled hexose. This feature is distinctive to the DPK pathway. Careful consideration of the consequences of metabolism of position-labelled substrates, in conjunction with enzyme analysis, allows distinction of the DPK pathway from other pathways of carbohydrate catabolism.

The citric acid cycle

The previously considered catabolic sequences may be regarded as alternative means of converting carbohydrates to pyruvate. Although all of the pathways provide energy and certain of the critical intermediates, and some provide reducing power for biosynthesis as NADPH, none of the previously discussed pathways provides *all* of the critical biosynthetic intermediates. Furthermore, the energy yield obtainable from carbohydrate catabolism to pyruvate is limited, compared to that obtainable if pyruvate is oxidized. The citric acid cycle, also known as the Krebs cycle and the tricarboxylic acid (TCA) cycle, provides the remaining critical intermediates and substantial amounts of additional energy not available to organisms that are devoid of the cycle. Although the TCA cycle in its complete form, is found in *oxybiontic* organisms (organisms that use oxygen in their metabolism), even organisms that do not use oxygen in energy metabolism display most, if not all, of the TCA cycle reactions. Obligate anaerobes, for example, typically contain all of the reactions of the TCA cycle with the exception of alpha-ketoglutarate dehydrogenase. Such a situation is physiologically necessary, since the critical intermediates for biosynthesis must be supplied for all organisms, irrespective of the way in which they obtain energy.

The reactions of the TCA cycle are shown in Figure 7.7. The cycle is amphibolic because it participates both in degradative and energy metabolism and in production of biosynthetic intermediates. The cycle literally begins where pyruvate formation ends. The cleavage of pyruvate by the pyruvic dehydrogenase complex yields acetyl CoA and carbon dioxide. Acetyl CoA is the connecting point between pyruvate and the TCA cycle. In the presence of condensing enzyme and oxaloacetate, citric acid is formed, which, by stereospecific water removal with *cis*-aconitate dehydratase, yields *cis*-aconitic acid. Stereospecific addition of water to the double bond of *cis*-aconitate, with aconitase, produces isocitric acid that through isocitric dehydrogenase, is converted first to oxalosuccinic acid, and then, with CO_2 evolution, to alpha-ketoglutarate.

The action of isocitrate dehydrogenase is interesting from several viewpoints. Its operation in many systems is NADP-specific. In this situation, it serves as a means of NADPH production, one of the few such reactions that do not result from the early steps of the hexose monophosphate pathways. Isocitric acid dehydrogenase is of further interest because it is multifunctional, in that it converts isocitrate to oxalosuccinate and oxalosuccinate to alpha-ketoglutarate (AKG). Finally, isocitrate dehydrogenase is of interest because the study of the source of the carbon dioxide (the middle carboxyl of isocitrate) formed during the

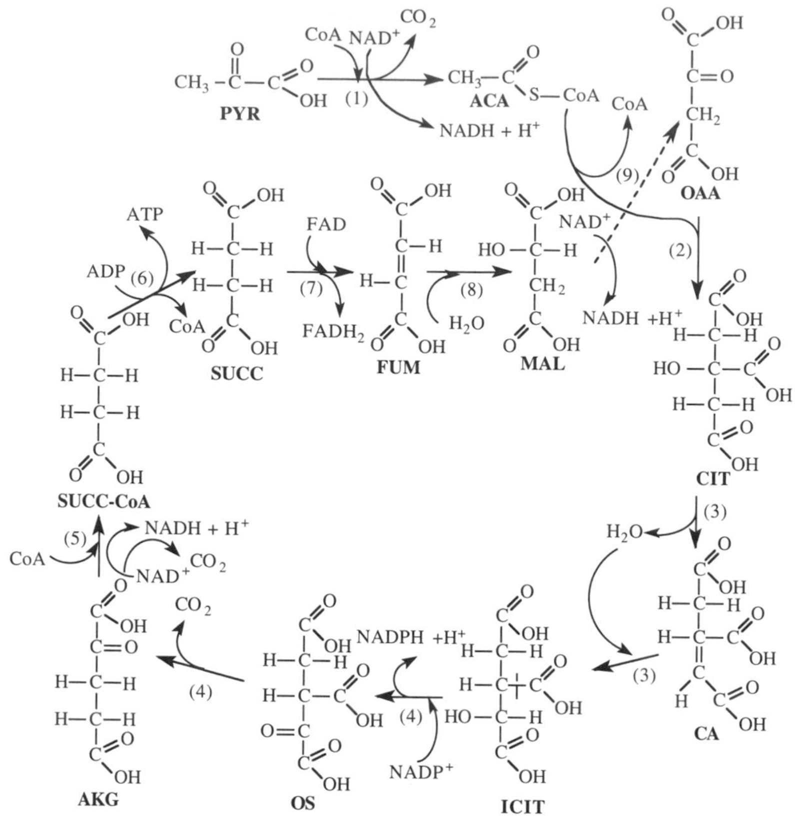

FIGURE 7.7 The tricarboxylic acid (TCA) cycle. (1) Formation of acetyl CoA (ACA) and carbon dioxide from pyruvate (PYR) by the pyruvic dehydrogenase complex. (2) Formation of citrate (CIT) from ACA and oxaloacetate (OAA) by citrate synthetase. (3) Formation of isocitrate (ICIT) from CIT with the intermediate formation of cis-aconitate (CA) by the action of CA dehydratase. (4) Conversion of ICIT to alpha-ketoglutarate (AKG) with the intermediate formation of oxalosuccinate (OS) by isocitric dehydrogenase. (5,6) Formation of succinic acid, with the concomitant formation of ATP, by the combined actions of the AKG dehydrogenase complex (5) and succinic thiokinase (6). (7–9) Regeneration of OAA by the sequential action of succinic dehydrogenase (7), fumarase (8), and malic dehydrogenase (9).

conversion of isocitrate to AKG led to general understanding of the stereochemical nature of enzyme action and to the formulation of the Ogston hypothesis, the hypothesis that the mechanisms of enzyme-substrate reactions depend upon the stereochemistry of the participating molecules.

Conversion of AKG to succinate with the intermediate formation of succinyl CoA is one of the most complicated enzyme-mediated processes in nature. It is accomplished by the multienzyme complex known as the alpha-ketoglutarate dehydrogenase complex that exhibits three activities: 1) the decarboxylation of AKG with the aid of reduced, enzyme-bound lipoate and enzyme-bound thiamine pyrophosphate to produce carbon dioxide and a succinyl moiety bound to lipoic acid; 2) trans-

fer of the succinyl moiety to a molecule of coenzyme A with the concomitant reduction of a second enzyme; and 3) the formation of NADH by reaction of reduced lipoate-combined enzyme with a third FAD-dependent enzyme. The mechanisms of these reactions are shown schematically in Figure 7.8 and are similar to the mechanisms of both the pyruvate dehydrogenase complex and phosphoketolase enzymes. The similarity of reaction mechanisms reflects a similarity in the nature of their substrates—the presence of an alpha carbonyl. All three processes involve thiamine pyrophosphate-associated enzymes and coenzyme A.

The conversion of succinyl CoA to succinate is mediated by succinic thiokinase and is associated with substrate-level, high-energy bound formation

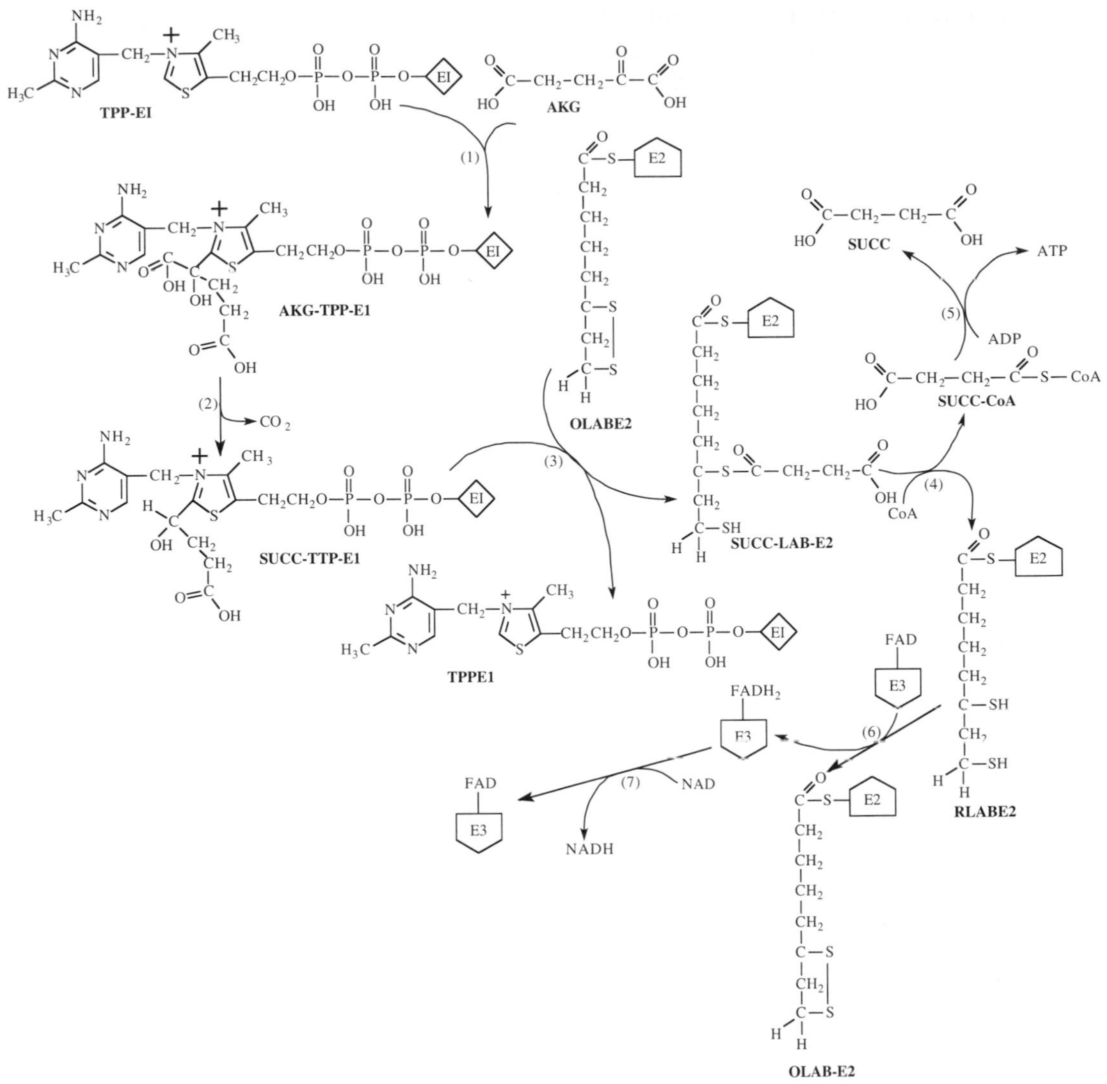

FIGURE 7.8 The nature of the reactions accomplished by the alpha ketoglutarate dehydrogenase complex. (1) Formation of a complex between thiamine pyrophosphate-bound alpha ketoglutarate dehydrogenase (E1) and alpha ketoglutarate (AKG). (2) Loss of carbon dioxide and formation of a TPP-E1-succinyl (SUCC) moiety complex (SUCC-TPP-E1). (3) Reaction of oxidized lipoic acid-bound dihydrolipoate transsuccinylase (OLABE2) with SUCC-TPP-E1 to regenerate TPP-E1 and form a succinate-LABE2 complex (SUCC-LABE2). (4) Regeneration of reduced (R) LABE2 and formation of succinyl CoA (SUCC-CoA). (5) Formation of ATP and free succinic acid (SUCC) by succinic thiokinase. (6) Regeneration of OLABE2 and formation of enzyme-bound reduced flavin adenine dinucleotide (FAD) by dihydrolipoate dehydrogenase (E3). (7) Regeneration of E3-bound oxidized FAD with the formation of NADH. The mechanisms of both the pyruvate dehydrogenase complex reactions and phosphoketolase reactions are similar or identical to the reactions of Figure 7.8. Although the reactions are shown separately, they occur via a multienzyme complex.

and the formation of free succinic acid. The energy of the CoA bond is used to form a high-energy phosphate bond, whose formation method varies from system to system. In microbes, ADP and inorganic phosphate combine to form ATP. In nonmicrobial systems, GDP or IDP may serve in the functions analogous to ADP. Succinyl CoA, in addition to serving as a substrate-level phosphorylation substrate, is a critical biosynthetic precursor for the synthesis of protoheme, the nonprotein substance essential for both cytochromes and a diversity of critical enzymes. The succinic acid formed from the previous reactions is the substrate for succinic dehydrogenase that converts free succinate, stereospecifically, to fumarate with the concomitant formation of reduced flavin adenine dinucleotide ($FADH_2$). Conversion of fumarate to L-malate is accomplished by fumarase. Conversion of L-malate to oxaloacetate by malate dehydrogenase, with the generation of a molecule of NADH, completes the cycle.

The amphibolic nature of the TCA cycle

The TCA cycle is both a catabolic and an anabolic cycle. Since it participates in both energy generation and biosynthesis, it is an amphibolic cycle. As a synthetic cycle, it provides succinyl CoA for heme synthesis, oxaloacetic acid as a precursor of the aspartic family of amino acids, and alpha-ketoglutaric acid as a precursor to the glutamic family of amino acids. In addition, it provides a large amount of energy through both electron transport-mediated and substrate-level processes. Each time the cycle operates completely, it oxidizes a molecule of acetyl CoA to carbon dioxide. The electrons released from oxidative processes are transferred to the carrier molecules NAD or FAD and are used for electron transport, with the concomitant generation of ATP and, in aerobes, reduction of oxygen to water.

The energetics of the TCA cycle

The precise amount of energy obtained by operation of the TCA cycle in consort with the EM pathway has been studied over the years, showing that a total of 38 ATP molecules may result from their joint operation, some derived from substrate-level phosphorylation and most resulting from electron transport. These calculations are based on the assumptions that 2 ATP molecules result from oxidation of a molecule of reduced FAD and 3 ATP molecules are formed per NADH molecule oxidized.

To better understand the energetic consequences of the TCA cycle and glycolysis, we will consider, separately, the respective contributions of hexose conversion to pyruvate and pyruvate oxidation to the total ATP yield obtained by their joint action. The conversion of glucose to pyruvate by the EM pathway yields a net of 2 ATP molecules per hexose by substrate-level processes via the operation of glyceraldehyde-3-phosphate dehydrogenase and pyruvate kinase. These two enzymes allow generation of a total of 4 ATPs per hexose mole, but since 2 ATPs are used in activation, the *net* yield is 2 ATPs by substrate-level processes. In addition, the operation of glyceraldehyde-3-phosphate dehydrogenase yields 2 molecules of NADH per hexose molecule. If these are oxidized, and the 3 ATP molecules per NADH molecule are formed, the total ATP yield per hexose molecule catabolized to pyruvate is 8—2 from substrate-level processes and 6 from electron-transport.

The oxidation of pyruvate to carbon dioxide and water allows formation of substantially more ATP than is obtainable from the conversion of hexose to pyruvate. The formation of acetyl CoA and carbon dioxide is associated with the formation of two molecules of NADH, one from each of the two pyruvates oxidized per hexose. The first energy-yielding step of the TCA cycle, per se, occurs during the conversion of isocitrate to α-ketoglutarate through the formation of NADPH. The second energy-yielding step is the formation of an NADH molecule during the conversion of AKG to succinyl CoA. Utilization of the succinyl CoA bond allows formation of ATP by substrate-level phosphorylation. Oxidation of free succinate to fumarate generates FADH, allowing the formation of additional ATP. The conversion of malate to oxaloacetate, with the attendant formation of an NADH molecule, is the final energy-generating step of the cycle.

The summative effects of the EM pathway and the TCA cycle in energy generation are the following: a total of 8 ATPs are formed from the conversion of hexose to pyruvate, 2 from substrate-level phosphorylation, and 6 from electron-transport by the oxidation of the two NADH molecules (3 ATP molecules each). Use of the 2 NADH molecules generated by pyruvate cleavage generates 6 more ATPs. Prior to entry into the TCA cycle, a total of 14 ATPs have been formed. Each turn of the cycle generates 12 additional ATP molecules. Since the cycle turns twice for each glucose molecule, the processing of a hexose molecule generates a total of 24 ATP molecules within the TCA cycle itself. For each turn of the cycle, 3 ATPs result from the use of NADPH produced from isocitric dehydrogenase, 3 more are formed from oxidation of the NADH

TABLE 7.2 The Summative Effects of the Embden-Meyerhof Pathway and the Tricarboxylic Acid Cycle in Energy Production

STEP OR ENZYME	TYPE OF PROCESS	ATP MOLECULES PER HEXOSE MOLECULE
3-phosphoglycerate kinase	Substrate-level phosphorylation	2
Glyceraldehyde-3-phosphate dehydrogenase	NAD reduction, electron transport	6
Pyruvate kinase	Substrate level phosphorylation	2
Pyruvate dehydrogenase complex	NAD reduction, electron transport	6
Isocitric acid dehydrogenase	NADP reduction, electron transport	6
α-ketoglutarate dehydrogenase complex	NAD reduction	6
Succinic thiokinase	Substrate-level phosphorylation	2
Succinic dehydrogenase	FAD reduction, electron transport	4
Malate dehydrogenase	NAD reduction, electron transport	6
Total ATP molecules per hexose molecule		40
Net ATP yield*		38

*Because 2 ATPs are expended to activate a glucose molecule, a net of only 38 ATPs is obtained, although a total of 40 may be produced.

formed by action of the alpha-ketoglutarate dehydrogenase complex, and 2 additional ATP moles result from reoxidation of the FADH formed from succinic dehydrogenase. Finally, through electron transport, 3 more ATP molecules are formed from the NADH produced when malate is converted to oxaloacetic acid. The total ATP yield, by electron transport, per turn of the TCA cycle is 11. The remaining ATP results from use of the succinyl CoA bond energy in the conversion of succinyl CoA to succinate, giving a total of 12 ATPs per turn of the cycle, 11 of which arise from electron transport and 1 of which results from substrate-level phosphorylation. The summative effects of the joint operation of the EM pathway and the TCA cycle in the production of energy are shown in Table 7.2.

The glyoxalate cycle

The TCA cycle, being amphibolic, participates both in energy-yielding and biosynthetic metabolism. It allows formation of substantial amounts of energy and, in addition, gives rise to the critical biosynthetic intermediates that are not formed by any of the carbohydrate degrading sequences. Since the TCA cycle participates in both energy generation and biosynthesis, there are occasions in which its involvement in these processes is unbalanced, that is, the rates of biosynthetic intermediates production and their use for biosynthesis are unequal. When this situation exists, a net drain on synthesis intermediates may occur, requiring mechanisms for their replenishment. The glyoxalate (G) cycle, often called the glyoxalate bypass cycle, is a major way in which depleted biosynthetic intermediates are replenished. The cycle is shown in Figure 7.9. It is, in some ways, similar to the TCA cycle, but is considered a bypass cycle because it converts isocitric acid to succinic acid directly, without the intermediate formation of α-ketoglutarate or succinyl CoA. The cycle begins with the conversion of acetyl CoA and oxaloacetate to citrate, which is converted, by TCA cycle reactions, to isocitrate. At this point, the sequence departs from the TCA cycle because isocitrate is cleaved directly to succinate and glyoxalate by isocitrate lyase rather than, as occurs in the TCA cycle, by isocitric dehydrogenase. The remaining reactions of the glyoxalate cycle include the conversion of succinate to fumarate, the hydration of fumarate to malate, and oxidation of malate to oxaloacetate. In addition, the glyoxalate formed from the operation of isocitric lyase is converted to malate by condensation with acetyl CoA via the critical enzyme malate synthetase.

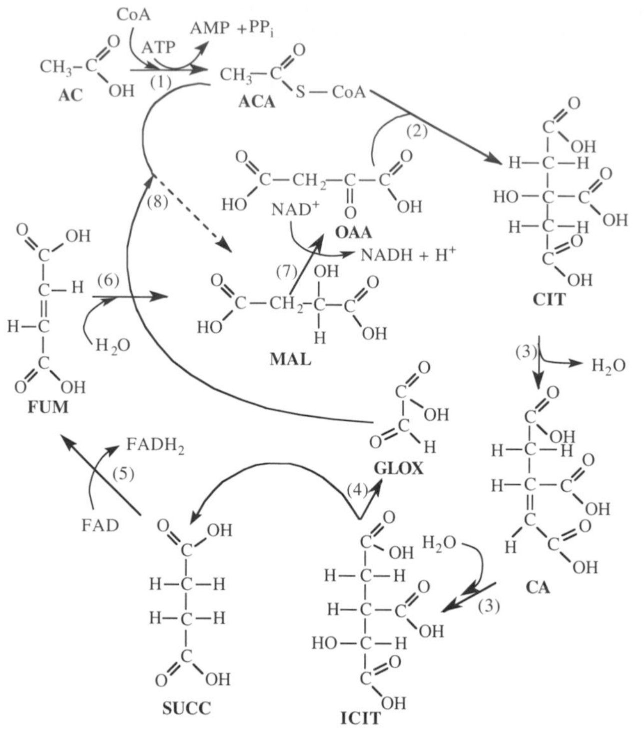

FIGURE 7.9 The glyoxalate bypass cycle. (1) Formation of acetyl CoA (ACA) from acetate (AC) by the action of acetyl CoA synthetase. (2) Formation of citrate (CIT) from ACA and oxaloacetate (OAA) by citrate synthetase. (3) Conversion of CIT to isocitrate (ICIT) with the intermediate formation of *cis*-aconitate (CA) by the action of CA hydratase. (4) Cleavage of ICIT to succinate (SUCC) and gloxalate (GLOX) by isocitrate lyase. Note the manner in which cleavage of ICIT occurs. (5) Conversion of SUCC to fumarate (FUM) by succinic dehydrogenase. (6) Formation of malate (MAL) from FUM by fumarase. (7) Formation of OAA from MAL by malate dehydrogenase. (8) Synthesis of malate from ACA and GLOX by malate synthetase.

Since it bypasses the reactions of the TCA cycle that involve stepwise decarboxylation of isocitric acid to α-ketoglutaric acid and of α-ketoglutarate to succinate, the glyoxalate cycle cannot be used for the oxidative metabolism of acetate. In addition, since it does not involve formation of succinyl CoA, the sequence does not generate energy by substrate-level phosphorylation as is possible in the TCA cycle, although energy production may result from oxidation of NADH or FADH. The major role of the glyoxalate cycle is replenishment of 4-carbon intermediates drained by biosynthesis.

Since it fails to synthesize 5-carbon biosynthetic intermediates, the glyoxalate cycle cannot serve as a substitute for the TCA cycle, but only as an adjunct to it. The glyoxalate cycle allows aerobic, but not anaerobic microbes to use acetate for synthetic purposes. Operation of the cycle is repressed, via catabolite repression, by the readily usable substrates. Certain anaerobes and photosynthetic bacteria use acetate by a different mechanism. These organisms, with the aid of the low potential electron carrier ferredoxin, can convert acetate to pyruvate. This mechanism allows anaerobes, as well as aerobes, to use acetate for biosynthesis.

The use of unusual carbohydrates

Although glucose is generally considered the preferred substrate for hexose catabolism for many microbes, many other hexoses and hexitols (6-carbon sugar-derived alcohols) may be used for energy production and biosynthetic intermediate formation. In addition, microbes exhibit substantial diversity in their ability to use a wide variety of 5-carbon sugars and their alcohol analogues. In many cases, the ability of microbes to use a wide variety of sugars and sugar alcohols reflects their ability to convert these substances into intermediates of central pathways. In other cases, unique sequences are found in organisms that degrade "unusual" carbohydrate substrates.

Pathways for galactose catabolism

Galactose is a sugar that may be catabolized by microbes, and is a hydrolysis product of lactose. Although galactose is used by several microbes, it is handled differently in various species. To a substantial extent, its utilization reflects the manner in which lactose is taken up by the cell. In members of the genus *Lactobacillus,* lactose uptake is mediated by the phosphotransferase system and the intracellular lactose is hydrolyzed in such a way as to yield galactose-6-phosphate. The latter is metabolized by the *tagatose pathway,* which is shown in Figure 7.10. In the tagatose pathway, galactose-6-phosphate is isomerized to tagatose-6-phosphate, which is converted by a kinase reaction involving ATP to tagatose-1,6-diphosphate. The tagatose-1,6-diphosphate is then split in an aldolase-like reaction, yielding dihydroxyacetone phosphate and

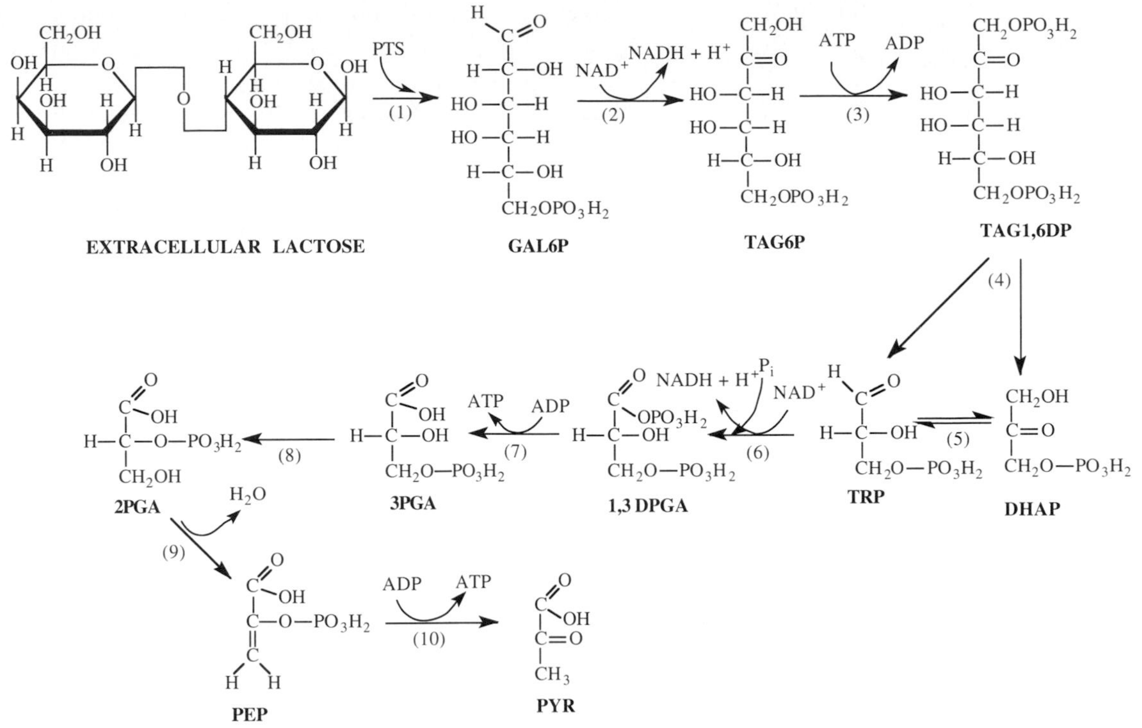

FIGURE 7.10 The utilization of lactose in *Lactobacillus* species by the tagatose pathway. (1) Transport of lactose into the cell with the aid of the phosphotransferase (PTS) system. In the process of transport, galactose-6-phosphate (GAL6P) is formed. (2) Isomerization of GAL6P to tagatose-6-phosphate (TAG6P). (3) Phosphorylation of TAG6P to form tagatose 1,6-diphosphate (TAG1,6DP) by an ATP-dependent kinase. (4) Cleavage of TAG 1,6DP to form dihydroxyacetone (DHAP) and triose phosphate (TRP). (5–10) Conversion of DHAP, via TRP, to pyruvate (PYR) by Embden-Meyerhof (EM) reactions. For names and abbreviations of the intermediates in the EM pathway and its enzymes see Figure 7.2.

glyceraldehyde-3-phosphate that are converted to pyruvate via EM pathway reactions.

Escherichia coli exhibits a different mode of lactose utilization. Transport is accomplished by an inducible transport system that converts the lactose to an intracellular mixture of D-glucose and D-galactose. This mixture is converted through a different kinase reaction than that involved in galactose utilization by *Lactobacillus* species, to galactose-1-phosphate, in contrast to the galactose-6-phosphate used in the tagatose pathway. The galactose-1-phosphate is converted to glucose-1-phosphate by a uridine diphosphate-mediated epimerase. Phosphoglucomutase then converts the glucose-1-phosphate to glucose-6-phosphate that by phosphohexose isomerase becomes fructose-6-phosphate. This is further metabolized by EM pathway reactions. The manner of galactose use by *E. coli* is commonly known as the *Leloir pathway* and is shown in Figure 7.11.

The methyl glyoxal pathway

The methyl glyoxal sequence is found in *Escherichia coli* and *Pseudomonas saccharophila*. It is unusual because it allows formation of pyruvate from dihydroxyacetone without involvement of typical EM pathway triose intermediates such as glyceraldehyde-3-phosphate, 1,3-diphosphoglyceric acid, 3-phosphoglyceric acid, 2-phosphoglyceric acid, and phosphoenol pyruvate. The sequence is shown in Figure 7.12. Fructose-1,6-bisphosphate is cleaved by aldolase to give glyceraldehyde-3-phosphate and dihydroxyacetone phosphate. The glyceraldehyde-3-phosphate is converted to pyruvate by classical EM pathway enzymes and intermediates. Dihydroxyacetone, however, is dephosphorylated and oxidized to methyl glyoxal by methyl glyoxal synthetase and converted to D-lactate by the combined action of glyoxalases I and II. The D-lactate is then converted to pyruvate by a flavin-dependent lactate oxidase.

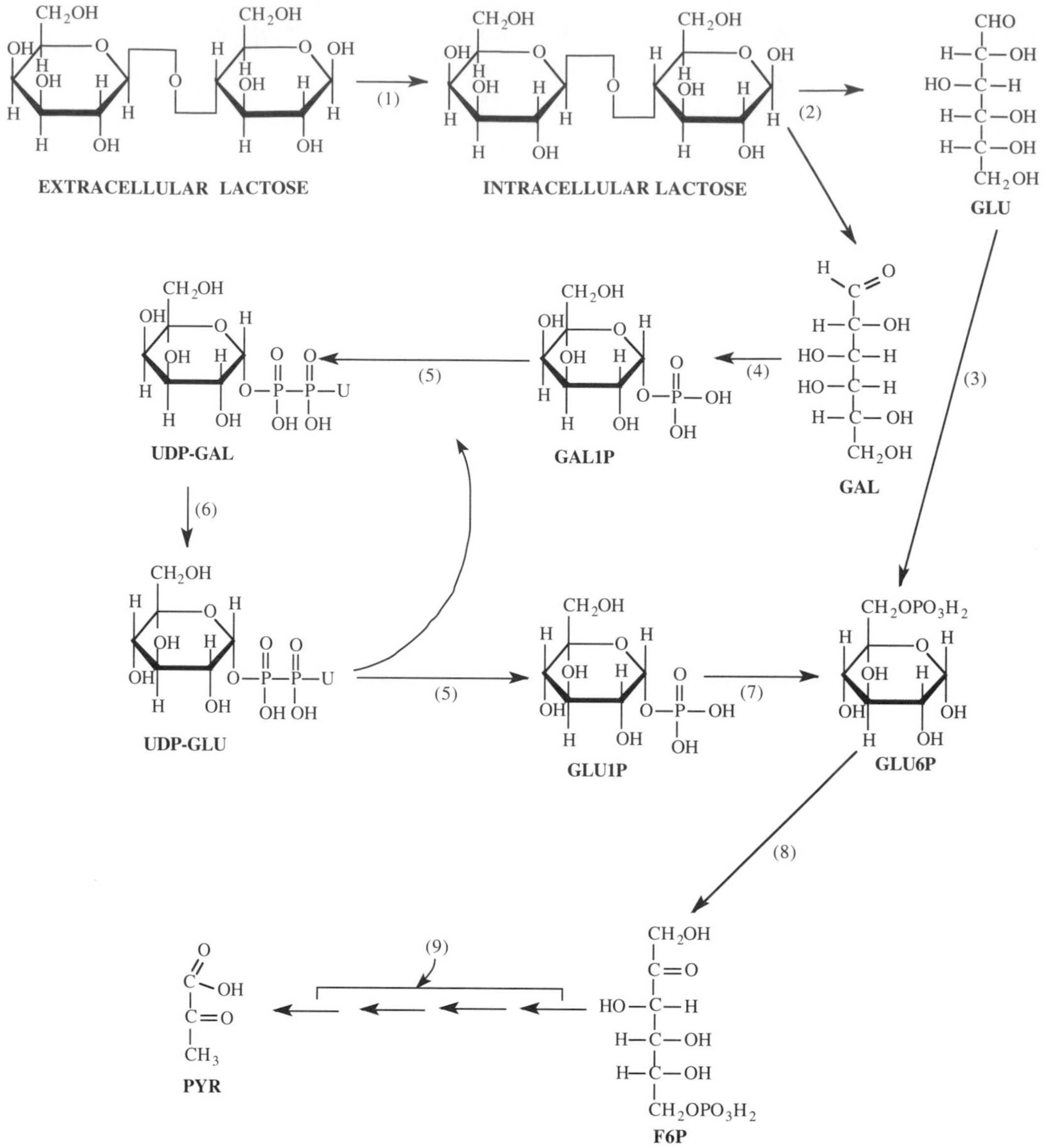

FIGURE 7.11 Lactose metabolism by the Leloir pathway. (1) Transport of lactose into the cell, without phosphoryation, by an inducible transport system. (2) Intracellular hydrolysis of lactose to D-glucose (GLU) and D-galactose (GAL) by beta-galactosidease. (3) Conversion of GLU to glucose-6-phosphate (GLU6P) by hexokinase. (4) Formation of galactose-1-phosphate (GAL1P) from GAL by galactokinase. (5) Formation of uridine diphosphate galactose (UDP-GAL) and glucose-1-phosphate by the action of GLU-GAL1P-uridylyl transferase. (6) Conversion of UDP-GAL to uridine diphosphate glucose (UDP-GLU) by UDP-GLU epimerase. (7) Formation of GLU6P from glucose-1-phosphate by phosphoglucomutase. (8) Formation of fructose-6-phosphate (F6P) from GLU6P by phosphohexose isomerase. (9) Conversion of F6P to pyruvate (PYR) by Embden-Meyerhof enzymes.

The use of dehydrosugars

Some microbes can use dehydrosugars such as fucose and rhamnose as carbon and energy sources. The use of these compounds by *E. coli* is shown in Figure 7.13. In *E. coli,* the L forms of fucose and rhamnose are used in a similar manner. Each of the substances enters the cell by a specific permease. Fucose is isomerized to its ketose analogue and subsequently phosphorylated by an ATP-dependent kinase to yield L-fuculose-1-phosphate. The latter is split by an aldolase enzyme to form a mixture of dihydroxyacetone phosphate and L-lactaldhyde.

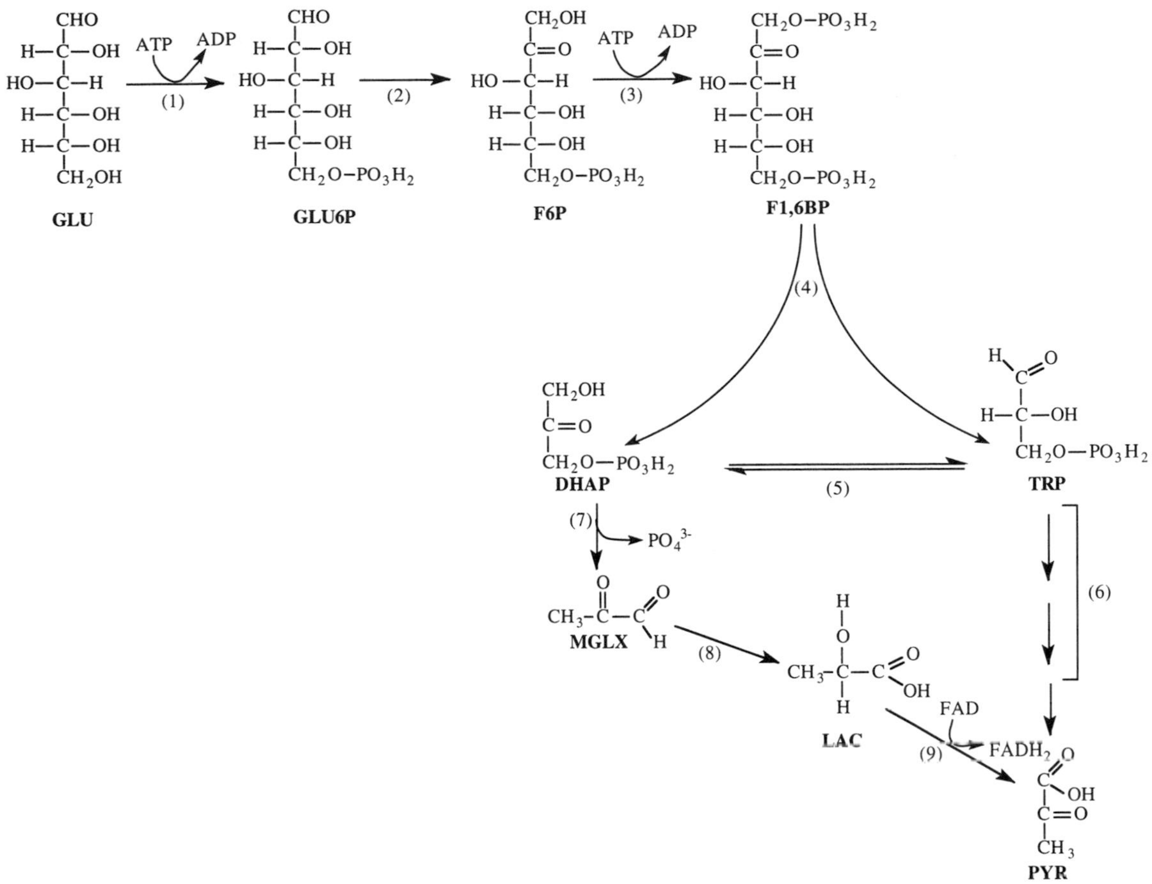

$$
\begin{array}{c}
\text{CHO} \\
\text{H-C-OH} \\
\text{HO-C-H} \\
\text{H-C-OH} \\
\text{H-C-OH} \\
\text{CH}_2\text{OH} \\
\textbf{GLU}
\end{array}
\xrightarrow[\text{(1)}]{\text{ATP} \quad \text{ADP}}
\begin{array}{c}
\text{CHO} \\
\text{H-C-OH} \\
\text{HO-C-H} \\
\text{H-C-OH} \\
\text{H-C-OH} \\
\text{CH}_2\text{O-PO}_3\text{H}_2 \\
\textbf{GLU6P}
\end{array}
\xrightarrow{\text{(2)}}
\begin{array}{c}
\text{CH}_2\text{OH} \\
\text{C=O} \\
\text{HO-C-H} \\
\text{H-C-OH} \\
\text{H-C-OH} \\
\text{CH}_2\text{O-PO}_3\text{H}_2 \\
\textbf{F6P}
\end{array}
\xrightarrow[\text{(3)}]{\text{ATP} \quad \text{ADP}}
\begin{array}{c}
\text{CH}_2\text{O-PO}_3\text{H}_2 \\
\text{C=O} \\
\text{HO-C-H} \\
\text{H-C-OH} \\
\text{H-C-OH} \\
\text{CH}_2\text{O-PO}_3\text{H}_2 \\
\textbf{F1,6BP}
\end{array}
$$

(4)

DHAP:
$$
\begin{array}{c}
\text{CH}_2\text{OH} \\
\text{C=O} \\
\text{CH}_2\text{O-PO}_3\text{H}_2 \\
\textbf{DHAP}
\end{array}
$$

(5)

TRP:
$$
\begin{array}{c}
\text{H-C=O} \\
\text{H-C-OH} \\
\text{CH}_2\text{O-PO}_3\text{H}_2 \\
\textbf{TRP}
\end{array}
$$

(7) → PO_4^{3-}

MGLX:
$$
\text{CH}_3\text{-C(O)-C(O)H} \qquad \textbf{MGLX}
$$

(8)

LAC:
$$
\text{CH}_3\text{-C(OH)H-C(=O)OH} \qquad \textbf{LAC}
$$

(9) FAD → FADH$_2$

(6)

PYR:
$$
\begin{array}{c}
\text{C(=O)OH} \\
\text{C=O} \\
\text{CH}_3 \\
\textbf{PYR}
\end{array}
$$

FIGURE 7.12 The methyl glyoxal pathway. (1) Conversion of glucose (GLU) to glucose-6-phosphate (GLU6P) by hexokinase or the phosphotransferase system. (2) Isomerization of GLU6P to fructose-6-phosphate (F6P) by phosphohexose isomerase. (3) Formation of fructose-1,6-bisphosphate (F1,6BP) from F6P by phosphofructokinase. (4) Cleavage of F1,6BP to form triose phosphate (TRP) and dihydroxyacetone phosphate (DHAP) by F1,6BP aldolase. (5) Interconversion of TRP and DHAP by triose phosphate isomerase. (6) Conversion of TRP to pyruvate (PYR) by Embden-Meyerhof enzymes. (7) Formation of methyl glyoxal (MGLX) by methyl glyoxal synthetase. (8) Conversion of MGLX to lactate (LAC) by the combined actions of glyoxalases 1 and 2. (9) PYR formation from LAC by a flavin-requiring lactate oxidase. The operation of this sequence allows PYR formation without TRP as an intermediate, and thus is particularly physiologically significant when phosphate limitation reduces TRP metabolism.

L-rhamnose is mobilized by an analogous series of reactions. After entrance into the cell by a specific permease, intracellular L-rhamnose is isomerized and phosphorylated in a single step to yield L-rhamnulose-1-phosphate. The latter, by its aldolase, is converted to dihydroxyacetone phosphate and L-lactaldehyde. Subsequent metabolism of dihydroxyacetone is accomplished by reactions of the EM pathway. Metabolism of L-lactaldehyde depends upon the conditions of growth. Under aerobic conditions, L-lactaldehyde is converted by an NAD-dependent oxioreductase to L-lactate which is oxidized by an FAD-mediated oxiore-ductase to pyruvate. Under anaerobic conditions, L-lactaldehyde is reduced by a NADH-requiring enzyme to 1,2-propane diol that is removed from the cell by facilitated diffusion.

Sugar alcohols

Many bacteria can use sugar alcohols, usually hexitols and pentitols, as carbon and energy sources. Although there are differences in the details of the mechanisms involved, the underlying mechanisms are often similar. Alcohol uptake is mediated by inducible transport systems, many of which involve

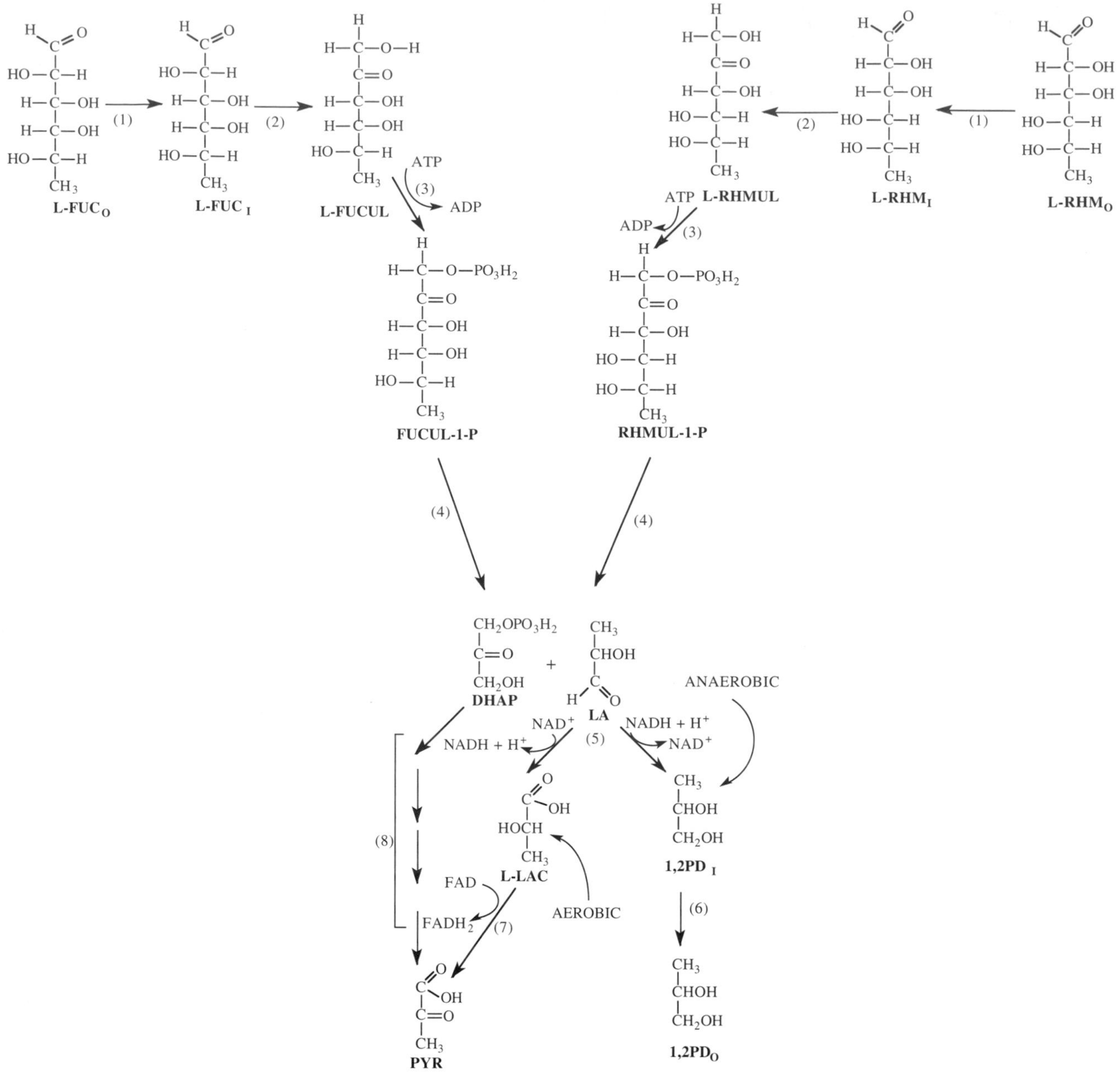

FIGURE 7.13 Utilization of L-fucose (L-FUC) and L-Rhomnose (L-RHAM) by bacteria. (1) Transport of the sugars from the outside to the inside of the cell by permeases. The O and I subscripts of the sugar abbreviations denote external and internal sugars, respectively. (2) Conversion of RHAM and FUC to rhamulose and fuculose (RHMUL and FUCUL) by isomerases. (3) Formation of ketose-1-phosphates (RHMUL-1-P and FUCUL-1-P) by ATP-dependent kinases. (4) Cleavage of FUCUL-1P and RHMUL-1-P by aldolases to form dihydroxy acetone phosphate (DHAP) and L-lactaldehyde. (5,7) Oxidation of lactaldehyde under aerobic conditions, via L-lactate, to pyruvate by the combined actions of an NAD-dependent oxioreductase (5) and a flavin-mediated lactate oxidase (7). (5,6) Anaerobic reduction of L-lactaldehyde by the same oxioreductase that mediates its oxidation (5) and removal of the resultant 1,2 propanediol (1,2PD) by facilitated diffusion (6). All of the permeases, isomerases, kinases, and aldolases are discrete and inducible. (8) Conversion of DHAP to pyruvate by Embden-Meyerhof reactions.

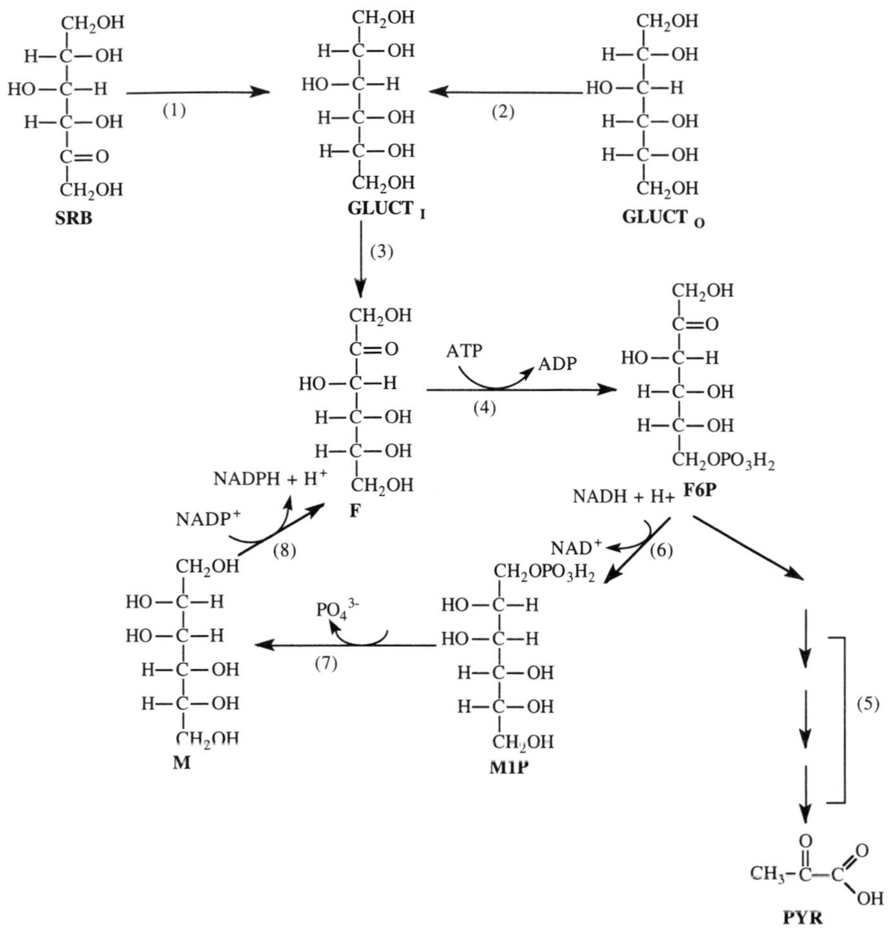

FIGURE 7.14 General mechanisms for use of unusual hexoses, pentoses, and their alcohol analogues. The mechanisms shown are for sorbose (SRB) and glucitol (GLUCT), but similar mechanisms apply for use of a variety of five- and six-carbon sugars and sugar alcohols. (1) Transport of sorbose to the cell interior by an inducible permease with reduction to glucitol (GLUCT). (2) Uptake of GLUCT without modification. The I and O subscripts of GLUCT denote the intracellular and extracellular materials, respectively. (3) Conversion of the intracellular GLUCT to fructose (F) by a dehydrogenase. (4) Conversion of F to fructose-6-phosphate (F6P) by an ATP-dependent kinase. (5) Conversion of F6P to pyruvate (PYR) by Embden-Meyerhof enzymes. (6) Conversion of F6P to mannitol-1-phosphate (M1P) by M1P dehydrogenase. (7) Conversion of M1P to free mannitol by a phosphorylase. (8) Conversion of free manitol to D-fructose by a mannitol dehydrogenase. Great diversity is found in various organisms in the nature of their mechanisms for hexitol and pentitol uptake, and in the nature and specificity of the reactions that convert them to central intermediates.

the PEP phosphotransferase system. The phosphorylated alcohols are then oxidized to their corresponding sugar phosphates with the aid of either NAD- or NADP-specific dehydrogenases and are converted to a sugar phosphate utilizable by central pathways. In some situations, alcohols are transported into the cell without conversion to alcohol phosphate. The transported alcohols may then be either oxidized to sugars and phosphorylated or phosphorylated first and then oxidized. Ultimately, the alcohols are converted to sugar phosphates that are metabolizable by central pathways. For hexitols, alcohols typically enter central metabolism as either glucose-6-phosphate or fructose-6-phosphate, while with pentitols, xylulose-5-phosphate and ribose-5-phosphate are the usual entry points. It is often the case that "unusual" sugars as well as their alcohol analogues are used by conversion to these intermediates. Representative mechanisms of sugar and sugar alcohol utilization are shown in Figure 7.14.

Aromatic compounds

A number of bacterial and other microbial species can degrade aromatic compounds. Although many organisms have been studied for their ability to degrade these substances because of the potential deleterious effects of aromatic compound accumulation for the environment, much of the work has focused on members of the genus *Pseudomonas.* The major intermediates recognized in aromatic compound degradation are catachol, protocatechuate, gentisate, and homogentisate. Degradation of aromatic compounds depends upon conversion of the material to one of these compounds and its subsequent conversion to intermediates of the TCA cycle or to metabolites utilizable by carbohydrate catabolism sequences.

It is difficult, if not impossible, to predict from its structure, the precise manner in which a particular aromatic compound may be catabolized. In some situations, several pathways may be used for a single compound. Such is the case of toluene utilization by *Pseudomonas putida,* which may convert toluene to catechol, protocatechuate, or 3-methyl catechol. Through ring cleavage, these materials are then converted to central intermediates. In other cases, many materials may be catabolized via a common intermediate. Thus, salicylate and compounds convertible to it, benzoate, and materials convertible to it, anthranilate and substances convertible to it, as well as benzene and phenol, may all be degraded by catechol. These situations are shown schematically in Figures 7.15 and 7.16.

Aromatic compound catabolism can be understood by understanding the nature of the central compounds to which such substances are converted, and the manner in which the central compounds are metabolized. Although gentisate and homogentisate are important intermediates in some situations, in most cases, the critical intermediates are catechol or protocatechuate. These structurally similar compounds are metabolized by two analogous series of reactions, ortho-cleavage and meta-cleavage. In ortho-cleavage, the ring structures of catechol and protocatechuate are ruptured by 1,2 catechol dioxygenase and protocatechuate 3,4 dioxygenase, respectively, producing *cis-cis* muconate (catechol) and β-carboxy *cis-cis* muconate (protocatechuate). The latter two compounds, by analogous enzymes, are converted to lactones and, subsequently, through isomerization (catechol) and decarboxylation (protocatechuate), to 4-oxoadipate enol lactone. Addition of water to

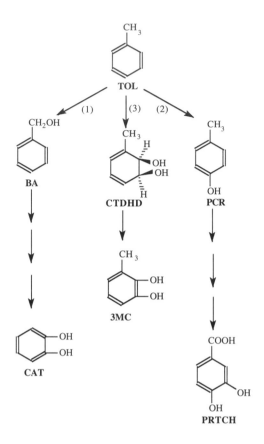

FIGURE 7.15 Degradation of aromatic compounds by alternate metabolic sequences as exemplified by toluene use of *Pseudomonas putida* and taxonomically similar organisms. (1) Conversion of toluene (TOL) to benzyl alcohol (BA) and its conversion, by several steps, to catechol (CAT). (2) Conversion of TOL to para-cresol (PCR) and its conversion, by several steps, to protochatechuate (PRTCH). (3) Metabolism of TOL to 3-methyl catechol (3MC) via *cis*-toluene dihydradiol (CTDHD). By whatever mechanisms TOL is converted to CAT, PRTCHC or 3MC, subsequent metabolism occurs via ortho- or meta-ring cleavage and conversion to central metabolites.

4-oxoadipate enol lactone by 4-oxiadipate hydrolase gives β-keto (3-oxo) adipate. Donation of coenzyme A from succinyl CoA, through the action of succinyl CoA transferase, forms β-ketoadipyl CoA, which with the aid of β-ketoadipate CoA thiolase and coenzyme A regenerates the succinyl CoA used to coacylate β-ketoadipate and allows formation of acetyl CoA. Thus, irrespective of structural differences between catechol and protocatechuate, the end products of ortho-cleavage are acetyl CoA and succinate, common intermediates of central metabolism.

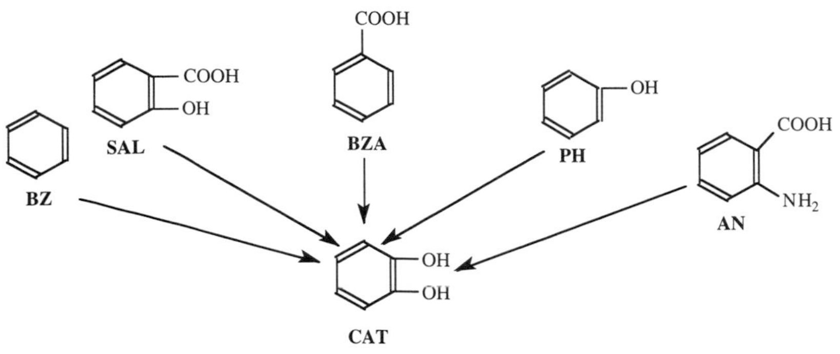

FIGURE 7.16 Degradation of aromatic compounds by a common intermediate as exemplified by conversion of benzene (BZ), salicylic acid (SAL), benzoic acid (BZA), anthranilic acid (AN), and phenol (PH) to catechol (CAT). Subsequent cleavage and metabolism of the CAT ring structure produces central intermediates.

Meta-cleavage of the aromatic ring structure results, once again, in a parallel series of reactions with catechol and protocatechuate. In meta-cleavage, in contrast to ortho-cleavage, the aromatic ring is broken between the carbon adjacent to the hydroxyl on carbon 2 (catechol) and carbon 4 (proto-catechuate) by monooxygenase enzymes. These reactions produce 2-hydroxymuconic semialdehyde from catechol and 2-hydroxy-4-carboxymuconic semialdehyde from protocatechuate. These compounds are acted upon by semialdehyde hydrolase enzymes, yielding 2-oxopent-4-enoate and 2-oxo-4-carboxypent-4-enoate. Hydrolysis of the latter compounds by hydrolase enzymes results in the formation of 4-hydroxy-2-oxovalerate and 4-hydroxy-4-carboxy-2-oxovalerate. In the final step of meta-cleavage, these last two compounds are converted by aldolases to pyruvate and acetaldehyde (4-hydroxy-2-oxovalerate) or 2 pyruvate molecules (4-hydroxy-4-carboxy-2-oxovalerate). The catabolism of catechol and protocatechuate by ortho- and meta- cleavage mechanisms is shown schematically in Figures 7.17 and 7.18. Regardless of whether the ortho- or meta- mode of metabolism is used, analogous reactions occur with catechol and protocatechuate.

Aromatic amino acid utilization

The degradation of aromatic amino acids by microbes differs substantially from aromatic hydrocarbon use. The pathway for phenylalanine and tyrosine degradation is shown in Figure 7.19. Phenylalanine is initially converted to tyrosine by phenylalanine hydroxylase. Tyrosine is then converted to p-hydroxyphenyl pyruvate by transamination. Para-hydroxyphenylpyruvate is oxidatively decarboxylated with attendant side-chain migration and hydroxylation, through p-hydroxyphenylpyruvate oxidase, to yield homogentisic acid. Rupture of the ring structure of homogentisate by homogentisate oxidase gives 4-maleylacetoacetate. Isomerization of this compound by maleylacetoacetate isomerase produces 4-fumarylacetoacetate. Finally, under the influence of fumarylacetoacetate hydrolase, fumarate and acetoacetate, compounds utilizable by central pathways, are formed.

Aliphatic hydrocarbon degradation

Aliphatic hydrocarbon degradation by microbes is substantially different than that of aromatic hydrocarbons. Hydrocarbon use is limited to organisms that have the ability to take up such compounds. The hydrophobic nature of the aliphatic hydrocarbons makes their utilization difficult. Uptake may involve modification of the cell wall, particularly by synthesis of trehalolipids, rhamnolipids, or other substances that allow dissolution of the hydrocarbons in the cell wall, as occurs in certain bacteria and yeasts, or the excretion of membranous particles that allow hydrocarbon dissolution, as occurs in the genus *Acinetobacter*. Intracellular compartmentalization is also an apparent feature of hydrocarbon degradation. The initial reactions involved in the use of hydrophobic materials normally occur in the membrane.

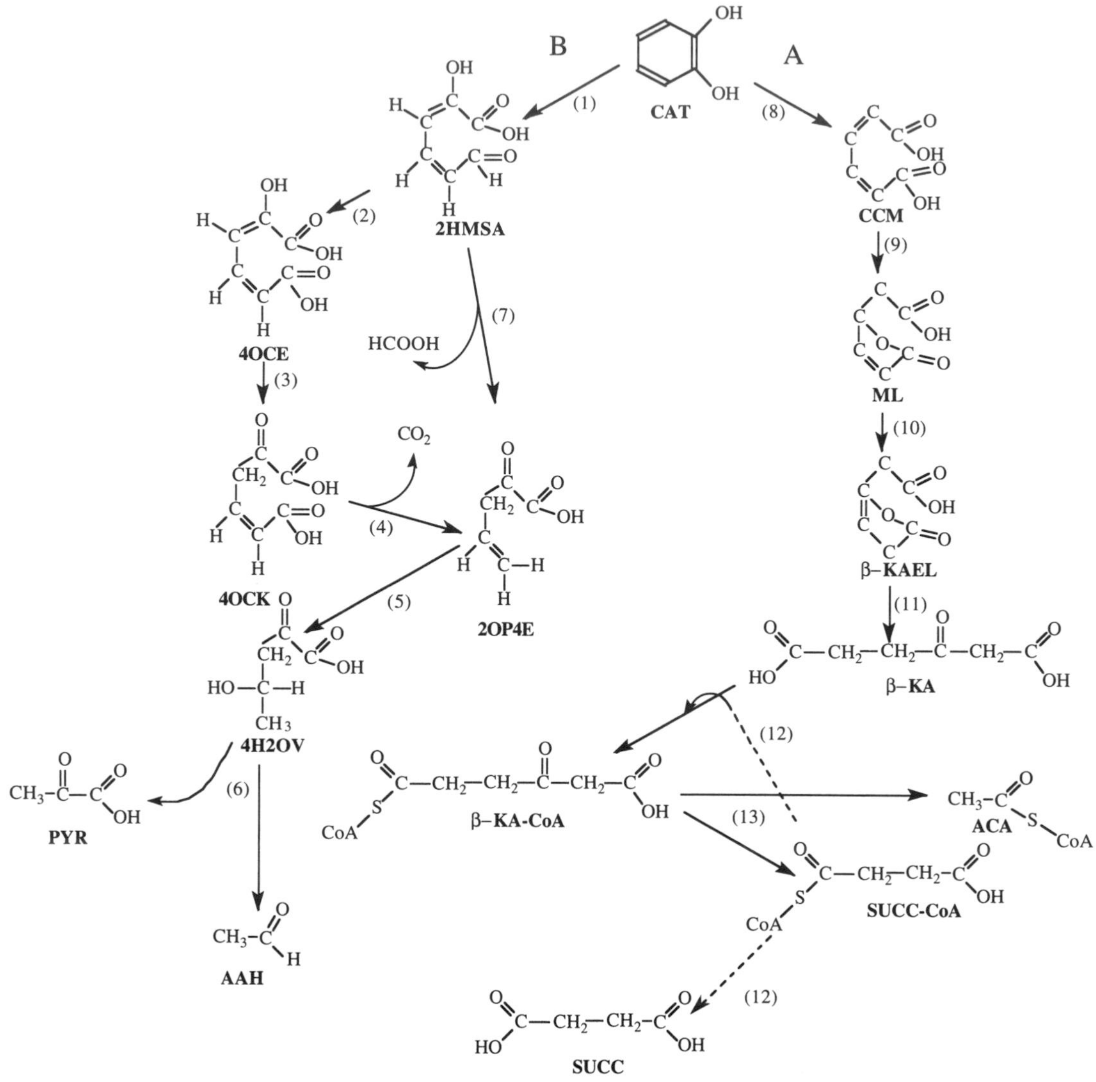

FIGURE 7.17 Catechol degradation by the (A) ortho- and (B) meta-pathways. (1) Conversion of catechol (CAT) to 2-hydroxymuconic semialdehyde (2HMSA) by CAT2,3 dioxygenase. (2) Conversion of 2HMSA to 4-oxalocrontonate enol (4OCE) by 2HMSA dehydrogenase. (3) Formation of the keto form of 4OC (4OCK) by 4OC tautomerase. (4) Formation of 2-oxy-pentanoate 4-enoate (2OP4E) from 4OCK by 4OCK decarboxylase or from 2HMSA by 2HMSA hydrolase (7). (5) Conversion of 2OP4E to 4-hydroxy-2-oxovaleric acid (4H2OV) by a hydrolase. (6) Cleavage of 4H2OV to pyruvate (PYR) and acetaldehyde (AAH) by 4H2OV aldolase. (8) Formation of *cis, cis* muconate (CCM) from CAT by CAT 1,2 dioxygenase. (9) Conversion of CCM to muconolactone (ML) by CCM lactonizing enzyme. (10) Formation of β-ketoadipate enol lactone (β-KAEL) from ML by ML isomerase. (11) Formation of β-ketoadipate (β-KA) from β-KAEL by β-KAEL hydrolase. (12) Formation of β-ketoadipyl CoA (β-KA-CoA) from β-KA and succinyl CoA (SUCC-CoA) by β-KA:SUCC-CoA transferase, with the concomitant formation of succinate (SUCC). (13) Cleavage of β-KA-CoA to form SUCC-CoA and acetyl CoA (ACA) by β-KA-CoA thiolase.

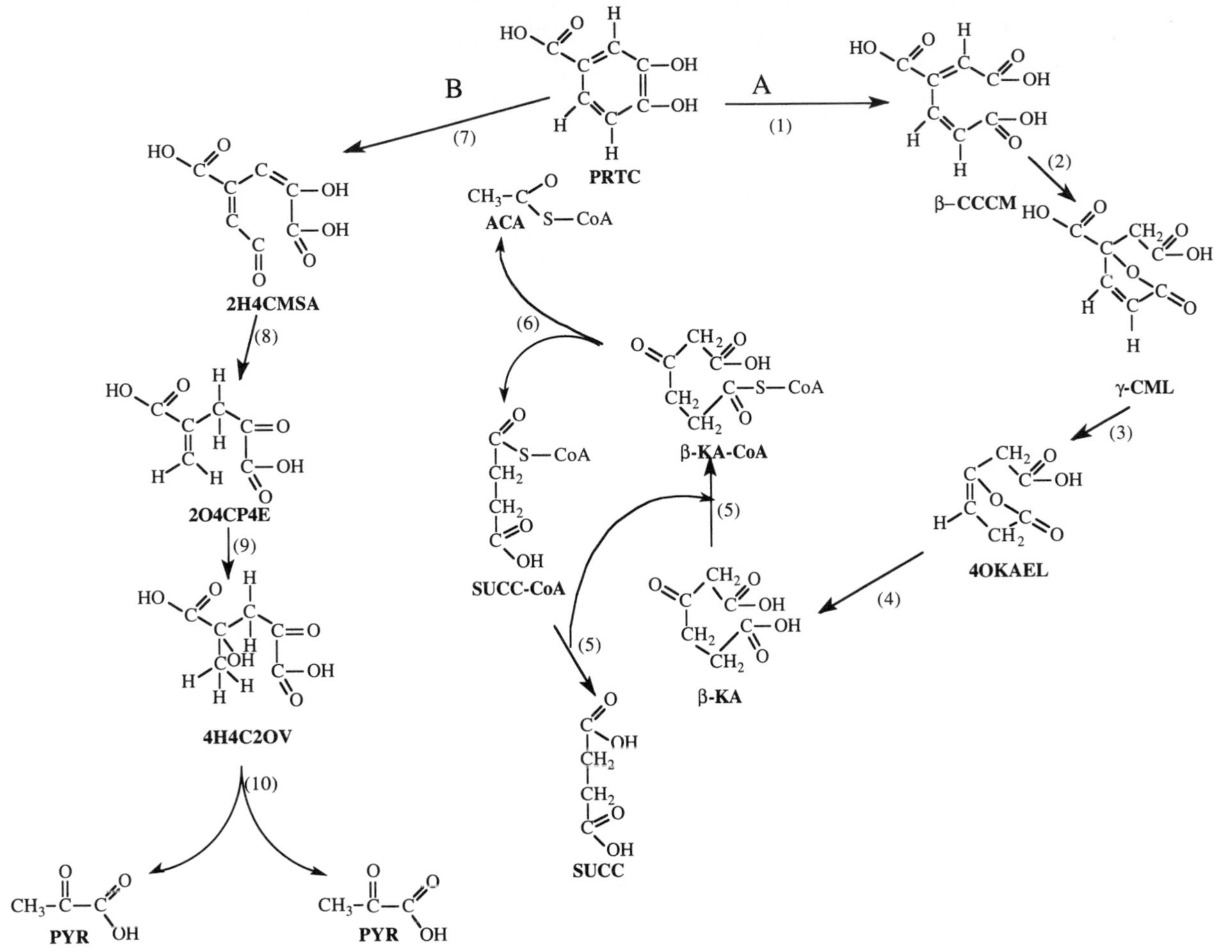

FIGURE 7.18 Protachatecuate (PRTC) degradation by the (A) ortho- and (B) meta-pathways. (1) Formation of beta-carboxy-*cis-cis* muconate (β-CCCM) from PRTC by PRTC 3,4 dioxygenase. (2) Conversion of β-CCCM to γ-carboxymuconolactone (γ-CML) by β-carboxymuconate lactonizing enzyme. (3) Formation of 4-oxoadipate-enol-lactone (4OKAEL) from γ-CML by γ-CML decarboxylase. (4) Conversion of 4OKAEL to β-ketoadipate (β-KA) by 4OKAEL hydrolase. (5) Formation of succinate (SUCC) and β-KA CoA from succinyl CoA (SUCC-CoA) and β-KA by β-KA:SUCC-CoA transferase. (6) Conversion of β-KA CoA to SUCC-CoA and acetyl CoA (ACA) by β-KA thiolase. (7) Formation of 2 hydroxy-4-carboxymuconic semialdehyde (2H4CMSA) from PRTC by PRTC 4,5-oxygenase. (8) Conversion of 2H4CMSA to 2-oxo-4-carboxypent-4-enoate (2O4CP4E) by 2H4CMSA hydrolase. (9) Formation of 4-hydroxy-4-carboxy-2-oxovalerate (4H4C2OV) from 2O4CP4E by 2O4CP4E hydrolase. (10) Conversion of 4H4C2OV to two molecules of pyruvate (PYR) by 4H4C2OV aldolase.

Two general mechanisms for aliphatic hydrocarbon degradation are currently known, one of which is similar, in some respects, to organic acid oxidation.

Terminal oxidation In many cases, aliphatic hydrocarbon oxidation occurs by terminal oxidation.

The general processes that occur in this mode of catabolism are shown in Figure 7.20. Oxidation by this mechanism requires participation of a cosubstance that is also oxidizable. In the case of hydrocarbons, this substance is the reduced, nonheme iron compound, rubredoxin. The reduced form of rubredoxin is formed from the oxidation of NADH

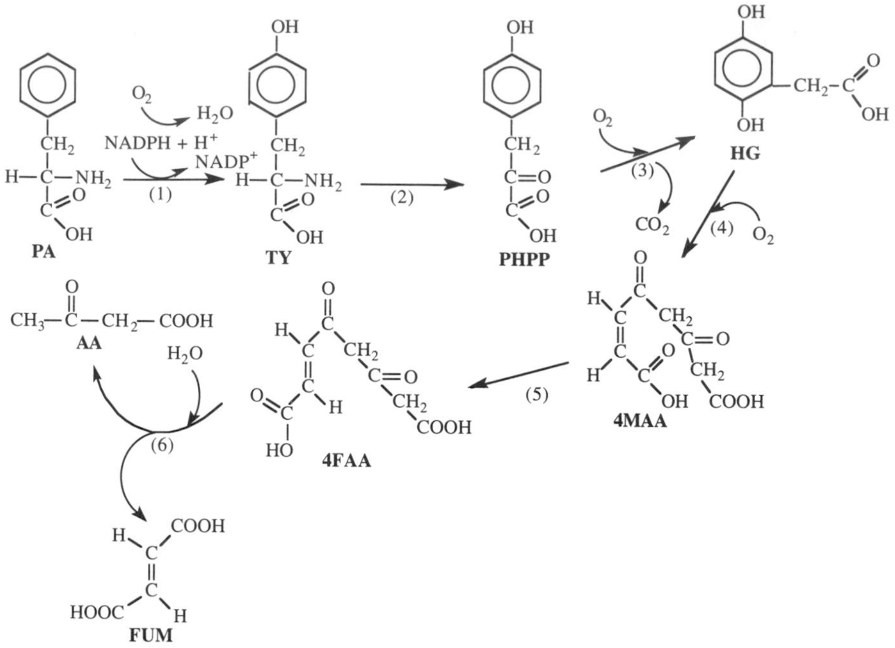

FIGURE 7.19 Degradation of phenylalanine (PA) and tyrosine (TY). (1) Conversion of PA to TY by PA hydrolase. (2) Formation of para-hydroxyphenyl pyruvate (PHPP) from TY by transaminse activity. (3) Decarboxylation of PHPP, and shift of the side chain with hydroxylation, to form homogentisate (HG) by the action of PHPP oxidase. (4) Formation of 4-maleylacetoacetate (4MAA) from HG by HG oxidase. (5) Conversion of 4MAA to 4-fumarylacetoacetate (4FAA) by 4MAA isomerase, a glutathione-requiring enzyme. (6) Formation of fumarate (FUM) and acetoacetate (AA) from 4FAA by the action of fumarylacetoacetate hydrolase. FUM and AA are metabolized by central metabolic pathways.

with the aid of rubredoxin: NADH oxioreductase. The next reaction, in the presence of an oxygen molecule and a proton and with the aid of n-alkane monooxidase, converts the terminal carbon of the alkane to a hydroxymethyl group, using one of the atoms of the oxygen molecule and reducing the other atom to water. The terminal hydroxymethylated n-alkane is then converted to an organic acid, with the intermediate formation of an aldehyde, by the actions of two NAD-requiring dehydrogenases.

Subterminal oxidation In certain circumstances, normal alkane oxidation proceeds via subterminal oxidation, as shown in Figure 7.21. Degradation by this mechanism begins by conversion of the n-alkane to a secondary alcohol by the action of a monooxygenase enzyme. The secondary alcohol is then converted by an NAD-mediated secondary alcohol dehydrogenase to a methyl ketone. A second monooxygenase reaction produces an acetic acid-

alcohol ester, by incorporation of one of the atoms of an oxygen molecule into the methyl ketone moiety and reduction of the other oxygen atom to water. Hydrolysis of the acetic acid-alcohol ester by acetyl ester forms free acetic acid and a primary alcohol two carbons longer than the initial one. The primary alcohol may then be further degraded by the terminal oxidation mechanism.

Organic acids

A substantial number of microbes use organic acids as carbon and energy sources. For the most part, they do so by the process of beta oxidation. The latter is accomplished by removal of acetyl CoA units. Thus, with even carbon acids, acetyl CoA is the sole degradation product, while with odd-chain acids, propionyl CoA is also a product. The basic mechanism for beta oxidation is shown in Figure 7.22. In the first reaction, the organic acid

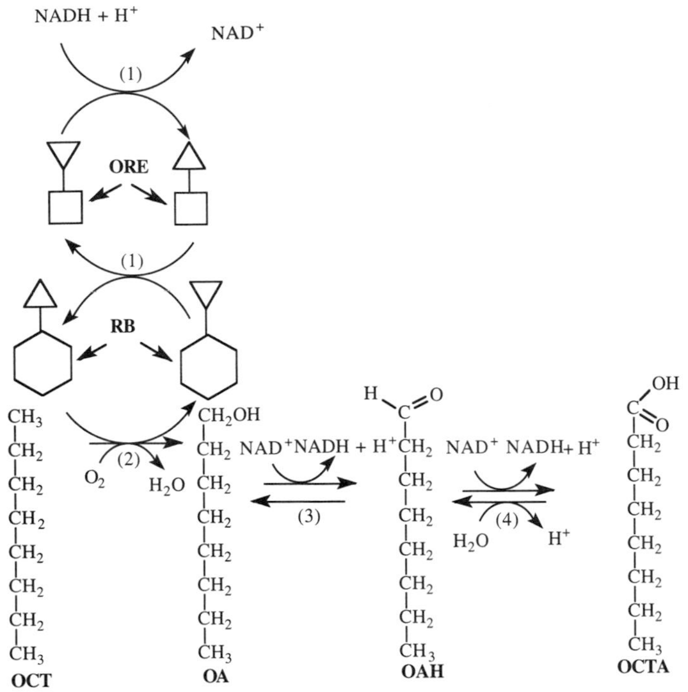

FIGURE 7.20 Terminal oxidation of normal alkanes. Octane is used as an example, but the general processes are the same, irrespective of differences between organisms and chain length. Because of the intrinsic hydrophobicity of hydrocarbons, the initial reactions of hydrocarbon utilization occur within the membrane. (1) Oxidation of NADH, with the formation of reduced rubredoxin, by the action of rubredoxin: NADH oxioreductase (ORE). Both the enzyme and rubredoxin (RB) may exist in oxidized and reduced forms that are denoted by normal (reduced) and inverted (oxidized) triangles. (2) Conversion of octane (OCT) to octyl alcohol (OA) by a monoxygenase enzyme, with reduced RB as the cosubstrate. (3) Formation of octyl aldehyde (OAH) from OA by alcohol dehydrogenase. (4) Conversion of OAH to octanoic acid (OCTA) by aldehyde dehydrogenase.

is converted to a coacylated form by an acyl CoA synthetase. The reaction requires expenditure of both CoA and ATP. The next reaction, catalyzed by an FAD-dependent fatty acid CoA dehydrogenase, inserts a double bond into the fatty acid chain between carbons 2 and 3. This reaction is followed by hydrolation of the double bond in such a manner that the hydroxyl portion of a water molecule is placed on carbon 3 of the organic acid. The reaction is mediated by the enzyme 3-hydroxyacyl-CoA-hydrolase. In the next reaction, the hydroxyl group on carbon 3 is oxidized to a carbonyl group by an NAD-dependent L-3-hydroxyacyl-CoA-dehydrogenase. The final reaction of oxidation is mediated by a β-thiolase. In the presence of coenzyme A, this enzyme cleaves the oxidized organic acid, giving rise to a molecule of acetyl CoA and an organic acid coenzyme A molecule two carbons shorter than the previous organic acid. Repetition of the reactions outlined above converts even-chain acids into acetyl CoA molecules. With odd-chain acids, analogous reactions give rise to acetyl CoA and a small amount of propionyl CoA as well. These two compounds are metabolized by central pathways.

Summary

This chapter illustrates the diversity of ways in which microbes may extract energy from a great variety of carbonaceous compounds and yet, do so with a high degree of "conservatism" in that, comparatively speaking, only a few metabolic sequences are required for use of a large diversity of

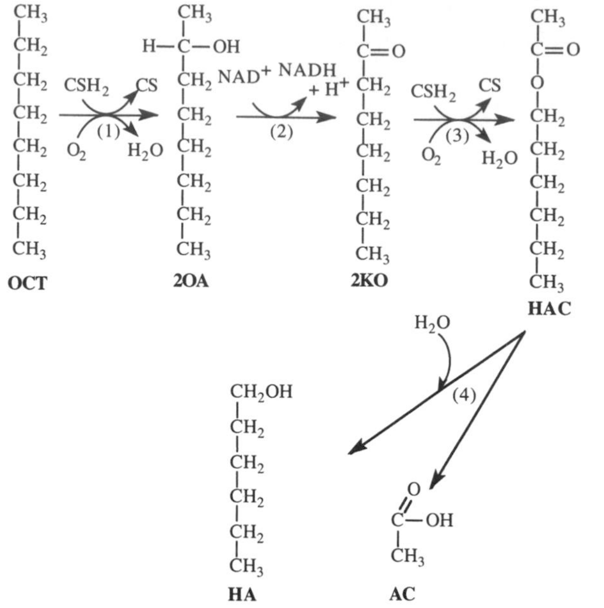

FIGURE 7.21 Normal alkane oxidation by the subterminal mechanism. Octane is used as an example, but analogous reactions occur with alcohols of other chain lengths. (1) Conversion of octane (OCT) to 2-octyl alcohol (2OA) by a monoxygenase enzyme that requires simultaneous conversion of a reduced cosubstrate (CSH_2) to its oxidized form (CS). (2) Formation of a methyl ketone, in this case 2-keto octane (2KO) from 2OA, by a secondary alcohol dehydrogenase, with the formation of NADH from NAD. (3) Formation of an alcohol ester of acetic acid from the methyl ketone (2KO), in this case hexyl acetate (HAC), by action of a second monoxygenase enzyme. (4) Hydrolysis of the acetic acid ester to form free acetate (AC) and a normal alcohol two carbons shorter than the chain length of the alkane, in this case hexyl alcohol (HA). The normal alcohol can then be metabolized by the sequence of reactions outlined in Figure 7.20. The acetate enters central metabolism.

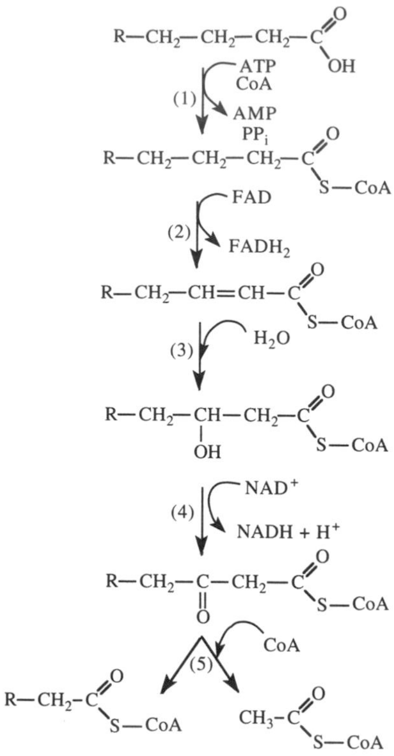

FIGURE 7.22 Fatty acid degradation by the β-oxidation pathway. (1) Mobilization of the fatty acid by conversion of the fatty acid to a coacylated form with expenditure of both CoA and ATP by acyl-CoA synthetase. The remaining degradative reactions occur on CoA-bound intermediates. (2) Formation of an α, β-unsaturated acid by the action of fatty acid acyl CoA dehydrogenase. The redox potential of the reaction is such that FAD is the electron acceptor. (3) Hydration of the α, β-double bond to form a β-hydroxyacyl-fatty acid acyl CoA derivative by the action of 3-hydroxyacyl-CoA hydrolase. (4) Oxidation of the 3-hydroxyacyl CoA derivative to a CoA-bound 3-keto derivative by the action of L-3-hydroxyacyl-CoA dehydrogenase. (5) Removal of a CoA-bound acetate residue from the fatty acid and formation of a CoA-bound fatty acid two carbons shorter than the previous acid. Repetition of these reactions allows complete degradation of the acid to acetyl CoA or to a mixture or acetyl and propionyl CoA, depending on the acid.

materials. In addition, the different pathways that often allow degradation of similar substances share a number of common intermediates. Although they differ in certain details, it is not necessary that each catabolic sequence involve an independently synthesized collection of substances in order to function. The EM and ED pathways, for example, share a number of common substances as do the TCA and glyoxalate cycles.

The necessity for synthesis of the twelve common biosynthetic precursors is a unifying aspect of catabolic metabolism. It is apparent that no single cycle allows formation of all of them and that a critical goal of catabolic metabolism is, in some manner, to allow formation of the entire spectrum of substances from which the critical biosynthetic molecules may be derived. Even for organisms that generate energy by noncatabolic mechanisms, formation of the twelve critical metabolites is essential. In the midst of metabolic diversity, there is a commonality.

Selected References

Abedal, A. T. 1979. Arginine catabolism by microorganisms. *Annual Review of Microbiology.* **33:** 139–168.

Cooper, R. A. 1984. Metabolism of methylglyoxal in microorganisms. *Advances in Microbial Physiology.* **38:** 49–68.

Dagley, S. 1978. Pathways for the utilization of organic growth substrates. *In* L. N. Ornston and J. R. Sokatch (eds.), *The Bacteria,* vol. 6. pp. 403–468. Academic Press, Inc., New York.

Doelle, H. W. 1975. *Bacterial Metabolism.* Academic Press, Inc., New York.

Fraenkel, Dan G. 1986. Mutants in glucose metabolism. *Annual Review of Biochemistry.* **55:** 317–338.

Fraenkel, D. G., and R. T. Vinopal. 1973. Carbohydrate metabolism in bacteria. *Annual Review of Microbiology.* **27:** 69–100.

Gottschalk, G. 1986. *Bacterial Metabolism,* 2nd edition. Springer-Verlag, New York, Berlin, Heidelberg, Tokyo.

Haddock, B. A., and C. W. Jones. 1977. Bacterial respiration. *Bacteriological Reviews.* **41:** 47–99.

Klug, M. J., and A. J. Markovetz. 1971. Utilization of aliphatic hydrocarbons by microorganisms. *Advances in Microbial Physiology.* **5:** 1–39.

Krampitz, L. O. 1962. Cyclic mechanisms of terminal oxidation. *In* I. C. Gunsalus and R. Y. Stanier (eds.), *The Bacteria,* vol. 2. pp. 209–256. Academic Press, Inc., New York.

Lessie, T. G. 1984. Alternative pathways of carbohydrate catabolism in *Pseudomonas. Annual Review of Microbiology.* **38:** 359–387.

Moat, A. G., and J. W. Foster. 1988. *Microbial Physiology,* 2nd edition. John Wiley and Sons, New York.

Mortlock, R. P. 1976. Catabolism of unusual carbohydrates by microorganisms. *Advances in Microbial Physiology.* **13:** 1–53.

Müller, M. 1988. Energy metabolism of protozoa without mitochondria. *Annual Review of Microbiology.* **32:** 465–488.

Opperdoes, F. R. 1987. Compartmentation of carbohydrate metabolism in trypanosomes. *Annual Review of Microbiology.* **41:** 127–152.

Perry, J. J. 1979. Microbial cooxidation involving hydrocarbons. *Microbiological Reviews* **43:** 59–72.

Sokatch, J. R. 1969. *Bacterial Physiology and Metabolism.* Academic Press, Inc., New York.

Stanier, R. Y., and L. N. Ornston. 1973. The β-ketoadipate pathway. *Advances in Microbial Physiology* **9:** 89–149.

Stouthamer, A. H. 1978. Energy-yielding pathways. *In* L. N. Orston and J. R. Sokatch (eds.), *The Bacteria,* vol. 6. pp. 389–462. Academic Press, Inc., New York.

van Poelje, P. D., and E. E. Snell. 1990. Pyruvoyl-dependent enzymes. *Annual Review of Biochemistry.* **59:** 29–60.

Wyatt, J. M. 1984. The microbial degradation of hydrocarbons. *Trends in Biochemical Sciences.* **9:** 20–23.

CHAPTER 8

Fermentation

The antiquity of interest in fermentation

Fermentation is a subject that has long been of interest to humankind. Indeed, it was largely through fermentation and disease that humankind first encountered microbes. The ancients were well aware of fermentation as a means of leavening bread and of producing alcoholic beverages. Over the years, our understanding of fermentation has substantially increased. We have rejected the suggestion of Liebig that fermentation was strictly a chemical process and, with the aid of Louis Pasteur and the Buchner brothers, understand that it is a biochemical process mediated by microbes. We understand further that specific microbes produce specific products because of their specific enzymes. We also understand that the fermentation end products formed by microbes reflect genetic differences between them, that is, their enzymes and other critical products are an expression of their genetic potential. Studies of fermentation and disease were, to a large extent, responsible for the recognition of the concept of microbial specificity and the delineation of microbial species. Fermentation has many implications.

The nature of fermentation

Although we recognize the importance of fermentation and the essentiality of microbes for its occurrence we do not, even today, fully understand its nature. The word fermentation implies different things to different people. To some it means any degradative process mediated by microbes. To others, fermentation implies any anaerobic, microbially-mediated degradative process. Still others regard fermentation as a degradative process in which energy is generated by substrate-level phosphorylation mechanisms only. Many people define fermentation as a process in which organic compounds serve as both electron donors and electron acceptors. Finally, some regard fermentation as an alternate means of reducing power disposal.

Each of the above definitions of fermentation has a certain amount of merit and yet none of them reflects a *complete* understanding of fermentation. To fully understand the nature of fermentation, we must understand both its chemical nature and its metabolic consequences. In particular, we must understand the relationship between end product formation and energy production. It is especially in this area that confusion concerning fermentation exists.

117

The relationship of oxygen to growth and fermentation

Oxygen is a critical substance for life. Its importance is such that we use it as a means of classifying microbes. We recognize certain organisms that require oxygen obligatorily for growth, that is, the *obligately aerobic* forms. We also recognize organisms for which oxygen or its metabolites, are toxic, the *obligate anaerobes*. In addition, we recognize that some organisms have the ability to grow both in the presence or in the absence of oxygen. For a period of time we regarded such organisms as *facultative anaerobes*, organisms that "preferred" to grow in the presence of oxygen, but had the ability to grow without it. Such terminology, although useful for certain purposes, is simplistic because it fails to consider the consequences of oxygen removal for microbial growth. For a long time it was believed that a correlation existed between the ability of an organism to grow in the presence or absence of oxygen and the nature of its energy generation mechanisms. It was believed that aerobic organisms produced energy by oxidative phosphorylation via electron transport and that fermentative organisms were those that required the absence of oxygen to grow and produced energy solely by substrate-level processes. We now know that such a dichotomy is untenable. We recognize certain organisms that are facultative in the sense that they grow either with or without oxygen, but that produce energy by substrate-level phosphorylation under *either* condition. *Streptococcus pyogenes* is such an organism. The organism is a facultative organism that is a biochemical anaerobe. It is *anoxybiontic,* that is, an organism that does not use oxygen in its metabolism. Under any growth condition, *S. pyogenes* is a fermentative organism, in the sense that it disposes of reducing power by the formation of fermentation end products with an organic compound as the electron donor.

Saccharomyces cerevisiae is an example of another facultative organism, but its relationship to oxygen is substantially different from that of *S. pyogenes*. In the presence of oxygen, *S. cerevisiae* produces energy both from substrate-level processes and by electron transport with oxygen as the electron acceptor. Under aerobic conditions, the organism produces water, the reduction product of oxygen, and carbon dioxide as major end products of glucose catabolism. When grown under these conditions, the organism may be considered as "energy rich and carbon poor." It produces an abundance of energy, primarily by electron transport, and it is difficult to provide the organism with sufficient materials to allow the organism to use its full biosynthetic potential.

When *S. cerevisiae* is grown under anaerobic conditions, it is a fermentative organism and produces carbon dioxide and ethanol as the major products of glucose catabolism. When the organism is grown under these conditions, it obtains energy solely from substrate-level processes. Since it cannot use oxygen as a terminal electron acceptor or "electron sink" compound, it uses acetaldehyde and produces ethanol. The major physiological differences between growth of the organism under aerobic conditions and its growth under anaerobic conditions are the ways in which the organism disposes of reducing power and the manner in which it obtains energy. In the case of *S. cerevisiae*, the difference between fermentative and oxidative metabolism reflects differences both in the manner of energy generation and in reducing power disposal.

Not all organisms grown in the absence of oxygen behave as does *S. cerevisiae*. Some organisms grow under anaerobic conditions and produce energy by both substrate-level phosphorylation and by electron transport. *Prevotella* (*Bacteroides*) *ruminicola* and *Vibrio succinogenes* are organisms that exhibit such behavior. During anaerobic growth, these organisms produce fermentation end products but obtain energy from both substrate-level and electron transport processes. The existence of organisms of this nature complicates our understanding of the nature of fermentation. Fermentative organisms cannot be regarded as those that obtain energy by substrate-level processes only, because some of them obtain energy by electron transport as well. Growth in the presence or absence of oxygen is also an inappropriate criterion by which to separate fermentative organisms from others because some organisms that are fermentative, in the sense that they produce large quantities of organic products during glucose catabolism, do so in the presence or absence of oxygen. It seems reasonable to suggest that fermentative organisms be considered to be those organisms that dispose of reducing power by the reduction of organic compounds and to recognize that aerobic microbes participate in the same process, reducing power disposal, but do so by reducing oxygen. Such a distinction ignores the nature of energy generation as a criterion by which to distinguish between fermentative and nonfermentative organisms and recognizes that irrespective of whether they are fermentative or nonfermentative, all organisms must dispose of reducing power and generate energy!

We may further understand that in the midst of metabolic diversity, there are only two ways in which biologically useful energy is formed, by *substrate-level phosphorylation* and by *electron transport*. Certain organisms accomplish reducing power disposal by reduction of molecular oxygen or forms of combined oxygen, such as nitrate, sulfate, or carbon dioxide. Others, like the fermentative organisms, accomplish the same process, reducing power disposal, by organic compound reduction. From this perspective, oxygen, sulfate, pyruvate, and fumaric acid are physiologically equivalent, since all of them may serve as terminal electron acceptors. Fermentative organisms are no longer distinguished by their manner of energy generation, but by the fact that they dispose of reducing power through organic compound reduction from organic sources of reducing power. Some fermentative organisms generate energy by a combination of substrate-level and electron transport processes and some by substrate-level processes only.

Fermentation end product formation

Irrespective of their means of forming energy, fermentative organisms, taken collectively, produce a wide variety of fermentation end products. Most of these products arise from pyruvate metabolism. The spectrum of products to which pyruvate may give rise is shown in Figure 8.1. The products produced by particular organisms are characteristic of them and may be used in their identification. Some microbes produce many products, and others produce only a single product, but fermentation end product formation, however it occurs, is a means of reducing power disposal.

Head-to-head and head-to-tail products

Pyruvate is a 3-carbon molecule. Some of the products formed from its metabolism are themselves 3-carbon compounds, but many products are formed by metabolic sequences in which pyruvate is cleaved. Fermentation end products are formed from further metabolism of the cleavage products. When pyruvate is cleaved, the cleavage occurs in a manner such that 1-carbon compounds arise from the carboxyl carbon of pyruvate, while 2-carbon compounds arise from pyruvate carbons 2 and 3. Many of the fermentation end products arise from acetyl CoA. Four-carbon products, and compounds derived from them, typically arise by condensation of two acetyl CoA molecules or by condensation of acetyl CoA with other two carbon-molecules derived from pyruvate cleavage. The condensations may occur in one of two ways, *head-to-head condensations* or *head-to-tail condensations*. In head-to-head condensations, the carbons corresponding to the 2-carbon moieties derived from pyruvate are joined to form a 4-carbon entity whose exterior carbons are equivalent to the methyl carbons of pyruvate or acetate, both of which arise from carbon 3 of pyruvate. The interior carbons of the 4-carbon entities formed by condensation in this manner are equivalent to the carboxyl carbons of the acetate moiety if it is involved, and, in any case, are derived from carbon 2 of pyruvate.

In head-to-tail condensations, acetyl CoA moieties or *equivalent* two-carbon units, are joined so that the methyl carbon of one acetate molecule is joined to a carboxyl carbon of a second acetate molecule. Condensations of this type produce 4-carbon molecules in which carbons 1 and 3 arise from the carboxyl carbons of the acetate moiety and carbons 2 and 4 arise from the methyl carbons. Since, irrespective of the manner of acetate condensation, the methyl carbons of acetate arise from the carbon 3 of pyruvate and the carboxyl carbons arise from pyruvate carbon 2, study of the radioactive label distribution in the end products formed from metabolism of labelled pyruvate, or the molecules that give rise to it, is of great value in determining precisely by which mechanism a particular end product arises.

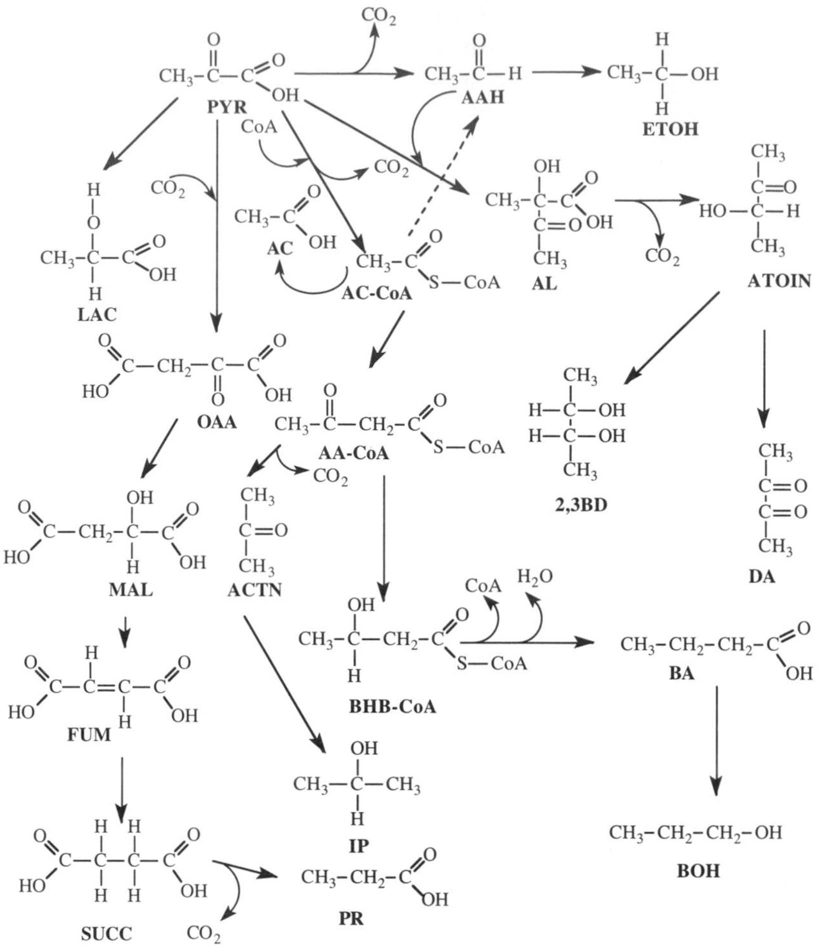

FIGURE 8.1 Typical fermentation products formed from pyruvate. The product abbreviations are the following: pyruvate (PYR); acetaldehyde (AAH); ethanol (ETOH); lactate (LAC); acetyl-CoA (AC-CoA); acetoacetyl-CoA (AA-CoA); β-hydroxybutyryl-CoA (BHB-CoA); butyric acid (BA); butanol (BOH); acetolactate (AL); acetoin (ATOIN); 2,3-butanediol (2,3BD); diacetyl (DA); acetic acid (AC); acetone (ACTN); isopropanol (IP); oxaloacetic acid (OAA); malic acid (MAL); fumaric acid (FUM); succinic acid (SUCC); propionic acid (PR). The precise nature of specific fermentations is discussed elsewhere.

Products formed by head-to-tail condensation

Fermentation products typically formed by the head-to-tail mechanism are shown in Figure 8.2. Not all of the products formed in this way are 4-carbon products. Initially, two molecules of acetyl CoA condense to form a molecule of acetoacetyl CoA. The products that result from further metabolism of acetoacetyl CoA depend upon whether the acetoacetyl CoA molecule is metabolized intact, or is cleaved into 3-carbon and 1-carbon molecules. If the acetoacetyl CoA molecule is metabolized as a unit, it is initially reduced, with the expenditure of

NADH, to β-hydroxybutyryl-CoA. This substance is then converted by dehydration between carbons 2 and 3 to crotonyl CoA. The crotonyl CoA is reduced to butyryl CoA, a pivotal molecule. In some organisms, butyryl-CoA is used as an energy source by the conversion of butyryl CoA to butyryl phosphate. The butyryl phosphate then reacts with ADP to give ATP and free butyric acid. Alternatively, butyryl CoA may be further reduced, via butyl aldehyde, to butanol.

If acetoacetyl CoA is metabolized by a mechanism in which the acetoacetyl CoA molecule is cleaved to 3- and 1-carbon entities, a different set of

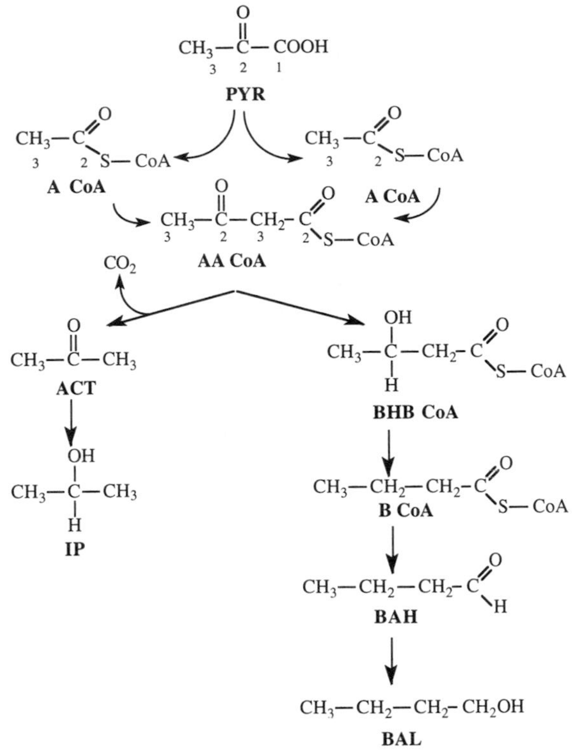

FIGURE 8.2 Typical products formed from pyruvate cleavage and condensation of two carbon moieties by a head-to-tail mechanism. The initial pyruvate cleavage forms acetyl-CoA (A CoA) from pyruvate carbons 2 and 3. Carbon dioxide is formed from pyruvate carbon 1. Two molecules of ACA condense to form a molecule of acetoacetyl-CoA (AA CoA). In all of the four carbon products, carbons 1 and 3 arise from pyruvate carbon 2, while carbons 2 and 4 are derived from pyruvate carbon 3. The three carbon products derived from pyruvate cleavage and head-to-tail condensation are formed from AA CoA in such a way that carbon dioxide is also formed from pyruvate carbon 2. The exterior carbons of three carbon products are derived from pyruvate carbon 3, while the middle carbon of the three-carbon products comes from pyruvate carbon 2. Aside from AA CoA, the products normally formed are β-hydroxybutyryl-CoA (BHB CoA), butyryl-CoA (B CoA), butyl aldehyde (BAH), butyl alcohol (BAL), acetone (ACT), and isopropanol (IP).

products may be formed. Cleavage of acetoacetyl CoA in a 3:1 manner yields carbon dioxide from the carboxyl carbon of the acetoacetyl CoA molecule. The remaining 3 carbons become, initially, acetone. Isopropyl alcohol may be formed by reduction of the acetone. In terms of the pyruvate carbons from which they arise, the exterior carbons

of both acetone and isopropanol are derived from pyruvate carbon 3 and the interior carbon originates from pyruvate carbon 2. Carbon dioxide formed in the sequence also results from pyruvate carbon 2.

Products formed from head-to-head condensation

Diacetyl, acetoin, and 2,3-butanediol are the major products formed from head-to-head condensation of two carbon moieties derived from pyruvate cleavage. Formation of these products may occur in one of three ways, all of which are shown in Figure 8.3. In mechanism 1, pyruvate reacts with a molecule of "active acetaldehyde" (hydroxyethyl thiamine pyrophosphate) to form acetolactate, which gives rise, by decarboxylation, to acetoin. Although the acetoin molecule does not result from the condensation of two 2-carbon moieties because acetolactate is a 5-carbon molecule, the carbons from pyruvate in the acetoin formed by this mechanism are such that the result is equivalent to that which *would* have occurred by condensation of two 2-carbon entities derived from pyruvate cleavage. The interior carbons of acetoin arise from the carbonyl carbons of pyruvate (pyruvate carbon 2) and the exterior carbons arise from pyruvate carbon 3. In mechanism 2, two molecules of pyruvate are converted to two molecules of active aldehyde that condense to give an acetoin molecule with carbons derived from pyruvate in the same manner as for mechanism 1. In mechanism 3, acetyl CoA condenses with active aldehyde to form diacetyl and it is converted to acetoin by the action of diacetyl reductase.

One-carbon products

Pyruvate cleavage occurs so that each pyruvate molecule gives rise to a 2-carbon moiety and a 1-carbon. The 1-carbon arises from the carboxyl carbon of pyruvate. A number of different systems for pyruvate cleavage are known (see Table 8.1).

Although they differ in mechanisms and actions, the various pyruvate-cleaving enzymes typically give rise to mixtures of acetyl CoA or acetyl phosphate and carbon dioxide, or formate. In addition, reduced electron carriers may be formed, such as, NADH, FADH, and reduced ferredoxin (FdH). The first two materials may be used, depending on the situation, to form fermentation end products or may be diverted to electron transport. Reduced

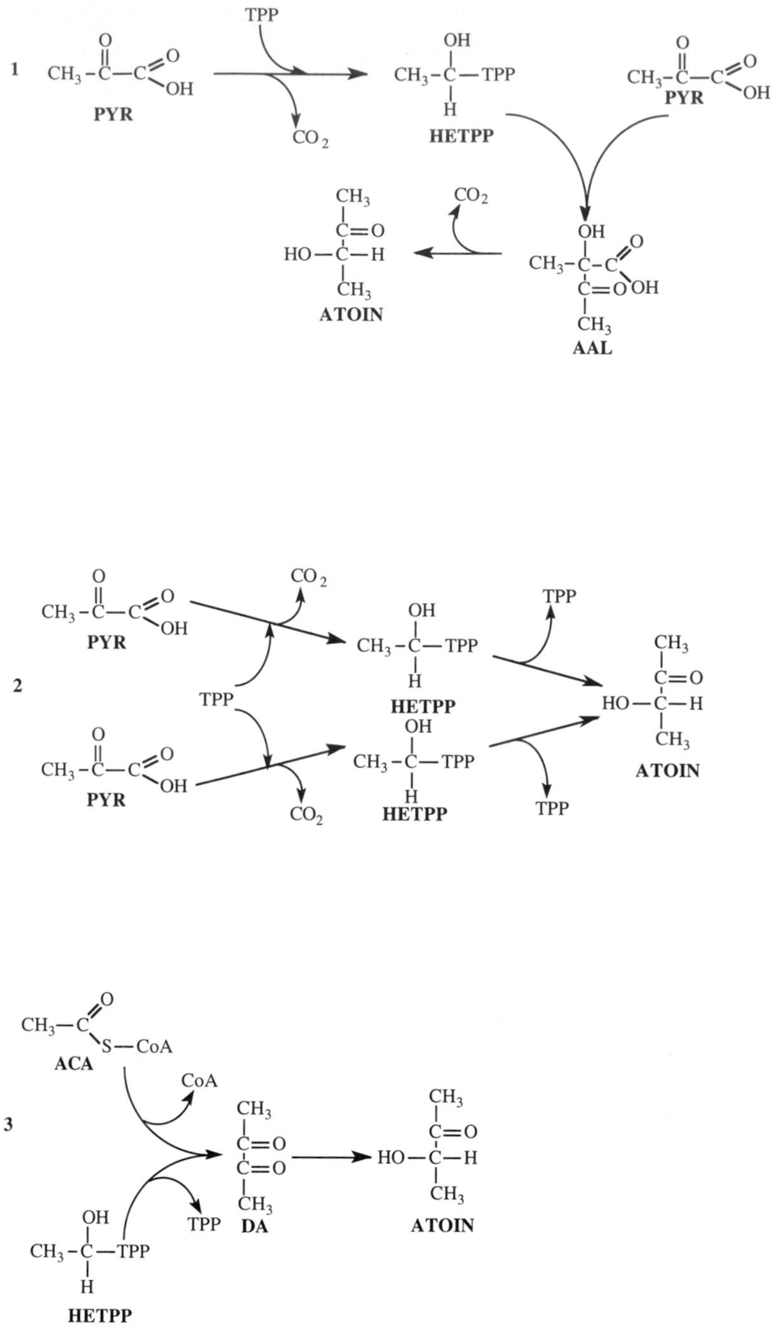

FIGURE 8.3 Three alternate mechanisms for acetoin formation from pyruvate. The abbreviations for compounds are the following: pyruvate (PYR); hydroxyethyl thiamine pyrophosphate (active aldehyde) (HETPP); alpha-acetolactate (AAL); acetoin (ATOIN); and acetyl-CoA (ACA). The details of specific fermentations are discussed in the text.

TABLE 8.1 Enzymes and Products of Pyruvate Cleavage

ENZYME	PRODUCTS
Pyruvate dehydrogenase	Acetyl CoA, CO_2, NADH
Pyruvate oxidase	Acetyl PO_4, CO_2, FADH
Pyruvate decarboxylase	Acetaldehyde, CO_2
Pyruvate-formate lyase	Acetyl CoA, formate
Pyruvate-ferredoxin oxioreductase	Acetyl CoA, CO_2, FdH

TABLE 8.2 Various Enzymes of Carbon Dioxide Fixation in Microbes

ENZYME	ACCEPTOR	PRODUCTS
1. Pyruvate carboxylase	Pyruvate	Oxaloacetate
2. Phosphoenol pyruvate carboxylase	PEP	Oxaloacetate
3. Malic enzyme	Pyruvate	Malic acid
4. Phosphoenolpyruvate carboxykinase	PEP	Oxaloacetate, ATP

ferredoxin, with the aid of the enzyme hydrogenase, may be converted to hydrogen. Formate seldom appears as an end product because it is usually converted to carbon dioxide and hydrogen by the formic hydrogenlyase system. This system was a subject of study for some time, but is currently considered as the combined action of formic dehydrogenase, that converts formate to carbon dioxide and reduced ferredoxin and hydrogenase that oxidizes reduced ferredoxin to produce hydrogen. Hydrogen produced during fermentation as a means of reducing power disposal may arise by at least two different mechanisms, oxidation of FdH produced from the action of pyruvate-ferredoxin oxioreductase and by formate degradation.

Products whose formation requires carbon dioxide fixation

Although carbon dioxide is a frequent fermentation end product, for certain organisms its fixation is required for end product formation. In some systems, succinate and propionate formation require carbon dioxide fixation. In these organisms, carbon dioxide may serve four physiological functions: 1) as a fermentation end product; 2) as an essential agent in energy generation through formation of a terminal electron acceptor; 3) as an agent in fermentation end product formation; and 4) in the formation of critical biosynthetic intermediates. The diversity of carbon dioxide fixation mechanisms is illustrated in Table 8.2.

The list in Table 8.2 is by no means an exhaustive one, but illustrates features of carbon dioxide fixation systems critical to their participation in fermentation end product formation. Enzymes 1–3 require energy or reducing power expenditure, while phosphoenolpyruvate carboxykinase allows the concomitant fixation of carbon dioxide and conservation of energy in the form of ATP. Mechanisms 1, 2, and 4 produce oxaloacetic acid, while mechanism 3 produces malic acid. Pyruvate is the acceptor molecule for mechanisms 1 and 3 and phosphoenolpyruvate (PEP) is the acceptor in mechanisms 2 and 4. In most organisms, more than one carbon dioxide fixing system is present and, irrespective of mechanism, precursors required for the synthesis of succinate and propionate via carbon dioxide fixation are formed.

The formation of succinate and propionate by carbon dioxide fixation is shown in Figure 8.4. The sequence begins with the formation of oxaloacetate by the fixation of carbon dioxide into pyruvate, or in some cases PEP, yielding oxaloacetate. This is reduced to L-malate by the action of malic dehydrogenase. The malic acid is then dehydrated by fumarase, acting in the reverse direction, to yield fumaric acid. The fumarate is converted to succinate by fumarate reductase.

Succinate is a critical intermediate in the sequence because it is a symmetrical molecule. With the aid of CoA transferase, succinic acid is converted to succinyl CoA. In the next reaction, mediated by (R)-methylmalonyl mutase, succinyl CoA is converted to (R)-methyl malonyl CoA. Because succinate is a symmetrical molecule, the CoA transferase enzyme may react with either end of the molecule. If the succinic acid is labelled in one of its interior carbons, isomerization of the suc-cinyl CoA formed from it results in two forms of (R)-methylmalonyl CoA, one in which the carbon adjacent to the coacylated carbon is labelled and an alternate form in which the methyl carbon of (R)-methylmalonyl CoA is labelled. After

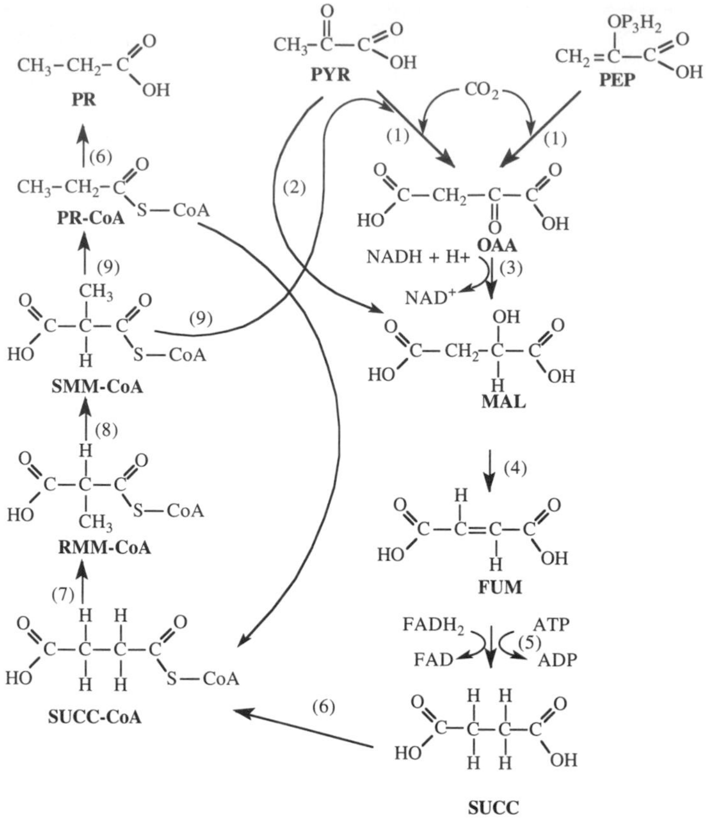

FIGURE 8.4 Formation of succinate and propionate by carbon dioxide fixation. (1) Formation of oxaloacetate (OAA) from pyruvate (PYR) or phosphoenolpyruvate (PEP) by a variety of CO_2-fixing enzymes. (2) Direct formation of L-malate (MAL) from pyruvate by malic enzyme. (3) Conversion of OAA to MAL with NADH expenditure by malic dehydrogenase. (4) Formation of fumarate (FUM) from MAL by fumarase. (5) Conversion of FUM to succinate (SUCC) by fumarate reductase, a flavin-containing enzyme, with the concomitant formation of ATP by electron transport.
(6) Formation of succinyl CoA (SUCC-CoA) from SUCC by the action of CoA transferase with propionyl Co-A (PR-CoA) as donor. In the process, free propionic acid (PR) is formed.
(7) Conversion of SUCC-CoA to R-methylmalonyl CoA (RMM-CoA) by RMM-CoA mutase, a vitamin B_{12}-requiring enzyme. (8) Formation of S-methylmalonyl CoA (SMM-CoA) from RMM-CoA by RMMA-CoA racemase. (9) Formation of PR-CoA and transfer of a carboxyl group to pyruvate by the action of S-methylmalonyl:pyruvate transcarboxylase.

conversion of the (R)-methylmalonyl CoA to the S form of the molecule, the carbon distal to the coacylated carbon is transferred by a biotin-mediated transcarboxylase reaction to a pyruvate molecule, forming a molecule of oxaloacetate and a molecule of propionyl CoA. The propionyl CoA, by CoA transferase, transfers its CoA moiety to succinate and becomes free propionic acid. However, because of the random coacylation of succinate, some of the molecules of propionate nate, some of the molecules of propionate

produced from labelled succinate contain label in the carbon adjacent to the carboxyl function of propionate and others are labelled in the methyl carbon of the molecule. For this reason, the carbon dioxide fixation pathway for succinate and propionate formation is sometimes referred to as the *randomizing* pathway, because the label in succinate is effectively randomized in the propionate formed from it. The transfer of a carboxyl function from (S)-methylmalonyl CoA is mediated by a transcarboxylase enzyme. The enzyme is sufficient for formation of oxaloacetate from pyruvate, but in some organisms, the enzymes listed in Table 8.2 are involved in either oxaloacetate or malate formation.

The direct reduction, or acrylate pathway, of propionate formation

In some organisms, propionate is formed without the intermediate formation of succinate. Such a sequence is found in *Megasphaera elsdenii, Clostridium propionicum,* and *Prevotella (Bacteroides) ruminicola. Prevotella ruminicola* produces succinate in addition to, or in place of, propionate in certain situations. However, it produces propionate by the direct reduction pathway. The details of this pathway are shown in Figure 8.5. In *M. elsdenii* and *C. propionicum,* lactate is the beginning point for propionate formation and acetate and carbon dioxide are also formed. The production of acetate and carbon dioxide is accomplished with D-lactate, while propionate is formed from L-lactate. However, either form of lactate may be used for both processes since they are interconvertible with a racemase.

Propionic acid formation by this sequence begins with the conversion of L-lactate to L-lactyl CoA by a CoA transferase reaction. The L-lactyl CoA is then converted to acrylyl CoA by a dehydratase reaction. The acrylyl CoA is reduced to propionyl CoA, with the reducing equivalents supplied by a reduced electron transferring flavoprotein. The propionyl CoA transfers its CoA moiety to a molecule of L-lactate, regenerating L-lactyl CoA and forming free propionic acid. This sequence is commonly referred to as the *direct reduction,* or *nonrandomizing* pathway, to contrast it to the sequence in which propionate formation requires carbon dioxide fixation. It is also called the nonrandomizing pathway because carbons in L-lactate have analogous carbons in the propionate formed from it.

In the direct reduction pathway, acetate and carbon dioxide formation accompany the formation of propionate. A portion of the available lactate is converted to its D-isomer. The latter, with the aid of a flavoprotein-dependent lactic dehydrogenase, is converted to pyruvate. Pyruvate is then cleaved by pyruvate-ferredoxin oxioreductase, giving carbon dioxide, acetyl CoA, and reduced ferredoxin. The ferredoxin is reoxidized by a transhydrogenase reaction, providing a portion of the

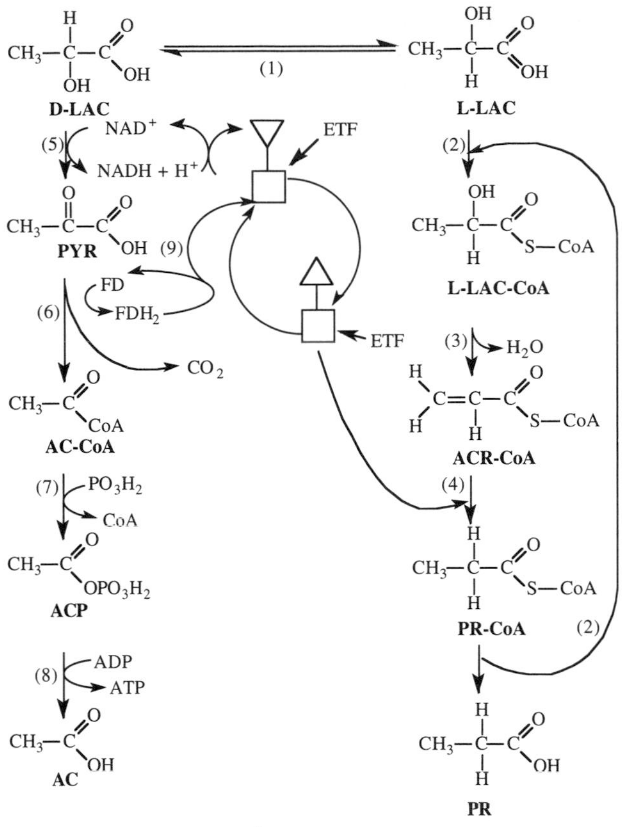

FIGURE 8.5 The acrylate or "direct reduction" pathway for propionate formation. (1) Interconversion of D- and L-lactate (D- and L-LAC) by a racemase. (2) Conversion of L-LAC to L-lactyl-CoA (L-LAC-CoA) by CoA transferase with propionyl-CoA (PR-CoA) as donor, with the formation of free propionate (PR). (3) Dehydration of L-LAC-CoA to form acrylyl-CoA (ACR-CoA). (4) Reduction of ACR-CoA to PRP-CoA with reducing power supplied by an electron transferring flavoprotein (ETF). The reduced and oxidized forms of ETF are indicated by the normal (reduced) and inverted (oxidized) triangles. The reducing power for ACR-CoA reduction is provided by D-LAC oxidation to pyruvate (PYR) by lactic dehydrogenase (5) and by the combined action of pyruvate-ferredoxin (FD) oxidoreductase (6) and transhydrogenase (9), producing reduced ferredoxin (FDH$_2$). Acetyl-CoA (AC-CoA) produced from the action of PYR-FD oxidoreductase is converted to free acetate (AC) and ATP by the combined actions of phosphotransacetylase (7) and acetyl kinase (8).

reducing equivalents required for conversion of acrylyl CoA to propionate. The acetyl CoA is converted by phosphotransacetylase to acetyl phosphate, which, through acetyl kinase, becomes free acetic acid. The last reaction also produces a molecule of ATP.

Ethanol and lactate formation

Ethanol and lactate are products produced by a number of microbes. Study of these two fermentations has been central to understanding carbohydrate fermentation. The formation of ethanol and lactate is shown in Figure 8.6. The demonstration by the Buchners that ethanol could be formed by a cell-free preparation was a major contribution to the recognition of fermentation as a biochemical process mediated by microbes, rather than simply a chemical one. Ethanol is normally formed in one of two ways, either by reduction of the acetaldehyde produced from the cleavage of pyruvate by pyruvate decarboxylase or from reduction of acetyl CoA formed from the action of the pyruvate dehydrogenase complex. Reduction of acetaldehyde is mediated by alcohol dehydrogenase, whereas reduction of acetyl CoA is accomplished by aldehyde dehydrogenase. If acetaldehyde reduction is inhibited by the interaction of acetaldehyde with bisulfite, the reducing power normally used for ethanol production may be diverted to the reduction of dihydroxyacetone phosphate and glycerol may accumulate as a substantial fermentation product.

Lactic acid may be formed as a mixture of D(−)- and L(+)-lactate or as either the D or the L isomer. In organisms that form a mixture, the nature of that mixture is reflective of the operation of stereospecific dehydrogenases, whether or not the organisms have a racemase. In organisms that have stereospecific enzymes but lack a racemase, the formation of D- and L lactate are independent processes. In cases in which a racemase is present, one isomeric form may be converted to the other.

The acetate fermentation

Acetate may be formed as the sole product of carbohydrate fermentation by some organisms. Such organisms are known as *acetogenic organisms*. Several clostridial species are of this nature. The formation of acetate as a sole fermentation end product by the acetogens is shown in Figure 8.7. The sequence begins with fructose that is converted via the EM pathway to pyruvate. Two pyru-

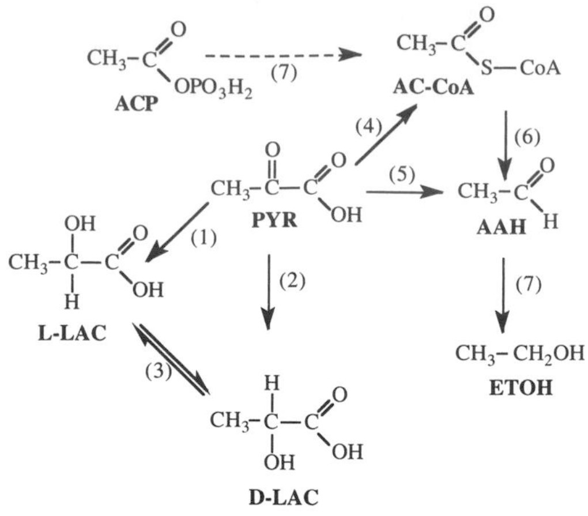

FIGURE 8.6 Mechanisms for ethanol (ETOH) and lactate (LAC) formation by microbes. (1,2) Production of D- (D-LAC) or L-lactate (L-LAC) by stereospecific reduction of pyruvate by lactic dehydrogenases. In some cases, a single isomeric form of LAC is formed while in other situations, a mixture of isomers is produced. It is often the case that a mixture of D- and L-LAC results from the initial formation of a single isomeric form followed by conversion of the latter to a mixture by an inducible racemase (3). Ethanol formation normally results from one of two mechanisms, reduction of acetaldehyde (AAH) by ethanol dehydrogenase (7) or by a combination of acetyl-CoA (AC-CoA) reduction to AAH by AAH dehydrogenase (6) and subsequent AAH reduction by ETOH dehydrogenase. AAH normally results from PYR by the action of a thiamine pyrophosphate-requiring pyruvate decarboxylase. AC-CoA is normally formed from pyruvate by the pyruvate dehydrogenase complex. Although most organisms produce ethanol by reduction of either AC-CoA or AAH produced from pyruvate cleavage, some organisms initially form acetyl phosphate (ACP) and convert the latter to AC-CoA by phosphotransacetylase. In heterofermentative organisms, it is often the case that the relative abundance of products is regulated by the stereospecificity of their dehydrogenase enzymes.

vate molecules are then converted to two molecules of acetyl CoA and two molecules of CO_2 by the action of pyruvate-ferredoxin oxioreductase. The acetyl CoA molecules are converted to free acetic acid with phosphotransacetylase and acetyl kinase, with the concomitant formation of ATP. The carbon dioxide molecules are converted to acetate by a complicated series of reactions (see Chapter 17 for details). One of the carbon dioxide molecules is converted to formate by the action of formate dehydrogenase. The formate is then con-

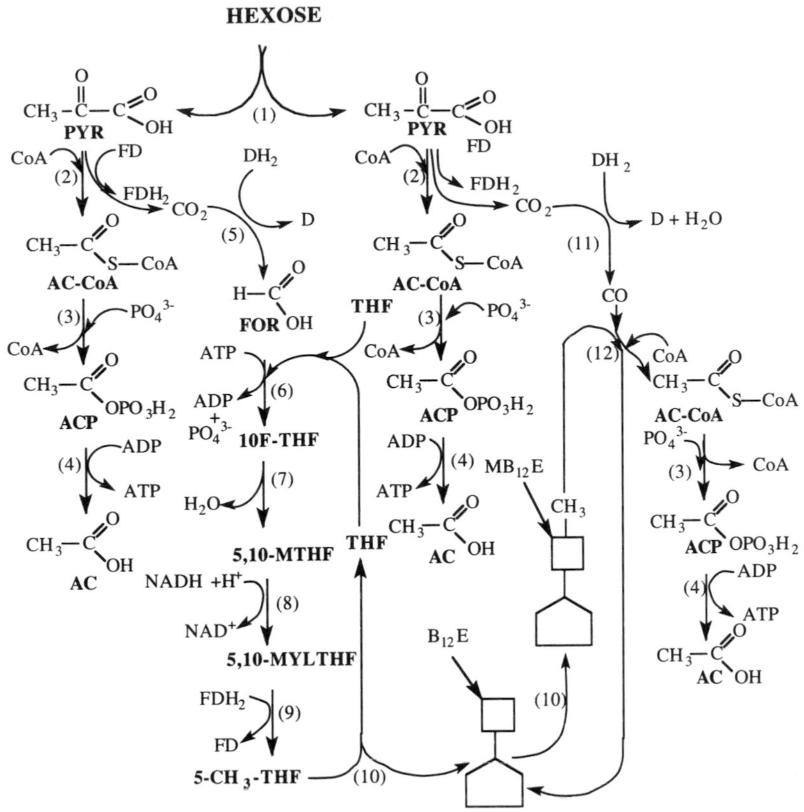

HEXOSE

(figure 8.7 pathway diagram with labels: PYR, CoA, FD, FDH₂, CO₂, AC-CoA, ACP, AC, FOR, 10F-THF, 5,10-MTHF, 5,10-MYLTHF, 5-CH₃-THF, THF, MB₁₂E, B₁₂E, CO, reaction numbers (1)–(12), ATP/ADP, NADH+H⁺/NAD⁺, etc.)

FIGURE 8.7 Formation of acetate as a sole fermentation end product. (1) Conversion of a hexose, normally glucose or fructose, to two molecules of pyruvate (PYR) by the Embden-Meyerhof pathway. This process yields ATP and NADH that are used in other reactions required for acetate formation. (2) Formation of acetyl-CoA (AC-CoA), CO_2, and reduced ferredoxin (FDH_2) from its oxidized form (FD) by the action of pyruvate:ferredoxin oxidoreductase. (3,4) Formation of free acetate (AC) from AC-CoA with the intermediate formation of acetyl phosphate (ACP), by the sequential actions of phosphotransacetylase (3) and acetyl kinase (4). (5) Synthesis of formate (FOR) from a molecule of CO_2 by the action of a selenium and tungsten-containing formate dehydrogenase. This reaction requires an electron donor (DH_2), whose nature varies from organism to organism. (6) Synthesis of 10-formyl-tetrahydrofolate (10F-THF) from FOR and tetrahydrofolate (THF) by formyl-THF synthetase. (7–9) Reduction of 10F-THF to 5-methyl-THF, with the intermediate formation of 5,10-methenyl-THF (5,10-MTHF) and 5,10-methylene-THF(5,10MYL-THF) by the actions of M-THF cyclohydrolase (7), MYL-THF dehydrogenase (8), and MYL-THF reductase (9). (10) Donation of a methyl group to a vitamin B_{12}-bound enzyme ($B_{12}E$) to produce a methylated enzyme ($MB_{12}E$) by the action of THF-B_{12} methyl transferase. (11) Reduction of CO_2 to CO by CO dehydrogenase. (12) Formation of AC-CoA from CO and $MB_{12}E$ and regeneration of $B_{12}E$ by AC-CoA-synthesizing enzyme. This reaction involves CO bound to a molecule of CO dehydrogenase.

verted, with ATP expenditure and reducing power supplied by NADH and FdH, to 5-methyl tetrahydrofolate. All of the reduction reactions occur as tetrahydrofolate intermediates. The methyl portion of the 5-methyltetrahydrofolate is transferred to an enzyme-vitamin B_{12} complex by methyltransferase to yield a methyl-B_{12}-enzyme complex. The second carbon dioxide molecule formed by the action of pyruvate-ferredoxin oxioreductase is reduced to carbon monoxide by carbon monoxide dehydroge-nase. The methyl moiety of the methyl-B_{12}-enzyme complex then combines with a carbon monoxide dehydrogenase-bound molecule of acetyl CoA-synthesizing enzyme to form acetyl CoA. ATP is derived from conversion of the acetyl CoA to free acetic acid by the actions of phosphotransacetylase and acetyl kinase. The net effect of all of these reactions is the conversion of a molecule of fructose to three molecules of acetate and three molecules of ATP.

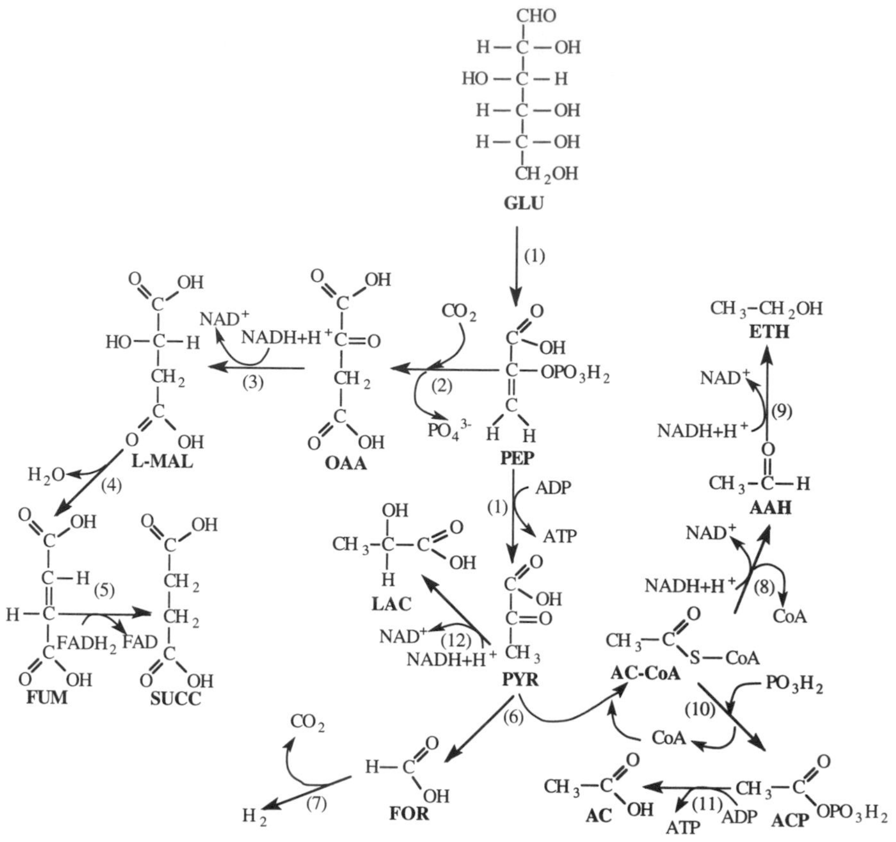

FIGURE 8.8 The mixed acid fermentation. (1) Conversion of glucose (GLU) to pyruvate (PYR) by Embden-Meyerhof reactions. (2–5) Conversion of GLU to succinate (SUCC) by carboxylation of phosphoenol pyruvate (PEP) by PEP carboxylase to yield oxaloacetate (OAA) (2), followed by the actions of malic dehydrogenase (3), fumarase (4), and fumarate reductase (5). During SUCC formation, malate (MAL) and fumarate (FUM) are formed as intermediates in addition to OAA. (6,7) Conversion of PYR to formate (FOR) and acetyl-CoA (AC-CoA) by the action of pyruvate-formate lyase (6) and conversion of FOR to CO_2 and H_2 by the combined actions of formate dehydrogenase and hydrogenase (7). (8,9) Reduction of AC-CoA to ethanol (ETH) with the intermediate formation of acetaldehyde (AAH) by the actions of acetaldehyde (8) and alcohol (9) dehydrogenases. (10,11) Formation of free acetate (AC) from AC-CoA, with the intermediate formation of acetyl phosphate (ACP) and ATP by the actions of phosphotransacetylase (10) and acetyl kinase (11). (12) Reduction of PYR to lactate (LAC) with LAC dehydrogenase.

The mixed acid fermentation

Some microbes are *homofermentative*, which means that they produce a single fermentation product, but many microbes are *heterofermentative*, producing more than one product. An example of hetero-fermentation, the mixed acid fermentation, is found in the enterobacteria and shown in Figure 8.8. The enterobacteria ferment glucose by the EM pathway and use EM intermediates for the formation of succinate, ethanol, lactate, formate, carbon dioxide, and hydrogen. Succinate is formed by the carboxylation of phosphoenol pyruvate (PEP) by PEP carboxylase to produce oxaloacetate

(OAA). The OAA is converted to succinate by the combined actions of malate dehydrogenase, fumarase, and fumarate reductase. The remaining products of the mixed acid fermentation arise from pyruvate metabolism: lactate is formed by direct pyruvate reduction with lactate dehydrogenase and formate and acetyl CoA are produced by the action of pyruvate-ferredoxin oxioreductase. Acetyl CoA is converted to free acetic acid, with the intermediate formation of acetyl-PO_4, by the actions of phosphotransacetylase and acetyl kinase. This sequence allows the CoA bond energy of acetyl CoA to generate ATP. A portion of the acetyl CoA is converted to ethanol by the actions

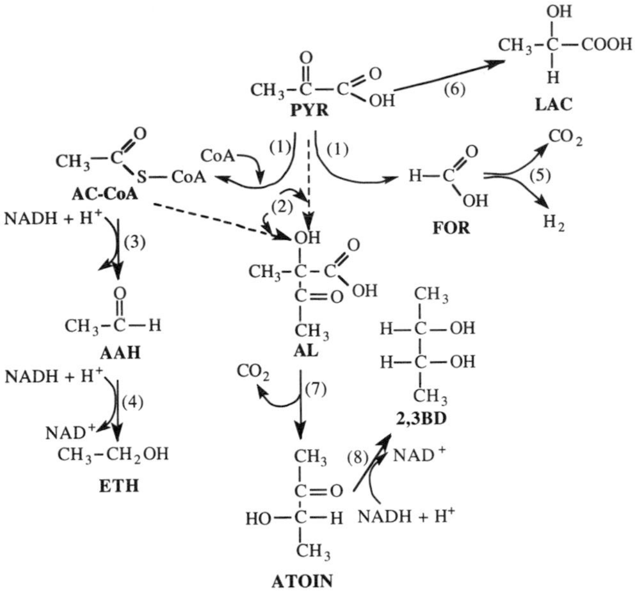

FIGURE 8.9 The butanediol fermentation. (1) Conversion of pyruvate (PYR), produced from glucose by Embden-Meyerhof reactions, to acetyl-CoA (AC-CoA) and formate (FOR) by PYR-FOR lyase. (3,4) Reduction of AC-CoA to ethanol (ETH) with the intermediate formation of acetaldehyde (AAH) by the actions of acetaldehyde (3) and ethanol (4) dehydrogenases. (5) Conversion of FOR to CO_2 and H_2 by the FOR hydrogen lyase system. (6) Reduction of PYR to lactate (LAC) with lactic dehydrogenase. (2,7,8) Butanediol synthesis from PYR and AC-CoA by the sequential action of α-acetolactate (AL) synthetase (2), AL decarboxylase (7), and 2,3-butanediol (2,3BD) dehydrogenase (8). Acetoin (ATOIN) is an intermediate in 2,3BD formation.

of aldehyde and alcohol dehydrogenases. Formate is converted to CO_2 and hydrogen by the joint operation of formic dehydrogenase and hydrogenase.

The butanediol fermentation

The butanediol fermentation is a second example of heterofermentation. Among the enterobacteria, *Enterobacter* and *Serratia* display this sequence, shown in Figure 8.9. As in the mixed acid fermentation, pyruvate is formed from glucose by the EM pathway. Lactate, ethanol formate, carbon dioxide, and hydrogen are formed by mechanisms identical to those of the mixed acid fermentation. A portion of the pyruvate is converted to acetolactate by acetolactate synthetase. The acetolactate is further converted to acetoin and carbon dioxide by acetoin

decarboxylase, and the acetoin is then reduced to butanediol with butanediol dehydrogenase.

The butanol-acetone fermentation

Clostridium acetobutylicum produces acetone and butanol from glucose fermentation as shown in Figure 8.10. Glucose is converted to pyruvate by the EM pathway reactions. Pyruvate, by the actions of pyruvate-ferredoxin oxioreductase and hydrogenase, yields acetyl CoA and hydrogen. The acetyl CoA is converted to acetoacetyl CoA by acetyl CoA:acetyltransferase and a portion of the acetoacetyl CoA is then converted to free acetoacetate by acetoacetyl CoA:acetate coenzyme A transferase. Decarboxylation of the free acetoacetate yields acetone and carbon dioxide. The remaining portion of acetoacetyl CoA is converted to β-hydroxybutyric acid by L(+)-β-hydroxybutyryl CoA dehydrogenase. The β-hydroxybutyryl CoA is converted to crotonyl CoA by L-3-hydroxyacyl-CoA hydrolase and the crotonyl CoA is reduced to butyryl CoA by butyryl CoA dehydrogenase. The final steps in butanol formation include reduction of butyryl CoA to butylaldehyde by butylaldehyde dehydrogenase and the formation of butanol from butylaldehyde by butanol dehydrogenase.

The ethanol-acetate fermentation

Ethanol and acetate are normally considered products of fermentation. In *Clostridium kluyveri,* these materials are used as fermentation substrates for the production of butyrate caproate and hydrogen. The sequence is shown in Figure 8.11. Ethanol is converted to acetaldehyde by the action of alcohol dehydrogenase. The acetaldehyde is further oxidized to acetyl CoA by aldehyde dehydrogenase. The former enzyme is NAD^+-mediated, whereas the latter requires $NADP^+$. Further metabolism of NADPH and NADH produces hydrogen. A portion of the acetyl CoA becomes free acetate and ATP, with the intermediate formation of acetyl phosphate by the actions of phosphotransacetylase and acetyl kinase. Additional acetyl CoA condenses with another acetyl CoA molecule in a reaction mediated by thiolase to produce acetoacetyl CoA. This compound is converted by NADPH-dependent L(+)-β-hydroxybutyryl CoA dehydrogenase to L(+)-β-hydroxybutyryl CoA. By removal of water with crotonase, the L(+)-β-hydroxybutyryl CoA becomes crotonyl CoA, which by NADH-mediated reduction with butyryl CoA dehydrogenase, becomes

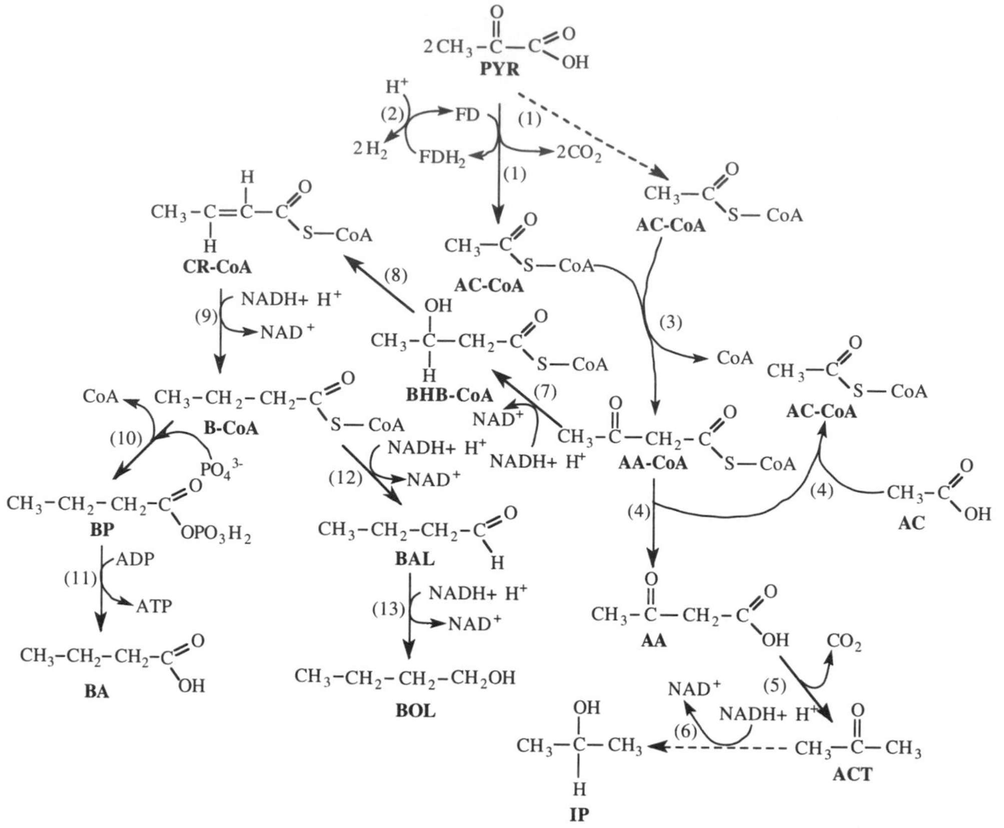

FIGURE 8.10 The butanol-acetone and butyric acid fermentations (1) Conversion of pyruvate (PYR), normally produced from Embden-Meyerhof reactions, to acetyl-CoA (AC-CoA), CO_2, and reduced ferredoxin (FDH_2) by the action of pyruvate-ferredoxin oxidoreductase. (2) Formation of H_2 and oxidized ferredoxin (FD) by hydrogenase. (3) Synthesis of actetoacetyl-CoA (AA-CoA) from two molecules of AC-CoA by AC-CoA acetyl transferase. (4) Formation of free acetoacetate (AA) from AA-CoA and free acetate (AC) by AA-CoA-acetyl transferase. (5,6) Conversion of AA to acetone (ACT) or isopropanol (IP) by the actions of AA decarboxylase (5) and IP dehydrogenase (6). (7–9) Reduction of AA-CoA to butyryl CoA (B-CoA) with the intermediate formation of β-hydroxy butyryl-CoA (BHB-CoA) and crotonyl-CoA (CR-CoA) by the sequential action of L(+) BHB-CoA dehydrogenase (7), L-3-hydroxy-CoA hydrolase (8) and B-CoA dehydrogenase (9). (10,11) Conversion of B-CoA to free butyric acid (BA) and ATP with the intermediate formation of butyryl phosphate (BP) by phosphotransbutylase (10) and butyrate kinase (11). (12,13) Reduction of B-CoA to normal butanol (BOL) with the intermediate formation of butyl aldehyde (BAL) by the actions of BAL (12) and BOH (13) dehydrogenases. Different organisms possess different various collections of enzymes and form various collections of products. Physiological conditions also affect product formation. The figure is intended to indicate the spectrum of possibilities.

butyryl CoA. Reaction of acetate with butyryl CoA, mediated by CoA transferase, forms free butyric acid and regenerates acetyl CoA. Caproic acid is formed by the reaction of butyryl CoA with a molecule of acetyl CoA to yield 3-oxocaproyl CoA, an analogue of the acetoacetyl CoA formed during butyrate formation. The remaining reactions of caproate formation are also analogous to those involved in butyrate, that is, reduction of 3-ketocaproyl CoA to 3-hydroxycaproyl CoA, and the dehydration and reduction of 3-hydroxycaproyl CoA to produce caproyl CoA.

Amino acid fermentation

Although many microbes ferment carbohydrates, and some ferment materials that are end products of other fermentation products, such as ethanol, acetate, and lactate, fermentation is not restricted to carbohydrate substrates. L-glutamate, for example is fermented by *Clostridium tetanomorphum* and *Peptostreptococcus asaccharolyticus*. The schemes for its degradation by these organisms are shown in Figure 8.12. The sequence in *C. tetanomorphum* is somewhat less complex than that of *P. asaccharolyticus*.

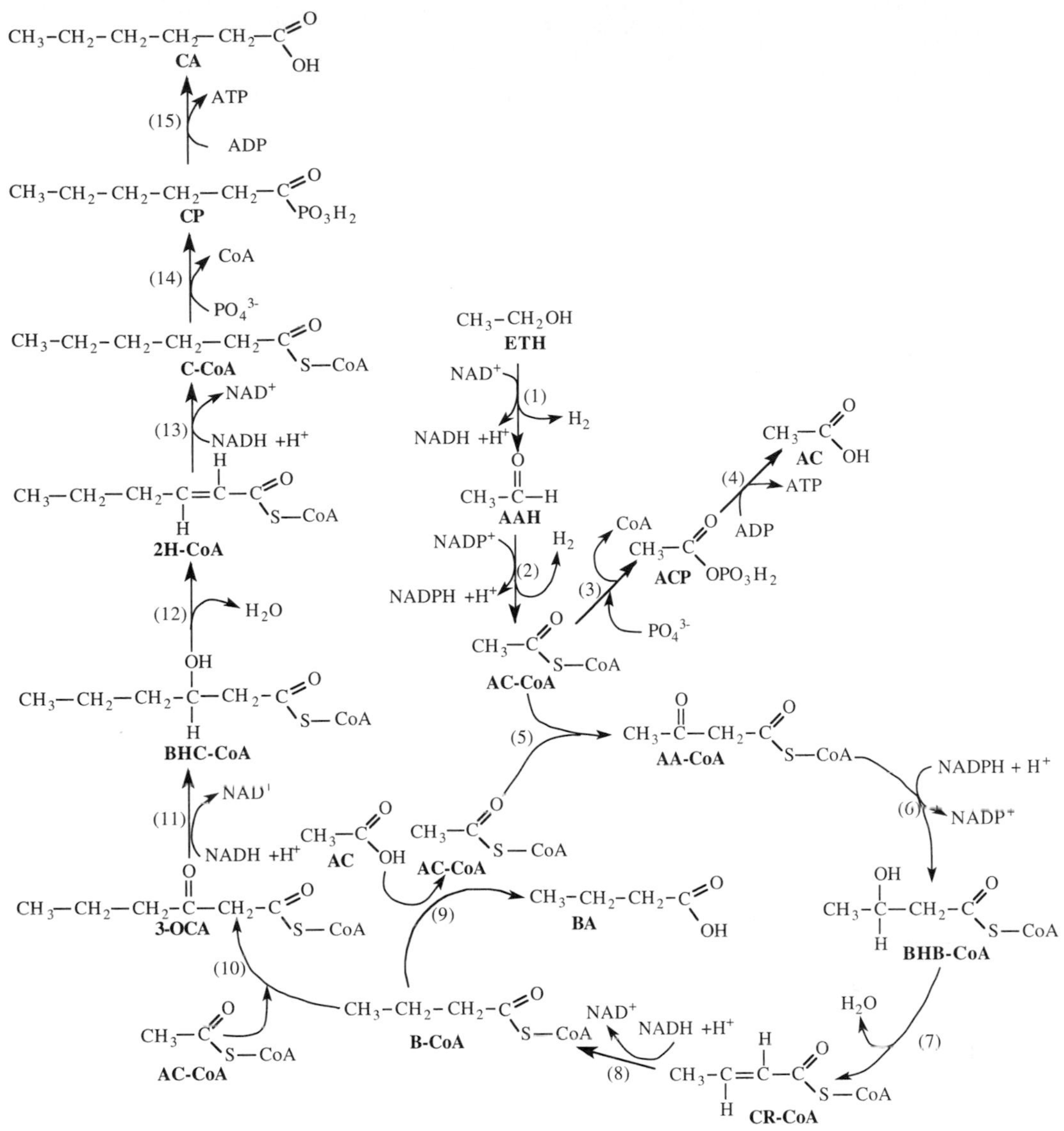

FIGURE 8.11 The ethanol-acetate fermentation of *Clostridium kluyveri*. (1,2) Conversion of ethanol (ETH) to acetyl-CoA (AC-CoA) with the intermediate formation of acetaldehyde (AAH) by the actions of ETH (1) and AC-CoA (2) dehydrogenases. H_2 is also formed by an incompletely characterized system during these reactions. (3,4) Formation of free acetate (AC) and ATP with the intermediate of acetyl phosphate (ACP) from a portion of the AC-CoA by the actions of phosphotransacetylase (3) and acetyl kinase (4). (5) Synthesis of acetoacetyl-CoA (AA-CoA) from AC-CoA by AA-CoA-acetyltransferase. (6–8) Conversion of AA-CoA to butryl-CoA (B-CoA), with the intermediate formation of β-hydroxybutyryl CoA (BHB-CoA) and crotonyl-CoA (CR-CoA) by the actions of BHB-CoA dehydrogenase (6), CR-CoA hydrolase (7), and B-CoA dehydrogenase (8). (9) Conversion of a portion of the B-CoA to free butyrate (BA), and formation of AC-CoA from AC, by CoA transferase. (10–15) Formation of caproic (CA) acid from B-CoA by reactions analogous to those involved in B-CoA synthesis from AA-CoA, except that the intermediates, 3-oxycaproate (3-OCA), β-hydroxycaproyl-CoA (BHC-CoA), 2-hexenyl-CoA (2H-CoA), caproyl-CoA (C-CoA), caproyl phosphate (CP), and caproic acid (CA), are two carbons longer. Differences are found in the specificities of the various dehydrogenases for NAD^+ and $NADP^+$. In the ETH-AC fermentation of *C. kluyveri*, BHB-CoA dehydrogenase specifically requires $NADP^+$.

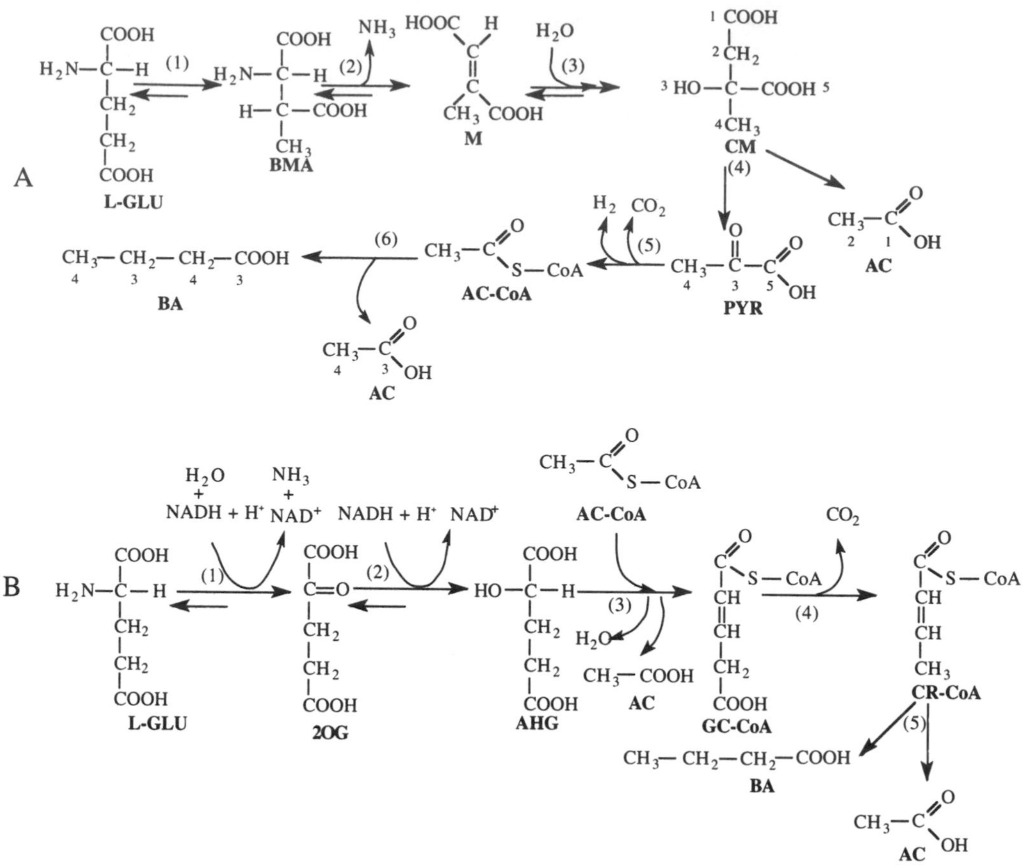

FIGURE 8.12 Alternate schemes for L-glutamate (L-GLU) degradation by (A) *Clostridium tetanomorphum* and (B) *Peptostreptococcus asaccharolyticus*. The enzyme numbers refer to the numbers *within* a particular sequence. For *C. tetanomorphum*: (1) Conversion of L-GLU to L-threo-β-methyl aspartate (BMA) by L-GLU mutase. (2) Deamination of BMA to form mesaconate (M) by β-methylaspartase. (3) Hydration of M to form citramalate (CM) by CM dehydratase. (4) Conversion of CM to pyruvate (PYR) and acetate (AC) by CM lyase. (5,6) Conversion of PYR to AC and butyrate (BA) via acetyl CoA by central metabolic pathways. The small numbers adjacent to the carbons of AC and BA are the carbons of CM that appear in these products. For *P. asaccharolyticus*: (1) Conversion of L-GLU to 2-oxoglutarate (2OG) by L-GLU dehydrogenase. (2) Reduction of 2OG to α-hydroxyglutarate (AHG) by AHG dehydrogenase. (3) Formation of glutaconyl-CoA (GC-CoA) by dehydration and coacylation of AHG with AC-CoA as CoA donor. This process involves two enzymes, a dehydratase and CoA transferase. Free AC is also formed. (4) Decarboxylation of GC-CoA to yield crotonyl-CoA (CR-CoA) by a Na^+-dependent GC-CoA decarboxylase. (5) Dismutation of CR-CoA to form butyrate (BA) and acetate (AC).

For *C. tetanomorphum*, L-glutamate is isomerized to L-threo-β methylaspartate by the action of L-glutamate mutase. The L-threo-β methylaspartate is deaminated to mesaconitate by β-methylaspartase. Hydration of mesaconitate by citramalate dehydratase, operating in the reverse direction, produces citramalate. Citramalate lyase then produces equimolar amounts of acetate and pyruvate. Further metabolism of pyruvate via central pathways yields additional acetate as well as butyrate, carbon dioxide, and hydrogen.

L-glutamate fermentation by *P. asaccharolyticus* occurs by a different set of reactions. Initially, L-glutamate is oxidatively deaminated by glutamate dehydrogenase, an NAD-mediated reaction, to form 2-oxoglutaric acid (2OG). The 2OG is then reduced to α-hydroxyglutaric acid by alpha hydroxyglutarate dehydrogenase. The combined action of a dehydratase and a CoA transferase on α-hydroxyglutaric acid produces glutaconyl CoA, which by decarboxylation with glutaconyl CoA decarboxylase, a sodium-requiring enzyme, becomes crotonyl CoA. Finally, crotonyl CoA dismutase produces acetate and butyrate in the ratio of 2:1, acetate to butyrate.

Perspectives

Fermentation is a process that involves many substrates and the formation of a diversity of end products. At first glance, it may seem difficult to develop a generalized understanding of the process. Further reflection reveals a substantial degree of commonality in the midst of diversity. All fermentations involve anaerobic oxidation and reduction processes in which organic compounds serve as both oxidants and reductants. In addition, although a diversity of apparently different products are formed, formation of many of them involves, for the most part, a common series of reactions and intermediates. Mechanistic differences in the products formed often reflect the functioning of only a few enzymes. For example, the difference between the production of butyric acid or butanol reflects whether butyryl CoA is oxidized and used for the production of ATP and butyrate or whether it serves as a vehicle for reducing power disposal and is thereby reduced to butanol. Similarly, whether acetyl CoA becomes ethanol or acetate is determined, once again, by whether acetyl CoA serves as a substrate for reducing power disposal or is oxidized for energy. In both cases, it is an acyl CoA molecule of some sort that is involved. Finally, in most cases, whether or not the fermentation substrate is carbohydrate in nature, pyruvate serves as the intermediate that allows fermentation product formation.

Study of fermentations

Determination of the nature of microbial fermentations is a difficult and challenging task. Fermentations may be studied in a variety of ways. It is useful, and often essential, to identify the characteristic enzymes that may be present, although it is often possible to determine the qualitative and quantitative importance of the sequence, or sequences, that may be operative in a particular organism by means other than enzyme analysis. In addition, the use of radioactivity is of particular

value. Its use has led to understanding of phenomena as diverse as the involvement of carbon dioxide in fermentation end product formation, the extent to which carbon in substrates is recovered in end products, the nature and quantitative importance of multiple pathways in the same organism, and the manner in which fermentation end products are formed.

Position labelling

Study of the metabolic fate of position-labelled substrates is a powerful tool in elucidation of fermentation mechanisms. We will illustrate its usefulness by consideration of "model" fermentations. Consider the following problem, shown graphically in Figure 8.13.

1. An organism is grown in the presence of 1-14 C-labelled glucose to form carbon dioxide, lactic acid, acetic acid, glycerol, and isopropanol.
2. Analysis of the fermentation end products shows that the carbon dioxide is not labelled, but the lactic acid is labelled in the methyl carbon, as is the acetic acid.
3. The isopropyl alcohol is labelled in carbons 1 and 3.
4. The glycerol is labelled in one, but not both, of the exterior carbons.
5. From this information, what is the probable degradative mechanism in the organism?

Consideration of the answer to the problem reveals the following:

1. Lactic acid is formed from direct reduction of pyruvate. Therefore, whatever carbon corresponds to the methyl carbon of pyruvate will correspond to the methyl carbon of lactate.
2. Since acetate arises from cleavage of pyruvic acid between carbons 1 and 2 of the pyruvate molecule, the methyl carbon of pyruvate will become the methyl carbon of acetate.
3. Carbon dioxide results from the carbon 1 of pyruvate.
4. Isopropyl alcohol results from the head-to-tail condensation of acetyl CoA, followed by decarboxylation. Therefore, carbons 1 and 3 of the isopropyl alcohol should arise from the methyl carbon of acetate.

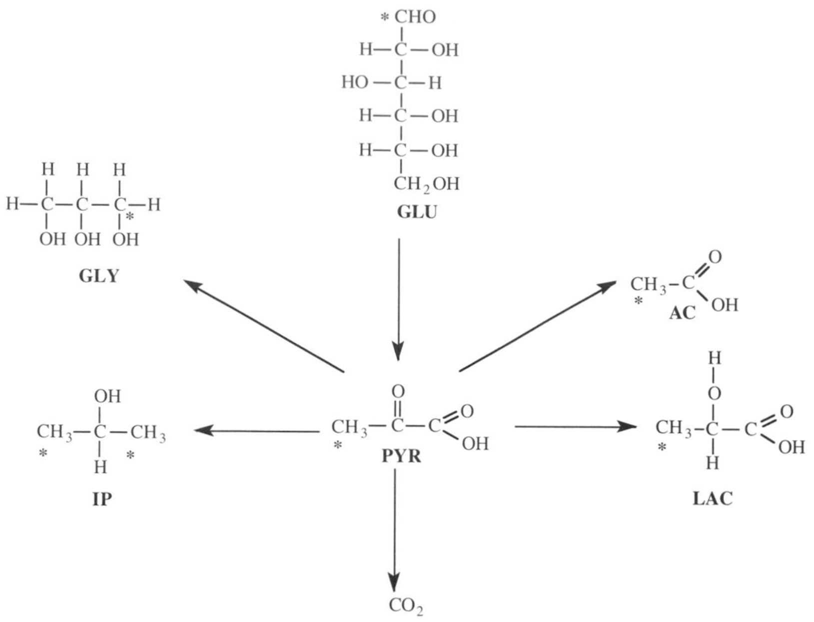

FIGURE 8.13 The label distribution obtained in acetate (AC), lactate (LAC), glycerol (GLY), isopropanol (IP), and carbon dioxide from fermentation of 1-labelled glucose (GLU) by the Embden-Meyerhof pathway. The carbons of the fermentation products that are labelled are indicated by the bold asteriks (*). For reasons discussed in the text, the labelling pattern displayed in the figure is indicative of the *sole* operation of the Embden-Meyerhof pathway.

5. Since it is labelled from carbon 1, glycerol must have been formed from metabolism of the top half of the molecule.

All of these results are compatible with the operation of the Embden-Meyerhof pathway:

1. The carbon dioxide is unlabelled, as it should be. If any of the pathways involving hexose monophosphate had been operative, carbon dioxide would have been labelled from carbon 1.

2. The labelling in acetate, lactate, and isopropanol are consistent with the EM pathway because carbon 1 of glucose degraded by the EM pathway is the methyl carbon of pyruvate. If the ED or any of the hexose monophosphate (HMP) pathways (see Chapter 7) were present, the label in acetate, lactate, and isopropanol would be different. In the case of the HMP pathways, acetate, lactate, isopropanol, *and* glycerol would all be unlabelled from 1-labelled glucose. They are not.

3. Lactate, from the ED pathway, would have been labelled in carbon 1, the carboxyl carbon, not carbon 3 as is the case here, and would be expected from the EM pathway.

4. Acetate would be unlabelled from 1-C-labelled glucose fermented by the ED pathway. The fact that it is labelled and in the methyl carbon, would be expected from the EM pathway.

5. Glycerol would not have been labelled from carbon 1 of glucose by a HMP type of pathway and also would not have been labelled from the ED pathway. All of the results are compatible with operation of the EM pathway, and *only* with it!

Consider another problem, which is shown graphically in Figure 8.14. An organism is grown in the presence of 1-14C-labelled glucose to produce the following end products:

1. Lactate labelled in the methyl carbon and also in the carboxyl carbon.

2. Acetate labelled in the methyl carbon.

3. Radioactive carbon dioxide.

4. Glycerol labelled in one of the exterior carbons.

5. Butyric acid labelled in carbons 2 and 4.

6. Acetoin (*not* acetone) labelled in carbons 1 and 4. What is the probable nature of the degradation pathway (or *pathways*) present?

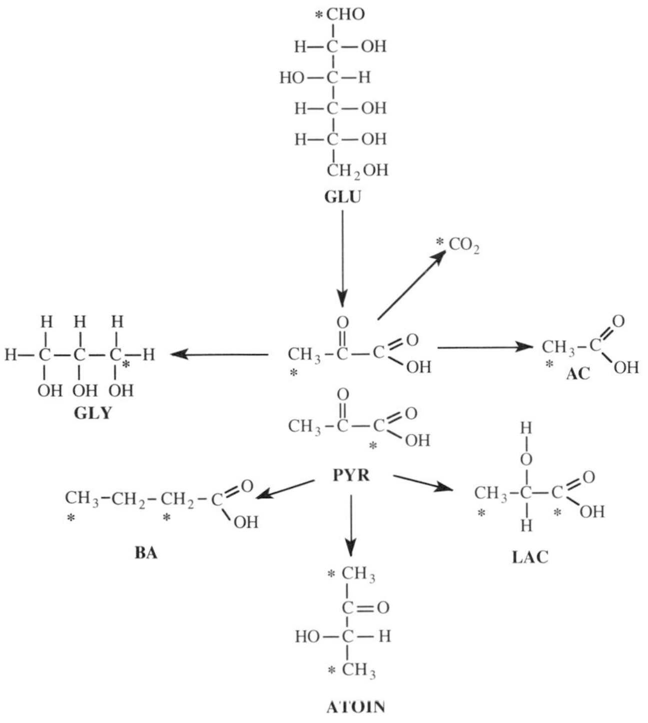

FIGURE 8.14 A model fermentation yielding acetate (AC), lactate (LAC), butyrate (BA), acetoin (ATOIN), glycerol (GLY), and carbon dioxide from 1-labelled glucose (GLU) by the joint operation of the Embden-Meyerhof and Entner-Doudoroff pathways. The labelled carbons are indicated by the bold asterisks (*). The diagram is not intended to indicate, in all cases, that the indicated carbons are labelled in the same molecule of a particular product but only that the operation of a metabolic pathway, or pathways, produces the labelling pattern observed. As discussed in the text, the pattern depicted in the figure is explained by the joint operation of the Embden-Meyerhof and Entner-Doudoroff pathways and is not possible if either of the pathways operates separately.

This situation is a bit more complicated than the first situation but we can, after reflection, establish the following:

1. The radioactivity in carbon dioxide could not arise from operation of the EM pathway. Therefore, if it is present, at least one additional pathway is also present.

2. The label in lactate also cannot be explained solely by the presence of the EM pathway since it could give rise to methyl labelled lactate, but it could *not* give rise to the lactate label in the carboxyl carbon.

3. The label pattern in the acetate, the butyrate, and the acetoin is consistent with operation of the EM pathway since from carbon 1, one would obtain methyl labelled acetate and lactate, as is found. Also, since butyrate arises from a head-to-tail condensation of units equivalent to acetate, one would expect, as is found, to get label in carbons 2 and 4 of

butyrate formed from acetate, which, in turn, arose from the EM pathway. Since acetoin arises from a head-to-head condensation of acetate units or functionally similar 2-carbon units, it is reasonable to find label in carbons 1 and 4 of the acetoin if it was formed from acetate arising from the EM pathway operating on 1-14C-labelled glucose.

4. The glycerol labelling is consistent with operation of the EM pathway. It could not arise from 1-14C-labelled glucose degraded by any of the other major, recognized pathways.

5. The lactate labelling and the label in carbon dioxide are the puzzles. However, the label found in the latter could be easily explained if the Entner-Doudoroff (ED) sequence were present, in addition to the EM pathway! If such were the case, one would obtain, by the ED pathway, label from carbon 1 in carbon dioxide, by pyruvate cleavage, and none of the

2-carbon products, or products formed from them, would be labelled. However, we have already explained the label pattern in the latter by the presence of the EM pathway. Also, by the ED sequence, label from 1-14C-labelled glucose would appear in the carboxyl carbon of lactate, as is observed. To detect label in various carbons of a fermentation end product, it is not necessary that both labelled carbons appear in the same molecule!

6. Label in glycerol from 1-14C-labelled glucose is impossible from the ED pathway, or any of the other commonly known major carbohydrate degradation sequences other than the EM pathway, because glycerol is formed from the reduction of dihydroxyacetone phosphate, which is not formed by the HMP pathways from 1-labelled glucose or by the ED pathway, since the ED pathway gives rise to pyruvate directly from the top half of the glucose molecule. However, we have already explained the glycerol label by postulating the existence of the EM pathway.

The labelling pattern in this situation is explainable by postulating the joint operation of the EM and ED pathways. This situation would account for all of the labelling observed because:

1. The label in glycerol, acetate, and all of the products formed from it could be attributed to the EM pathway.

2. The label in carbon dioxide could be explained by the ED pathway from pyruvate cleavage.

3. The label in lactate could not arise from 1-14C glucose by any other of the HMP pathways, except the ED pathway, because the major product from 1-14C-labelled glucose, produced by all other HMP pathways is radioactive carbon dioxide. The apparent finding of label in both the methyl carbon and the carboxyl carbon of lactate results from the fact that while some of the lactate arises from the EM pathway, giving methyl-labelled lactate from carbon 1, some of it also arises from the ED pathway, which gives rise, from 1-labelled glucose, to carboxyl labelled lactate. The two types of lactate mix with each other, giving rise to label detection in both carbons, although the actual label occurs in different molecules of the same product.

Carbon balances

The problems in the previous section illustrate the complexities of attempting to decipher the nature of fermentation sequences and the usefulness of position-labelled substrates in seeking to understand them. In addition to understanding of the nature of fermentation sequences and the manner in which end products are formed, it is necessary to determine the extent to which supplied carbon is recovered both in end products and in cellular material. In this situation, it is appropriate to use uniformly labelled radioactive substrates and to assess the extent to which the radioactivity in substrates is recovered in products. If the radioactivity in cells is also analyzed, the extent to which the fermentation substrate is used for biosynthesis may also be assessed. Measurement of the specific activities of the end products and the cells, on a carbon basis, provides information concerning the extent to which both cells and fermentation end products may arise from the substrate supplied or from other *nonradioactive* materials.

Although radioactivity may be used to determine carbon balances, it is often possible to assess carbon balance by the use of nonradioactive methods. These are less hazardous than radioactivity, but the answers regarding carbon obtained by radioactive and nonradioactive methods should agree. Provided that proper analytical methods are available, it is often preferable to initially study carbon recovery without the use of radiochemicals and to confirm the nonradioactive results with radioactivity as appropriate. The ideas and strategies concerning carbon balance calculations are illustrated by the following problem:

An organism metabolizes 100 millimoles of glucose to produce 50 millimoles of lactic acid, 200 millimoles of acetic acid, 25 millimoles of methane, and 25 millimoles of carbon dioxide. What is the carbon recovery?

To answer this question, we determine the millimoles of carbon that must be recovered. To do this, we multiply the millimoles of substrate by the number of carbons per molecule of substrate. In the present problem, we multiply 100 millimoles of the substrate, glucose, by the number of carbons per molecule, in this case, 6. The result of this calculation tells us that we must account for 6×100, or 600, millimoles of carbon. Next we consider each end product, on the same molar basis as the substrate, and calculate the number of molar units of carbon that are attributable to it. For both substrate and end products, we are free to select any convenient molar unit, provided the same is used for both substrate and product. In the present case, all of the information is in millimolar units. In our problem, 50 millimoles of lactate are produced from 100 millimoles of glucose. Since lactate is a 3-carbon molecule, the millimoles of carbon attributable to 50 millimoles of lactic acid are 150 (50 × 3). A quarter of the total end product carbon is attributable to lactate. Consideration of the acetate reveals that 400 millimoles of carbon are attributable to it (200 millimoles of acetate multiplied by the carbons per molecule, 2). Two-thirds of the end product carbon is attributable to acetate. Thus far, we have accounted for 550 millimoles of carbon—150 in the form of lactate and 400 as acetate. Since methane and carbon dioxide are 1-carbon molecules, and 25 millimoles of each of them is produced from 100 millimoles of glucose, the carbon recovery is complete. We have recovered 600 millimoles of substrate carbon in the end products—150 millimoles from lactate, 400 millimoles from acetate, and 25 millimoles each from carbon dioxide and methane. In this case, the materials for cell synthesis must arise from a different source. Any fermentation could be analyzed in an identical manner. It is rare that complete carbon recovery is obtained and, contrary to our example, it is often the case that small amounts (only 1 to 2 percent) of substrate carbon are used for cell material.

Oxidation-reduction balances

Carbon balances provide information regarding conservation of matter. Oxidation-reduction balances assess the extent to which energy is conserved. Since fermentations involve anaerobic oxidations and reductions, in any fermentation, the extent of oxidation must equal the extent of reduction. How do we determine whether this is the case? To answer this question, we compare the oxidation state and millimolar amounts of fermentation products with those of the substrates from which they are formed. The ideas underlying the calculations are best explained from consideration of a "model" carbohydrate fermentation. Carbohydrates, for example, glucose, derive their name from the fact that they may be considered to be composed of "hydrated" carbons. On the average,

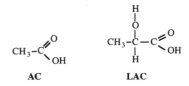

FIGURE 8.15 Lactate (LAC) and acetate (AC) as "model" fermentation end products. Although the two acids differ in the number of carbons per molecule, their formation as products of carbohydrate fermentation produces the same result electrically. Irrespective of differences in the number of carbons per molecule, the ratio of hydrogens to oxygen atoms in the two molecules is identical, 2:1, the same ratio that occurs in the carbohydrates whose fermentation produces them.

each carbon is associated with 2 hydrogen atoms and 1 oxygen atom. The ratio of hydrogens to oxygens is the same as that of water. Any product that has hydrogens to oxygens in the ratio of 2:1 has the same oxidation state as does a carbohydrate. Consider the cases of lactic acid and acetic acids, whose formulae are shown in Figure 8.15. Lactic acid contains three carbons, three oxygens, and six hydrogens, while acetic acid contains two carbons, four hydrogens, and two oxygens. Irrespective of differences in the numbers of hydrogens, oxygens, and carbons between the two compounds, however, the ratio of hydrogens to oxygens is the same, 2:1, and the ratio of hydrogens to oxygens in both lactate and acetate is the same as that in carbohydrate. Thus, production of *either* acetate or lactate as a sole product of carbohydrate fermentation is an electrically equivalent process. In the case of lactic acid, the methyl carbon is substantially more reduced than carbohydrate since the methyl carbon is associated with three hydrogens and no oxygens. However, the carboxyl carbon is substantially more oxidized than a carbohydrate. The degree of "excess" reduction of the methyl carbon of lactate is compensated for by the "excess" oxidation of the carboxyl carbon. The middle carbon has hydrogens and oxygens in the ratio of 2:1. Thus, the *entire compound* has an oxidation state identical to the material from which it was produced. Acetate, although it is only a 2-carbon compound, also has a ratio of hydrogens to oxygen of 2:1. From the viewpoint of electrical balance, the number of carbons that comprise a fermentation product is unimportant. What is critical is the extent to which

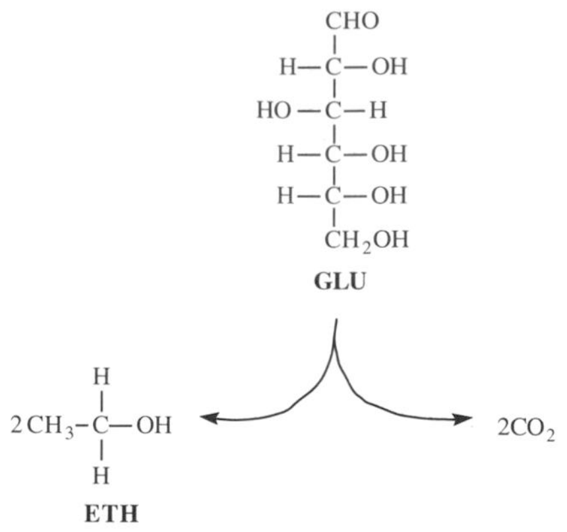

FIGURE 8.16 The general nature of the ethanol fermentation of *Saccharomyces cereviseae*. A molecule of glucose (GLU) is converted to two molecules of CO_2 and ethanol (ETH). Electrical equivalence results from the fact that ETH is reduced, relative to GLU, to the same extent that CO_2 is oxidized. The joint production of CO_2 and ETH thus results in an overall ratio of hydrogens to oxygen of 2:1. In this situation, electrical equivalence is obtained by formation of two products, while acetate or lactate production, as a sole product may lead to the same electrical result.

the material is oxidized or reduced in comparison to the material from which it was produced. The homofermentation of carbohydrate to lactate or acetate is such that the production of a single material allows electrical balance.

Formation of a single end product during fermentation is the exception rather than the rule. In most cases, two or more products are formed, some of which are more oxidized than their substrate and some of which are more reduced. It thus becomes necessary to study the relationship between the production of oxidized and reduced products. Initially, we must calculate the number of molar units of product formed per unit of substrate catabolized. Next we assign an *oxidation number* to each product. To do this, we consider that each oxygen atom in a molecule is equivalent to two hydrogens. For the purposes of accounting, we assign a value of +1 to each oxygen and a value of −1 to every *two* hydrogens. In this way, by counting the numbers of hydrogens and oxygens in a molecule we can determine whether or not the product is more or less oxidized or reduced than the material from which it arises, and the extent to which this is the case. This procedure allows us to compare the production of oxidized and reduced products and to determine whether or not their production is equal. To illustrate these ideas, consider the

ethanol fermentation of *S. cerevisiae* (Figure 8.16). This organism produces two molecules of ethanol and two molecules of carbon dioxide per molecule of glucose fermented. Since ethanol contains six hydrogens and one oxygen atom per molecule, it has an oxidation number of −2 (+1[O] −3[H] = −2). Carbon dioxide is devoid of hydrogens and contains two oxygens, so it has an oxidation number of +2 (1[O]+1[O] = +2). Production of an ethanol molecule is thus equivalent to production of a molecule of carbon dioxide. In the ethanol fermentation of *S. cerevisiae*, ethanol and carbon dioxide are formed in equal molar amounts, so the fermentation is electrically balanced. An organism that produced equal molar amounts of carbon dioxide and methane would similarly be balanced. In this situation, although only 1-carbon compounds would be produced, electrical balance would occur because a molecule of methane requires the input of reducing power to the same extent that electrons are lost in the production, from glucose, of a molecule or carbon dioxide. All fermentations can be analyzed for electrical balance in this manner, and should be.

General rules for oxidation-reduction balance calculation

The general rules for oxidation-reduction balance calculations are the following:

1. Assign an oxidation number to each of the products of the fermentation by comparing the number of oxygen and hydrogen atoms in the molecule, assuming that each oxygen is numerically equal to +1 and two hydrogens together are equal to −1.

2. Determine the number of molecules of products formed, on a molar basis, per unit of substrate catabolized.

3. Multiply the oxidation number of each product by the number of molar units per unit of substrate catabolized that the product represents.

4. Sum up the resulting positive and negative numbers and divide the *total* of the positive numbers by the total of the negative numbers. In a perfect balance, one should obtain a number numerically equal to 1.0. The sign (+ or −) of the calculated number is not important, only its absolute value. A number less than 1.0 indicates that oxidized products are missing, and a number greater than 1.0 indicates that reduced products are missing.

5. Compare the ratio of C2 products and C1 products. If the products result from pyruvate cleavage only, the ratio should be 1.0.

Precautions in fermentation balance studies

Study of fermentation balances is a complicated and challenging process. It is particularly fruitful, and often essential, to study both carbon and oxidation reduction balances. The joint study of carbon and oxidation-reduction balances is often useful in eliminating confusion. For example, the identification of the missing products in carbon balances may help to balance an otherwise unbalanced oxidation-reduction balance. Particular sources of confusion regarding fermentations include: 1) the possibility of simultaneous fermentation of more than one substrate; 2) the loss of volatile products; 3) carbon dioxide fixation for growth or end product production; 4) inaccuracies in product measurement; and 5) incomplete product recovery. With all of the difficulties outlined above, critical studies of fermentation are most instructional.

Relationships between fermentation, energy production, and protoplasm

Comparatively speaking, fermentation allows substantially less energy release than that which occurs when carbonaceous compounds are oxidized to carbon dioxide, the most oxidized form of carbon. Thus fermentation is an inherently less energy-efficient process than is aerobic metabolism. This concept is illustrated by the fact that 38 ATP molecules are formed per glucose molecule degraded aerobically via respiration, while a typical fermentation yields no more than 2 ATP molecules per glucose molecule catabolized. The relationship between catabolism, energy production, and the growth of fermentative organisms has been a subject of keen interest over an extended period of time. It has become of increased interest with the recognition that certain fermentative organisms obtain energy by both substrate level and electron transport-mediated processes, whereas others are restricted to substrate level processes only. In the early 1950's, Monod and coworkers studied the relationship between energy supply and growth of *Saccharomyces cerevisiae*. During

anaerobic growth of the organism in medium under conditions of glucose-limitation, a linear relationship was found between the amount of cell dry weight and glucose concentration. Calculations revealed that between 20 and 22 g of cells were formed per glucose mole degraded. Since the organism was known to contain the EM pathway and therefore derived a net of 2 ATP molecules per glucose mole catabolized, Monod suggested that the yield of cells per ATP mole was 10 to 11 grams. The work of Monod and colleagues was extended by the studies of Elsden and Bauchop, who studied the relationship between growth and energy supply for *Streptococcus faecalis* and found that during growth with glucose, 20 to 22 grams of cells were obtained per glucose mole degraded, a finding consistent with the fact that *S. faecalis* also contains the EM pathway. Additional work showed that when the organism was grown in the presence of glucose and arginine, 30 g of cells were formed per glucose mole degraded. Since arginine catabolism was known to provide a single ATP molecule per molecule of arginine metabolized, the results were consistent with the suggestion that arginine catabolism provided additional energy and thus allowed more abundant growth of the organism. The work of Bauchop and Elsden supported the work of Monod and associates, indicating that the amount of growth of an organism was a function of its energy supply and that a constant amount of cells, 10 to 11 g, was obtained per mole of ATP. Furthermore, Elsden and Bauchop's work suggested that the relationship between energy and growth of an organism was independent of the nature of the energy source and reflected *only* the amount of ATP obtainable from it. The work from both laboratories suggested that the amount of growth obtained per ATP unit was constant.

Anomalous growth yields

Monod suggested a term, "molar growth yield," to describe the relationship between the substrate used and the cell yield obtained. It was suggested that if one knew the metabolic sequence used for degradation of a particular substrate, and the number of ATPs per substrate molecule degraded, one could predict the cell yield, in dry weight, per mole of substrate degraded. It was further reasoned that the nature of an unknown fermentation could be determined, to a great extent, by study of the relationship between substrate catabolism and growth.

Organisms that possessed the EM pathway, which yields 2 ATP moles per glucose mole degraded, would display 20–22 g of cells per mole, while organisms that possessed the Entner-Doudoroff sequence, which produces a single ATP mole per mole of glucose degraded, would obtain only 10–11 g.

In many cases, the expectations predicted by early workers have been confirmed, however, in some cases they have not. To a certain extent, anomalous results may reflect the presence of multiple pathways in the same organism, but sometimes this suggestion is inadequate to explain the relationship between growth and energy supply. A number of reports indicate that some organisms obtain extremely large cell yields per glucose molecule degraded by fermentative metabolism. *Selenomonas ruminantium, Prevotella (Bacteroides) ruminicola, Ruminobacter amylophilus,* and *Ruminococcus albus* have all been shown to obtain cell yields greater than or equal to 60 g per mole of hexose catabolized. All of these organisms contain the EM pathway and may obtain only 2 ATPs per hexose by substrate level processes. It is difficult to explain the manner in which these organisms obtain such elevated molar growth yields. A portion of the explanation may reflect the presence of electron transport systems. *Prevotella ruminicola* displays a functional ATP-generating, cytochrome-containing electron transport system. It appears that the elevated yield of the organism results from the generation of more energy per glucose by a *combination* of substrate level and electron transport mediated processes. The explanation may, on the other hand, reflect the ability of the organism to form more cells per unit of ATP than is the case with other organisms. Studies with fermentative organisms from a variety of sources suggest that this is possible.

Yield per ATP mole

If the chemical composition of a microbe is known, the amount of ATP required to synthesize a known weight of the organism may be calculated on a molar basis from consideration of established synthetic pathways, and the maximum possible amount of growth per ATP mole may be determined. Such calculations depend on reasonable assumptions about the weight of a "building block" for major cell polymers and information concerning the number of ATP molecules required to add a monomeric unit to a growing polymer. Calcula-

tions of this nature are useful for three reasons: 1) They allow us to compare observed cell yields with what is theoretically possible; 2) they allow us to determine the extent to which, in a given situation, biosynthetic energy is used for the addition of small or large monomers; and 3) they allow us to determine the extent to which polymerization involves expenditure of large amounts of energy for addition of few monomers or, relatively speaking, small amounts of energy for the addition of many monomers.

The amount of energy required for synthesis of a constant amount of cell dry weight reflects its chemical composition. The total cell dry weight obtained per ATP mole also reflects chemical composition. Consider the consequences of chemical composition for biosynthetic energy utilization by the organism depicted in Table 8.3.

Although the information in Table 8.3 is entirely hypothetical and would seldom, if ever, occur in nature, consideration of the "data" of Table 8.3 yields interesting information. The total energy requirement for 100 mg of cells is 1,872 micromoles of ATP. Of this total, 62.2% is used to synthesize nucleic acid of one type or another, while only 21.3% of the total is used to form protein. Approximately 13.5% is used for lipid synthesis and only about 3% of the total is used for carbohydrate synthesis. The most striking aspect of the information in Table 8.3, however, is the amount of cells that *could* be formed from the ATP required to synthesize 100 mg of cells. Since it takes 1,872 micromoles to synthesize 100 milligrams of cells, each micromole of ATP could potentially give rise to 100,000/1,872 = 53.4 micrograms per micromole! This yield is about 5 times the yield actually observed for many microbes! It appears, in this situation, that the cells are only about 20% efficient in the use of ATP for synthesis. Perhaps it is not too surprising that some organisms are more efficient than was originally thought possible. If the cells in this example obtained 20 g/mole, they would still be only about 40% efficient in energy utilization for biosynthesis. There are many situations in which biological systems are at least 40% efficient. The number of monomers required for synthesis of 100 mg of cells is a final interesting aspect of the data in Table 8.3. Only 478 µmoles of monomers are required. The percentage of nucleic acids and protein is high. However, relative to the other monomers, the monomers for nucleic acid are heavy. Although the cells are high in proteins and nucleic acids, the amount of total energy required

TABLE 8.3 The Micromoles of ATP Required to Synthesize 100 mg of Cells Composed of 40% DNA, 30% RNA, 20% Protein, 5% Lipid, and 5% Carbohydrate

COMPONENT	PERCENTAGE	MONOMER WEIGHT	MICROMOLES MONOMER (100 mg)	ATP PER MONOMER	MICROMOLES ATP (100 mg cells)
DNA	40	300	133	5	665
RNA	30	300	100	5	500
Protein	20	150	133	3	399
Lipid	5	60	84	3	252
Carbohydrate	5	180	28	2	56

Total ATP micromoles = 1,872
Total monomers per 100 mg = 478
Total potential micrograms per ATP micromole = 53.4

Monomer weights and ATPs per monomer are based on classical assumptions and known metabolic sequences. Slight differences in assumptions about monomer weights do not appreciably affect the calculations.

TABLE 8.4 The Micromoles of ATP Required to Synthesize 100 mg of Cells Composed of 5% DNA, 5% RNA, 20% Protein, 40% Lipid, and 30% Carbohydrate

COMPONENT	PERCENTAGE	MONOMER WEIGHT	MICROMOLES MONOMER (100 mg)	ATP PER MONOMER	MICROMOLES ATP (100 mg cells)
DNA	5	300	17	5	85
RNA	5	300	17	5	85
Protein	20	150	133	3	399
Lipid	40	60	667	3	2,001
Carbohydrate	30	180	167	2	334

Total ATP micromoles = 2,904
Total monomers = 1,001
Total potential micrograms per ATP micromole = 34.4

for their synthesis (3 or 5 ATPs per monomer) is to some extent compensated for by the fact that a nucleic acid monomer is substantially heavier than are the other monomers. Therefore, not as many nucleic acid monomers are required to constitute a large portion of the cell dry weight as is the case for macromolecules, whose monomers are relatively light.

It is helpful to compare the information in Table 8.3 with that of Table 8.4, which displays cells of substantially different chemical composition than the cells of Table 8.3.

The cells of Table 8.4 are composed, by weight, of 5% each of DNA and RNA, 20% protein, 40% lipid, and 30% carbohydrate. Consideration of Table 8.4 reveals a number of things. To begin with, synthesis of 100 mg of cells requires 2,904 micromoles of ATP, an approximate 50% increase in

the total energy required compared to the energy required for the cells described in Table 8.3. In addition, for the cells of Table 8.4, 68% of the total energy is used for lipid synthesis, while in Table 8.3, 62% of the total biosynthetic energy is used for nucleic acid formation. The amounts of energy required for protein synthesis are identical in both cases, but the fraction of the total biosynthetic energy used for protein synthesis in the cells of Table 8.4 is smaller than that of the cells in Table 8.3. The majority of the synthetic energy used for the cells in Table 8.4 is involved in the addition of relatively small building blocks to monomers, while in Table 8.3, most of the biosynthetic energy is used to incorporate relatively large monomers into nucleic acids. The energy cost per small monomer addition to the macromolecules of the cells in Table 8.4 is substantially less than the cost of adding nucleic acid precursors to the cells described in Table 8.3. This fact allows addition of a substantially greater number of monomers to the polymers of the cells described in Table 8.4 without a proportionate increase in energy. To a substantial extent, the increased energy for lipid synthesis for the cells in the cells of Table 8.4 is compensated for by decrease in the amount of energy required for nucleic acid formation. The amount of energy required to form the carbohydrates in the cells of Table 8.4 is approximately six times that of Table 8.3, but in both cases, the total energy expended for carbohydrate synthesis is a rather small fraction of the total biosynthetic energy. Finally, the difference in total energy required for synthesis of 100 mg of cells, 1,872 for the cells of Table 8.3 as opposed to 2,904 for those of Table 8.4, is associated with a difference in the maximum possible cell weight per molar unit of ATP—53.4 mg per millimole for the cells in Table 8.3 as opposed to "only" 34.4 for the cells in Table 8.4. Irrespective of the differences between the two yields, on an ATP molecule basis, the potential cell yield per ATP mole in either situation is substantially greater than that proposed by Monod. It is likely that there are, indeed, organisms that are more efficient in biosynthesis than those that obtain 10 to 11 g of cells per unit ATP, but further work is required to delineate the mechanisms by which anomalous growth yields may be achieved.

Summary

Compared to our original understanding, our knowledge of the nature and consequences of fermentation is substantial. Although a diversity of specific fermentations are known, and more are being discovered as we explore organisms from previously unstudied environments, a commonality emerges amidst the diversity of fermentations that we currently understand. Just as with conversion of carbonaceous substances to pyruvate, there is a commonality among organisms in the ways in which pyruvate is used for energy metabolism and for fermentation end product formation. Most of the unusual fermentations are variations on a common theme, that is, they involve a few unusual intermediates and enzymatic reactions in addition to those known to be involved in similar processes of other, more frequently encountered, organisms. To be fully understood, fermentation must be considered not only on the basis of its energetic consequences, but also with regard to the manner in which degradative processes provide intermediates that serve as beginning points for biosynthesis. In addition, we must consider the ways in which degradative and synthetic processes are coupled. It is only when both degradation and synthesis, and the connections between them, are considered that we will fully appreciate the nature and implications of fermentation.

Selected References

Barker, H. A. 1961. The fermentation of nitrogenous organic compounds. *In* I. C. Gunsalus and R. Y. Stanier (eds.) *The Bacteria*, vol. 2. pp. 151–207. Academic Press, Inc., New York.

Barker, H. A. 1981. Amino acid degradation by anaerobic bacteria. *Annual Review of Biochemistry*. **50:** 23–40.

Berry, D. F., A. J. Francis, and J. M. Bollag. 1987. Microbial metabolism of homocyclic and heterocyclic aromatic compounds under anaerobic conditions. *Microbiological Reviews*. **51:** 43–59.

Doelle, H. W. 1975. *Bacterial Metabolism.* Academic Press, Inc., New York.

Gottschalk, G. 1986. *Bacterial Metabolism,* 2nd edition. Springer-Verlag, New York, Berlin, Heidelberg, Tokyo.

Gottschalk, G., and J. R. Andreesen. 1979. Energy metabolism in anaerobes. *International Review of Biochemistry*. **21:** 5–115.

Moat, A. G. 1988. *Microbial Physiology*, 2nd edition. John Wiley and Sons, New York.

Morris, J. G. 1975. The physiology of obligate anaerobiosis. *Advances in Microbial Physiology*. **12:** 169–233.

Sleat, R., and J. P. Robinson. 1984. The bacteriology of anaerobic degradation of aromatic compounds. *Journal of Applied Bacteriology*. **57:** 381–394.

Thauer, R. K., K. Jungermann, and K. Decker. 1977. Energy conservation in chemotrophic anaerobic bacteria. *Bacteriological Reviews* **41:** 100–180.

Thauer, R. K., and J. G. Morris. 1984. Metabolism of chemotrophic anaerobes: Old views and new aspects. *In* D. P. Kelley and N. G. Carr (eds.) *The Microbe. 1984.* pp. 123–168. Society for General Microbiology Symposium, Warwick, England.

Wood, W. A. 1961. Fermentation of carbohydrates and related compounds. *In* I. C. Gunsalus and R. Y. Stanier (eds.) *The Bacteria*, vol. 2. pp. 59–149. Academic Press, Inc., New York.

CHAPTER 9

Energy Generation

The universal requirement for energy

Energy is essential for all living things. In one way or another, all microbial physiology may be regarded as being composed of processes that produce or utilize energy. The ability of a microbe to survive and reproduce is, in large measure, a function of its ability to obtain energy, store it, and use it to accomplish the critical processes of life. The bacteria display tremendous versatility and efficiency in their ability to form energy and to couple energy generation with the other processes of life. The bacteria can extract biologically useful energy from sources as diverse as radiant energy, inorganic ions, carbohydrates, and hydrocarbons. In consort with their ability to accomplish rapid transport and to carefully regulate their critical processes, it is largely the ability of microbes, particularly the bacteria, to extract and store energy that allows organisms so physically small to exert such an effect on the totality of living things. Within the microbes, we find examples of all of the ways in which biologically utilizable energy may be obtained. Study of energy generation in the bacteria and other microbes allows us a generalized understanding of energy generation in all living things.

Diversity and unity

Although microbes obtain energy from a variety of sources, the mechanisms by which energy is obtained are relatively few. Although differences are found in the details of particular processes, all living forms obtain energy in one of two ways, by substrate-level phosphorylation or by electron transport. Each of these processes has unique characteristics. Substrate-level phosphorylation occurs in the cytoplasmic area of the cell, requires a separate and distinct enzyme or enzymes for each mechanism, and produces a single, high-energy-bond molecule per molecule of energy-yielding substrate degraded. Electron transport phosphorylation, in contrast, occurs in membranes or membranous organelles, uses a common series of carrier molecules and enzymes, and normally produces more than one high-energy molecule per unit of electrons processed.

Mechanisms of substrate-level phosphorylation

Although microbes obtain energy at the substrate level from a variety of materials, they accomplish the phosphorylation process by only a few mechanisms. The ability of microbes to obtain energy

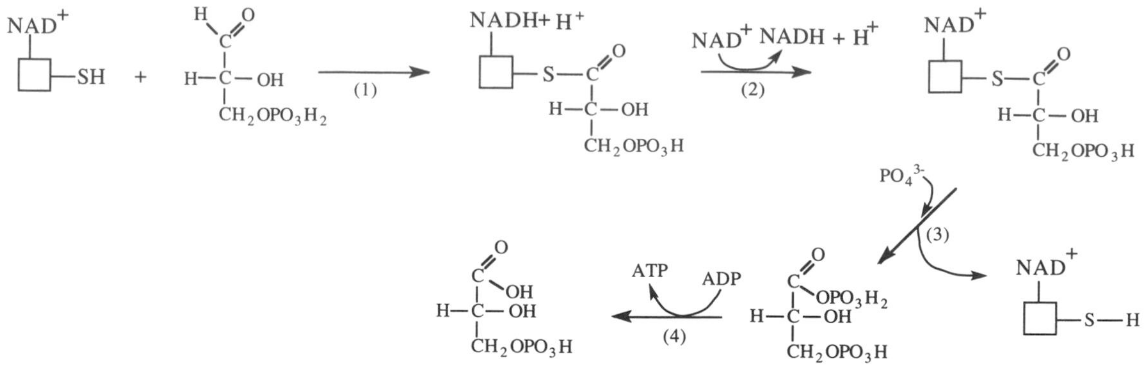

FIGURE 9.1 Generation of ATP at the substrate level by oxidation of glyceraldehyde-3-(triose) phosphate. (1) The substrate molecule interacts with a molecule of NAD^+-bound glyceraldehyde-3-phosphate dehydrogenase to form a complex between enzyme and substrate and to oxidize the carbonyl function of the substrate while simultaneously reducing the enzyme-bound NAD^+: (2) The enzyme-bound NADH is reoxidized by interaction with a free molecule of NAD^+, producing NADH free in solution and an oxidized NAD^+-enzyme-substrate complex. (3) Inorganic phosphate interacts with the tricomplex, producing NAD^+-bound enzyme and liberating a molecule of a 1,3-bisphosphoglyceric acid. (4) With the mediation of 3-phosphoglycerate kinase, 1,3-bisphosphoglyceric acid transfers its acid phosphate residue to ADP to yield ATP and 3-phosphoglyceric acid. Depending upon the organism, the NADH is reoxidized by fermentation end product formation or channeled into electron transport by NADH oxidoreductase.

from a wide variety of substances reflects the microbes' ability to channel diverse materials into central pathways that allow substrate-level phosphorylation to occur. The precise mechanisms by which high-energy phosphate bonds are formed differ from each other substantially, both mechanistically and in their physiological consequences. It is useful to compare the various substrate-level mechanisms for ATP formation.

The glyceraldehyde-3-phosphate dehydrogenase complex

Glyceraldehyde-3-phosphate is a critical intermediate of central metabolism and a major intermediate in most of the common carbohydrate degradation pathways. The oxidation of glyceraldehyde-3-phosphate (GA3P) to 3-phosphoglyceric acid (3PGA) is, therefore, a major mechanism for substrate-level phosphorylation. A summation of the processes involved in the conversion of GA3P to 3PGA, with the formation of reducing power and ATP, is shown in Figure 9.1. In the early studies of the action of GA3P dehydrogenase, the enzyme was regarded as a multifunctional, single enzyme. We now recognize that oxidation of GA3P is a multistep process involving the action of at least three proteins, GA3P dehydrogenase, 3PGA kinase, and NADH oxioreductase. Initially, a mole-

cule of GA3P reacts with a molecule of GA3P dehydrogenase bound to a molecule of NAD^+. This process produces an enzyme-bound molecule of NADH and a bound molecule of GA3P. Labelling studies have shown that the enzyme-bound NADH molecule is reoxidized by interaction with an NAD^+ molecule that is free in solution, producing a free molecule of NADH. In the next step of the process, inorganic phosphate is introduced into the glyceraldehyde molecule, resulting in the formation of 1,3-diphosphoglyceric acid (1,3DPG). This molecule dissociates from the enzyme, regenerating an NAD^+-enzyme complex that is available for further reaction. The final step in the process involves addition of a phosphate to ADP by the reversal of 3-PGA kinase, yielding ATP and 3PGA.

Pyruvate kinase

Pyruvate kinase is another central mechanism for substrate-level ATP formation. Its mechanism is much simpler than that for GA3P dehydrogenase. In the pyruvate kinase reaction operating in the reverse direction, phosphoenol pyruvate (PEP), normally formed from carbohydrate catabolism, donates a phosphate group to ADP, forming ATP and pyruvate. Normally, the direction of this reaction is substantially in the direction of pyruvate formation. Because this is the case, pyruvate kinase is a major regulatory enzyme in carbohydrate degradation sequences that yield pyruvate.

Acetyl kinase

Some of the systems for pyruvate metabolism, as well as other systems, produce acetyl phosphate, which by transfer of its acid phosphate to ADP, becomes free acetic acid and ATP. Although pyruvate cleavage is the major way in which acetyl phosphate is formed, acetyl phosphate may also arise by the action of phosphoketolase enzymes. Furthermore, energy generation by use of the acid phosphate bond is not restricted to *acetyl* phosphate. The acid phosphate bond is intrinsically energetic, so propionyl and butyryl phosphate, among others, may generate ATP by a mechanism analogous to that of acetyl phosphate.

The CoA bond as a source of energy for substrate-level phosphorylation

Coacylated forms of a variety of acids may be formed via central pathways of metabolism. Although acid CoA derivatives are often used for synthetic purposes, in certain situations, the CoA bond energy is used to generate high-energy phosphate bonds, normally in the form of ATP. The acid CoA is converted to an acid phosphate by a phosphotransacetylase reaction or an analogous enzyme, and the acid phosphate, by transfer to ADP, becomes ATP, with the liberation of free acid. This type of mechanism operates in processes as diverse as pyruvate cleavage and the generation of ATP from succinyl CoA in the TCA cycle. In some circumstances, a purine or pyrimidine nucleotide other than ADP, for example, GDP, may serve as the acceptor for high-energy bond formation, resulting in formation of a high-energy bond other than ATP.

Phosphoenol pyruvate carboxykinase

In certain anaerobes, oxaloacetate is intimately involved in fermentation end product formation and in the generation of acceptors for electron transport. Generation of energy by substrate-level phosphorylation is coupled with carbon dioxide fixation. In the phosphoenol pyruvate carboxykinase reaction, PEP reacts with carbon dioxide to form oxaloacetic acid and, at the same time, transfers its phosphate to ADP to form ATP.

Phosphoenol pyruvate carboxytransphosphorylase and pyruvate-phosphate orthodikinase

Some microbes have substrate mechanisms for generation of high-energy phosphate bonds other than a nucleotide phosphate bond. The propionibacteria contain two such enzymes, phosphoenol pyruvate carboxytransphosphorylase and pyruvate-phosphate orthodikinase. The action of both enzymes results in energy conservation as inorganic pyrophosphate (PP_i), which may be used to phosphorylate critical molecules such as fructose-6-phosphate and serine.

Arginine desmidase and ureidase

The majority of substrate-level phosphorylation systems involve enzymes normally associated with carbohydrate metabolism. However, some organisms generate ATP at the substrate level by degradation of noncarbohydrate substances such as arginine. Arginine is degraded to citrulline and ammonia by the action of arginine desmidase. The citrulline, in the presence of phosphoric acid and ADP is converted to ornithine, CO_2, NH_3, and ATP by action of the enzyme ureidase. The formation of ATP by substrate-level mechanisms other than the GA3P system are shown in Figure 9.2.

Electron transport

Throughout the microbial world, a number of mechanisms exist for substrate-level phosphorylation, but electron transport is an equally prominent mechanism for ATP formation. Electron transport was originally thought to be restricted to oxybiontic organisms, that is, organisms that use molecular oxygen in their metabolism. Thus, initially, organisms were regarded as being of one of two types, respiratory or fermentative. Respiratory organisms were regarded as those organisms that generated energy, obligatorily, by electron transport, with oxygen as the terminal electron acceptor. Fermentative organisms, by contrast, were those organisms that could not generate energy by electron transport, could not use molecular oxygen in energy metabolism, and were thus restricted to substrate-level mechanisms for energy generation. It is now known that such an understanding was simplistic and that many organisms that may use oxygen as a terminal electron acceptor, may also

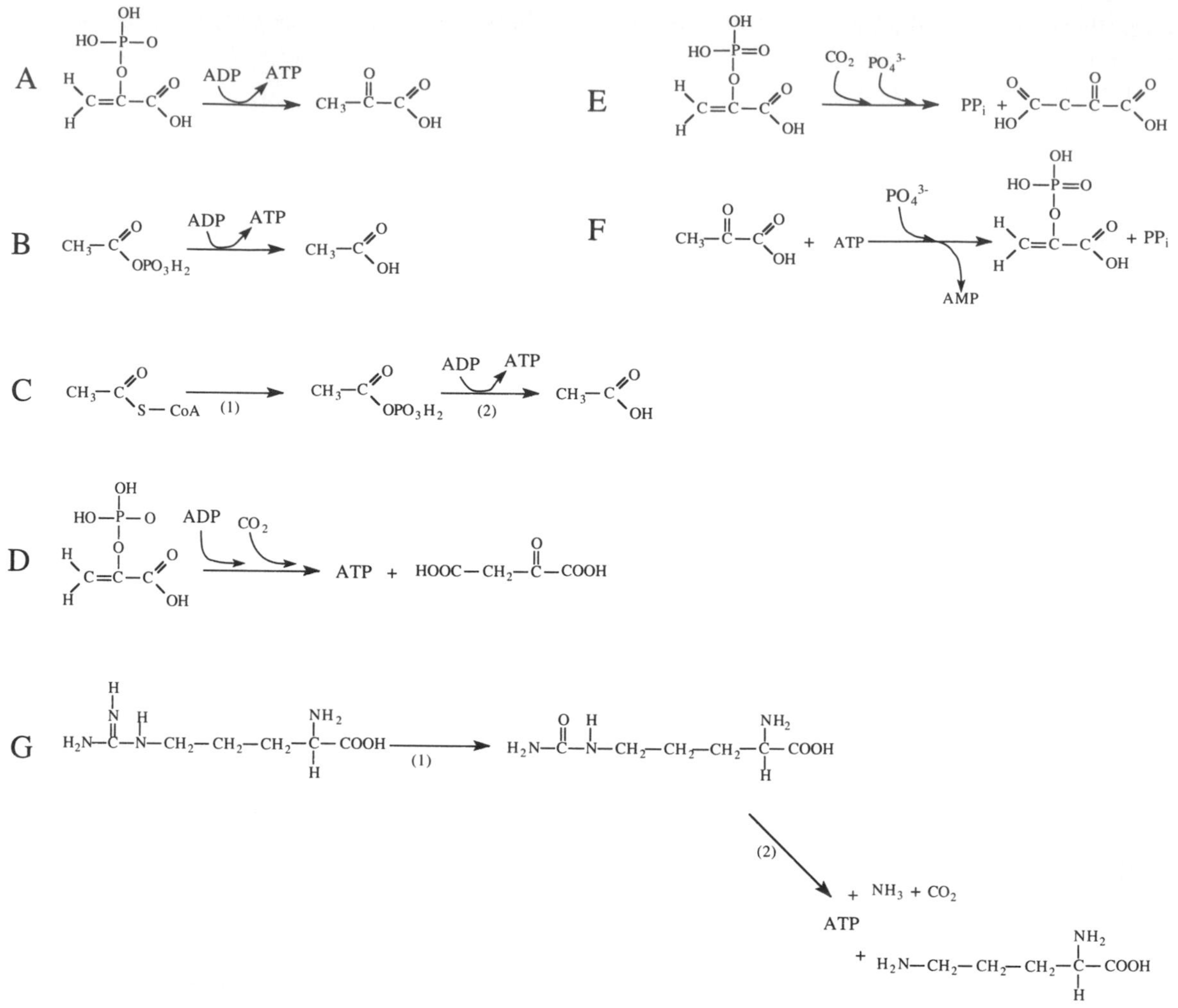

FIGURE 9.2 Various mechanisms of substrate-level phosphorylation. The different mechanisms are denoted by capital letters. Within each mechanism, individual reactions are identified by the numbers in parenthesis. (A) Phosphoenol pyruvate kinase. The enzyme transfers a phosphate from phosphoenol pyruvate (PEP) to ADP to form ATP and pyruvate. (B) Acetyl kinase. The enzyme transfers a phosphate group to ADP from an acid-phosphate bond to yield ATP and free acetic acid. This type of reaction, although shown for acetic acid, is a general mechanism and is widely applicable because the acid-phosphate bond is intrinsically energetic. Thus, other acid-phosphate bonds may be exploited for ATP formation. (C) Conversion of acetyl CoA to ATP with free acetic acid formation by the combined actions of phosphotransacetylase (1) and acetyl kinase (2). (D) ATP formation with CO_2 fixation by the action of PEP carboxykinase. (E) Pyrophosphate (PP_i) and oxaloacetate formation by the action of PEP carboxytransphosphorylase. (F) PP_i synthesis by the action of pyruvate-phosphate orthodikinase. (G) ATP formation from arginine by the combined actions of arginine desmidase (1), which converts arginine to citrulline, and ureidase (2), which dismutates citrulline to ornithine, ATP, NH_3, and CO_2.

use nitrate or sulfate as alternate electron acceptors when oxygen is not available. In addition, it is now recognized that a substantial number of obligately anaerobic microbes generate ATP by electron transport with the use of organic compounds as electron-accepting compounds. The obligatory distinction between respiratory and fermentative organisms is no longer tenable, if electron transport, or the lack of it, is the criterion. Electron transport is widely distributed among both respiratory and fermentative organisms.

Electron transport is an intrinsically complex subject both because of the diversity of ways in which it occurs and because of its intimate relationship to other cellular activities, particularly in procaryotic organisms. In these organisms, electron transport serves not only to allow formation of ATP but also to allow substance transport, maintenance, and motility, among other processes. Although electron transport is complex, there is unity in the midst of diversity. Although it is connected to the totality of cell physiology in various ways, depending on the system, electron transport, as a process, has common requirements.

The common processes of electron transport

Irrespective of the system in which it occurs, whether in a mitochondrion or in a bacterial membrane, electron transport requires a source of reducing power. The latter may be as diverse as radiant energy, electrons derived from oxidation of the ferrous ion, or electrons derived from organic compound oxidation. The electrons derived initially from either radiant energy or from chemical oxidation are used to reduce a primary carrier molecule, which is then reoxidized by reduction of a second, more electropositive carrier molecule. The second carrier is itself reoxidized by a third carrier. This process continues, by mechanisms that differ from system to system in detail, but that are identical in principle. Finally, electron transport is terminated in one of two ways. In cyclic electron transport, which is characteristic of anoxygenic, (i.e., nonoxygen-forming) photosynthetic procaryotes, the electron removed from the chlorophyll molecule that allows initiation of the electron transport process in these organisms is returned to the chlorophyll molecule from which it arose. The process is thus regarded as cyclic because there is no net electron loss from the system.

In chemotrophic electron transport, the terminal step of electron transport is different from the one just described. Oxidation of the last electron carrier in the chain is used to reduce a terminal electron acceptor, sometimes known as an "electron sink compound." Under these conditions, a net loss of electrons from the system occurs and a reduced compound is formed. In aerobic organisms, the compound is water, formed by the reduction of a half a molecule of molecular oxygen with two protons and two electrons derived from the transport process. In some anaerobic organisms, organic compounds, particularly fumarate and acrylate, may serve functions analogous to molecular oxygen. Their reduction produces reduced products, succinate and propionate, from relatively oxidized precursors, just as water is formed from the reduction of oxygen. In addition, some organisms may use nitrate or sulfate as terminal electron acceptors when oxygen is not available, leading to the formation of reduced forms of nitrogen and sulfur. Representative procaryotic electron transport systems are shown in Figure 9.3.

The carriers in electron transport systems

Consideration of Figure 9.3 reveals that the precise nature of the electron transport systems of procaryotic microbes differ substantially among organisms. However, the carrier molecules involved in most electron transport systems are both structurally and functionally similar. The similarity reflects the fact that the electron transport process obligatorily requires that the members of the electron transport chain be reversibly oxidizable and reducible. As is the case with catabolic metabolism, a diversity of processes is accomplished with similar or identical materials.

In phototrophic electron transport, the original light-absorbing substance is one or more chlorophyll molecules, in association with carotenoid pigments and proteins. The chlorophylls differ both structurally and in their light-absorbing properties. The carotenoids serve primarily, or exclusively, as protective agents against photochemical damage. In chemotrophic electron transport, the first molecule in the electron transport chain is typically NAD or NADP. Reduced forms of these substances are normally reoxidized by transfer of their electrons to a molecule of a flavin-containing compound, typically a flavoprotein. The flavoprotein is reoxidized by electron transfer to a quinone or napthoquinone compound. In some systems, this transfer is mediated by a nonheme-containing iron sulfur protein, most frequently a ferredoxin or a structurally similar compound. Transfer of electrons from reduced quinone compounds to a series of cytochromes is the next step in electron transport. A diversity of cytochromes is found in electron transport systems, but amidst this diversity, there is structural similarity. All of the cytochromes are heme proteins that differ in the nature of both their proteins and the side chains of their heme moieties. Representative structures of the major carrier molecules in electron transport are shown in Figure 9.4. Various electron transport systems may use slightly different chemical substances from those depicted in Figure 9.4, but the structures are physiologically equivalent.

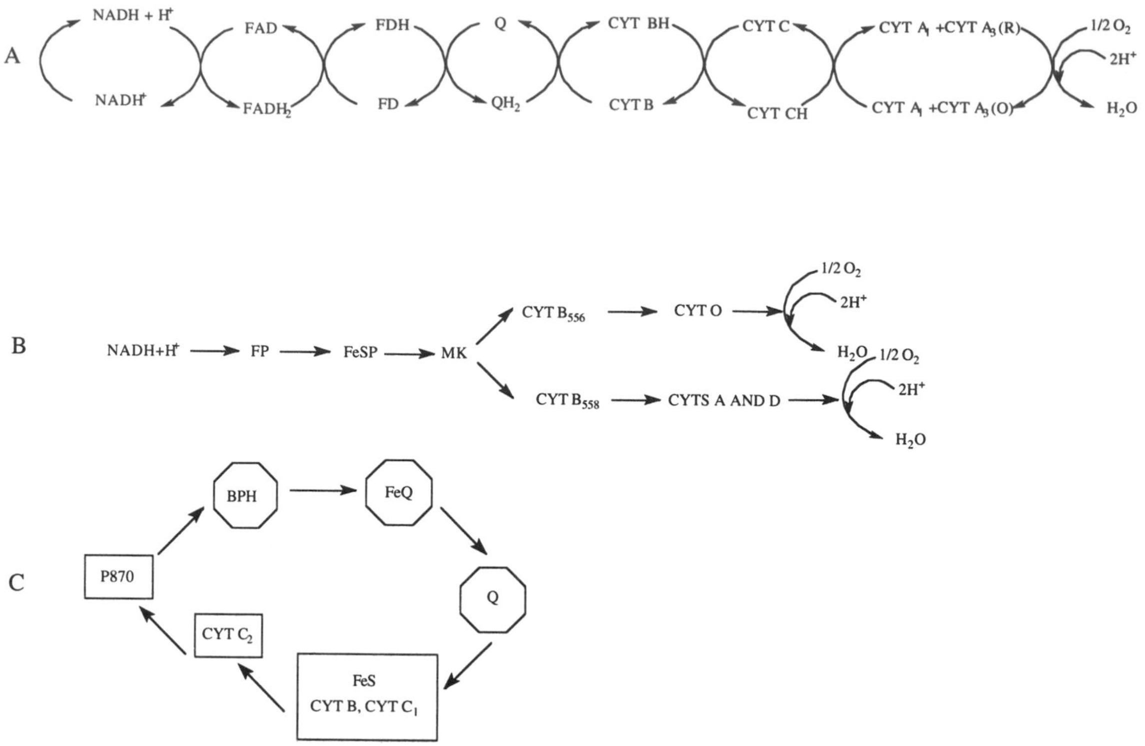

FIGURE 9.3 Various types of procaryotic electron transport systems. Throughout the diagram, a carrier with an H represents the reduced form of the material, whereas the same substance without the H represents the oxidized form. (A) A generalized system beginning with NADH and proceeding to flavin adenine dinucleotide (FAD), an iron-sulfur protein; ferredoxin (FD), a coenzyme containing a quinone (Q); cytochromes (CYT) of the *b, c,* and *a* types, and oxygen, with the formation of water as a reduced compound. (B) A branched linear system in which menaquinone (MK) replaces the quinone in system (A) and the branches result from the operation of two *b*-type cytochromes. In the example shown, one of the cytochromes absorbs maximally at 556 nm and the other absorbs maximally at 558 nm. The branches are also distinguished by additional cytochromes between cytochromes of the *b*-type and oxygen. The 556 nm branch uses cytochrome O, while the 558 nm branch uses a combination of cytochromes *a* and *d.* The 556 nm branch operates in an abundant oxygen environment, while the 558 nm branch operates during oxygen limitation. A system of this type is found in *E. coli,* (C). The cyclic system in the purple bacteria. Light energy extracts an electron from the pigment complex 870 and the electron in transferred to a series of carriers. Finally, the electron is returned to the P870 complex.

Oxidation-reduction potentials

The ability of electron transport carrier molecules to participate in electron transport reflects the ability of particular carriers, on the one hand, to be reduced and, on the other hand, to be oxidized. In a particular electron transport chain, each component of the chain occupies its position in the chain because it is reduced by the carrier that precedes it and is oxidized by the carrier that succeeds it. The relative tendency of particular substances to exist in oxidized and reduced forms can be measured under standard conditions and the substances may be ranked in order. Table 9.1 displays the standard oxidation potentials of the major carrier molecules in typical electron transport systems.

Examination of Table 9.1 reveals that the components of typical electron transport systems are arranged in an order so that each component oxidizes the carrier that precedes it. The amount of energy released during a particular oxidation step in electron transport can be calculated by consideration of the difference in standard oxidation poten-

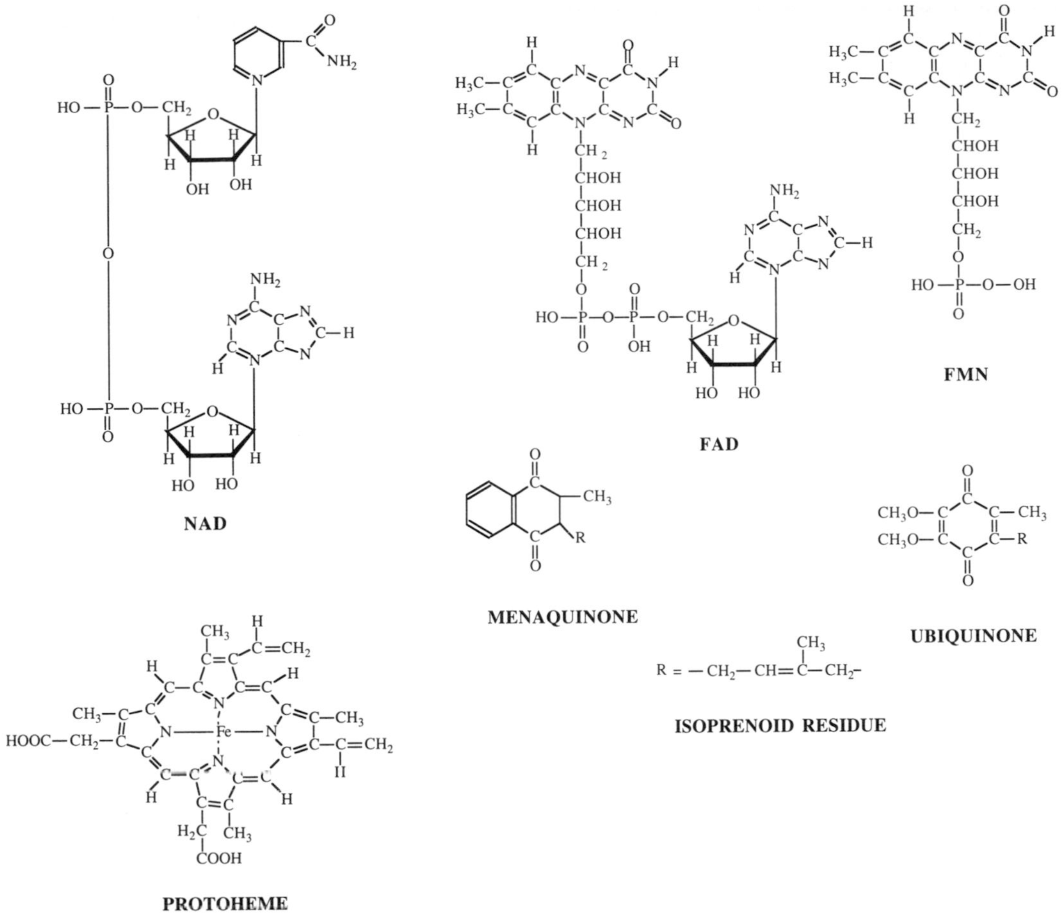

FIGURE 9.4 The structures of the major functional elements of electron transport systems. NADP differs from NAD (nicotinamide adenine dinucleotide) by the presence of a phosphate group attached to carbon 2 of the ribose ring structure. Hydrogen is stereospecifically donated and received at carbon 1 of the pyridine ring structure. For flavin compounds, oxidation and reduction occur at the unsaturated nitrogen atoms of the ring structures. The sites of oxidation for flavin mononucleotide (FMN) are identical to those for flavin adenine dinucleotide (FAD). Ubiquinones and menaquinones are structurally and functionally similar molecules, except that the menaquinone structure contains two rings, whereas the ubiquinone structure contains only one ring. The carbonyl functions of both molecules are the sites of oxidation and reduction. The various quinones and menaquinones differ from each other by the number of R units attached to the carbonyl-containing rings. Protoheme, or heme B, is a prototype for all of the prosthetic groups of the cytochromes. Hemes associated with cytochromes other than cytochromes of the b-type have side chains other than those shown for heme B, allowing differing interactions between the various hemes and their associated proteins, resulting in differing oxidation-reduction potentials.

tial between the system that is oxidized and the system that oxidizes. The equation that allows this calculation is shown below:

$$\Delta G^{0\prime} = -nF\Delta E_0{}^\prime$$

In the above equation, $G^{0\prime}$ = the standard free energy; Δ = change in; F = the Faraday (96,500 coulombs); n = the number of electrons trans-ported; and $E_0{}^\prime$ = the standard electron potential. Application of the above equation to substances whose standard oxidation potentials are known allows prediction of the direction of the reaction and the amount or energy released from it. Furthermore, the equation allows determination of whether or not the energy released from a particular oxidation is sufficient to allow formation

TABLE 9.1 The Standard Oxidation Potentials of Typical Electron Carriers

COMPONENT		E_0' (VOLTS)
NAD/NADH	$+2H^+ + 2e^-$	-0.32
FAD/FADH$_2$	$+2H^+ + 2e^-$	-0.22
FMN/FMNH$_2$	$+2H^+ + 2e^-$	-0.19
Fumarate/Succinate	$+2H^+ + 2e^-$	$+0.03$
Flavoproteins	$+2H^+ + 2e^-$	-0.45 to 0.00
FeS-proteins	$+2H^+ + 2e^-$	-0.40 to $+0.20$
Menaquinone	$+2H^+ + 2e^-$	-0.07
Ubiquinone	$+2H^+ + 2e^-$	$+0.11$
2 Cyt b_{ox}/ Cyt b_{red}		$+0.07$
2 Cyt c_{ox}/ Cyt c_{red}		$+0.25$
2 Cyt a_{ox}/ Cyt a_{red}		$+0.38$
$1/2O_2$/ H_2O	$+2H^+ + 2e^-$	$+0.82$
SO_4^{2-}/SO_3^{2-}	$+2H^+ + 2e^-$	$+0.20$
$2NO_3^-$/N_2O_4	$+2H^+ + 2e^-$	$+0.80$

NAD = Nicotinamide adenine dinucleotide; FAD = flavin adenine dinucleotide; FMN = flavin mononucleotide; FeS = nonheme iron proteins, most frequently ferredoxin. The range of redox potentials for FeS and flavoproteins reflects varying interactions between the protein and nonprotein moieties of these materials leading to differences in oxidation-reduction potentials. Cyt = cytochrome. The subscripts ox and red refer to the oxidized and reduced forms, respectively. A particular substance will be oxidized by a substance with a standard oxidation potential more positive than itself and reduced by a substance whose standard electron potential is more negative than its own.

of a molecule of ATP. Consider the following example: NADH is oxidized by FAD. How much energy is released in the process? Reference to Table 9.1 reveals that the standard oxidation potential (E_0') value for NAD is –0.32, whereas that of FAD is –0.22. Also, the processes involve 2 electrons, so the amount of energy released per mole of NADH oxidation is:

$$\Delta G^\circ = \frac{-2 \times 96,5000\,[-.22 - (-0.32)]}{4.18} = -4,617\ cal = -4.67\ Kcal$$

In this equation, the number 4.18 is the conversion factor between coulomb-volts (Joules) and calories. In the example, only 4.67 Kcal of energy are released per mole of NADH oxidized, an amount insufficient to produce an ATP molecule, since formation of an ATP molecule requires 7.5 to 8 Kcal per mole. Calculations of this nature, although useful, must be used with caution because the actual amount of energy released from oxidative processes may be different than that obtained under standard conditions.

Structural aspects of electron transport

Electron transport is an obligatorily membrane-associated process. If the components of electron transport systems were free in solution, the primary product of their reactions would be heat rather than the proton motive force (PMF). Although electron transport requires membrane-association, a variety of membranes and membranous organelles participate in the process. In most procaryotes, electron transport, both chemotrophic and phototrophic, typically occurs in membranes or membranous invaginations, although in a few photosynthetic procaryotes, members of the genus *Chlorobium,* and the cyanobacteria the process occurs in membranous organelles.

Photosynthetic electron transport in the cyanobacteria (Figure 9.5) is accomplished by intracytoplasmic membranous organelles known as thylakoids. Thylakoids are covered with phycobilisomes that contain the proteinaceous pigments phycocyanin, phycoerythrin, and allophycocyanin. These pigments, in consort with tetrapyrroles, absorb energy in the vicinity of 680 nm and collectively constitute photosystem two (PSII). Photosystem I is composed of a chlorophyll of the a type and proteins and absorbs at 700 nm and above. The existence of two photocenters in the cyanobacteria allows oxygenic photosynthesis.

The green bacteria of the genus *Chlorobium* also conduct photosynthetic electron transport in a specialized structure discrete from the cytoplasmic membrane. The structure is known as a chlorosome or vessicle and contains a chlorophyll of the c, d, or e type, plus carotenoids and proteins. Interaction occurs between the chlorosomes and the cytoplasmic membrane in the photosynthetic processes of these organisms. It appears that the primary function of chlorosomes is light energy absorption and that the majority of the events of photosynthetic electron transport are functions of the membrane. The details of the reactions involved in photosynthesis by members of the genus *Chlorobium* are shown in Figure 9.6.

Photosynthetic electron transport in the purple bacteria is a structurally less complicated process than in either the green sulfur bacteria or the cyanobacteria. In the purple bacteria, photosynthetic electron transport is accomplished by invaginations of the cytoplasmic membrane. Light energy is absorbed by *light-harvesting centers* that contain

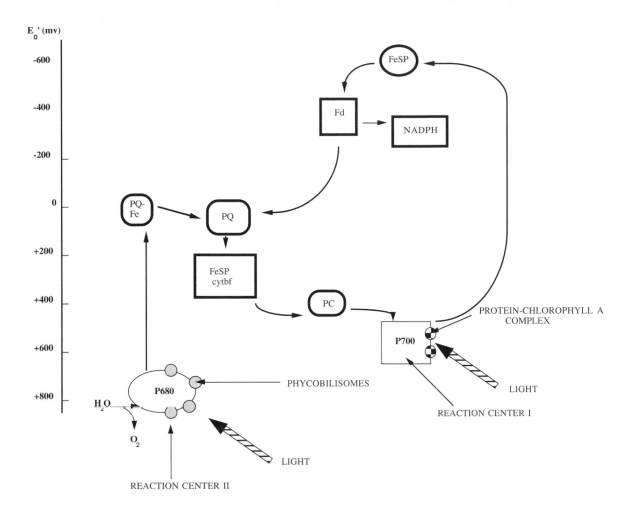

FIGURE 9.5 Electron transport in the cyanobacteria. The organisms possess two reaction centers that allow the trapping of radiant energy over a broad-energy spectrum. Phycobilisomes, which contain phycocyanin, phycoerythrin, allophycocyanin, and tetrapyrroles are the light-harvesting elements of reaction center II, while protein-chlorophyll *a* complexes serve the same function in reaction center I. Light energy trapped by either photosystem is used to eject an electron that reduces a low-oxidation potential carrier molecule, an iron-sulfur protein (FeSPN) for reaction center I, and a plastoquinone-iron complex (PQ Fe) for reaction center II. Electrons from FeSPN are transferred sequentially to ferredoxin (Fd), plastoquinone (PQ), an iron sulfur-cytochrome *b*-cytochrome *c* complex, plastocyanin (PC), and, finally, back to reaction center I. Electrons from PQ Fe are fed into PQ and allow net formation of NADPH with Fd as electron donor, providing reducing power for carbon dioxide reduction.

chlorophyll, carotenoids, and proteins. The two light-harvesting centers of the purple bacteria are normally known as P_{850} and P_{875} and contain the vast majority of the chlorophyll in the purple bacteria. The light-harvesting centers interact with reaction centers in the membrane, in which the bulk of the photosynthetic events occur. The reaction center is designated as P_{870}, denoting the absorption wavelength maximum of the chlorophyll associated with it.

Relatively speaking, the structural nature of the entity involved in photosynthesis in the purple bacteria is substantially simpler than are the entities involved in photosynthesis for both the green sulfur bacteria and the cyanobacteria, but photosynthesis in the purple bacteria involves complex structural organization of the components of both the light-harvesting centers and reaction centers within the membrane. The general nature of relationships of the components of the photosynthetic

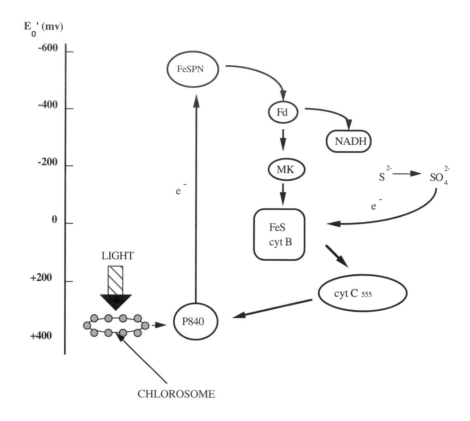

FIGURE 9.6 Electron transport in the green bacteria. Light energy is trapped in a chlorosome vessicle adjacent to the cytoplasmic membrane. The trapped energy is transferred to a reaction center, P840, in the cytoplasmic membrane. Electrons ejected from P840 are transferred sequentially to an iron-sulfur protein (FeSPN), ferredoxin (Fd), menaquinone (MK), an iron sulfur-cytochrome b complex (FeS cyt B), a cytochrome c (cyt C), and are finally returned to P840. Sulfide oxidation to sulfate, with donation of electrons to FeS-cyt B, provides reducing power for carbon dioxide reduction, with Fd serving as the immediate donor of electrons to NAD.

apparatus in the purple bacteria is shown in Figure 9.7. The light-harvesting centers are physically adjacent to the reaction center, which is physically oriented toward the exterior of the membrane and adjacent to an iron-quinone complex that is oriented toward the cell interior. A second component of the photosynthetic apparatus of the purple bacteria that contains ferredoxin or a ferredoxin-like protein, cytochrome b, and cytochrome c_1, is physically separated from both the reaction center and the iron-quinone complex. Quinones and cytochrome c_2, both of which are mobile within the membrane, mediate communication between the other components of the photosynthetic apparatus. The essential reactions of photosynthesis in the purple bacteria are shown in Figure 9.8.

In eucaryotes, both chemotrophic and phototrophic electron transport are accomplished by complex double-membraned organelles, the mito-

chondrion and the chloroplast. These two organelles have both structural and functional similarities. Both of them contain 70S ribosomes and circular DNA, properties that suggest that they are vestigial procaryotic cells that are no longer capable of existence as free-living entities. In addition to these structural similarities, for both electron transport-mediated and chloroplast-mediated phosphorylation, high-energy bond formation is a function of the inner membrane of the structure.

The mitochondrion

The structure of a mitochondrion is shown schematically in Figure 9.9. It contains an outer membrane and an invaginated inner membrane. The inner membrane contains the complexes involved in electron transport and membrane-bound ATP synthetase, the enzyme that forms ATP. The

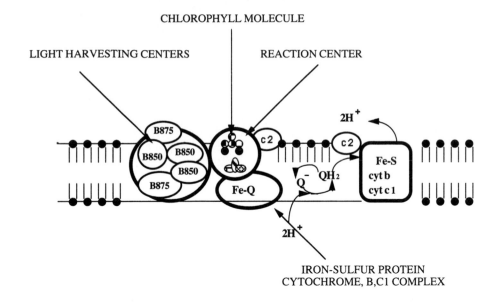

FIGURE 9.7 Structural relationships among the electron transport components of the purple photosynthetic bacteria.

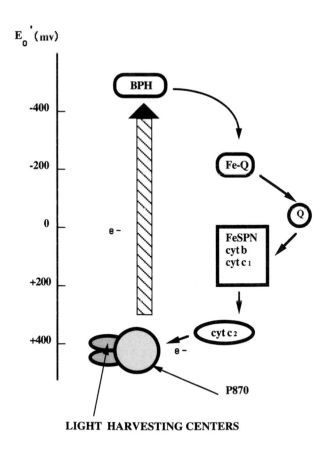

FIGURE 9.8 The reactions of electron transport in the purple bacteria.

mitochondrion contains three major complexes that accomplish electron transport and generation of the proton motive force (PMF) that allows generation of ATP. The nature and function of these complexes are shown in Figure 9.10. The NADH-coenzyme Q reductase complex contains NADH, coenzyme Q (CoQ), flavin mononucleotide (FMN), nonheme iron proteins (FeS), commonly known as ferredoxins, and NADH-Q reductase, a protein that contains 25 polypeptide chains. The components of this complex operate so that electrons from NADH, produced from chemotrophic metabolism, are transferred to a molecule of FMN, producing a reduced flavin mononucleotide molecule, FMNH2. Electrons from reduced FMNH2 are then transferred to FeS proteins, and from them to CoQ to produce a reduced form of CoQ ubiquinol.

The ubiquinol-cytochrome c reductase complex accomplishes the next events of mitochondrial electron transport. The complex contains FeS proteins, cytochrome b, the polypeptides of cytochrome c reductase, and cytochrome c_1. Mechanistically, the complex operates in the following way: Initially, ubiquinol transfers one of its electrons to an FeS protein. The FeS transfers its electron sequentially to cytochrome c_1 and from there to cytochrome c. Cytochrome c is a water-soluble cytochrome and removes the donated electron from the complex. However, the processes just described result in

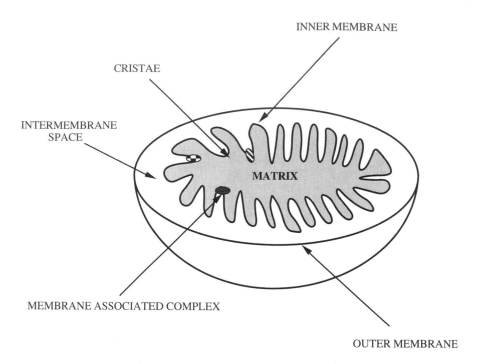

INNER MEMBRANE

CRISTAE

INTERMEMBRANE
SPACE

MATRIX

MEMBRANE ASSOCIATED COMPLEX

OUTER MEMBRANE

FIGURE 9.9 A schematic representation of a mitochondrion. The structure is composed of two membranes, an outer membrane and an inner membrane. Invaginations of the inner membrane are known as cristae. The mitochondrial matrix is the mitochondrial region interior to the inner membrane in which TCA cycle reactions occur. Electron transport and generation of the proton motive force required for ATP formation are accomplished by stationary complexes associated with the inner side of the inner membrane. The mitochondrial membranes are separated by a space. The outer membrane is generally permeable to cytosolic metabolites, but the inner membrane is impermeable to many products of cytosolic metabolism, particularly NADH. This apparent difficulty is overcome by formation, via cytosolic oxidation of NADH, of reduced compounds permeable to the mitochondrial inner membrane. Reoxidation of the transported compounds within the mitochondrion generates intramitochondrial NADH and other reduced substances that channel reducing power to the mitochondrial electron transport system and allow intramitochondrial ATP formation. Additional features of mitochondrial electron transport and its physiological consequences are shown in Figure 9.10 and discussed in the text.

production of a ubiquinol molecule with a single electron, that is, a semiquinone. Cytochrome b allows the semiquinone to be converted to a molecule of ubiquinol. The electron from a semiquinone molecule is accepted by one of the heme groups of cytochrome b, heme B562, which transfers its electron to a second molecule of semiquinone, converting the second molecule back to the ubiquinone state. The net effect of these processes is to allow a two electron-accepting compound, ubiquinol, to transfer a single electron to a cytochrome. The donation of a single electron from one semiquinone molecule to another, mediated by cytochrome b, allows a single electron transport to cytochrome b and the simultaneous formation of a molecule of oxidized quinone and a molecule of ubiquinol.

The cytochrome oxidase complex mediates the final steps of mitochondrial electron transport. The complex contains cytochrome oxidase composed of approximately eight subunits, some of which are formed within the mitochondrion and some that result from cytoplasmic protein synthesis. The cytochrome oxidase molecule has two heme groups, heme A and heme A3, distinguishable by differences in their electrical environments. In addition, the complex contains two copper ions, one of which (CuA) is associated with heme A and the other that is associated with heme A3. The reduced cytochrome c molecule initially transfers its electron to heme A that subsequently transfers its electron to the heme A3 moiety and, from there to oxygen.

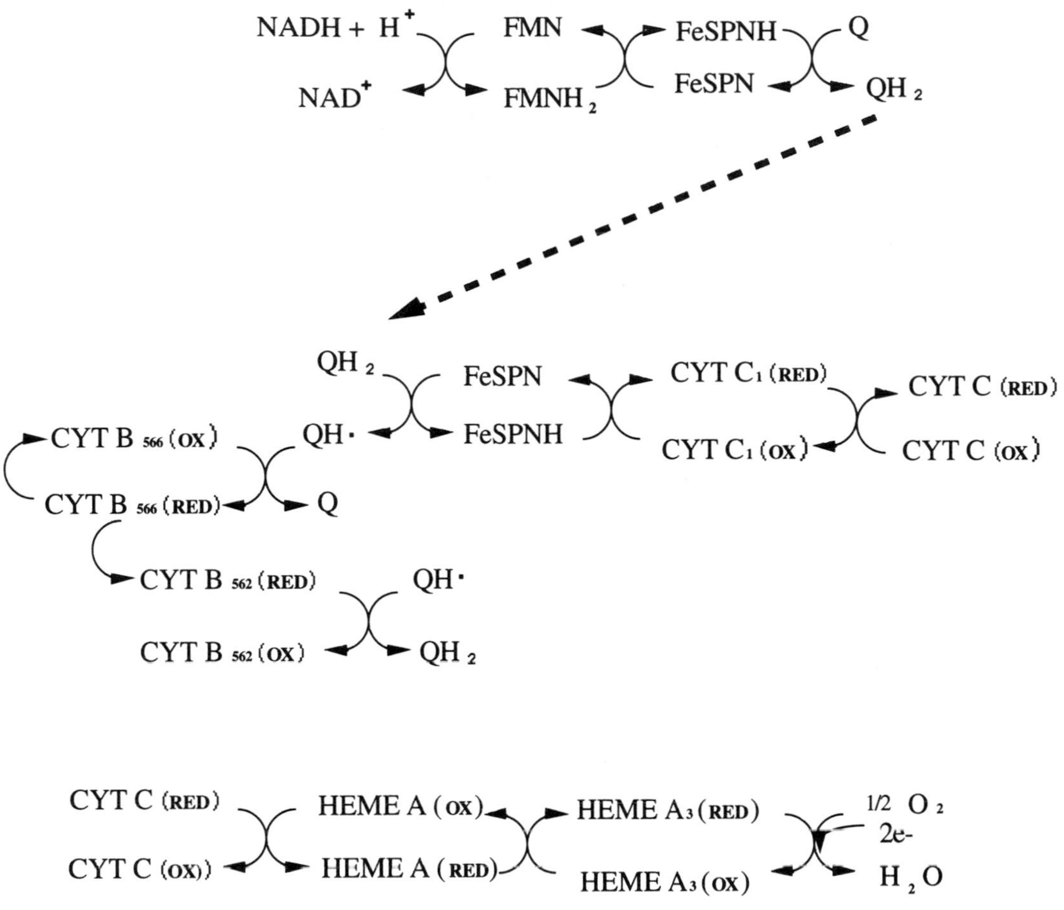

FIGURE 9.10 The electron transport reactions in the major functional complexes of the mitochondrion.

Succinate as a donor of electrons Although NADH is the most frequent donor of electrons to the mitochondrial electron transport system, there are situations in which electrons enter the system by succinate oxidation. This process typically results in formation of reduced flavoprotein. Entry of electrons from succinate oxidation into electron transport is mediated by the succinate-CoQ reductase complex. Reduced FAD donates its electrons of molecules to FeS proteins that, in turn, donate their electrons to CoQ.

Structural enigmas of the mitochondrion NADH is the most abundant and versatile donor of electrons to the mitochondrion, but paradoxically, the bulk of NADH arises in the cytoplasm as a result of glycolysis and similar processes. Furthermore,

NADH is most impermeable to the inner mitochondrial membrane, a difficulty overcome by use of the reducing power of NADH to produce substances that are permeable to the membrane. Two mechanisms by which this process occurs are known: the reduction of dihydroxyacetone phosphate to α-glycerol phosphate by an NADH-dependent cytoplasmic enzyme and the conversion of cytoplasmic oxaloacetate to malate by a second NADH-requiring enzyme. The latter mechanism is best characterized in mitochondria obtained from liver cells. In the "glycerol phosphate" shunt, glycerol phosphate produced from cytoplasmic reduction of dihydroxyacetone phosphate traverses the exterior mitochondrial membrane and is reoxidized within the membrane by an FAD-mediated enzyme. The reducing power thus generated as

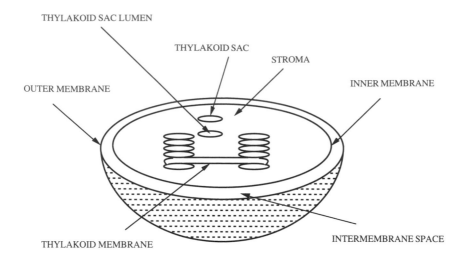

FIGURE 9.11 The general structure of a chloroplast. As is the case with the mitochondrion, the basic structure is composed of two membranes, an outer, highly permeable membrane and an inner, relatively impermeable membrane. Thylakoid sacs within the double-membraned structures are the site of photosynthetic and electron transport events. The components involved in photosynthetic processes are within the sac membranes. Stacked sacs, interconnected by the thylakoid membranes, are known as grana and contain most of the photosynthesis components. ATP synthetase and photosystem I are located in unstacked sacs, which have immediate access to the chloroplast stroma.

FADH is channelled into electron transport and the intramitochondrial dihydroxyacetone phosphate returns by diffusion to the cytoplasm. In the "malate shuttle," cytoplasmic oxaloacetate is reduced to malate, which crosses the exterior mitochondrial membrane and is reoxidized by an FAD-mediated enzyme to oxaloacetate. The latter is not permeable to the mitochondrial membrane and is converted by a transaminase reaction to aspartate, which returns to the cytoplasm.

Chloroplast-mediated electron transport

Electron transport in chloroplasts has substantial similarity to the process in mitochondria. The structure of a model chloroplast is shown in Figure 9.11. Like the mitochondrion, the chloroplast is a double-membraned structure. Within the inner chloroplast membrane are the stroma and the membranous thylakoid sacs. It is within the membranes of the thylakoid sacs that the major events of photosynthetic electron transport occur. The sacs are stacked into structures known as grana that are connected by the thylakoid membrane.

Chloroplast-mediated electron transport is accomplished by the cooperative action of two photosystems, photosystem I (PSI) and photosystem II

(PSII). The manner in which these two entities participate in photosynthesis is shown in Figure 9.12. PSI absorbs radiant energy maximally at 700 nm and is commonly designated P_{700}. It participates in generation of reducing power as NADPH and in the generation of a proton gradient. In addition, by a cyclic process, it may generate ATP. Photosystem II absorbs maximally at 680 nm. It is involved in the photolysis of water, a Mn^{2+}-dependent process that results in liberation of O_2. The apparent function of Mn is to allow charge accumulation without formation of toxic materials. In addition, PSII donates electrons to PSI and facilitates formation of a proton gradient by the cytochrome bf complex.

The cooperative action of PSII and PSI occurs in the following way. Light energy absorbed by PSII gives rise to activated pigment 680 ($P_{680}*$), which donates electrons successively to pheophytin (PPh), plastoquinone-binding proteins QA and QB, and reduced plastoquinone (QH_2). With mediation of the cytochrome bf complex, QH_2 reduces plastocyanin (PC) with the concomitant release of protons. The cytochrome bf complex contains two cytochromes, cytochrome f and cytochrome b, and a FeS protein. Reduction of plastocyanin is mediated by the FeS protein that facilitates electron transport from a two electron

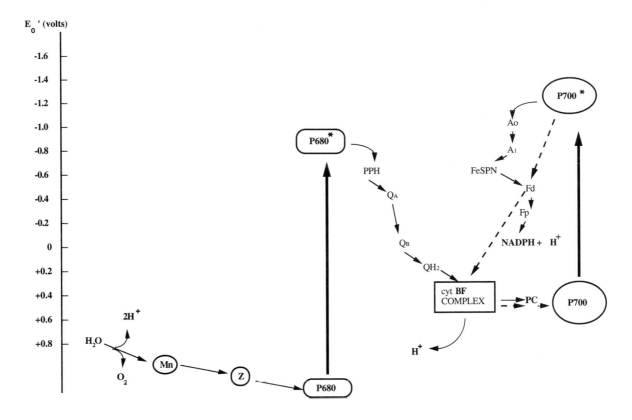

FIGURE 9.12 The cooperative participation of photosystems I and II in electron transport in eucaryotic photosynthesis. See text for commentary.

compound, plastoquinone, to a one electron compound. The function of the cytochrome b of the cytochrome bf complex is apparently similar to that of the cytochrome b of the mitochondrial system. The reduced plastocyanin donates reducing power to P_{700}, activating it so that it may transfer its electrons to plastoquinone-binding protein A0. Donation of electrons from protein A0 to plastoquinone-binding protein A1 and from protein A1 via FeS, ferredoxin (Fd), and flavoprotein (Fp) generates reducing power as NADPH and contributes to the overall proton gradient.

In chloroplasts, generation of the proton gradient depends upon the joint action of two photosystems, P_{680} and P_{700}. Light absorbed by P_{680} is inherently more energetic than is that absorbed by P_{700}. Critical studies have shown that the efficiency of photosynthesis is enhanced when wavelengths at or below 680 nm are used as energy sources, compared to its efficiency when the system is illuminated with wavelengths at or above 700 nm. The stronger reductant power obtained with the P_{680}

system, compared to the P_{700} system, facilitates both NADP reduction and formation of the proton gradient.

The mechanism of ATP formation

Irrespective of organism, the membranous system in which electron transport occurs, or the nature of the carriers involved, the major function of electron transport is generation of the proton motive force (PMF) so that it may be used to accomplish critical processes of life, such as, transport, motility, and, above all, formation of ATP. The PMF is composed of two components, the proton gradient (ΔpH) and the membrane potential. These elements are interconvertible. For the most part, the PMF may be considered to be a function of the proton gradient. The general manner in which PMF is generated is depicted in Figure 9.13. In the process of electron transport, reduction of an electron carrier is coupled with extrusion of a proton to the cell exterior, resulting in an increase in hydrogen ion

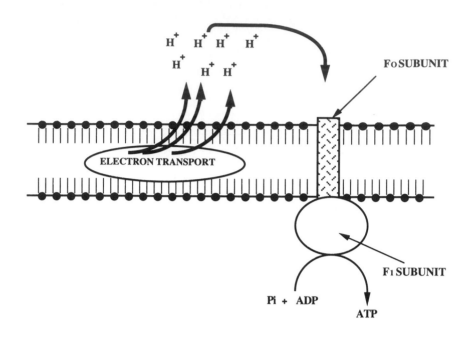

FIGURE 9.13 The basic mechanism of electron transport-mediated ATP formation. Electron transport leads to selective extrusion of hydrogen ions to the exterior of the cell, generating a proton motive force, largely as a function of the difference in hydrogen ion concentrations between the cell exterior and the cytoplasm. Hydrogen ions returning to the cell interior via the F_0 portion of ATP synthetase serve as the driving force for cytoplasmic ATP formation from ADP and inorganic phosphate by the F_1 portion of the enzyme. An analogous process occurs during photosynthesis. Photosynthetic electron transport produces a hydrogen ion excess in the lumen of nonstacked thylakoid sacs compared to the hydrogen ion concentration in the stroma. This hydrogen ion excess serves as the driving force for ATP synthesis in the stromal region of the chloroplast.

concentration in the extracellular environment and a corresponding decrease in the intracellular proton concentration. The difference in proton concentration between the intracellular and extracellular environment serves as the driving force for ATP synthesis by the action of the membrane-bound enzyme F_0F_1-ATPase, which may serve in two ways. When a proton gradient is required, as in the case of certain anaerobes devoid of electron transport systems, the gradient can be generated by ATP hydrolysis. Alternatively, and more frequently, the proton gradient generated by selective extrusion of protons from the cell serves as the driving force for ATP synthesis. F_0F_1-ATPase is composed of two subunits, the F_0 portion and the F_1 portion. The F_0 portion is within the membrane and serves as a "pore" through which protons may flow through an otherwise fundamentally, hydrophobic layer. The F_1 portion of the F_0F_1-ATPase contains subunits denoted as alpha, beta, gamma, delta, and epsilon. Each F_1 portion contains three alpha and beta subunits plus one unit each of gamma, delta, and

epsilon. Isolated alpha and beta subunits have been shown to hydrolyze ATP, but net ATP synthesis requires that the alpha and beta subunits be attached to the F_0 portion by the delta, gamma, and epsilon moieties.

The precise way in which the PMF is formed has been a subject of much study. It is generally conceded that selective extrusion of protons requires stereospecific arrangement of the components of the electron transport system, but the precise nature of that arrangement and the manner in which its components participate in the extrusion process are imperfectly understood. Two general mechanisms have been suggested, the loop mechanism and the quinone cycle. According to the loop mechanism, electrons and hydrogens at the cytoplasmic side of the membrane reduce a hydrogen carrier that moves to the exterior of the membrane. At this location, the hydrogen carrier reduces an electron carrier by a two-electron transfer and releases two protons to the extracellular environment for every pair of electrons used to reduce the electron carrier. According to the quinone cycle, quinones are reduced by hydrogen on the cytoplasmic side of the membrane and transfer a single electron to an electron carrier at the outer edge of cell membrane while, at the same time, extruding two protons to the cell exterior.

Although ATP synthesis is undoubtably the most prominent physiological function of the PMF, the PMF is intimately involved in other critical activities, particularly in procaryotes. It is the form of energy used in most mechanisms of procaryotic motility (see Chapter 2) and in addition, serves as the force for active transport.

Reverse electron transport

It is usually the case that ATP is generated by the proton motive force produced by oxidation of reduced materials. In particular, NADH is a frequent source of the reducing power used for generation of the PMF. However, in all organisms, reduced NADH and NADPH are required for biosynthesis. Because of substantial differences in the oxidation-reduction potentials of certain substances, particularly NO_2^- and Fe^{2+}, their oxidation by the chemolithotrophic bacteria does not allow formation of NADH and NADPH. To obviate this difficulty, a portion of the PMF derived from NO_2^- and Fe^{2+} oxidation is used to reduce NADH.

Summary

Energy generation in microbes occurs in a variety of ways, but in comparison to the potential energy sources, the mechanisms for energy production are relatively few. The ability of the bacteria to form energy from a diversity of carbonaceous materials reflects, in large measure, the channeling of diverse materials into central phosphorylated metabolites so that these materials may participate in substrate-level phosphorylation. In an analogous manner, electron transport occurs in a diversity of systems, although the principles and processes underlying its operation in various membranous systems, and the molecules involved, are similar. By studying the diversity of mechanisms for energy production at both the substrate level and in electron-transport mediated processes of microbes, we may understand the spectrum of processes involved in energy generation by all cell life forms. Wherever energy generation occurs, and by whatever mechanism, neither the processes involved nor their consequences may be fully understood without considering the relationship of energy generation to the remaining activities of the cell.

Selected References

Anzel, L. M., and P. L. Pedersen. 1983. Proton ATPases : Structure and mechanism. *Annual Review of Biochemistry.* **52:** 801–824.

Blankenship, R. E., and W. W. Parson. 1978. The photochemical electron transfer reactions of photosynthetic bacteria and plants. *Annual Review of Biochemistry.* **47:** 635–654.

Glazer, A. N. 1983. Comparative biochemistry of photosynthetic light-harvesting systems. *Annual Review of Biochemistry.* **52:** 125–158.

Haddock, B. A., and C. W. Jones. 1977. Bacterial respiration. *Bacteriological Reviews.* **41:** 47–99.

Harold, F. M. 1986. Vectorial metabolism. *In* L. N. Ornston and J. R. Sokatch (eds.) *The Bacteria,* vol. 6. pp. 463–521. Academic Press, Inc., New York.

Ingeledew, W. J., and R. K. Poole. 1984. The respiratory chains of *Escherichia coli. Microbiological Reviews.* **48:** 181–198.

Jones, C. W. 1982. *Bacterial respiration and photosynthesis.* Thomas Nelson and Sons Ltd., Walton on Thames, Surrey, England.

Krampitz, L. O. 1962. Cyclic mechanisms of terminal oxidation. *In* I. C. Gunsalus and R. Y. Stanier (eds.) *The Bacteria,* vol. 2. pp. 209–256. Academic Press, Inc., New York.

Krulwich, T. A., P. G. Quirk, and A. A. Guffanti. 1990. Uncoupler-resistant mutants of bacteria. *Microbiological Reviews.* **54:** 52–65.

Lacy, D., R. Spaulding, and G. D. Sprott. 1984. The bioenergetics of methanogenesis. *Biochemica Biophysica Acta.* **768:** 113–163.

Moodie, A. D., and W. J. Ingledew. 1990. Microbial anaerobic respiration. *Advances in Microbial Physiology.* **31:** 225–265.

Odom, J. M., and H. D. Peck, Jr. 1984. Hydrogenase, electron-transfer proteins and energy coupling in the sulfate-reducing bacteria *Desulfovibrio. Annual Review of Microbiology.* **38:** 551–592.

Parson, W. W. 1974. Bacterial photosynthesis. *Annual Review of Microbiology.* **28:** 41–59.

Racker, E. 1980. From Pasteur to Mitchell: a hundred years of bioenergetics. *Federation Proceedings.* **39:** 210–215.

Schlegel, H. G., and B. Bowien. 1989. *Autotrophic Bacteria.* Springer-Verlag, New York, Berlin, Heidelberg, Tokyo.

Stanier, R. Y., J. L. Ingraham, M. L. Wheelis, and P. R. Painter. 1986. *The Microbial World.* 5th edition. Prentice Hall, Englewood Cliffs, New Jersey.

Stewart, V. 1988. Nitrate respiration in relation of facultative metabolism in enterobacteria. *Microbiological Reviews.* **52:** 190–232.

Thauer, R. K., and W. Badziong. 1980. Respiration with sulphate as electron acceptor. *In* C. J. Knowles (ed.) *Diversity of Bacterial Respiratory Systems.* pp. 65–85. CRC Press, Inc., Boca Raton, Florida.

Wolin, M. J., and T. L. Miller. 1985. Methanogens. *In* A. L. Demain and N. A. Solomon (eds.) *Biology of Industrial Microorganisms.* Benjamin Cummings, Menlo Park, CA.

CHAPTER 10

Small Molecules

The critical role of small molecules as connecting substances

Small molecules are essential to the life processes of microbes. As we have briefly seen in Chapter 7, all of biosynthetic metabolism can be regarded as a study of the fate of twelve molecules generated during catabolism. From a particular perspective, all of life may be regarded as dependent upon the successful coupling of processes that generate energy and small molecules with processes that utilize energy and small molecules, so that the fundamental processes of life may be accomplished. In several chapters of this book, we consider how energy is generated and how polymeric molecules are formed. In this chapter, we shall consider the diversity of ways by which the critical molecules formed by catabolism are converted to the small molecules that the microbial cell uses as building blocks for macromolecular synthesis. In a very real sense, it is small molecules that constitute the connecting points between energy generation and energy use. For the most part, our discussion will focus on the formation of building blocks for proteins, nucleic acids, and lipids, since these substances constitute

three of the four major, quantitatively significant, organic components of protoplasm and since synthesis of the fourth component, the critical carbohydrate entities of cells has been, for the most part, considered elsewhere (see Chapter 3).

Formation of nucleotide precursors

Formation of nucleotide precursors is a critical process for cells. Without appropriate precursors, cells cannot form either DNA or the molecular vehicles for its expression. Synthesis of both DNA and RNA requires formation of the purine and pyrimidine ring structures and in addition, the specific formation of adenine, guanine, uridine, thymine, and cytosine. Additional purines and pyrimidines are required in some situations. However, simple formation of the appropriate purines and pyrimidines is insufficient to allow their functioning in cellular physiological processes. We must also understand the manner in which the pentoses, to which the rings are attached, are formed and how the various purines and pyrimidines are bound to pentose sugars. Finally,

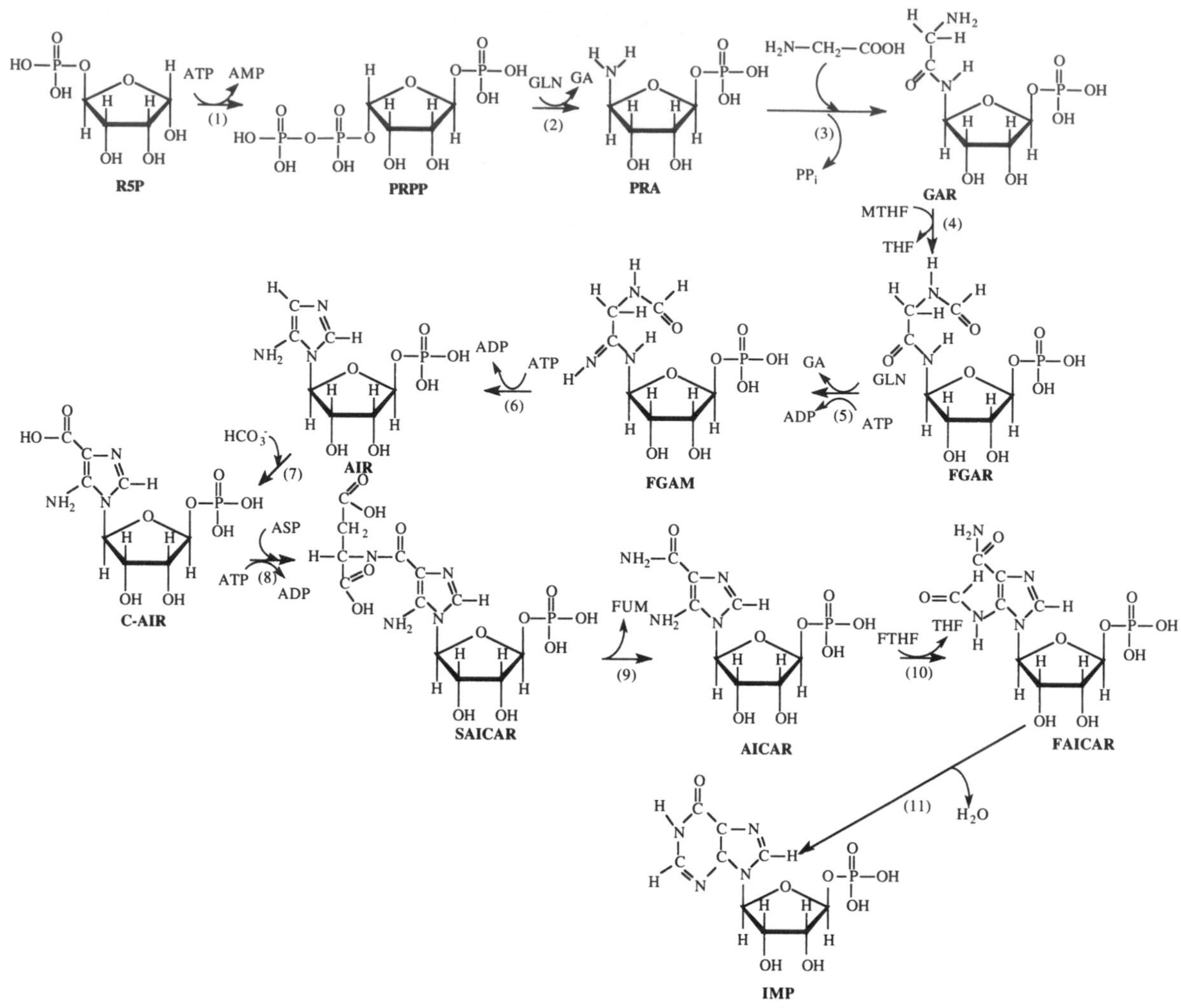

FIGURE 10.1 Synthesis of the purine ring structure from ribose-5-phosphate (R5P). The enzyme names and the names and abbreviations of the various intermediates between R5P and inosine monophosphate (IMP) are discussed in the text. The names and abbreviations of substances other than intermediates that are required for the process are: adenosine triphosphate (ATP); 5,10-methenyl tetrahydrofolic acid (MTHF); glutamine (GLN); glutamic acid (GA); aspartate (ASP); N^{10}-formyl tetrahydrofolic acid (FTHF); tetrahydrofolic acid (THF); and fumaric acid (FUM).

we must understand the ways in which incipient nucleic acid building blocks are phosphorylated, so that they may serve as precursors for synthesis.

Formation of the purine ring structure

The steps leading to formation of the purine ring structure are shown in Figure 10.1. Purine ring formation begins by the reaction of ribose-5-phosphate (R5P), normally derived from pentose phosphate pathways, with ATP by the action of ribose-5-PO_4 pyrophosphokinase, yielding phosphoribosyl pyrophosphate (PRPP). In this reaction, ATP is converted to AMP, and rather than

serving as an energy compound, serves as an *activating agent*, preparing the pentose ring for the subsequent events of purine synthesis. In purine synthesis, the purine ring is built on a pentose molecule, while in pyrimidine ring synthesis, the initial stages of ring structure synthesis occur in the absence of the pentose moiety. In pyrimidine synthesis, the pentose ring structure is added during the process of synthesis rather than serving as the entity upon which the ring structure is formed. For both purines and pyrimidines, however, PRPP is a critical substance in formation of the biologically active molecules.

In purine ring synthesis, PRPP is converted by aminophosphoribosyl transferase to 5-phosphoribosylamine (PRA), with glutamine as the amino donor. The PRA is subsequently converted by the addition of a glycine residue to glycineamide ribonucleotide (GAR) through the action of phosphoribosylglycineamide synthetase. By donating a carbonyl moiety with the aid of phosphoribosylglycineamide formyltransferase, GAR is converted to formylglycineamide ribonucleotide (FGAR). Addition of an amino group to the carbonyl function in FGAR, attributable to its glycine component, converts FGAR to alpha-N-formylglycineamide ribonucleotide (FGAM). The first ring of the purine structure is formed by the conversion of FGAM to 5-aminoimidazole ribonucleotide (AIR) with the aid of phosphoribosylimidazole synthetase.

Initiation of the formation of the second ring component of the purine ring structure begins by addition of a carboxyl function, via HCO_3^-, to the carbon adjacent to the amino group of the ring structure. This reaction, mediated by phosphoribosylaminoimidazole carboxylase, produces 5-amino-4-carboxyimidazole ribonucleotide (CIAR). An amino group is then added to CIAR, with ATP expenditure and aspartate as amino donor, to produce 5-amino-4-imidazole-(N-succinylo-)carboxamide ribonucleotide (SAICAR). This reaction is accomplished by phosphoribosylaminoimidazole succinocarboxamide synthetase. The carbon skeleton of aspartate is removed from SAICAR by the action of adenylosuccinate lyase, allowing donation of the amino group of aspartate to SAICAR and forming 5-amino-4-carboxamide ribonucleotide (AICAR) with the concomitant formation of fumarate. AICAR is further converted to 5-formamidoimidazole-4-carboxamide ribonucleotide (FAICAR) by the action of phosphoribosylaminoimidazolecarboxamide formyltransferase. Finally, dehydration of FAICAR by IMP cyclohydrolase forms inosine monophosphate (IMP), the ring structure from which all of the purines are formed.

The complexity of purine ring formation

Formation of the purine ring structure is an extremely complicated process. It involves eleven enzymatically catalyzed steps. It is mediated by three amino acids (i.e., glycine, glutamine, and aspartate). It involves three separate donations of one-carbon entities. In addition, the ring structure is built upon ribose-5-phosphate, an intermediate in carbohydrate metabolism. The system is no less complex at the genetic level. Eleven genetic loci participate in the reactions of purine ring synthesis. Regulation of purine formation occurs at two general locations in the sequence: at the initial conversion of PRPP to PRA and at the first steps of the branches that lead to the formation of adenine and guanine from IMP. The primary mechanisms of regulation of this system are apparently feedback inhibition and repression.

Pyrimidine ring synthesis

Synthesis of the pyrimidine ring structure is shown in Figure 10.2. It begins with the reaction of CO_2, glutamine, and two molecules of ATP to form carbamoyl phosphate (CP), by the action of carbamoyl phosphate synthetase. Next, under the influence of carbamoyl phosphate transferase (aspartate transcarbamoylase), carbamoyl phosphate reacts with aspartate (ASP) to form carbamoyl aspartate (CA). Dihydroorotase forms dihydroorotic acid (DOR) by removal of a water molecule from carbamoyl aspartate. Dihydroorotate is further converted to orotic acid (OR) by dihydroorotate dehydrogenase, with NAD as an acceptor. At this point, phosphoribosyl-1-pyrophosphate (PRPP) reacts with orotic acid under the influence of orotate phosphoribosyl transferase to form orotidine-5'-phosphate (O5P). The latter, by decarboxylation with orotidine monophosphate decarboxylase, becomes uridine monophosphate (UMP). The primary point of regulation of pyrimidine ring formation is the first reaction in the sequence, the conversion of carbamoyl phosphate and aspartate to carbamoyl aspartate under the influence of aspartate transcarbamoylase.

Formation of active nucleotides

The various phosphorylated forms of nucleotides are formed by further metabolism of ring structures. The formation of cytosine, uridine, thymidine, and their various phosphated forms from UMP is shown in Figure 10.3. Uridine triphosphate (UTP) arises from UMP by separate, ATP-mediated, phosphorylation reactions. Cytosine triphosphate (CTP) is formed by ATP-dependent amination of UTP. The analogous processes, that is,

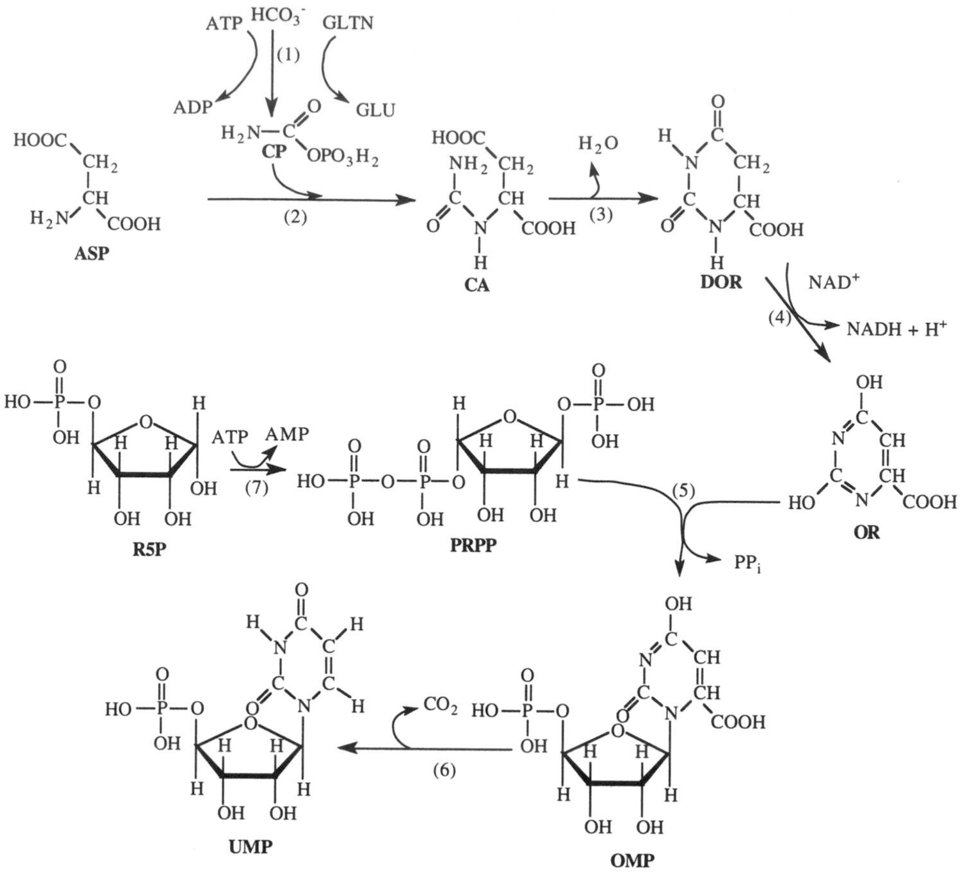

FIGURE 10.2 Synthesis of the pyrimidine ring structure. (1) Formation of carbamoyl phosphate (CP) by the action of CP synthetase. (2) Condensation of aspartate (ASP) with CP to form carbamoyl aspartate (CA) by the action of aspartate transcarbamoylase. (3) Formation of 4,5 dehydroorotate (DOR) from CA with the loss of water by dehydroorotase. (4) Oxidation of DOR to orotate (OR) with NAD^+ reduction by DOR dehydrogenase. (5) Formation of orotidine monophosphate (OMP) from phosphoribosyl-1-pyrophosphate (PRPP) and OR by OR-PRPP transferase, with loss of inorganic pyrophosphate (PP_i). (6) Decarboxylation of OMP to yield uridine monophosphate (UMP) by orotidine-5-phosphate (OMP) decarboxylase. (7) Synthesis of PRPP from ribose-5-phosphate (R5P) by R5P pyrophosphorylase, with adenosine monophosphate (AMP) release.

formation of adenosine, guanosine, and their phosphated forms are shown in Figure 10.4. Adenosine monophosphate is formed from IMP by reaction of aspartate and guanosine triphosphate (GTP) with inosine monophosphate (IMP) to form adenylosuccinate (AS) by the action of adenylosuccinate synthetase. Concomitantly, AMP and fumarate are formed from AS by the action of adenylosuccinate lyase. The di- and triphosphorylated modifications of AMP are formed by adenylate kinase and ATP formed from either electron transport or substrate level phosphorylation. The di- and triphosphorylated modifications of GMP are formed from the action of nucleoside mono- and dikinases.

Deoxynucleotides

With the exception of deoxythymidine triphosphate (dTTP), the deoxy forms of nucleotides are formed from reduction of diphosphorylated nondeoxynucleotides. The flavoprotein, thioredoxin, serves as the immediate source of reducing power

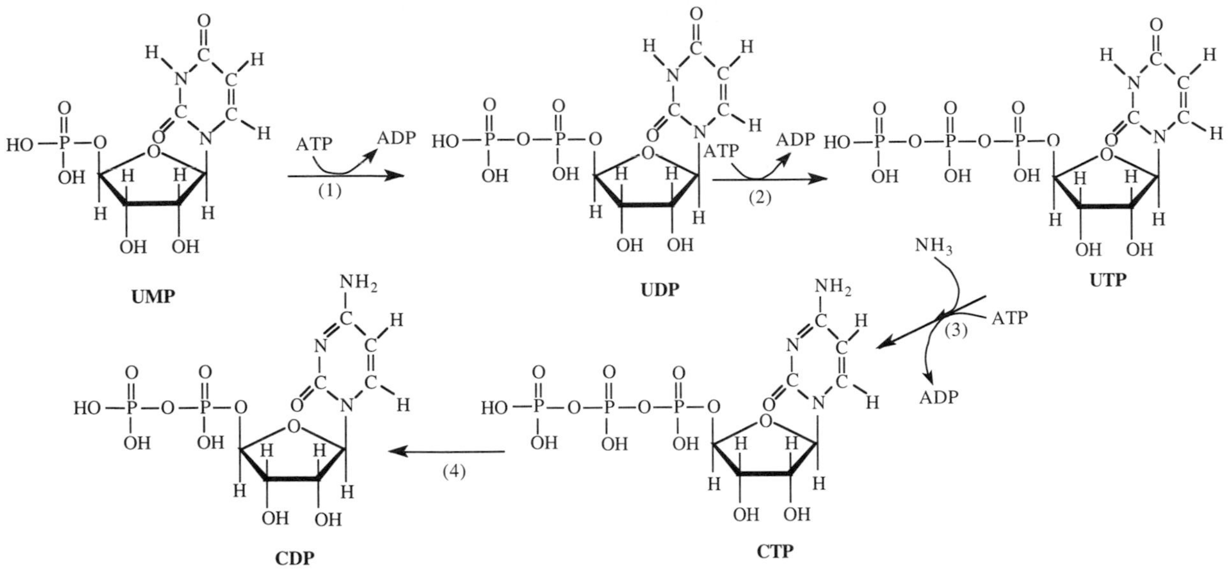

FIGURE 10.3 Formation of di- and triphosphated forms of uridine and cytidine nucleotides from uridine monophosphate (UMP). (1) Conversion of UMP to uridine diphosphate (UDP) by nucleoside monophosphate kinase. (2) Conversion of UDP to uridine triphosphate (UTP) by nucleoside diphosphate kinase. (3) Amination of UTP to form cytosine triphosphate (CTP) by CTP synthetase. (4) Dephosphorylation of CTP to form cytosine diphosphate (CDP) by a phosphatase enzyme.

and NADPH regenerates the reduced thioredoxin. Deoxythymidine triphosphate formation occurs in a different way than the other deoxynucleotide triphosphates. Deoxythymidine triphosphate arises from NADPH-directed reduction of UMP, producing dUMP. This reaction is followed by methylation of dUMP by the action of thymidylate synthase. Methyl tetrahydrofolate serves as both reductant and methyl donor in this reaction. The dihydrofolate formed as a consequence of dUMP reduction is regenerated with NADPH. The reactions involved in deoxynucleotide triphosphate formation are shown in Figure 10.5.

Regulation of nucleotide synthesis

The formation of nucleotides is a complicated process that intimately reflects other cellular processes, particularly amino acid synthesis. Pyrimidine synthesis involves use of both glutamine and aspartate while purine synthesis involves glutamine, aspartate, and glycine. In addition, PRPP is a critical intermediate in the formation of both the purine and pyrimidine ring structures. PRPP is normally derived from carbohydrate me-

tabolism and is also involved in histidine and tryptophan formation. Because of its intimate interconnection to other cellular processes, regulation of nucleotide base formation is both physiologically and genetically complex.

Amino acid synthesis

Microbial cells require an array of amino acids for protein synthesis. Fortunately, amino acid synthesis may be considered in terms of *families*—groups of amino acids formed from common precursors. In addition, as with all small molecules, the precursors arise from catabolic metabolism or an equivalent process. Within the amino acids, we can identify six families: the pyruvate family, the 3-phosphoglycerate or serine family, the aspartate or oxaloacetate family, the glutamic or alpha-ketoglutarate family, the aromatic family, and as a separate family histidine. For the most part, amino acid synthesis in microbes is functionally identical to that in higher life forms although some differences are found between microbial and nonmicrobial synthetic pathways and even within microbes.

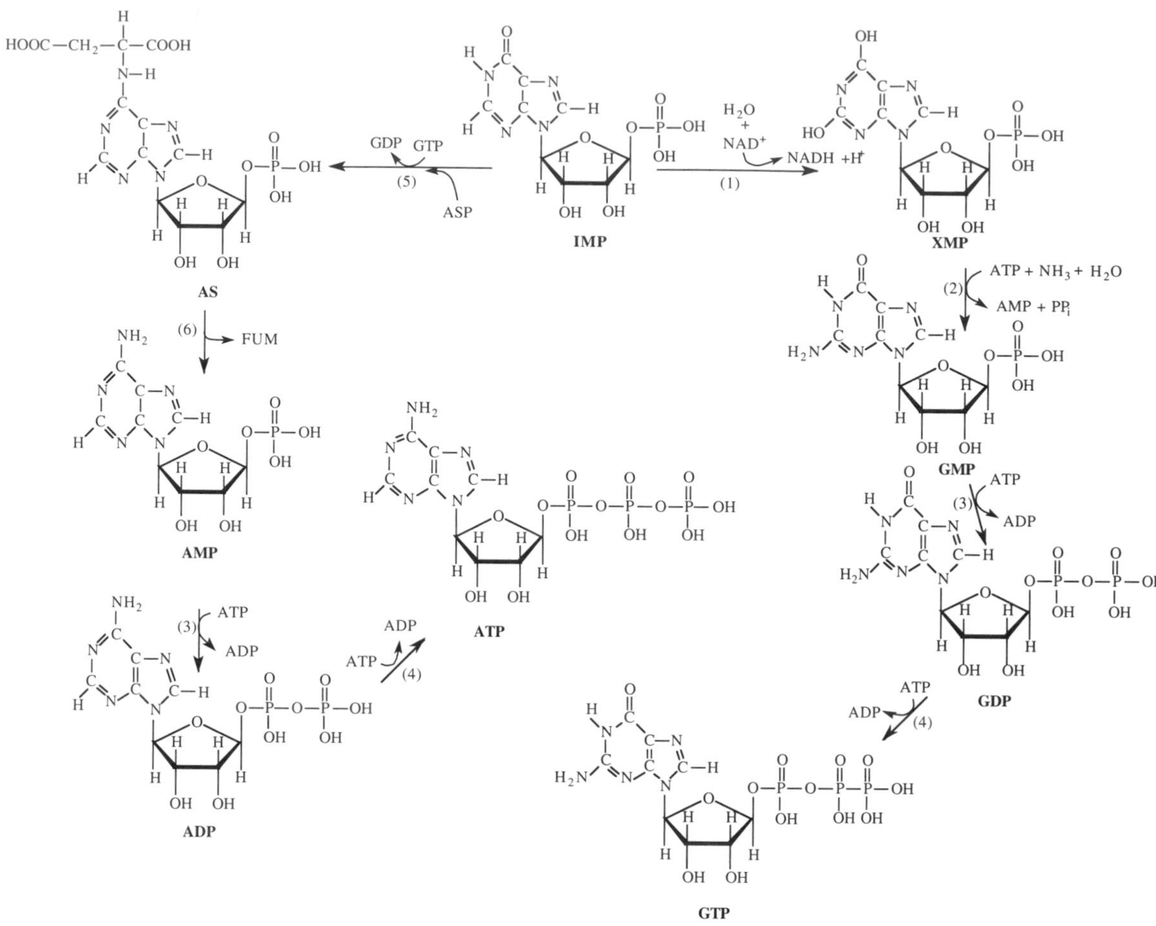

FIGURE 10.4 Formation of phosphorylated purines from inosine monophosphate (IMP). (1) Conversion of IMP to xanthine monophosphate (XMP) by XMP dehydrogenase. (2) Formation of guanosine monophosphate (GMP) from XMP by GMP synthetase. (3,4) Conversion of GMP to guanosine diphosphate (GDP) and guanosine triphosphate (GTP) by the sequential actions of nucleoside monophosphate (3) and nucleoside diphosphate (4) kinase reactions. (5) Conversion of IMP to adenylosuccinate (AS) by reaction of IMP with aspartate (ASP) under the influence of AS synthetase. (6) Formation of adenosine monophosphate (AMP) from AS with fumarate (FUM) release by the action of AS lyase. (3,4) Conversion of AMP to adenosine diphosphate (ADP) and adenosine triphosphate (ATP) by kinase reactions.

The pyruvate family

The pyruvate family of amino acids includes alanine, valine, and leucine. The synthetic pathways involved are shown in Figure 10.6. Alanine is formed from pyruvate by transamination with glutamate as donor and the concomitant formation of α-ketoglutarate (AKG). Valine and leucine are formed by a different reaction sequence. Initially, two molecules of pyruvate condense, with the loss of CO_2, to form α-acetolactate (AAL). NADPH-dependent reduction of AAL produces α, β-dihydroxy-isovalerate (ABIV). ABIV is sub-

sequently dehydrated, yielding α-keto-isovalerate (AKIV) that becomes valine by transamination with glutamate. Leucine is also formed from AKIV, but initially, AKIV reacts with acetyl CoA to form 2-isopropyl malate (2-IPM). The 2-IPM is dehydrated to form *cis*-dimethylcitraconate (DMC). Hydration of DMC produces 3-isopropyl malate (3-IPM). By NAD-dependent oxidation and decarboxylation, 3-IPM becomes α-ketoisocaproate (KIC), the structural analogue of AKIV. Leucine is formed from KIC by transamination with glutamate.

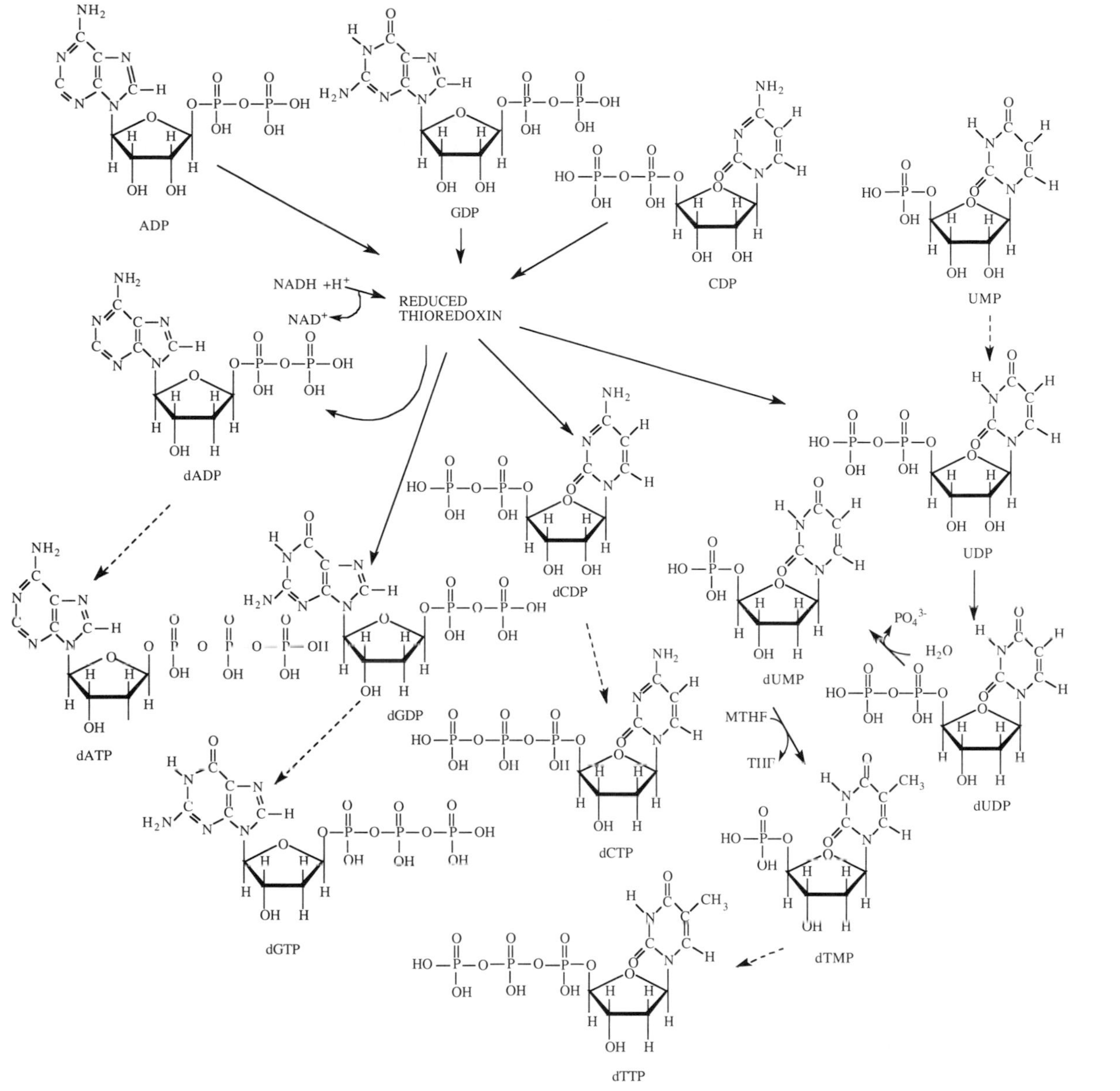

FIGURE 10.5 Formation of deoxy (d) nucleotides from non-deoxy precursors. For adenosine, guanosine and cytosine, deoxytriphosphates arise from reduction of nondeoxydinucleotides to yield deoxydinucleotides by the flavoprotein thioredoxin. Reduced thioredoxin is provided by NADH oxidation. The deoxydinucleotides are converted to triphosphodeoxynucleotides by kinase reactions. Deoxythymidine triphosphate is formed from deoxyuridine *mono*phosphate (dUMP). The latter arises by conversion of nondeoxy-uridine monophosphate to uridine diphosphate (UDP), thioredoxin-mediated reduction of UDP to yield dUDP and hydrolysis of the latter to yield dUMP. Methylation of the dUMP by methenyl tetrahydrofolate, produces deoxythymidine monophosphate (dTMP), which becomes deoxythymidine triphosphate by kinase reactions. Kinase reactions are indicated by dashed arrows.

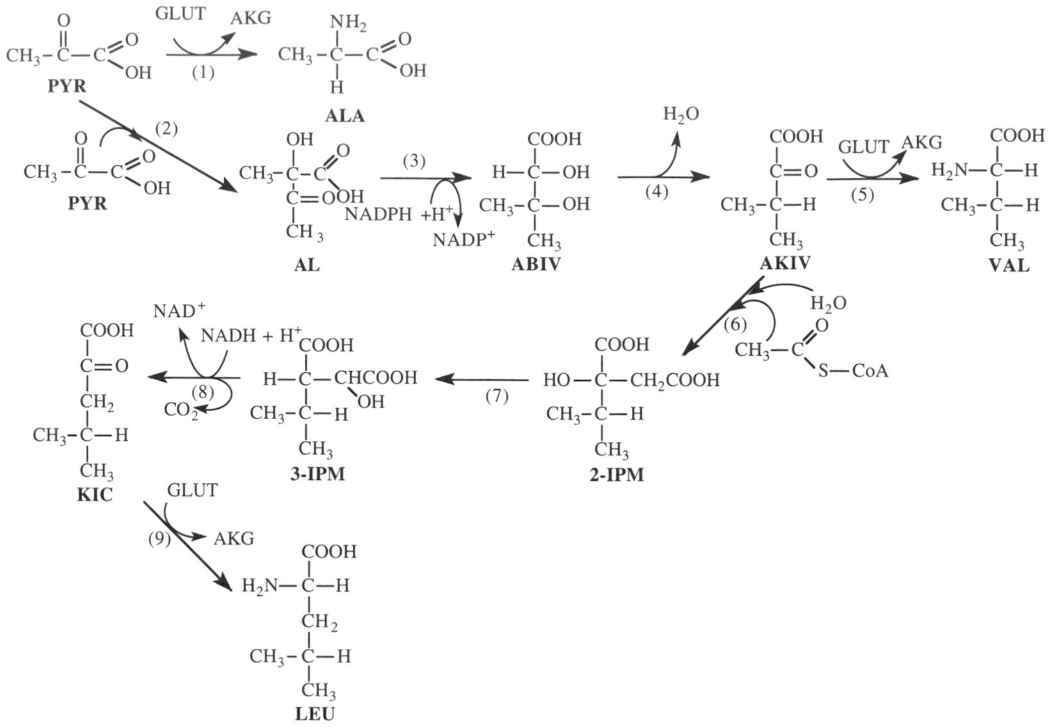

FIGURE 10.6 Synthesis of the pyruvate amino acid family. (1) Conversion of pyruvate (PYR) to alanine (ALA) by the action of glutamic (GLUT)-PYR transaminase. Alpha-ketoglutarate (AKG) is formed in this reaction. (2–5) Synthesis of valine (VAL) by the sequential action of acetolactate (AL) synthetase (2), acetohydroxy acid isomeroreductase (3), dihydroxy acid dehydratase (4), and a transaminase (5). Intermediates in this process are α, β-dihydroxyisovalerate (ABIV) and α-ketoisovalerate (AKIV). (6–9) Conversion of AKIV to leucine (LEU). Initially, AKIV is converted to 2-isopropylmalate (2-IPM) by 2-IPM synthetase (6). The 2-IPM is converted to 3-isopropylmalate (3-IPM) by IPM isomerase (7). The 3-IPM is oxidatively decarboxylated to yield α-ketoisocaproate (KIC) by 3-IPM dehydrogenase (8) and the KIC is then converted to LEU by transaminase activity (9).

The 3-phosphoglycerate family

The 3-phosphoglycerate (serine) pathway includes serine, cysteine, and glycine. Its reactions are shown in Figure 10.7. Initially, 3-phosphoglycerate (3PGA) is oxidized to 3-phosphohydroxypyruvate (3PHP). The 3PHP is then converted to 3-phospho-serine (3PS) by amination with glutamate. Serine (SER) is then formed by dephosphorylation of 3PS. Serine is a branchpoint of the pathway. By transfer of its hydroxymethyl residue to tetrahydrofolate, serine is converted to glycine (GLY). Alternatively, serine may react with acetyl CoA to form O-acetyl serine (OAS). Conversion of OAS to cysteine (CYS) results from reaction of OAS with H_2S, often derived from assimilatory sulfate reduction. Serine formation from OAS is accompanied by the release of free acetate.

The glutamic acid family

The glutamic acid family includes the linear amino acids glutamine, ornithine, citrulline, and arginine. In practice, among the linear glutamic family amino acids, only glutamic acid, glutamine, and arginine are normally of major synthetic consequence. In addition to linear amino acids, the glutamic family includes the cyclic amino acid, proline. The reactions of the glutamic acid family are shown in Figure 10.8. Glutamic acid (GLUT) arises from α-ketoglutaric acid in one of two ways, either by NH_4^+ fixation with NADH or NADPH-dependent glutamic dehydrogenases or by the combination of ATP and NH_4^+-dependent L-glutamine formation from L-glutamate by glutamine synthetase and glutamate synthesis by reaction of

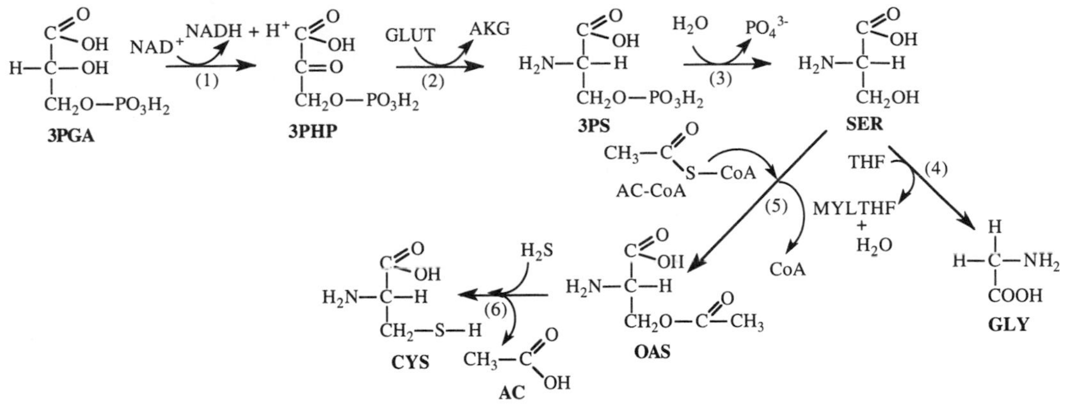

FIGURE 10.7 Synthesis of the serine amino acid family. (1) Conversion of 3-phosphoglycerate (3PGA) to 3-phosphohydroxypyruvate (3PHP) by 3PGA dehydrogenase. (2) Amination of 3PHP with phosphoserine aminotransferase to form 3-phosphoserine (3PS). (3) Dephosphorylation of 3PS to produce serine (SER) by the action of phosphoserine phosphatase. (4) Transfer of the hydroxymethyl group of SER to tetrahydrofolate (THF) to form glycine (GLY) and N^5,N^{10} methylene tetrahydrofolate (MYLTHF) by serine hydroxymethyltransferase. (5) Acetylation of SER to yield O-acteylserine (OAS) by serine transacetylase. (6) Conversion of SER to cysteine (CYS), with acetate release, by O-acetylserine sulfhyrdylase, with H_2S generated by assimilatory sulfate reduction.

L-glutamine with α-ketoglutarate to yield two molecules of L-glutamate. Which of the two mechanisms operates depends upon the ammonia concentration to which the microbe is exposed. At ammonia concentrations below 1 mM, the glutamine synthetase is operative, while at NH_3 concentrations above 1 mM, the dehydrogenase mechanism functions. Collectively, the two mechanisms constitute the major mechanisms of ammonia assimilation by microbes.

Irrespective of the origin of L-glutamate, its further metabolism produces the amino acids of the glutamic acid family. For proline formation, L-glutamate reacts with ATP, forming γ-glutamyl phosphate (GGP). Reduction of GGP with NADPH, coupled with phosphate loss, yields glutamic γ-semialdehyde (GSA). Spontaneous loss of water from GSA produces Δ'-pyrroline-5-carboxylate (P5G). Proline (P) is formed from NADPH-dependent reduction of P5G.

Arginine (ARG) formation occurs by a different series of reactions. To begin with, GLUT reacts with acetyl CoA to form N-acetyl glutamate (NAGL). By ATP expenditure, the NAGL is converted to N-acetylglutamyl phosphate (NAGLP). NADPH-dependent reduction of NAGLP produces N-acetylglutamyl-γ-semialdehyde (NAGLA). Glutamate-mediated amination of NAGLA yields N-acetyl ornithine (NAO), which by loss of acetate, becomes ornithine (ORN). ORN is converted to citrulline (CIT) by reaction with carbamoyl phosphate. ARG is formed from CIT by aspartate and ATP-mediated formation of argininosuccinate (ARSUC) and release of fumarate from the ARSUC.

The aromatic family

The aromatic amino acid family includes phenylalanine, tyrosine, and tryptophan. Common reactions of the aromatic amino acid pathway are shown in Figure 10.9. The pathway begins with intermediates from carbohydrate fermentation, erythrose-4-phosphate (E4P) and phosphoenolpyruvate (PEP). These compounds are condensed to give 3-deoxy-D-arabino-heptulosonate-7-phosphate (DHP). DHP, by the loss of phosphate and cyclization, becomes the first ring intermediate of the pathway, 5-dehydroquinate (5DQ). Loss of water from 5DQ produces 5-dehydroshikimate (5DS). NADPH-dependent reduction of 5DS yields shikimate (SK). SK, with the aid of ATP, is converted to 5-phosphoshikimate (5PSK). By reaction of 5PSK with phosphoenol pyruvate (PEP), 3-enolpyruvylshikimate-5-phosphate (3EPSK5P) is formed, and by loss of phosphate from 3EPSK5P, chorismic acid (CH) is generated.

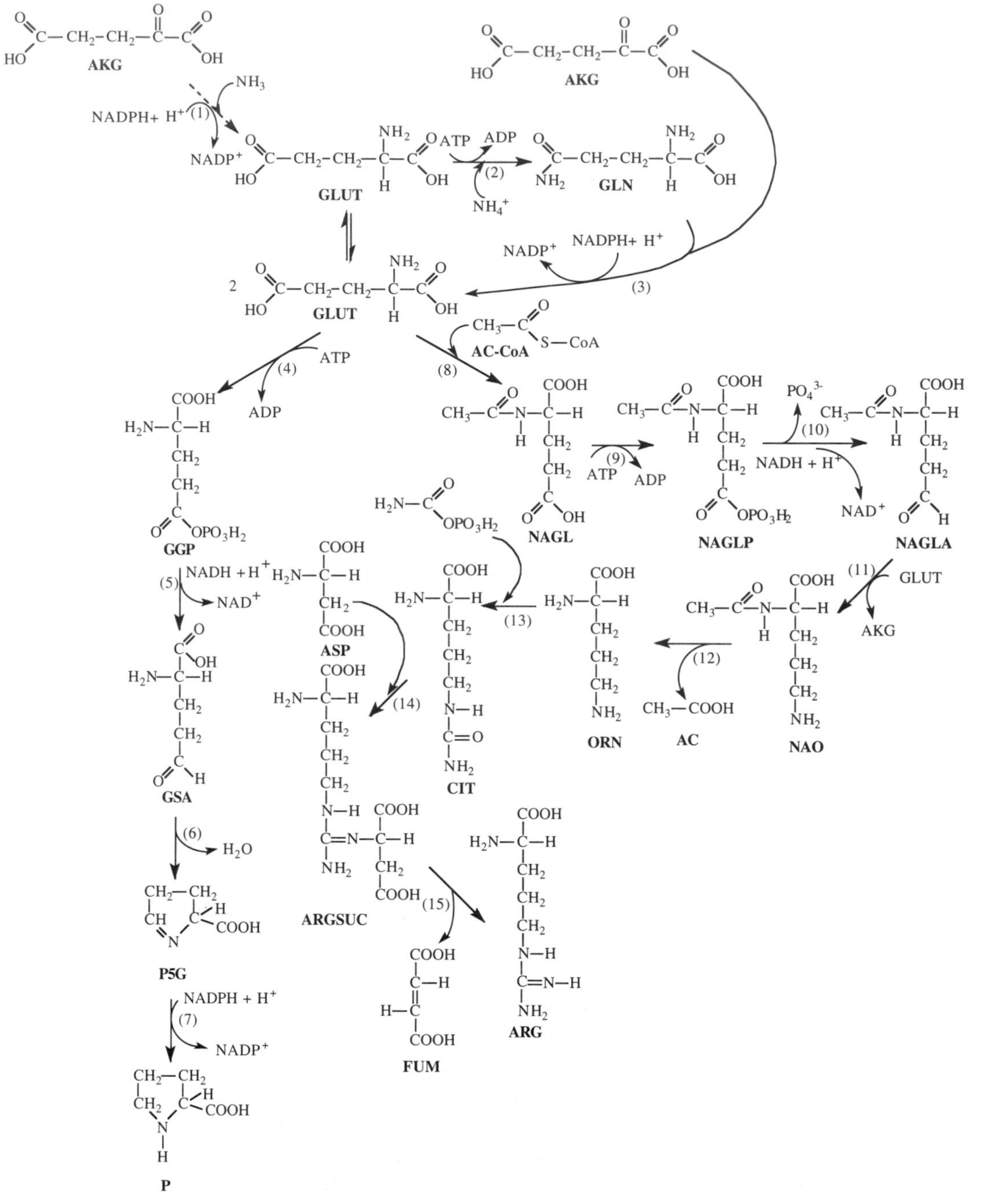

FIGURE 10.8 Synthesis of the glutamic amino acid family. The major amino acids of the family are glutamic acid (GLUT), glutamine (GLN), arginine (ARG), and proline (P). (1–3) Formation of glutamate, with ammonium ion fixation, by the actions of glutamate dehydrogenase (1), glutamine synthetase (2), and glutamate synthetase (3). (4–7) Conversion of GLUT to P by the actions of GLUT kinase (4), to form γ-glutamyl phosphate (GGP), glutamic semialdehyde (GSA) dehydrogenase to form GSA from GGP (5), the spontaneous loss of water from GSA to form Δ-1-pyrroline-5-carboxylate (P5C) (6) and conversion of P5C to P by P5C reductase (7). (8–15) Conversion of GLUT to ARG by the sequential actions of amino acid acetyl transferase (8) and N-acetyl glutamate (NAGL) kinase (9), to form N-acetylglutamyl phosphate (NAGLP). NAGLP is then converted to N-acetylglutamate semialdehyde by (NAGLA) dehydrogenase (10). NAGLA is aminated by GLUT to form N-acetyl ornithine (NAO) by NAO transaminase (11). NAO is deacetylated by NAO deacetylase (12), and the ornithine (ORN) is converted to ARG by the actions of ORN transcarbamoylase (13) to form citrulline (CIT), argininosuccinate (ARGSUC) synthetase (14), and ARGSUC lyase (15).

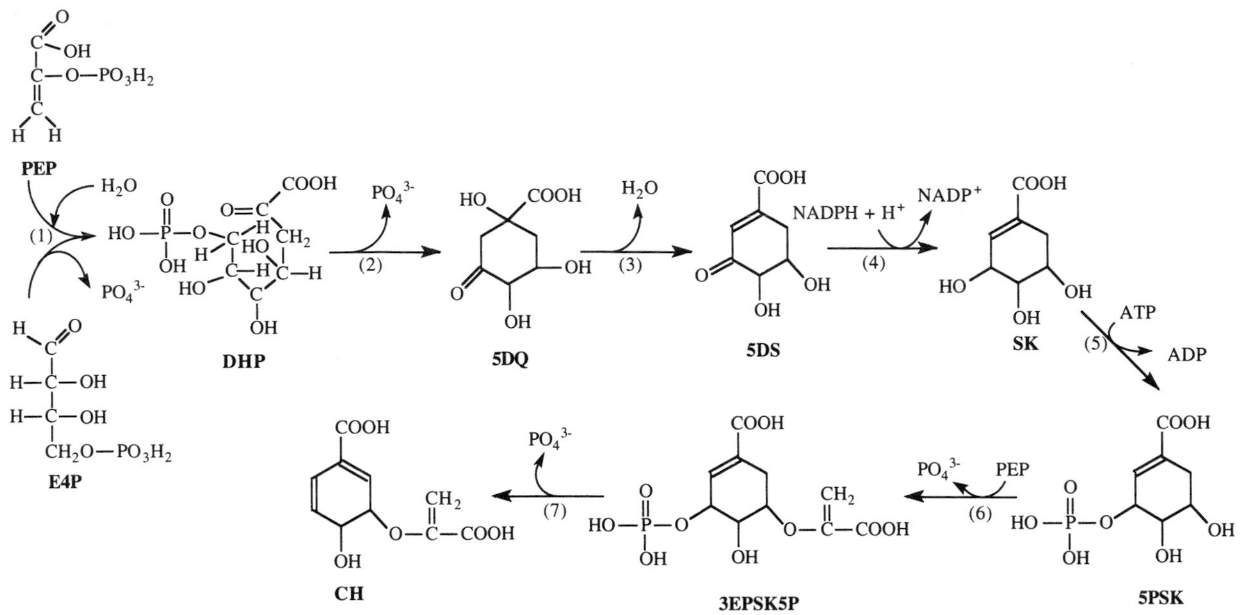

FIGURE 10.9 The common reactions of aromatic amino acid synthesis. (1,2) Formation of the first ringed-structured intermediate, 5-dehydroquinate (5DQ), from erythrose-4-phosphate (E4P) and phosphoenol pyruvate (PEP) with the intermediate formation of 3-deoxy-D-arabinoheptulosonate-7-phosphate (DHP) by the actions of DHP synthase (1) and 5DQ synthase (2). (3,4) Conversion of 5DQ to shikimate (SK) by dehydration and reduction through the actions of 5DQ dehydratase (3) and 5-dehydroshikimate (5DS) dehydrogenase (4). (5) ATP-dependent phosphorylation of SK to form 5-phosphoshikimate (5PSK). (6) Donation of a PEP residue to 5PSK to form 3-enolpyruvylshikimate-5-phosphate (3EPSK5P) by 3EPSK5P synthetase. (7) Dephosphorylation of 3EPSK5P to yield chorismate (CH) by the action of CH synthetase.

Chorismic acid is a critical intermediate in the aromatic amino acid family. Its conversion, by anthranilate synthetase, to anthranilic acid (AN) leads to formation of tryptophan. Alternatively, conversion of chorismate to prephenate (PRP) leads to the formation of phenylalanine and tyrosine. The formation of L-tryptophan from chorismate is shown in Figure 10.10. Amino group donation to chorismate by glutamine, with concomitant loss of pyruvate, forms anthranilate (AN). Reaction of AN with phosphoribosyl-1-pyrophosphate (PRPP) forms N-(5'-phosphoribosyl)-anthranilate (N5PRA). Opening of the ring structure of N5PRA yields enol-1-(o-carboxyphenylamino)-1-deoxyribulose-5-phosphate (CPADR). Dehydration and decarboxylation of CPADR produces the tryptophan ring structure in the form of indole-3-glycerol phosphate(IGP). Displacement of the IGP residue and its replacement by a serine (SER) moiety produces tryptophan (TRYP).

The formation of tyrosine and phenylalanine from chorismate (Figure 10.11) is accomplished by conversion of chorismate (CH) to prephenate (PRP) by the action of chorismate mutase, a key regulatory enzyme in aromatic amino acid synthesis. In this reaction, the 3-carbon moiety from carbon 3 of CH is transferred to carbon 1 of the ring. Phenylpyruvate (PPY) is formed from PRP by removal of water and CO_2 by the action of prephenate dehydratase, and phenylalanine (PA) is formed from PPY by transamination with glutamate.

Tyrosine (TY) arises from PRP by a reaction sequence analogous to those required for PA synthesis. By the action of prephenate dehydrogenase, another key enzyme of aromatic amino acid synthesis, PRP is converted to para-hydroxyphenyl pyruvate (PHPP). This reaction preserves the para-hydroxyl group of PRP by removing hydrogen with NAD as an acceptor. Finally, TY is formed by transamination with glutamate.

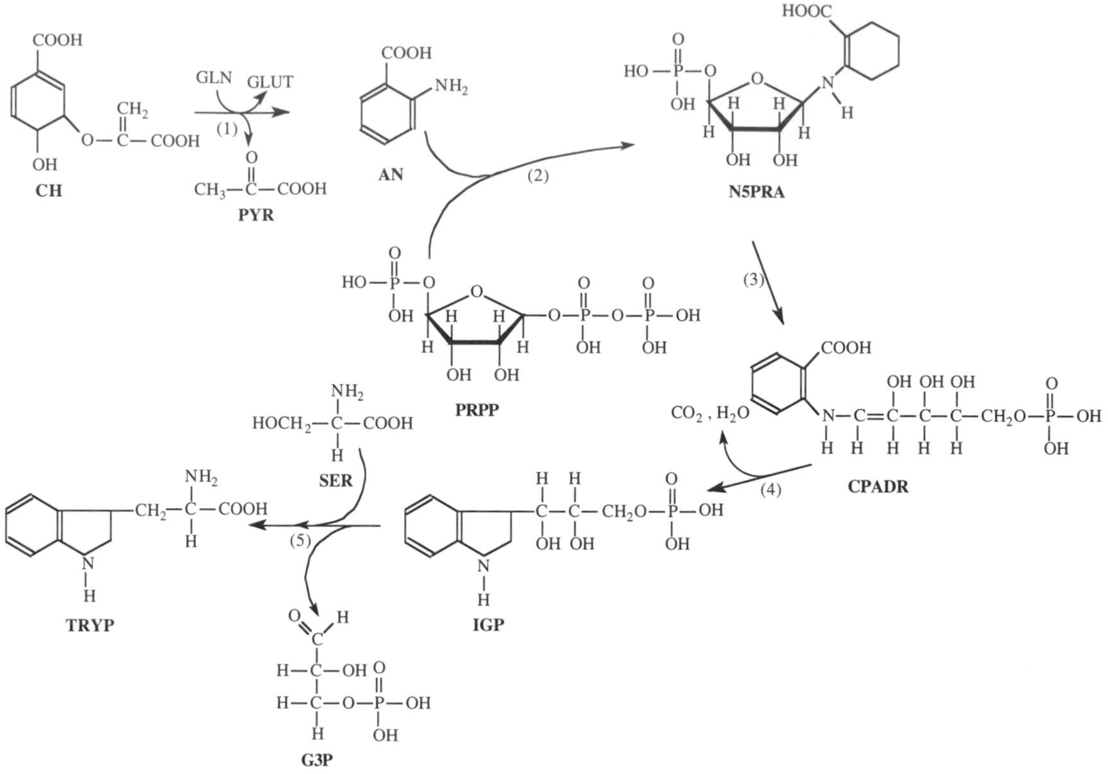

FIGURE 10.10 Synthesis of L-tryptophan (TRYP) from chorismate (CH). (1) Amination of CH, with glutamine (GLN) as amino donor, to form anthranilate (AN) and glutamate (GLUT) by the action of AN synthase. (2) Formation of N-5′-phosphoribosyl anthranilate (N5PRA) by reaction of AN and phosphoribosyl-1-pyrophosphate (PRPP), with the mediation of anthranilate phosphoribosyl transferase. (3) Conversion of N5PRA to enol-1-(o-carboxyphenylamino)-1-deoxyribulose-5-phosphate (CPADR) by phosphoribosylanthranilate isomerase. (4) Decarboxylation and dehydration of CPADR to form indole-3-glycerol phosphate (IGP) by IGP synthase. (5) Formation of TRYP from IGP by replacement of the glycerol phosphate moiety of IGP with the a serine (SER) residue and the simultaneous release of a molecule of glyceraldehyde-3-phosphate (G3P) by the action of TRYP synthase.

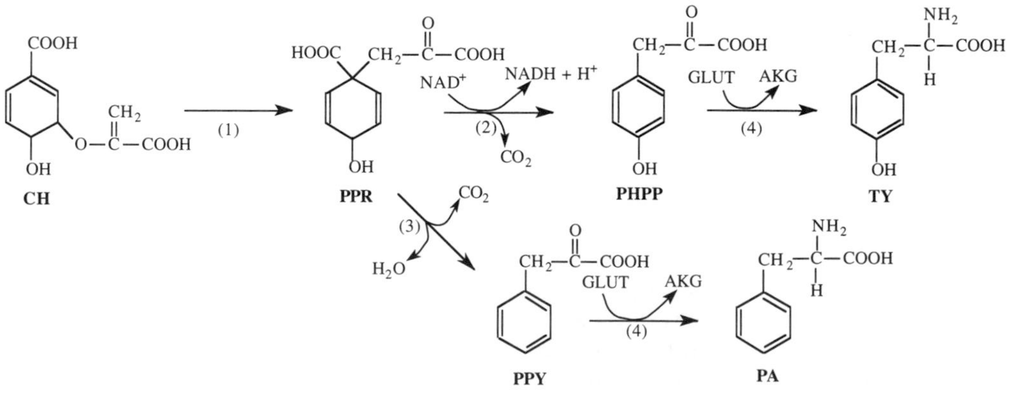

FIGURE 10.11 Synthesis of phenylalanine (PA) and tyrosine (TY) from chorismate (CH). (1) Formation of prephenate (PRP) from CH by CH mutase. (2,3) Conversion of PRP to phenyl pyruvate (PPY) by (PRP) *dehydratase* (2) or to para-hydroxyphenyl pyruvate (PHPP) by PRP *dehydrogenase* (3). (4) Amination of PHPP or PPY, with glutamate (GLUT) as amino donor, to produce TY or PA and alpha-ketoglutarate (AKG).

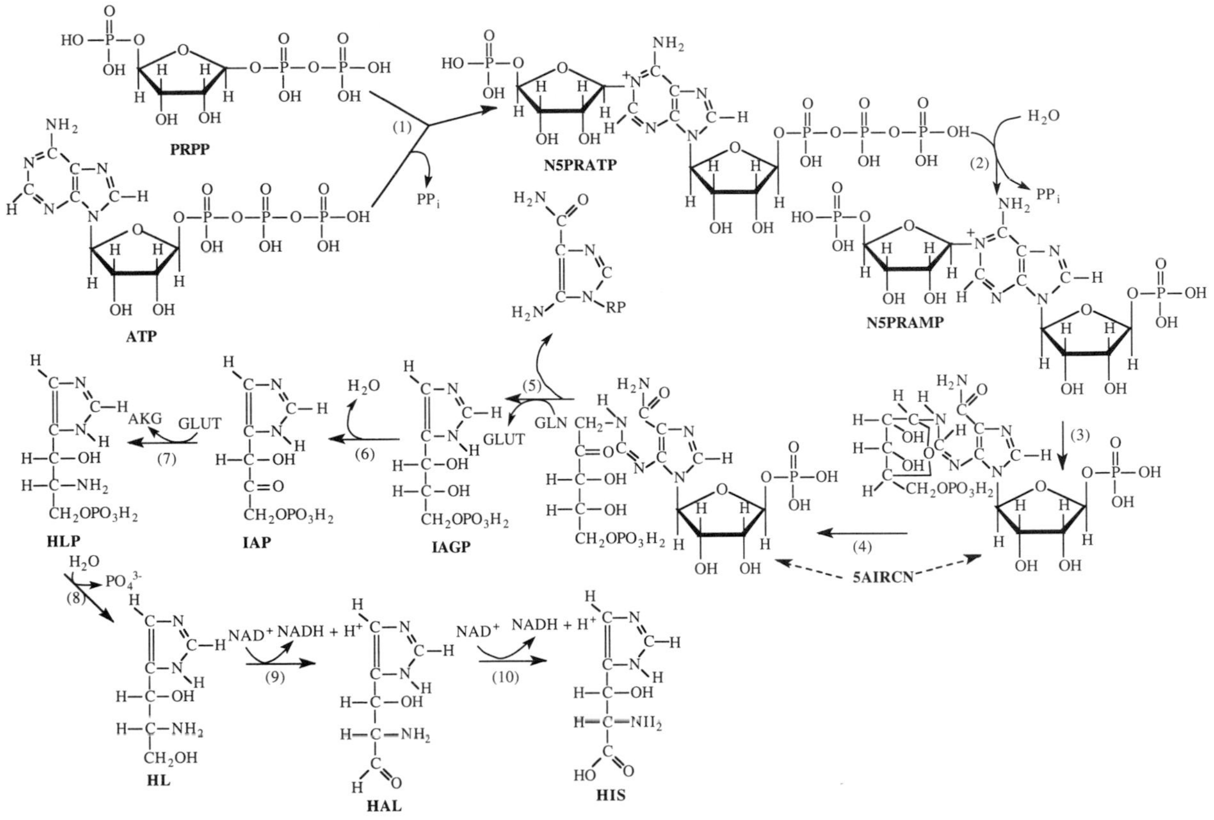

FIGURE 10.12 The synthesis of histidine. (1) Conversion of adenosine triphosphate (ATP) and phosphoribosyl-1-pyrophosphate to *N*1(5′ phosphoribosyl ATP) (N5PRATP) by phosphoribosyl:ATP-pyrophosphorylase. (2) Formation of *N*1-(5′-phosphoribosyl AMP) (N5PRAMP) from N5PRATP by phosphoribosyl-ATP pyrophosphohydrolase. (3) Conversion of N5PRAMP to phosphoribosyl-formimino-5-amino-imidazole carboxyamide (5AICRN) by N5PRAMP cyclohydrolase. (4) Isomeration of 5AICRN by 5AICRN isomerase. (5) Amination and displacement of one of the ring structures of ATP to form imidazole glycerol phosphate (IAGP) by the action of 5AICRN:glutamine aminotransferase. (6) Removal of the elements of water from IAGP to form imidacetazole phosphate (IAP) by IAGP dehydratase. (7) Amination of IAP to yield histinol phosphate (HLP) by HLP transaminase. (8) Dephosphorylation of HLP to form histidinol (HL). (9,10) Successive NAD⁺-mediated oxidations of HL to produce histidinal (HAL) and histidine (HIS) by the action of HL dehydrogenase. The same enzyme is involved in both reactions (9) and (10).

Histidine

Histidine (H) synthesis (Figure 10.12) begins by reaction of phosphoribosyl-1-pyrophosphate (PRPP) with ATP. In this reaction, pyrophosphate is lost and the ring structure of adenine is attached to the ribose moiety, producing, transitorily, N^1-(5′-phosphoribosyl)-AMP. Hydration of the latter opens the ring structure of both the second adenine function and the ribose moiety of PRPP, producing phosphoribosyl-formimino-5-amino-imidazole carboxamide ribonucleotide (5AICRN).

Reaction of 5AICRN with glutamine displaces the ribose-5-bound-adenine ring and yields imidazoleglycerol phosphate (IAGP). The imidazole ring results from the N-C-N portion of the nonribose phosphate-bound adenine ring and the top two carbons of PRPP. Dehydration of IAGP forms imidazolelacetol phosphate (IAP). Transamination of IAP with glutamate produces histidinol phosphate (HLP), which by hydration, becomes histidinol (HL). Stepwise, successive NAD-mediated oxidations of HL produce histidinal (HAL), and finally, L-histidine (H).

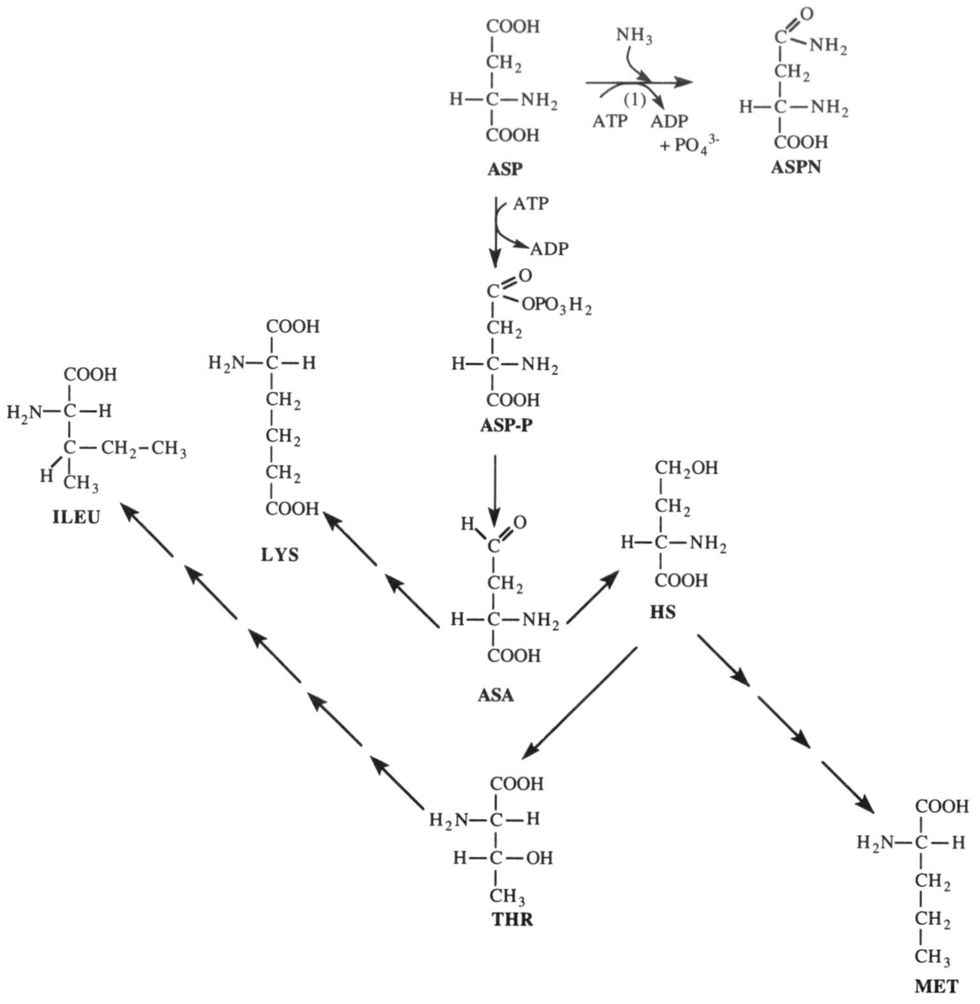

FIGURE 10.13 The general nature of the synthetic processes in the aspartic acid family of amino acids. Aspartate (ASP), formed from oxaloacetate by transamination, is converted to asparagine (ASPN) by ASPN synthetase (1), with ATP expenditure. The remaining amino acids of the aspartate family are formed by multistep processes. ASP is converted to aspartic semialdehyde (ASA) via aspartyl phosphate (ASP-P). ASA serves as a branchpoint, giving rise to either lysine (LYS) or to homoserine (HS). HS, in turn, also serves as a branchpoint, giving rise to either threonine (THR) and isoleucine (ILEU) or to methionine (MET). The details of the MET, ILEU, and LYS branches of the aspartic acid amino acid synthetic family are shown in Figures 10.14–10.16.

The aspartic acid family

The aspartic acid (oxaloacetate) family is the most complex of all of the amino acid families. Operation of the various sub-pathways in the family allows formation of aspartate, asparagine, methionine, threonine, isoleucine, and lysine.

The aspartic acid family contains several branched pathways. The general nature of the pathways within the aspartic acid family is shown in Figure 10.13. Details of the lysine, methionine, and isoleucine branch pathways are shown in Figures 10.14, 10.16, and 10.17. The family of pathways begins with oxaloacetate (OAA), which by transamination with glutamate, becomes aspartate (ASP). By ATP-dependent NH_3 fixation, aspartic acid is converted to asparagine. This reaction, in some systems, serves as a significant mechanism for NH_3 fixation. The remaining amino acids of the aspartic acid family are synthesized from aspartate metabolism.

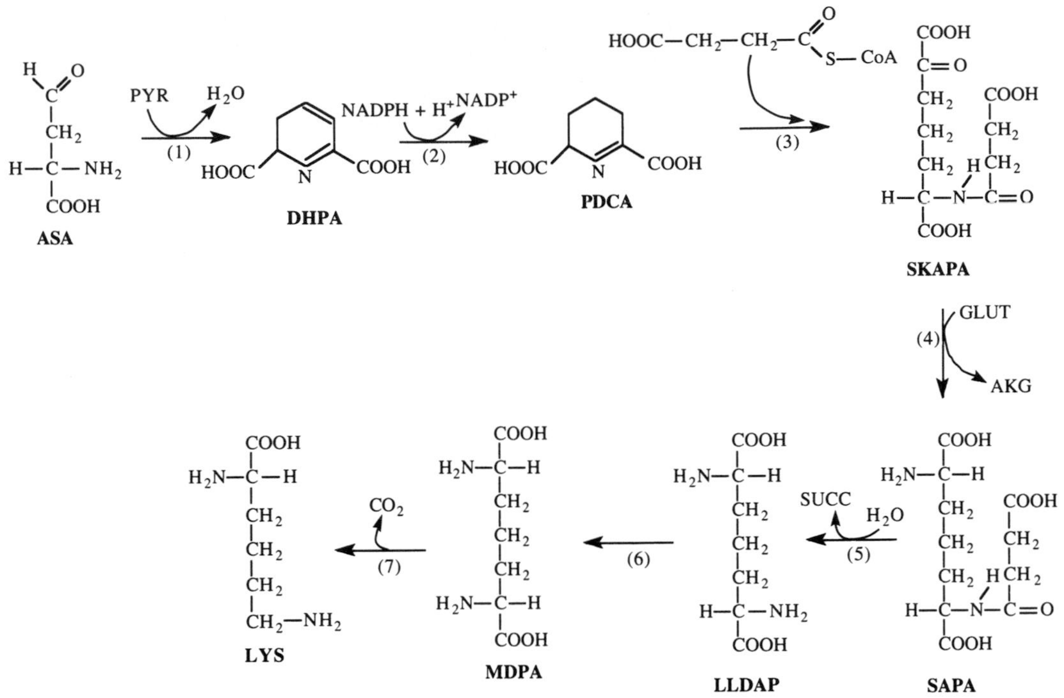

FIGURE 10.14 The usual pathway for lysine (LYS) synthesis in bacteria and algae. (1) Formation of 2,3-dihydropicolinic acid (DHPA) from aspartic-β-semialdehyde (ASA) by reaction of ASA with pyruvate (PYR) under the influence of DHPA synthase. (2) NADPH-mediated reduction of DHPA to piperideine-2,6-dicarboxylic acid (PDCA) by DHPA reductase. (3) Succinylation of PDCA to form N-succinyl-ε-keto-L-α-amino-pimelic acid (SKAPA) by PDCA succinylase. (4) Amination of SKAPA, with glutamate (GLUT) as donor, to form N-succinyl-L,L-α,ε-diaminopimelic acid (SAPA) and α-ketoglutrate (AKG) by the action of GLUT:succinyl-diaminopimelate amino transferase. (5) Succinate removal from SAPA to form L,L-α,ε-diaminopimelate (LLDAP). (6) Isomerization of LLDAP to form meso-α,ε-diaminopimelate (MDPA) by diaminopimelate epimerase. (7) Decarboxylation of MDPA to form lysine (LYS) by diaminopimelate decarboxylase.

The various sub-pathways of the aspartic acid family may be perceived as a "tree," with several branches. Aspartic-β-semialdehyde is the first branch, produced from aspartic acid in a two-step process. By reaction with ATP, ASP is converted to β-aspartyl phosphate (ASP-P) that by NADPH-dependent reduction and dephosphorylation, becomes aspartic-β-semialdehyde (ASA). The lysine synthetic pathway diverges from the rest of the aspartic acid pathways at this point.

By NADPH-mediated reduction, ASA is converted to homoserine (HS). HS is the second major branching compound of the aspartic acid family. Depending upon its metabolism, it may be converted to methionine (ME) or to threonine (THR) and isoleucine (IL). ATP-dependent phosphorylation of HS forms homoserine phosphate (HSP) and dephosphorylation and rearrangement of HSP produces threonine (THR).

Lysine formation

Lysine (LYS) is formed in procaryotes and most algae by metabolism of ASA. This sequence is shown in Figure 10.14. ASA is initially converted by reaction with pyruvate to a 7-carbon ringed structure, dihydropicolinic acid (DHPA). NADPH-mediated reduction of DHPA produces piperideine-2,6-dicarboxylic acid (PDCA). Succinylation of PDCA breaks open its ring structure, forming N-succinyl-ε-keto-L-α-aminopimelic acid (SKAPA). SKAPA is aminated with glutamate to form N-succinyl-LL-α,ε-diaminopimelic acid (SAPA). Removal of succinate from SAPA produces LL-α,ε-diaminopimelic acid (LLDAP). LLDAP is isomerized to produced meso-α, ε-diaminopimelic acid (MDAP). Decarboxylation of MDAP yields lysine.

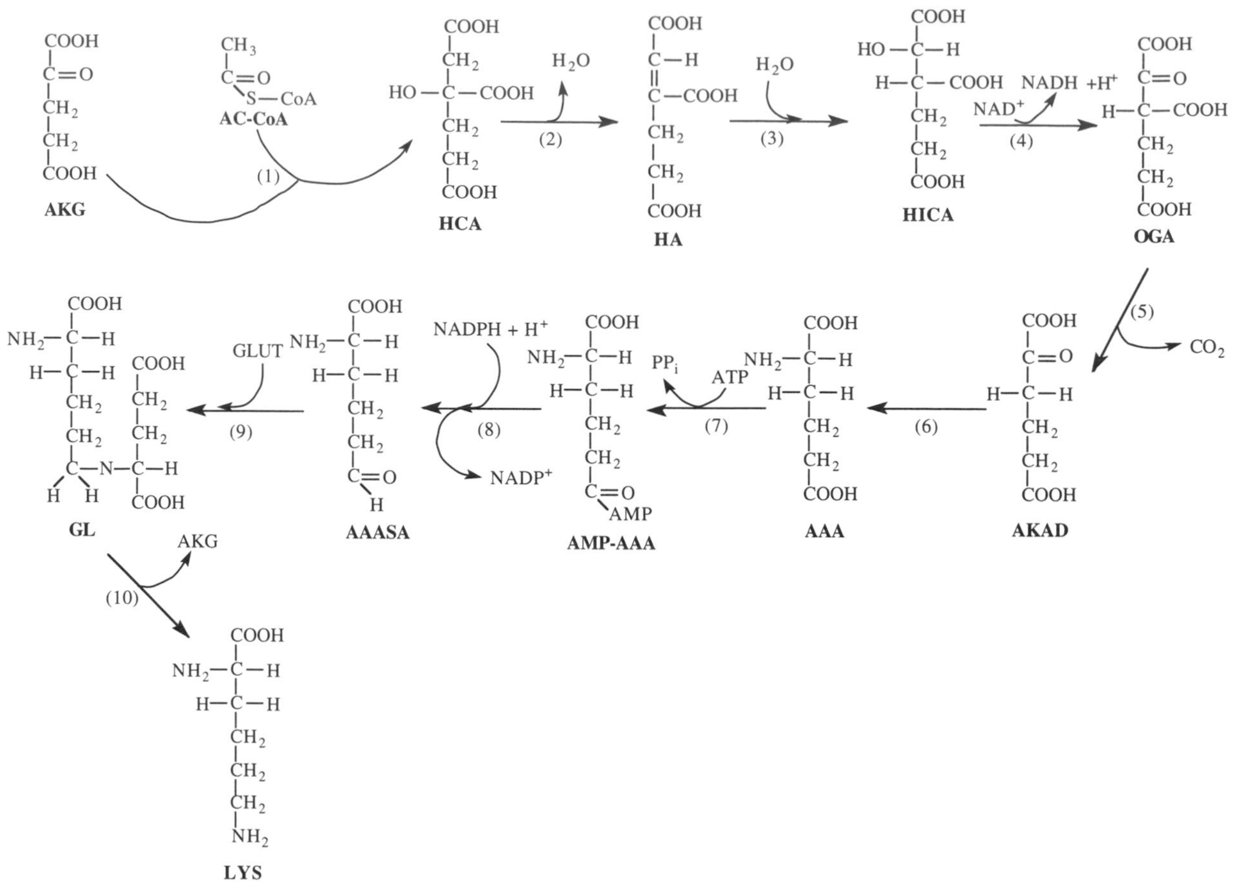

FIGURE 10.15 The alpha aminoadipate (AAA) pathway of lysine synthesis. (1) Condensation of α-ketoglutarate (AKG) and acetyl-CoA (AC-CoA) to form homocitric acid (HCA). (2) Dehydration of HCA to form *cis*-homoaconitic acid (HA). (3) Hydration of HA to produce homoisocitric acid (HICA). (4) NAD^+-mediated oxidation of HICA to form oxaloglutaric acid (OGA). (5) Decarboxylation of OGA to produce α-ketoadipic acid (AKAD). (6) Transamination of AKAD to produce α-aminoadipic acid (AAA), the unique intermediate of the sequence. (7) ATP-mediated adenylation of AAA to yield AMP-bound AAA. (8) NADPH-requiring reduction of AMP-AAA to form α-aminoadipyl semialdehyde (AAASA). (9) Glutamate (GLUT) interaction with AAASA to produce ε-*N*(L-glutaryl-2)-L-lysine (GL). (10) Lysine (LYS) formation from GL with the release of α-ketoglutarate (AKG). The sequence has been studied most thoroughly in *Saccharomyces* and *Neurospora* species. The formation of LYS from AAA is unusual in that semialdehyde formation involves a nucleotide-bound intermediate and the amination process involves formation of a glutamate-containing intermediate.

Certain algae, and most fungi, synthesize lysine by the pathway shown in Figure 10.15. The sequence differs markedly from the pathway just described. It begins with α-ketoglutarate (AKG). AKG reacts with acetyl-CoA, forming homocitric acid (HCA). By dehydration, HCA is converted to homoaconitic acid (HA) and the latter, by hydration, becomes homoisocitric acid (HICA). NADP-mediated oxidation of HICA forms oxaloglutaric acid (OGA) and decarboxylation of OGA produces α-ketoadipic acid (AKAD). By glutamate-dependent amination, AKAD is converted to α-aminoadipic acid (AAA), the intermediate from which the sequence derives its name. By NADPH and ATP-dependent reduction, the omega carboxyl function of AAA is reduced to an aldehyde, α-aminoadipic-ε-semialdehyde (AAASA). Reductive amination of AAASA yields lysine.

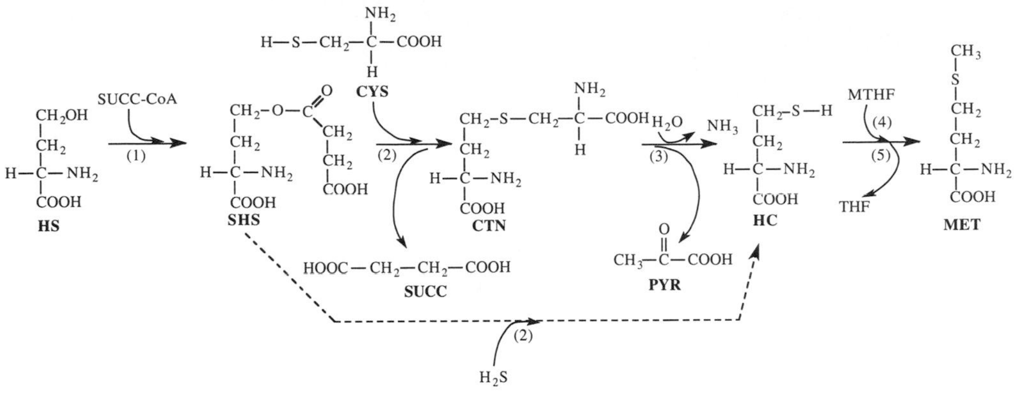

FIGURE 10.16 The reactions of the methionine branch of the aspartic amino acid family. (1) Formation of O-succinyl homoserine (SHS) from homoserine (HS) by the action of HS acetyltransferase. Succinyl CoA is used for this reaction. In many organisms, the action of this enzyme produces the O-succinyl derivative of HS, but in certain cases, the O-acetyl analogue is formed. (2) Reaction of SHS with cysteine (CYS), with succinate (SUCC) release, to form cystathione (CTN) by the action of CTN-γ-synthase. The latter enzyme is multifunctional and, in addition to facilitating CTN formation, may catalyze the direct formation of homocysteine (HC) by reaction of SHS, or its acetate analogue, with H_2S. (3) Conversion of CTN to HC, with the release of pyruvate (PYR) and NH_3, by the action of CTN lyase. (4,5) Formation of methionine (MET) from HC by methylation. Depending on the system, a vitamin B_{12}-requiring (4) or an B_{12}-independent (5) enzyme may be involved. In either case, 5-methyl tetrahydrofolate (MTHF) is formed from serine metabolism, with tetrahydrofolate (THF) as the donor and receiver of methyl groups.

The methionine and isoleucine branches of the aspartic acid pathway

Methionine (Figure 10.16) is formed by metabolism of homoserine (HS). The first step in the process involves reaction of succinyl CoA with the hydroxyl group of HS, producing succinyl homoserine (SHS). Reaction of SHS with cysteine forms cystathione (CTN). Removal of a pyruvate residue and NH_3 from CTN yields homocysteine (HC). Methionine (ME) is formed from HC by methylation with methylene tetrahydrofolate (MTHF) or with MTHF plus vitamin B_{12}.

Isoleucine (ILEU) and threonine (THR) synthesis (Figure 10.17) begins from HS. HS is phosphorylated with ATP to form homoserine phosphate (HSP). THR is formed from HSP by dehydration, isomerization, and dephosphorylation. THR, in turn, gives rise to ILEU. The THR is converted by deamination to α-ketobutyrate (AKB). AKB reacts with pyruvate (PYR) in decarboxylation, forming α-aceto-α-hydroxy-butyrate (AAAHB). AAAHB is a chemical analogue of acetolactate, a key intermediate in synthesis of valine, a branched-chain amino acid one carbon shorter than ILEU. AAAHB

reduction by NADPH produces α,β-dihydroxy-β-methylvalerate (ABHMV). Loss of water from ABHMV forms α-keto-β-methylvalerate (AKBMV) and amination of AKBMV with glutamate produces ILEU.

Common threads in amino acid synthesis

The descriptions in the previous sections illustrate the complexity of both the nature and consequences of amino acid synthesis and the interrelationships in the syntheses of particular amino acids. Interrelationships between amino acid synthesis and nucleotide synthesis are also abundant and complex. However, in the midst of complexity, commonality is found. Synthesis of purines, pyrimidines, and amino acids all begin with intermediates formed by catabolism. In addition, *common* intermediates are used for many processes, but in *different* ways. PRPP, for example, serves as a starting point for purine and pyrimidine formation and for synthesis of both histidine and tryptophan. As another example, acetyl-CoA, is a central participant in amino acid synthesis processes, but the manner of its utilization varies. In some cases, it is

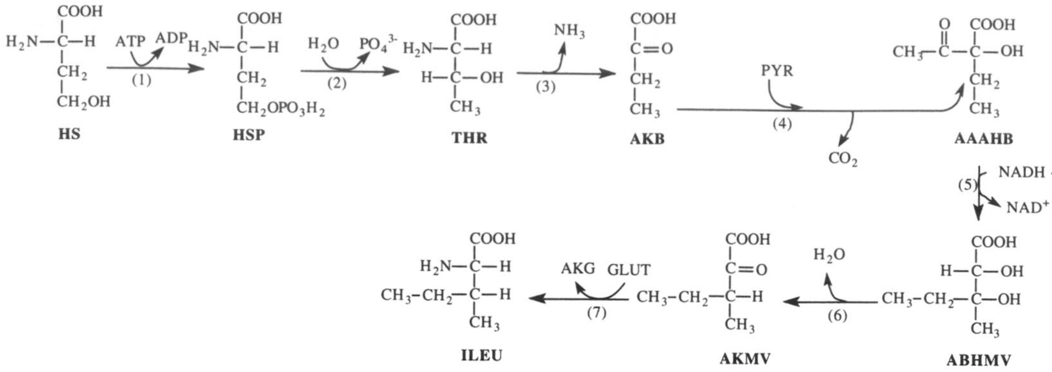

FIGURE 10.17 The reactions of the isoleucine (ILEU) branch of the aspartic amino acid family. (1) Phosphorylation of homoserine (HS) to form homoserine phosphate (HSP) by homoserine kinase. (2) Dehydration, dephosphorylation, and isomerization of HSP to form threonine (THR) by the action of THR synthase. (3) Conversion of THR to α-ketobutyrate (AKB) by THR deaminase. (4) Condensation of AKB and pyruvate (PYR), with CO_2 release, to form α-aceto-α-hydroxybutyrate (AAAHB) by acetohydroxybutyrate synthase II. (5) NADH-reduction and isomerization of AAAHB to form α,β-dihdroxy-β-methyl valerate (ABHMV) by the action of acetohydroxy acid isomeroreductase. (6) Dehydration of ABHMV to form α-keto-β-methylvalerate (AKMV). (7) Amination of AKMV, with glutamate (GLUT) as donor, to form ILEU by the action of branched-chain amino acid aminotransferase. Alpha-ketoglutarate (AKG) is formed in this reaction. Many of the reactions in the ILEU branch of the aspartic acid pathway, particularly reactions (4–7), are analogous to those involved in valine synthesis, although valine is usually considered a member of the pyruvate amino acid family. Much evidence indicates that, at least in many systems, the enzymes involved are identical and multifunctional.

used as a carbon source and in others it functions, by *N*-acetylation, to prevent ring formation and to allow selective phosphorylation. Glutamate and aspartate use is another example of the commonality and diversity of metabolic events in amino acid and nucleic acid synthesis. Both glutamate and aspartate serve as structural entities for the formation of many amino acids. At the same time, in both nucleotide and amino acid formation, glutamate and aspartate or the compounds derived from them, serve as amino donors.

To a substantial degree, the reactions of amino acid synthesis are parallel. In both the glutamic and aspartic families, *N*-acetylation, phosphorylation, reduction, and amination of central intermediates occur by analogous processes. In addition, the central intermediates of both pathways may be considered structural analogues, differing from each other only by a single carbon. A similar analogy is found in many of the intermediates involved in valine and isoleucine synthesis. Finally, certain of the intermediates of isoleucine synthesis bear both structural and metabolic similarity to citric acid cycle intermediates.

Regulation of amino acid synthesis

The complexity and branched nature of the metabolic sequences involved in amino acid formation is associated with an equally complex array of regulatory mechanisms for their synthetic processes. Virtually every recognized metabolic and physiological regulatory mechanism currently known is found within amino acid synthetic processes: repression and induction; catabolite activation; corepression; concerted, polyvalent, and sequential inhibition and repression; attenuation; and isoenzyme formation, all of which are discussed elsewhere (see Chapters 13 and 14).

Lipid synthesis

Lipid synthesis is a critical process for microbes. Although lipids normally constitute only a small portion of the cell's weight, they are universal components of the membranes and membranous organelles of the cell. Particularly in procaryotes, membranes participate in a diversity of physiological processes. The critical role of membranes in the physiology of procaryotes is a reflection of the relative morphological simplicity of the procaryotic cell. The centrality of lipids in membrane function in all cells requires that we understand the nature of the major lipid components of the cell and the manner of their synthesis.

Significant functional differences are found in the lipids found in procaryotic and eucaryotic microbes. The presence of sterols in the membranes of eucaryotes, and, with the exception of the mycoplasma, the absence of these materials in procaryotes, is the basis for the selective mode of action of many membrane-directed antimicrobial chemicals. Other lipid differences between procaryotic and eucaryotic cells are known. Thus poly-β-hydroxybutyrate, a storage substance formed from repeating units of β-hydroxybutyric acid, is found in many bacteria but not in eucaryotes. Conversely, neutral lipids, lipids in which organic fatty acids are esterified to glycerol, are characteristic storage compounds of eucaryotes, but are absent as storage compounds in procaryotes. Phospholipids are common to both types of organisms since they are integral components of all membranes. The occurrence of glycolipids, lipids containing a sugar residue, is more restricted than that of phospholipids. Among procaryotes, glycolipids are found as components of the membranes of certain photosynthetic organisms. In eucaryotes, similar substances are found in the chloroplast membrane, presumably a reflection of the fact that chloroplasts may be regarded as vestigial procaryotic cells. Lipids esterified to an amino acid are found in some procaryotes, such as the gram-negative bacteria since such lipids are a major and essential part of the gram-negative cell envelope, but these lipids are absent in eucaryotes.

Polyisoprenoid lipids and compounds containing them are found in both procaryotes and eucaryotes. Carotenoids may participate in light-harvesting in both procaryotic and eucaryotic phototrophs and, in addition, may exert a protective effect against excessive radiation. Bactoprenol (undecaprenol) is an essential lipid carrier molecule for procaryotes and is intimately involved in wall synthesis in both gram-negative and gram-positive bacteria. Although undecaprenol is absent in eucaryotes, dolichol, a chemical analogue, participates in synthesis of the cell wall polymers of

certain fungi. Finally, quinones and isoprenoid compounds participate in both photosynthesis and nonphototrophic electron transport in both procaryotic and eucaryotic cells.

When considered in all of their diversity, microbes contain a wide variety of lipids. However, in the midst of this diversity, the synthetic mechanisms for lipids are, relatively speaking, few. Aside from lipid A formation, the major components of lipid synthesis in microbes are: 1) the formation of saturated and unsaturated fatty acids, 2) the formation of isoprenoids, and 3) the formation of complex lipids.

Fatty acid synthesis

Although differences are found in the chain lengths of the fatty acids in microbial lipids, and in the proportions that various acids constitute of the total cellular fatty acid, the basic mechanism by which fatty acids are formed is universal in all cellular life forms and is shown in Figure 10.18. For even-chain acids, the process begins with acetyl-CoA. A molecule of acetyl-CoA combines with a molecule of a special protein, acyl carrier protein (ACP). A second acetyl-CoA molecule reacts with CO_2 and biotin to form malonyl-CoA, which in turn, reacts with ACP, giving rise to an ACP-bound malonyl residue and releasing a molecule of CoA. Subsequent events of saturated fatty acid synthesis involve repeated condensations of malonyl-ACP and acetyl-ACP residues, with the release of a CO_2 molecule and an ACP molecule at each condensation event. These processes produce a sequential series of ACP-bound β-keto compounds that are reduced to β-hydroxy compounds by NADPH-dependent enzymes. The β-hydroxy compounds then are dehydrated, forming α, β-unsaturated intermediates. A second NADPH-mediated reduction forms saturated ACP-bound intermediates two carbons longer than their predecessors. Repetition of these processes allows sequential addition of two-carbon moieties and produces a supply of even-chain saturated fatty acids. Odd-chain fatty acids are formed by an analogous process except that a propionyl-ACP residue and a malonyl-ACP participate in the initial reaction.

In most, if not all, organisms, a certain portion of the fatty acid residues are unsaturated, so mechanisms for unsaturated acid formation must exist. The mechanisms fall into two major categories. The

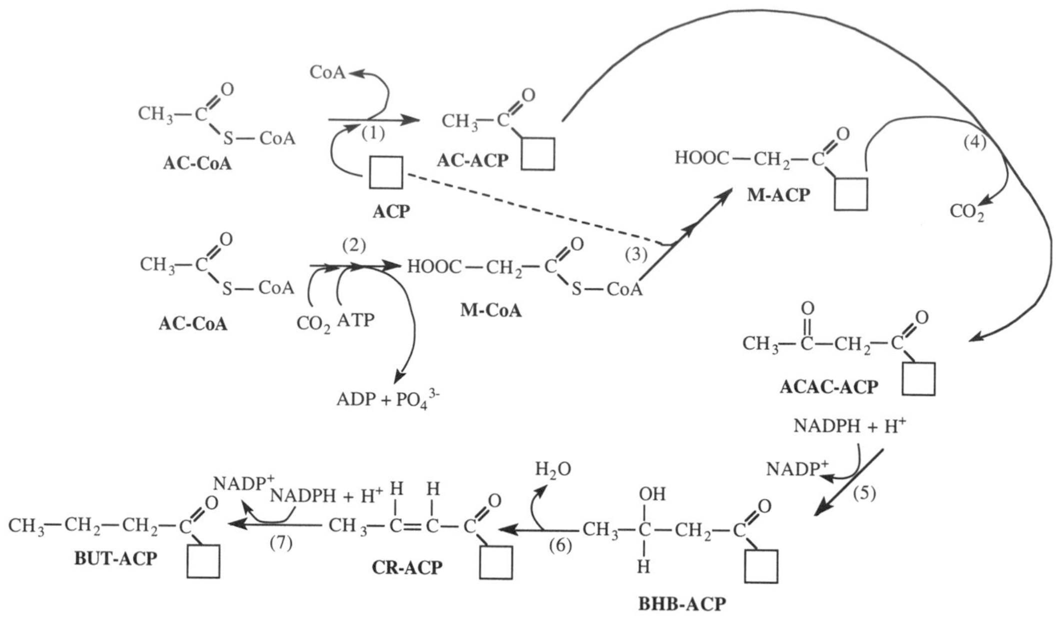

FIGURE 10.18 The general pathway for microbial saturated fatty acid synthesis. (1) Synthesis of acetyl-acyl carrier protein (AC-ACP) from acetyl-CoA (AC-CoA) and free acyl carrier protein (ACP) with CoA release by the action of acetyl transacetylase. (2) Formation of malonyl-CoA (M-CoA) from AC-CoA and CO_2 by AC-CoA carboxylase. This enzyme is ATP- and biotin-dependent. (3) Transfer of the malonyl portion of M-CoA to ACP to form malonyl-ACP. The remaining reactions occur as ACP-bound intermediates. (4) Formation of acetoacetyl-ACP (ACAC-ACP) from AC-ACP and M-ACP, with CO_2 loss by 3-ketoacyl-ACP synthase. (5) NADPH-mediated reduction of ACAC-ACP to β-hydroxybutyryl-ACP (BHB-ACP) by 3-keto-ACP reductase. (6) Dehydration of BHB-ACP to crotonyl-ACP(CR-ACP) by β-hydroxyacyl-ACP dehydratase. (7) NADPH-requiring reduction of CR-ACP to butyryl-ACP (BUT-ACP) by enoyl-ACP reductase. Repetition of these steps allows formation of a diversity of ACP-bound saturated fatty acids by sequential addition of two-carbon residues. If AC-CoA is the sole starting material, only even-chain acids are produced, but odd-chain acids may be formed if propionyl-CoA, in addition to AC-CoA, is a starting material. In general, saturated fatty acid-synthesizing enzymes are relatively nonspecific in their substrates. In some cases, the individual enzymes are soluble but in others they function as a multienzyme complex.

anaerobic pathway involves unsaturation in the process of synthesis, while the *aerobic* pathway involves desaturation of saturated fatty acids after synthesis. Both schemes are shown in Figure 10.19. In the aerobic pathway, an ACP-bound palmitic acid (16C) molecule reacts with oxygen to form a 16-carbon monounsaturated molecule, palmitoleic acid, with the unsaturated bond between carbons 9 and 10. The anaerobic mechanism, which may also occur in aerobes, depends upon selective dehydration of β-hydroxydecanoyl-ACP. This acid may be dehydrated either between the α and β carbons or between the β and γ carbons. In the first case, subsequent addition of two-carbon residues produces saturated fatty acids, and in the second case, the

unsaturated bond is retained, and by further C2 unit addition, results in the formation of *cis*-vaccenic acid, the characteristic 18-carbon, mono-unsaturated (Δ11) compound of the anaerobic pathway.

Complex lipid formation

At some point, the fatty acids synthesized on ACP-bound intermediates are converted to the phospholipids of the membrane. The general ways in which this process occurs are shown in Figure 10.20. Basically, ACP-fatty acids interact with molecules of glycerol-3-phosphate (α-glycerol phosphate) to produce a phosphatidic acid—a molecule of glycerol esterified to two fatty acid residues and a

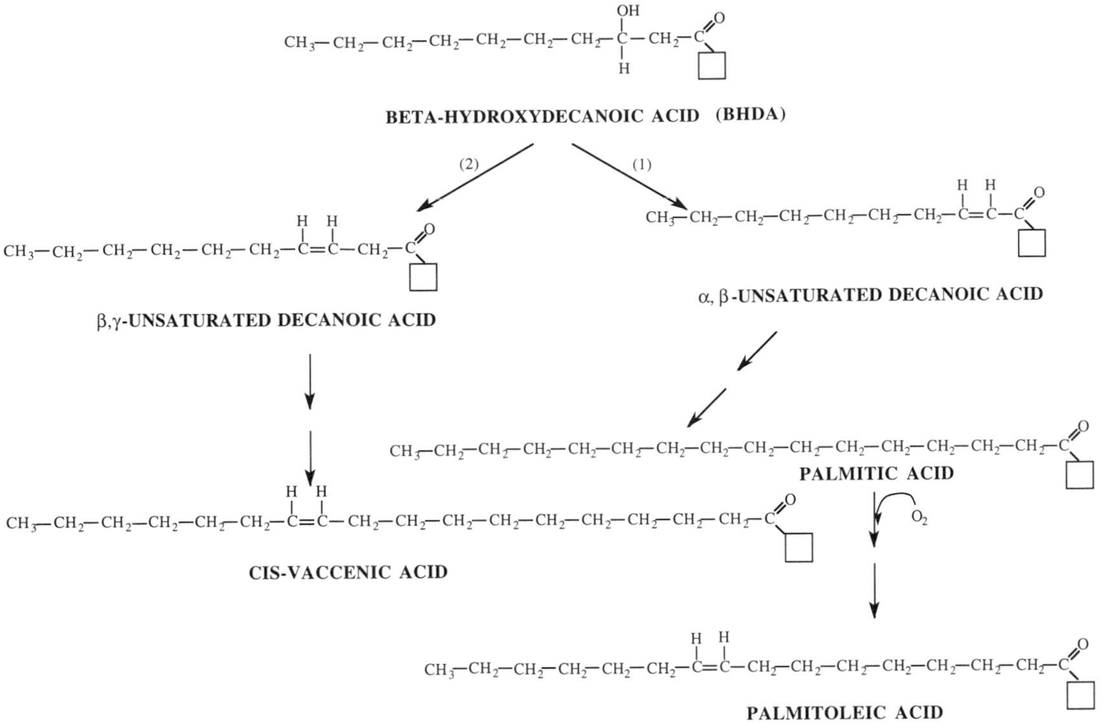

FIGURE 10.19 The aerobic (1) and anaerobic (2) mechanisms of unsaturated fatty acid synthesis. The critical intermediate is β-hydroxydecanoic acid (BHDA). Alpha-beta unsaturation of BHDA allows the product to serve as a substrate for enoyl-ACP reductase. Thus, with α, β-unsaturation of BHDA, saturated acids are formed by further synthetic reactions. In this situation, unsaturated acids result from oxygen-requiring desaturation of previously formed saturated acids. In the anaerobic mechanism, β, γ-unsaturation of BHDA by a special dehydratase precludes the product from serving as a substrate for enoyl-ACP reductase. Thus, the unsaturated bond is retained during further metabolism, leading to formation of cis-vaccenic acid, an 18-carbon monounsaturated acid with a double bond between carbons 11 and 12. The aerobic mechanism produces unsaturation between carbons 9 and 10. Although cis-vaccenic and palmitoleic acids are shown as examples, unsaturated acids of different chain lengths may also be formed by similar reactions. The anaerobic pathway is not restricted to anaerobes alone, occurring in some aerobes as well.

phosphate moiety. Modifications of the phosphatidic acid structure produce the various phospholipids of the membrane.

Phosphatidic acid formation is a stepwise process, involving the sequential interaction of ACP-bound fatty acids with a glycerol-3-phosphate residue derived from reduction of dihydroxyacetone phosphate. The initial fatty acid addition, which may involve either of the free hydroxyl groups of glycerol, forms a lysophosphatide. The second fatty acid–α-glycerol phosphate interaction produces a phosphatidic acid. The subsequent fate of phosphatides reflects the nature of the substances with which they interact. By reaction with cytosine triphosphate (CTP), phosphatidic acids

are converted to CDP-bound diglycerides. Reaction of the CDP-bound diglycerides with serine produces phosphatidyl serine and the decarboxylation of phosphatidyl serine yields phosphatidyl ethanolamine. Methylation of the phosphatidyl ethanolamine with S-adenosylmethionine forms phosphatidyl choline.

As an alternative to the processes just described, a CDP-diglyceride may react with a second molecule of glycerol-3-phosphate, initially forming phosphatidyl glycerol phosphate, whose dephosphorylation forms phosphatidyl glycerol. If phosphatidyl glycerol phosphate reacts with a second molecule of CDP-diglyceride, diphosphatidyl glycerol (cardiolipin) is produced.

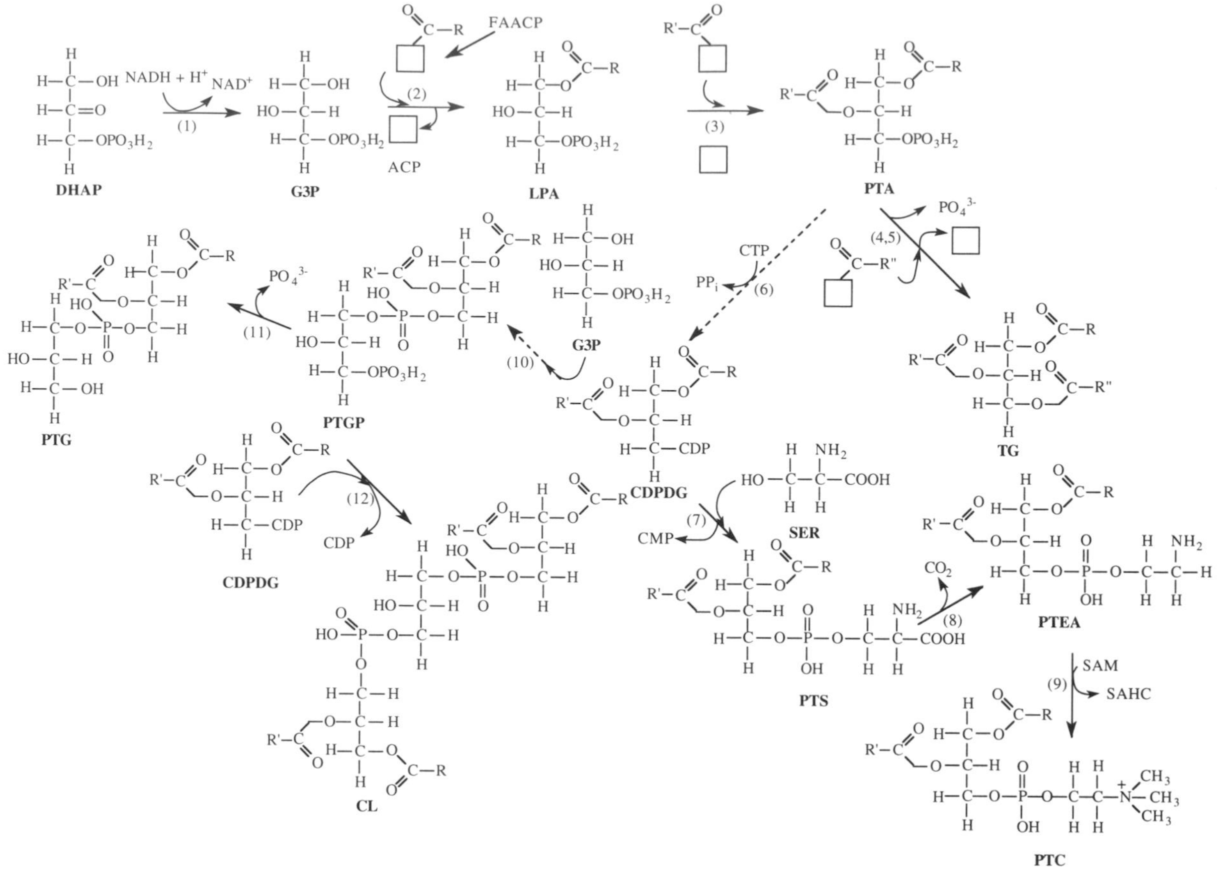

FIGURE 10.20 The major modes of complex lipid formation in microbes. (1) Formation of glycerol-3-phosphate (G3P) from dihydroxyacetone (DHAP) by G3P dehydrogenase. (2) Reaction of a molecule of acyl carrier protein-bound fatty acid (FAACP) with G3P to form a lysophosphatidic acid (LPA) with the release of a molecule of acyl carrier protein (ACP) by the action of glycerol phosphate acyltransferase. (3) Addition of a second fatty acid residue to G3P to form a phosphatidic acid (PTA). (4,5) Dephosphorylation of G3P and addition of a third fatty acid moiety to form a triglyceride (TG). (6) Reaction of PTA with cytosine triphosphate (CTP) to form a cytosine diphosphate-bound diglceride (CDPDG) by the action of phosphatidate cytidyl transferase. (7–9) Conversion of CDPDG to phosphatidyl serine (PTS) (7), decarboxylation of PTS to form phosphatidyl ethanolamine (PTEA) (8), and methylation of PTEA with S-adenosylmethionine (SAM) to form phosphatidyl choline (PTC) (9). In the latter reaction, S-adenosyl homocysteine (SAHC) is formed. (10–12) Reaction of CDPDG with G3P to yield phosphatidylglycerol phosphate (PTGP). The PTGP may be dephosphorylated (11) to yield phosphatidyl glycerol (PTG) or converted to cardiolipin (CL) (12). CL formation in the bacteria normally occurs by reaction of two PTGP molecules with release of one molecule of G3P. Alternatively, a PTGP molecule may react with a molecule of CDPDG.

Contrasts between fatty acid synthesis and degradation   Superficially, fatty synthesis and degradation might appear as reverse processes, but this is not the case. Net oxidation of fatty acids involves CoA-bound intermediates, while, with the exception of β-hydroxybutyrate, synthesis is mediated by ACP-bound entities. In addition, oxidative enzymes typically use NAD as an acceptor, whereas NADPH is the reductant in synthetic processes. Finally, synthesis requires biotin, but oxidation is biotin-independent.

Isoprenoid synthesis

Unsaturated, branched, linear lipids, such as the isoprenoids, arise from CoA intermediates. The general mechanism for isoprenoid synthesis is shown in Figure 10.21. Initially, two molecules of acetyl CoA condense in a head-to-tail fashion (see Chapter 8) to form acetoacetyl-CoA. Acetoacetyl-CoA is also an intermediate in β-hydroxybutyrate synthesis. Reaction of an additional molecule of

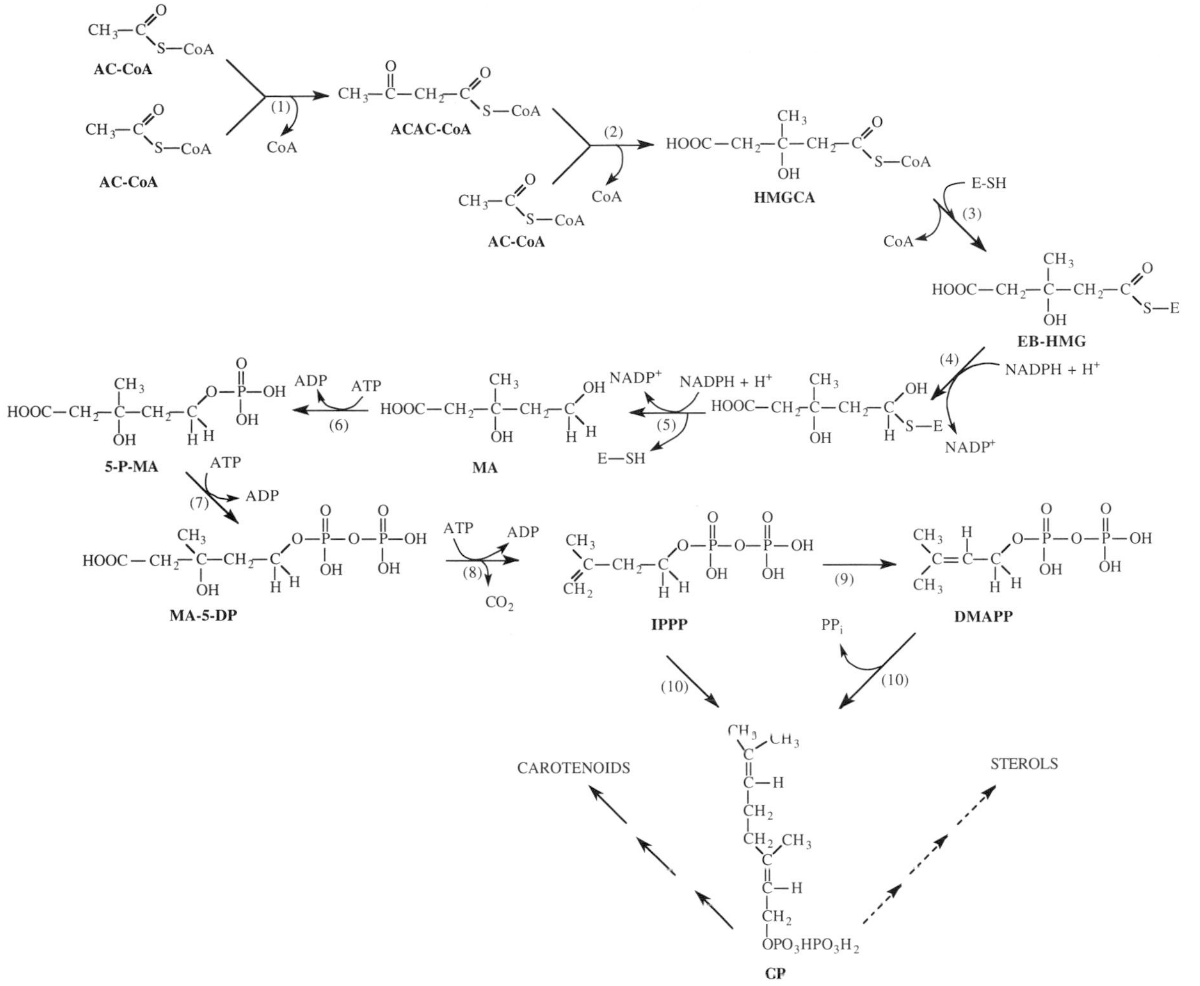

FIGURE 10.21 Microbial synthesis of isoprenoid compounds. (1) Formation of acetoacetyl-CoA (ACAC-CoA) from two molecules of acetyl-CoA (AC-CoA). (2) Conversion of ACAC-CoA to β-hydroxymethyl glutaryl-CoA (HMGCA) by reaction of AC-CoA and ACAC-CoA. (3) Conversion of HMGCA to an enzyme-bound form of EB-HMG with CoA release. (4,5) NADPH-mediated reduction of EB-HMG, with enzyme release, to form mevalonic acid (MA). (6,7) Sequential ATP-mediated phosphorylation of MA to yield mevalonic acid 5-diphosphate (MA-5-DP), with intermediate formation of mevalonic acid-5-phosphate (5-P-MA). (8) ATP-dependent decarboxylation of MA-5-DP to form isopentyl pyrophosphate (IPPP). (9) Isomerization of IPPP to yield dimethylallyl pyrophosphate (DMAPP). (10) Formation of geranyl phosphate (GP) by condensation of molecules of IPPP and DMAPP. Subsequent metabolism of GP may lead to the formation of carotenoids or sterols.

acetyl-CoA with acetoacetyl-CoA with release of a CoA molecule, forms hydroxymethyl glutaryl CoA (HMGCA), the first committed step in isoprenoid synthesis. NADPH-dependent reduction of HMGCA, with CoA release, produces mevalonic acid (MA) and two ATP-mediated phosphorylations produce 5-diphosphomevalonic acid (MA-5-DP). Decarboxylation and unsaturation of

MA-5-DP by dehydration produces isopentyl pyrophosphate (IPPP). By isomerization of IPPP, dimethylallyl pyrophosphate (DMAPP) is formed. Additional chain elongation occurs by reaction between a DMAPP molecule and a varying number of IPPP residues. These reactions allow formation of the various carotenoid pigments and, in certain cases, the sterol ring as well.

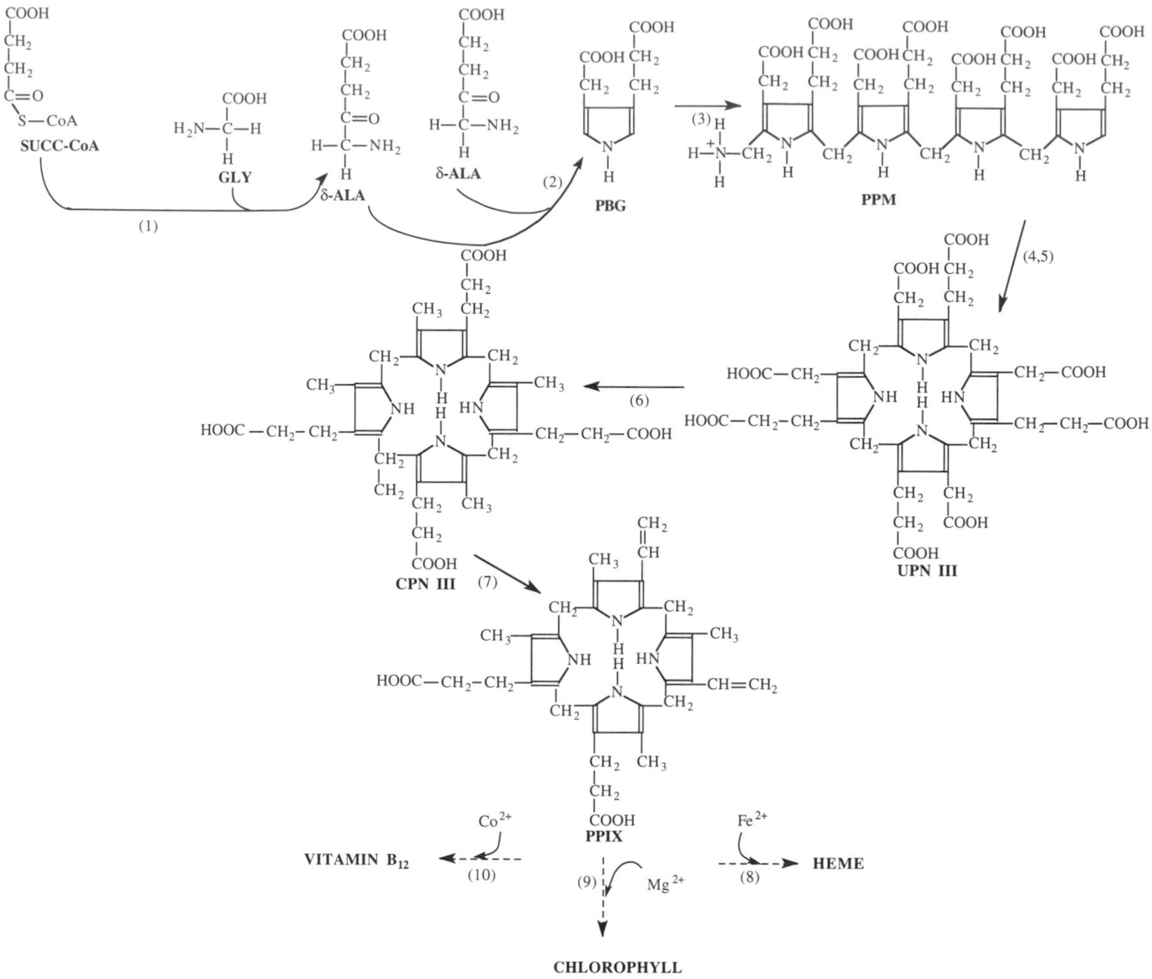

FIGURE 10.22 Processes involved in synthesis of tetrapyrrolic compounds. (1) Synthesis of Δ-aminolevulinic acid (Δ-ALA) from succinyl CoA (SUCC-CoA) and glycine (GLY) by the action of Δ-ALA synthase. (2) Conversion of two molecules of Δ-ALA to porphobilinogen (PBG) by Δ-ALA dehydratase. (3–5) Formation of polypyrrl methane (PPM) from 4PBG molecules by PBG deaminase (3) and conversion of PPM to uroporphyinogen III (UPN III) by the joint action of UPN III synthase and cosynthase (4,5). In the absence of the cosynthase, UPN in the I, rather that the III, configuration is formed. (6) Conversion of the acetyl side chains of UPN III to methyl side chains by decarboxylation to form copropophyrinogen III (CPN III). (7) Conversion of the propionyl side chains of CPN III to vinyl groups by decarboxylation and dehydrogenation. (8–10) Insertion of Fe^{2+} (8), Mg^{2+} (9), or Co^{2+} (10) into the tetrapyrrole ring structure to form heme, chlorophyll, or vitamin B_{12}. Substantial modification of the ring structure of the protoporphyrin IX (PPIX) produced by CPN III decarboxylation is required to allow chlorophyll and vitamin B_{12} formation.

Tetrapyrrole formation

Tetrapyrrolic compounds are central to all of biology. They are components of various chlorophylls and are involved in oxygen transport in animals as heme-proteins. In addition, throughout nature, they are components of oxidative enzymes such as peroxidase, lactoperoxidase, verdoperoxidase, and catalase. Finally, they are components of the cytochrome pigments in a diversity of electron transport systems and are thus intimately involved in energy metabolism.

Synthesis of the tetrapyrrole nucleus is a critical physiological process in both microbial and nonmicrobial systems. The universal pathway of heme synthesis is shown in Figure 10.22. It begins with the condensation of succinyl CoA and glycine

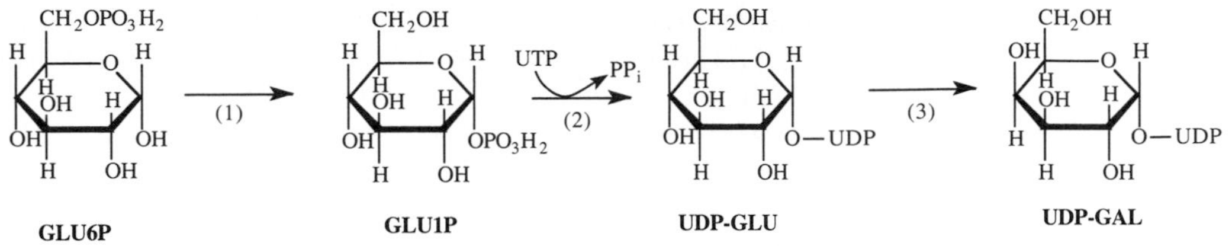

FIGURE 10.23 The formation of galactose from glucose. (1) Conversion of glucose-6-phosphate (GLU6P) to glucose-phosphate (GLU1P) by phosphoglucomutase. (2) Activation of glucose by reaction with uridine triphosphate (UTP) to produce uridine diphosphate glucose (UDP-GLUC) by UDP-GLU pyrophosphorylase. (3) Isomerization of UDP-GLU to form UDP-galactose (UDP-GAL) by an epimerase.

to form delta-aminolevulinic acid (DAL). Condensation of two molecules of DAL produces porphobilinogen (PBG). Condensation of four molecules of PBG produces uroporphyrinogen III (UP), the first tetrapyrrole intermediate of the pathway. By a series of reactions, UP becomes protoporphyrin IX (PPIX). Modification of the PPIX ring structure and metal insertion produces metallotetrapyrrole compounds as diverse as chlorophylls, hemes, and vitamin B_{12}.

Carbohydrates

Since carbohydrate metabolism is discussed extensively in other chapters of this book, only brief attention to the subject will be given here. A great number of microbes ferment glucose and many are also able to degrade polymers of glucose, such as starch and cellulose. In addition, many microbes use other sugars. In general, carbohydrate polymers are degraded to small molecules, often dimers, which are taken into the cell, converted to monomers, and used by central pathways. Galactose is a particularly important monosaccharide since it is used by certain organisms as the sole carbon and energy source and is an obligatory intermediate in lipopolysaccharide synthesis. Galactose is often formed by isomerization of UDP-bound glucose that may have been formed from isomerization of glucose-6-phosphate to glucose-1-phosphate and the reaction of the latter with UTP. These reactions are shown in Figure 10.23. In procaryotes, carbohydrate polymers normally are formed from ADP glucose, which may be formed from glucose-1-phosphate. UDP-bound intermediates are normally the carbohydrate polymer precursors in eucaryotic animal cells.

Summary

Small molecule formation and utilization are critical processes for both procaryotic and eucaryotic cells. Without the formation of small molecules, synthesis of critical cell macromolecules and structures would not be possible. It is through the formation and utilization of small molecules that the metabolic and physiological processes essential to life are accomplished and that the connection is made between formation and utilization of energy.

In the midst of synthetic diversity, nature is fundamentally conservative. All of the small molecules required for biosynthesis are derived from only twelve compounds. These twelve compounds typically arise from the degradative metabolic sequences that provide the energy for their use. It is truly remarkable that the use of twelve intermediates, or the substances readily derived from them, allows formation of all critical cell substances and structures.

The ability to form a diverse spectrum of cell substances and structures from a relative paucity of starting materials reflects the fact that critical synthetic molecule formation is an interconnected process. In other words, the formation of a particular synthetic molecule is biochemically related to the formation of others. Interconnection occurs not only *within* a particular molecule class (e.g., amino acids), but also *between* classes (e.g., amino acids and nucleotides).

The interconnectedness of anabolism is, in some respects, similar to that for catabolism. We cannot fully appreciate catabolism or anabolism without considering them together. From the viewpoints of both energy and synthesis, the two processes may be considered as inverses of each other, but both processes are comprised of a series of interconnected biochemical events, which allows a lot to be achieved from the operation of relatively few, but central, processes. Because of the interconnectedness of metabolism, its study allows understanding, not only of its chemical consequences, but also of the diverse ways in which it is regulated at both the biochemical and genetic level.

Selected References

Barker, H. A. 1978. Explorations of bacterial metabolism. *Annual Review of Biochemistry.* **47**: 1–34.

Braus, G. H. 1991. Aromatic amino acid biosynthesis in the yeast *Saccharomyces cereviseae:* A model system for the regulation of a eukaryotic biosynthetic pathway. *Microbiological Reviews.* **55**: 349–370.

Campbell, Iain M. 1984. Secondary metabolism and microbial physiology. *Advances in Microbial Physiology.* **25**: 2–60.

CIBA Foundation. Purine and pyrimidine metabolism. 1977. *CIBA Foundation Symposium.*

Cole, J. A. 1976. Microbial gas metabolism. *Advances in Microbial Physiology.* **14**: 1–92.

Cooper, A. J. L. 1983. Biochemistry of sulfur-containing amino acids. *Annual Review of Biochemistry.* **52**: 187–222.

Dalton, H. 1979. Utilization of inorganic nitrogen by microbial cells. *International Review of Biochemistry.* **21**: 227–266.

Danson, Michael J. 1988. Archaebacteria: The comparative enzymology of their central metabolic pathways. *Advances in Microbial Physiology.* **29**: 166–231.

Dawidowicz, E. A. 1987. Dynamics of membrane lipid metabolism and turnover. *Annual Review of Biochemistry.* **56**: 43–62.

Doelle, H. W. 1975. *Bacterial Metabolism,* 2nd edition. Academic Press, Inc., New York.

Finnerty, W. R. 1978. Physiology and biochemistry of bacterial phospholipid metabolism. *Advances in Microbial Physiology.* **18**: 177–233.

Fraenkel, D. G., and R. T. Vinopal. 1973. Carbohydrate metabolism in bacteria. *Annual Review of Microbiology.* **27**: 69–100.

Gottschalk, G. 1986. *Bacterial Metabolism,* 2nd edition. Springer-Verlag, New York, Berlin, Heidelberg, Tokyo.

Herrman, K. M., and R. L. Somerville (eds.). 1983. *Amino acids: Biosynthesis and genetic regulation.* Addison-Wesley, Reading, Massachusetts.

Knowles, C. J., and A. W. Bunch. 1986. Microbial cyanide metabolism. *Advances in Microbial Physiology.* **27**: 73–112.

Kulaev, I. S., and V. M. Vagabov. 1983. Polyphosphate metabolism in micro-organisms. *Advances in Microbial Physiology.* **24**: 83–171.

Lovley, D. R. 1991. Dissimilatory Fe(III) and Mn(IV) reduction. *Microbiological Reviews.* **55**: 259–287.

Moat, A. G., and J. W. Foster. 1988. *Microbial Physiology,* 2nd edition. John Wiley and Sons, New York.

Mohn, W. W., and J. M. Tiedje. 1992. Microbial reductive dehalogenation. *Microbiological Reviews.* **56**: 482–507.

O'Brian, M. R., and R. J. Maier. 1988. Hydrogen metabolism in *Rhizobium:* Energetics, regulation, enzymology and genetics. *Advances in Microbial Physiology.* **29**: 2–52.

Truffa-Bachi, P., and G. N. Cohen. 1973. Amino acid metabolism. *Annual Review of Biochemistry.* **42**: 113–134.

Umbarger, H. E. 1978. Amino acid biosynthesis and its regulation. *Annual Review of Biochemistry.* **47**: 533–606.

van den Bosch, H. 1974. Phosphoglyceride metabolism. *Annual Review of Biochemistry.* **43**: 243–278.

Vignais, P. M., A. Colbeau, J. Willison, and Y. Jouanneau. 1985. Hydrogenase, nitrogenase, and hydrogen metabolism in the photosynthetic bacteria. *Advances in Microbial Physiology.* **26**: 156–235.

Wakil, S. J., J. K. Stoops, and V. C. Joshi. 1983. Fatty acid synthesis and its regulation. *Annual Review of Biochemistry.* **52**: 537–579.

Weinberg, Robert A. 1973. Nuclear RNA metabolism. *Annual Review of Biochemistry.* **42**: 329–354.

Zeikus, J. G. 1983. Metabolism of one-carbon compounds by chemotrophic anaerobes. *Advances in Microbial Physiology.* **24**: 215–299.

CHAPTER 11

Protein Synthesis

The centrality of protein synthesis

Protein synthesis is a central physiological process that occurs in all cellular life forms. Its importance is underscored by the fact that it is a process accomplished not only by cellular, but also by noncellular, life forms—the viruses, who divert host cell machinery from its "normal" cellular activities. In most, if not all, cases, proteins constitute the bulk of the weight of the cells. In addition, no cellular process is understandable or controllable without consideration of proteins. Proteins mediate all of the critical aspects of cellular physiology: the generation of energy, the formation of cell building blocks, the formation of macromolecules, the formation of structures, the uptake and removal of substances, the regulation and coordination of cellular activities, and last but by no means least, their own formation. For all of these reasons, it is essential that a microbial physiologist have a fundamental understanding of the process of protein synthesis.

Protein synthesis is a multicomponent process

From the physiological viewpoint, the protein synthesis process may be considered from a number of perspectives. For example, it may be considered on the basis of its components, transcription and translation. Although *transcription,* the process of forming the message for protein, is not, by itself, protein synthesis, transcription is essential for the protein synthesis process. Although it occurs at the genome, transcription dictates the nature and order of the amino acids that are linked together to form the protein and also provides molecules of transfer ribonucleic acid (tRNA), the molecule by which activated amino acids may be recognized and added to the growing polypeptide. Furthermore, transcription provides molecules of ribosomal RNA (rRNA) from which the ribosomal particle is made. The intimate interactions between genome, ribosome, and other cell areas are such that it is useful to consider protein synthesis as an example of a multicomponent process that involves the coordinated functioning of various cell components. In addition to these perspectives, protein synthesis may be regarded as a process whose accomplishment depends on the formation and function of particulate entities, the ribosomes. We may further consider it as a model for understanding the transmission of information and as a model for understanding the interaction of macromolecules. Finally, protein synthesis may be regarded as the mechanism of genetic expression, because it is through the formation of proteins that the potential abilities of all living things are expressed.

Each of these perspectives are entirely valid ways of regarding the protein synthesis process and none of them are mutually exclusive. The complexity and implications of protein synthesis are so profound that the process cannot be understood without, at the same time, understanding a number of other phenomena and processes. These processes are not, by themselves, protein synthesis, but protein synthesis cannot occur without them and their interaction.

Requirements for the process

The general nature of the protein synthesis process is shown in Figure 11.1. To start, a code must be formed—a molecule of messenger ribonucleic acid (mRNA) that tells the ribosome the nature and order of addition of the amino acids that will constitute the protein. Formation of the message requires that we have both an entity, DNA, which specifies the nature of the message, and a mechanism for code formation, that is, a means of forming mRNA. The mRNA code is of no physiological value, however, unless a functioning ribosome is available. This ribosome is composed of subparticles that participate intimately in the process of joining, by peptide bonds, the amino acids specified by a mRNA molecule to form a particular protein. However, the subparticles of the ribosome cannot operate as discrete individuals to accomplish protein formation. Instead, ribosomal subparticles react in specific ways, and in a particular sequence, so that intact ribosomes and proteins may be formed. Normally ribosomes act in coordinated collections known as *polysomes* that are attached at various locations to a mRNA molecule, allowing the simultaneous formation of several proteins from a single messenger molecule.

Building blocks

Functioning ribosomes depend upon the availability of building blocks, the activated amino acids formed at cellular locations other than the genome or the ribosomes. The availability of activated amino acids materially affects the protein synthesis process. Limitations on amino acid availability may result from alterations of the synthetic sequences for particular amino acids, limitation of formation of their activated derivatives, the amino acid adenylates, or limitation of conversion of these derivatives to forms that are recognized by message-attached ribosomes (molecules of amino acyl tRNA). In addition, for those organisms deficient in synthesis of particular amino acids, factors affecting uptake of pre-formed amino acids from the external environment can exert serious effects, not only on protein formation but on the rate of message synthesis as well.

Message formation

Unless message formation, that is, the formation of mRNA by transcription, occurs, potentially functional ribosomes are of little or no value since they have no "instructions" on how to proceed in the formation of protein. It is only if a code is available that protein can be formed. Formation of the mRNA code is a complicated process that occurs in a substantially different way in procaryotes than in eucaryotes, although, in principle, the processes are similar.

Message formation in procaryotes

In procaryotes, message formation is dependent on a single multicomponent enzyme, DNA-dependent RNA polymerase. The manner in which this enzyme functions in message formation is depicted in Figure 11.2. The enzyme contains five subunits, two denoted alpha, and the others as beta, beta prime, and sigma. Each polymerase molecule contains two alpha subunits, whose functions are not entirely clear; a single beta unit that forms phosphodiester bonds; a single beta prime unit that binds the DNA template; and a single sigma unit whose function is to recognize the sites, called the promoter regions, along the DNA to which the polymerase enzyme may attach. The entire enzyme is referred to as a *holoenzyme*, while the enzyme with the sigma factor removed is known as the *core*

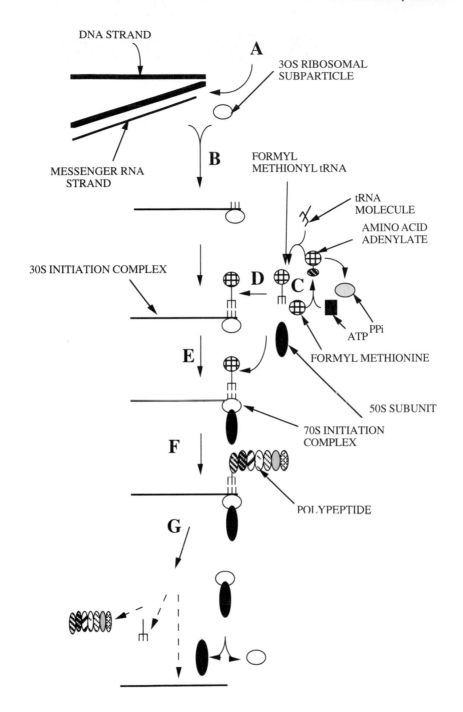

FIGURE 11.1 The general nature of protein synthesis in procaryotes. (A) Transcription of a messenger RNA molecule. (B) Interaction of a mRNA molecule with a 30S ribosomal subparticle. (C) Mobilization of a molecule of formyl methionine so that it can be used to initiate protein formation. Initially, the formyl methionine reacts with ATP, by pyrophosphate release, to form an adenylate. This reaction is followed by transfer of formyl methionine to a molecule of transfer RNA. These general processes are repeated when subsequent amino acids are mobilized for addition to the growing polypeptide. (D) Interaction of the tRNA-attached formyl methionine molecule with the 30S-messenger RNA complex to produce the 30S initiation complex. (E) Addition of a 50S ribosomal subparticle to the resultant of (D), to form the 70S initiation complex. (F) Repetition of processes (C) and (D), and repeated transpeptization and translocation to allow protein formation. (G) Dissociation of the system so that 50S and 30S ribosomal subparticles may participate in additional protein formation. The intricacies of the protein synthesis process are discussed in the text and depicted graphically in other figures.

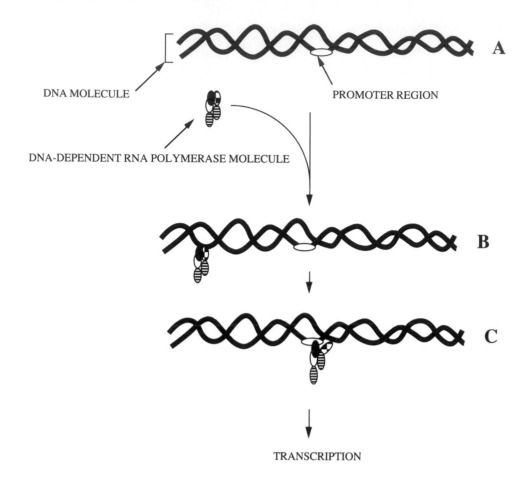

FIGURE 11.2 The initial interaction of DNA-dependent RNA polymerase and a DNA molecule. (A) At the outset, the polymerase and DNA are separate entities. DNA-dependent RNA polymerase contains two alpha units (horizontal lines), a beta unit (clear), a beta prime unit (checkered), and a sigma unit (black). The beta prime unit binds to the DNA template. The beta unit catalyzes phosphodiester linkages. The sigma unit specifically recognizes promoter regions so that transcription may occur nonrandomly. (B) After an original nonspecific attachment to a double-stranded DNA molecule, the polymerase migrates along the double-stranded molecule until a promotor region is encountered (C).

enzyme. That the sigma portion of the enzyme is involved in recognition is shown by experiments in which the properties of the holoenzyme were compared with the properties of the core enzyme. These studies showed that the core enzyme attaches randomly to DNA and gives rise to random mRNA transcripts, but the holoenzyme gives rise to specific transcripts.

Under normal circumstances, when the complete enzyme is functioning, transcription occurs in the following way. DNA-dependent RNA polymerase initially attaches to a double-stranded DNA molecule and moves down it until it encounters a promoter region. The promoter region is a region

upstream from the *structural region,* which codes for structural proteins as well as transfer RNA (tRNA) and ribosomal RNA (rRNA). Studies have shown that, in procaryotes, certain nucleotide sequences are found in many, if not all, promoters and at identical locations relative to the point at which transcription begins. A sequence of TATAAT is found at a location 10 bases upstream from the transcription start site, the so-called –10 region. In addition, a sequence of TTGACA is found at a location 35 bases upstream from the transcription start site, the so-called –35 region. The sigma factor recognizes these, and other sequences within the promoter region and facilitates attachment of the polymerase enzyme to the promoter region so that transcription may be initiated. The totality of the bases within various promoters differs among organism and from region to region along a particular DNA molecule. Different promoters interact with the polymerase enzyme with different degrees of affinity. So-called "strong" promoters facilitate more frequent enzyme attachment than do "weak" promoters, the promoters that bind the polymerase less avidly. The relative production of various transcripts reflects the relative strength of different promoters.

Mere attachment of the polymerase is insufficient to allow transcription, since the initial attachment is made to a double-stranded DNA molecule that cannot serve as a template for the polymerase. The polymerase enzyme facilitates the unwinding of the double-stranded DNA molecule. The extent of unwinding was deduced from studies of the effects of various amounts of the polymerase, in the presence of a known amount of DNA, on the action of topoisomerase I, an enzyme that facilitates the unwinding and resealing of DNA. These studies showed that each polymerase enzyme molecule allowed unwinding of a 17 base-pair region, a region corresponding to 1.6 turns of the DNA helix.

Transcription begins with the incorporation of an adenine or a guanosine triphosphate. This fact was recognized by radioactive experiments with gamma-labelled nucleotide triphosphates. Gamma-labelled nucleotides are nucleotides labelled in the phosphate most distal from the ribose component of the nucleotide. It was found that radioactivity was incorporated into incipient RNA from gamma-labelled adenine or guanosine 5′triphosphate, but not from the corresponding radiolabelled molecules of thymine or cytosine. This result could only occur if the adenine and guanosine residues were at the end of the growing chain, since only alpha (proximally)-labelled entities would occur within the chain. The fact that the label was at the 5′ end and not the 3′ end of the molecule was shown by the findings that the specific activity of newly synthesized RNA decreased with time, but that the total amount of radioactivity did not. These two facts indicated that the radioactivity was added early in the synthesis process, that is, at the 5′ end.

Elongation of the RNA transcript is mediated by a transcription bubble, a short segment of unwound DNA attached to a molecule of polymerase

enzyme and the newly forming RNA transcript. The size of the bubble remains constant during the transcription process, indicating that the rate of unwinding ahead of the transcription site is equal to the rate of DNA sealing after transcription has occurred. Transcription proceeds by phosphodiester bond formation, in the 5′ to 3′ direction (the formation of a bond involving interaction of the phosphate attached to the 5-carbon of the material to be added with the 3-hydroxyl group of the existing polymer) until such time as a termination signal is encountered. During this time, there is an intimate physical interconnection between the coding strand of DNA (the DNA strand that dictates the nature of the RNA message), the DNA-dependent RNA polymerase molecule, the template strand of DNA (the strand complementary to the coding strand), and the newly synthesized RNA transcript. The essential events of transcription are shown in Figure 11.3.

Termination of transcription

As is the case with transcription initiation, termination of transcription is a carefully regulated process that involves cessation of phosphodiester formation, rewinding of unwound DNA, dissociation of the transcribed RNA from the template DNA strand, and dissociation of the DNA-dependent RNA polymerase from the DNA molecule. All of these processes are mediated by "stop signals," sequences within the DNA that, by transcription, give rise to RNA transcripts that form loop structures. These structures exist, physically, as loops because certain regions of the transcript are compatible and associate by hybridization, while intervening areas are not compatible. It is often the case that loop-forming sequences contain *palindromes*, sequences that are the same in regard to their base composition, except that these identical regions have sequences in which the bases, although chemically identical, occur in reverse order and are thus complementary. When such a situation exists and the transcripts contain substantial amounts of guanosine and cytosine, tightly bound loops may occur. When DNA codes for tightly bound loops and a series of adenine residues follow the loop region, transcription of the adenine residues produces a transcript that contains a series of uridine residues. Uridine residues form inherently weak bonds with the transcription complex and facilitate

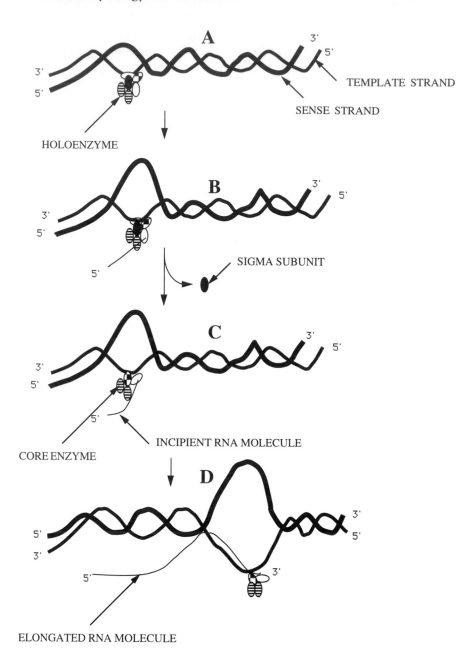

FIGURE 11.3 The essential events of procaryotic transcription. The heavy DNA strand is the sense strand and the strand of intermediate thickness is the template strand. The thinnest strand is messenger RNA. The checkered subunit of DNA-dependent RNA polymerase is the beta prime subunit, while the black subunit is the sigma factor, the clear unit is the beta subunit, and the horizontal-lined subunits are alpha subunits. (A) The holoenzyme locates the promoter region. (B) Unwinding of the double-strand DNA molecule begins, exposing the template strand so that a small amount of transcription may occur as a function of the holoenzyme. (C) The sigma factor is lost from the DNA-dependent RNA polymerase molecule. (D) Elongation continues as the enzyme moves along the template strand in the 3′ to 5′ direction, while synthesizing RNA in the 5′ to 3′ direction. After transcription of a small portion of the DNA, the DNA is rewound. The rate of rewinding is equal to the rate of the unwinding of DNA ahead of the transcription site. For this reason, the size of the transcription "bubble" remains constant through the process.

dissociation of messages from the transcribing system. The net effect of all of these factors is to cause the polymerase enzyme to pause, allowing dissociation of the DNA-RNA hybrid, resealing of the DNA in the transcription bubble, and reassociation of the sigma factor with the core enzyme so that transcription may be reinitiated.

Rho protein-mediated transcription termination

When a DNA template contains a region allowing stable loop formation followed by a series of bases, most notably adenine, whose transcription leads to weak bonds with hybrid strands, transcription termination is relatively easy. In many cases, however, energy is required to terminate transcription because the bases in transcripts form relatively strong bonds with hybrids. The energy is provided by the rho protein, which has ATPase activity. However, its activity is mediated by a single RNA strand and it is inactive with a DNA or DNA-RNA hybrid. Rho protein contains six subunits, each of which binds twelve bases. It thus interacts with a 72 base region of single-stranded RNA. The rho protein, by hydrolyzing ATP, provides energy that allows dissociation of DNA-RNA hybrids and facilitates the other events associated with termination. Rho dependent and rho-independent transcription terminations are shown in Figure 11.4.

Transcript modification

In procaryotes, transcription is mediated by a single DNA-dependent RNA polymerase. The original transcript is a large molecular weight molecule that contains RNA precursors for the rRNA and tRNA required for translation, as well as messenger molecules for critical cell proteins. Therefore, substantial processing of the initial transcript containing rRNA, tRNA, and mRNA is required. In procaryotes, mRNA is typically polycistronic in that it contains information for formation of several proteins that are functionally related, in a single message molecule. In procaryotes, mRNA molecules are normally processed intact, but considerable modification of both rRNA and tRNA molecules occurs. The modifications are of three major types. Initially, the various rRNA and tRNA molecules are removed from a common precursor molecule. This process, for rRNA molecules in *E. coli*, is mediated by RNAase III and RNAase E. RNAase P removes tRNA molecules and modifies

their 5' ends. RNAase D modifies tRNA molecules by removing nucleotides distal to the CCA residues at the 3'OH end of functional tRNAs. In this manner, it converts precursor tRNA molecules to an active form. Transfer RNA nucleotide transferase is yet another enzyme that may modify tRNA molecules. It adds CCA residues to nonfunctional tRNA molecules, allowing them to be active. In addition to these enzymes, a number of enzymes modify RNA transcripts by changing the nature of the bases that they contain. Methylases, thiolases, and pseudouridylating enzymes are examples of these types of enzymes. The major physiological consequence of the action of these enzymes is to allow the organism to respond to environmental stress. To reiterate, transcript-modifying enzymes in procaryotes exert three major actions: 1) removal of particular RNA molecules from precursors, 2) modification of the transcripts to make them functional, and 3) modification of the bases in transcripts. The actions of the various transcript-modifying enzymes in procaryotes are shown in Figure 11.5.

Transcription in eucaryotes

The complexity of transcription in eucaryotes is substantially greater than that of the analogous process in procaryotes. In procaryotes, transcription is accomplished by a single enzyme, while the process in eucaryotes involves three enzymes: RNA polymerase I that forms most rRNA molecules, RNA polymerase II that forms mRNA, and RNA polymerase III that forms the 5S ribosomal component of ribosomes and tRNA molecules. Transcription in eucaryotic systems is biochemically compartmentalized, in the sense that separate enzymes participate in formation of the various RNA types. In addition, the process is physically compartmentalized, since polymerase I is located in the nucleoli and both polymerases II and III are located within the nucleoplasm. Furthermore, in eucaryotes, mRNA sequences that code for functional genes associated with related processes are usually separated physically from each other on the chromosome as *exons* that are interspersed with nonfunctional regions known as *introns*. This situation contrasts sharply with the procaryotes, in which mRNA molecules are polycistronic, containing adjacent messages that code for proteins associated

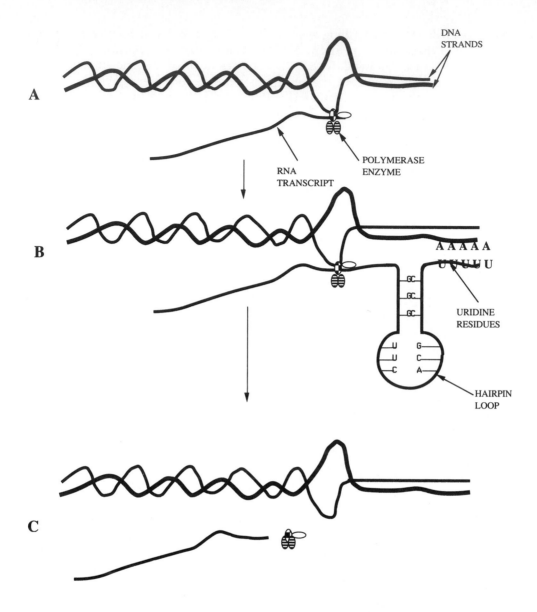

FIGURE 11.4 Termination of transcription. (A) Active transcription is occurring. (B) As transcription continues, a region of DNA rich in guanosine is transcribed, yielding a transcript that forms a hairpin loop. A DNA region containing adenine residues is also transcribed to produce an RNA region that contains uracil residues. (C) The hairpin loop impedes movement of the polymerase enzyme, slowing transcription. In addition, uracil forms weak bonds with the template DNA strand. The combination of these factors allows dissociation of the polymerase enzyme and the RNA from the transcription complex. The core enzyme may then combine with a new sigma factor so that transcription may be reinitiated. Although not shown in the figure, after the release of the enzyme and the transcript, the DNA in the transcription bubble rewinds. When rho protein participates in termination, it hydrolyzes ATP to provide energy for dissociation of the transcription complex in situations where uridine residues do not follow the hairpin. The ATPase activity of rho protein requires the presence of single RNA strands.

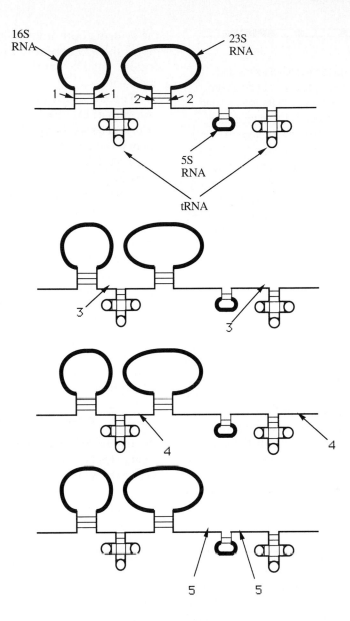

FIGURE 11.5 The modes of action of various RNA-processing enzymes. (1,2) The sites of action of RNAase III. The enzyme cleaves the stems of both 16S and 23S RNA but is inactive on 5S RNA. (3) The site of action of RNAase P. The enzyme cleaves the transcript on the 5′, but not the 3′, end of tRNA molecules. (4) The site of action of RNAase F. The enzyme cleaves the transcript on the 3′, but not the 5′, end of tRNA molecules. (5) The sites of action of RNAase E. This enzyme cleaves 5S-RNA on both the 5′ and 3′ ends of the transcript. Collectively, these enzymes, and others discussed in the text, allow conversion of the original large RNA transcript into physiologically functional forms.

FIGURE 11.6 A schematic representation of exons and introns. The dark areas are exons and the clear areas are introns. Exons code for genes and are both transcribed and translated, while introns are transcribed but not translated. When introns are present, original RNA transcripts are modified to remove introns from primary transcripts and to combine exons into functional mRNA molecules. Primary transcript modification, by cutting and splicing, is an obligatory event for eucaryotic microbes and to a lesser extent, for procaryotes. When introns are present, it is often the case that exons that code for related aspects of a common process occur in a linear order on the chromosome. Splicing and cutting of primary transcripts allows formation of a greater variety of messages than would otherwise be possible.

with various aspects of a common process. The physical orientation of exons and introns and of primary mRNA transcripts in eucaryotes is shown in Figure 11.6 The nature of eucaryotic messenger molecules allows for a higher degree of control in the formation of functional eucaryotic messages than may occur in procaryotes and also dictates a higher level of processing of primary transcripts. In eucaryotes, extensive processing of all three types of RNA occurs, while in procaryotes, only tRNA and rRNA are extensively processed.

Although significant differences are found in the transcription and processing events in procaryotes and eucaryotes, the processes in both types of organisms are, in many respects, similar. In both systems, the polymerizing enzymes are recognized by upstream regions of the genome, relative to the site of initiation of transcription. In both systems, consensus sequences are involved, although the nature of consensus sequences in eucaryotes is somewhat different from those found in procaryotes. At the same time, certain similarities exist. The proximal recognition region (TATA) in many eucaryotes is slightly more distant, approximately 25 bases upstream from the transcription start site, than in procaryotes, but it is physically similar to the TATAAT recognition site at −10 in procaryotes. Even in systems in which the primary recognition site is more distant from the transcription initiation site than most eucaryotes, its nature, TATAAA, is similar to the analogous location in procaryotes. In addition to the presence of a proximal recognition region for RNA polymerase action in both systems,

the systems both involve secondary recognition regions, the −35 region in procaryotes and the CAAT region typically found in the −40 to −110 region of eucaryotes.

The presence of regions that influence transcription at locations far distant from the transcription start site is apparently a unique aspect of eucaryotic transcription. Proteins known as *transcription factors* enhance recognition and attachment of polymerases in eucaryotes. These proteins apparently are not components of the polymerase enzymes per se, but facilitate recognition and attachment of polymerases to recognition regions and allow formation of a diversity of transcripts. In addition to transcription factors, proteins known as *transcription-enhancing proteins* that may be located upstream, downstream, or in the middle of a transcribed region and may be far distant from the transcribed regions, are found in eucaryotes. The manner in which these proteins act is not entirely understood, but it is possible that they serve to facilitate assembly of complexes associated with recognition.

Eucaryotic transcript processing

The extent and nature of transcript processing in eucaryotes is much more extensive than in procaryotes, particularly in regard to mRNA. In both systems, transcription occurs in the 5′ to 3′ direction, and begins with either adenine or guanine, but in eucaryotes, the 5′ end of the transcript is modified by hydrolysis of its terminal phosphate, producing an adenine or guanosine-5′-diphosphate. The latter reacts with a guanosine triphosphate molecule to form a 5′-5′-triphosphate linkage and to yield a cap at the 5′ end of the transcript. The N7 entity of the capping guanosine molecule is then modified by reaction with an S-adenosylmethionine molecule to produce a "cap O." In some cases, further modification of the cap occurs. Two functions of capping are recognized: protection of transcripts from degradation by phosphatases and nucleases and facilitation of translation.

In addition to capping, tailing is a frequent, if not invariant, aspect of eucaryotic mRNA processing. Tailing involves two subprocesses, cleavage of the transcript by an endonuclease that recognizes an AAUAAA sequence and the addition, by poly A polymerase, of about 250 adenine nucleotides to the 3′ end of the transcript.

Splicing

Extensive modification of primary transcripts is a characteristic feature of eucaryotic RNA metabolism. Not only are primary mRNA transcripts modified by capping and tailing, but multiple messages for proteins are made from primary transcripts by the removal of introns and the combining of various mRNA molecules into new message molecules. This process allows formation of a tremendous diversity of proteins from a finite amount of DNA. For multiple messages to be formed from primary transcripts, *splicing* must occur. Splicing allows the removal of introns and is also essential for multiple message formation. Splicing requires a mechanism to determine precisely where in the primary transcript a cut should be made so that an intron may be removed without, at the same time, destroying or altering an exon. Study of the sequences of many introns, and the regions on either side of them, has shown that introns have characteristic sequences at their ends. The splicing enzymes recognize these sequences and remove the introns in a specific manner. In addition, within the introns, regions known as branch sites are found. These sites are critical to many splicing phenomena.

Spliceosomes   Splicing, in most cases, is mediated by mRNA-associated entities known as *spliceosomes.* These particulate entities result from combination of mRNA molecules with aggregates of ribonucleoproteins and mediate the majority of slicing events. The mechanism by which spliceosomes appear to function is diagrammed in Figure 11.7. The splice site of the upstream (5′) exon is attacked by a 2′OH from an adenine residue in the branch site. The 5′ phosphate of the downstream (3′) exon is then attacked by the 3′ end of the upstream exon, forming a hybrid exon product and a "lariat" form that contains the branch site and the intron, joined by a phosphodiester linkage. Splicing accomplished in this fashion is achieved by transesterification reaction rather than by reactions involving hydrolysis.

Translation

The formation and processing of RNA molecules is a critical component of the protein synthesis process, but is only a portion of it. Processing of RNA provides the three types of RNA in forms that allow the formation of protein, but much more is required before protein synthesis is accomplished. To begin with, functioning ribosomes must be formed. Functional ribosomes are assembled from sequential interaction of proteins and other materials to form ribosomal subparticles and the subsequent aggregation of these subparticles to form intact protein-synthesizing ribosomes. Various subparticles of ribosomes are denoted by their sedimentation values in the *Svedberg system.* The Svedberg system classifies particles on the basis of differences in their rate of sedimentation in a centrifuge that are a function of particle density. Although the nature of ribosomal subparticles differs somewhat, depending upon whether the completed ribosome is a 70S procaryotic or an 80S eucaryotic cytoplasmic ribosome, in either system, formation of functional ribosomal particles requires aggregation of subparticles. The manner in which functioning ribosomes are formed is shown schematically in Figure 11.8.

Although the details differ in the two systems, the processes involved in 70S and 80S ribosome formation are similar. For the 70S ribosome, formation begins by the interaction of its 30S subunit with protein initiation factors, known as IF1, IF2, and IF3, and with the aid of GTP, formation of a 30S subparticle attached to initiation factors. The latter complex, with the aid of GTP, binds to a molecule of mRNA, and subsequently to a molecule of formyl methionyl tRNA, whose TAC anticodon is compatible with the AUG or GUG codon on the molecule of mRNA attached to the 30S ribosomal particle. This attachment occurs at the P, or peptide, site of the 30S ribosomal subparticle. The junction of mRNA and formylmethionyl tRNA to the 30S subparticle is associated with release of IF3. Collectively, the processes just described allow formation of the 30S initiation complex. Formation of the 70S initiation complex results from combination of a 50S ribosomal subparticle with the 30S initiation complex. During this process, IF1 and IF2 are lost and GTP hydrolysis occurs. The result of all these processes is formation of a functional 70S ribosome attached to a molecule of tRNA-bound formylmethionine at the P site. In this condition, the 70S ribosome is prepared to participate in the actual formation of protein by peptide bond formation.

Formation of a functioning 80S ribosome occurs in a manner similar to that of the 70S ribosome, although the smaller subparticle of the

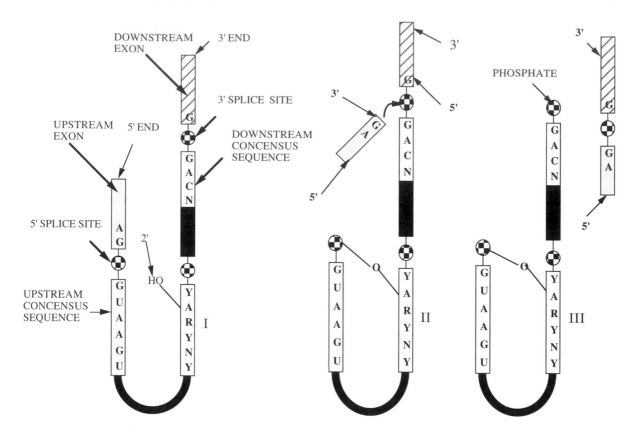

FIGURE 11.7 The mechanism of splicing of eucaryotic mRNA precursors by spliceosomes. A messenger RNA precursor containing an upstream exon, an intervening intron, and a downstream exon (I) is converted to a spliced product in which the 3′ end of the upstream exon is joined to the 5′ end of the downstream exon, producing a new message molecule and a "lariat" that contains the intron (III). This process is accomplished because the 2′ hydroxyl of an adenine residue within the "branch site" of the intron interacts with the 5′ phosphate on the upstream end of the intron, freeing a 3′ hydroxyl of the upstream exon. (II) The latter interreacts with the 5′ phosphate end of the downstream exon to produce the spliced product and a lariat-containing intron. The 5′ and 3′ "slice sites" are delineated from the intervening intron by concensus sequences of nucleotides, but the precise length of introns varies from situation to situation. This fact is indicated by the black region of the diagram between the branch site and the 3′ splice site of the intron. The black region contains differing numbers of purine residues. Abbreviations: adenine (A), pyrimidine (R), purine (Y), nucleotide (N), uridine (U), guanine (G), and phosphate residues (stippled circles).

80S ribosome, the 40S particle, is substantially larger than its 70S analogue. Formation of the 40S initiation complex begins by interaction of eucaryotic initiation factor four (eIF4), a cap-binding protein, a molecule of capped mRNA, a complex composed of eucaryotic initiation factor two (eIF2), GTP, methionyl tRNA, and an initiation factor-bound 40S subunit to form a complex composed of mRNA, bound by its 5′ end to a 40S subparticle that is bound to a molecule of methionine at the P site. However, initially, the bound methionine molecule is not paired with the message, so translation may not be initiated. Pairing of the methionine tRNA complex with its codon on the message is accom-

plished by movement of the message, in a "scanning" activity, facilitated by ATP hydrolysis, allowing alignment of the methionyl tRNA complex and its appropriate codon on the message. Normally, the first methionine codon (AUG) on the eucaryotic message is the codon that determines translation initiation. In procaryotic systems, specification of the formylmethionine codon that determines the initiation of translation is accomplished by compatibility between a purine rich region on the message and a pyrimidine region on the 16S component of the 30S portion of the 70S ribosome. A comparison of the major events in formation of functioning 70S and 80S ribosomes is shown in Figure 11.8.

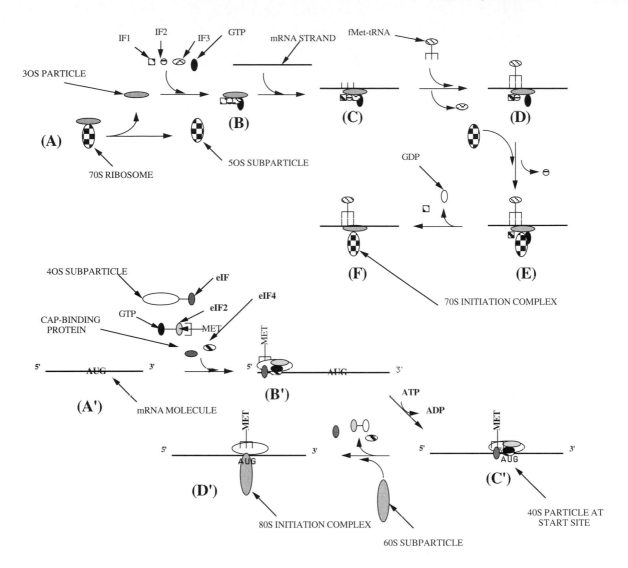

FIGURE 11.8 A comparison of the processes involved in formation of functional 70S and 80S ribosomes. The upper diagram depicts the events for 70S ribosomes and the lower diagram describes the events for the 80S system. For the 70S system, initially, an inactive ribosome (A) dissociates into 50S and 30S subparticles with the aid of initiation factors (IF1, IF2, and IF3) plus guanosine triphosphate, giving rise to a 30S particle bound to proteins and GTP. (B) The bound 30S particle complexes with an mRNA molecule (C). The mRNA-bound 30S particle then interacts with a molecule of tRNA bound to formyl methionine to produce a complex of 30S subunit bound to IF1, IF2, GTP, mRNA, and an f-Met residue with the loss of IF3 (D). During formation of complex E, the 50S subparticle combines with complex D, releasing IF2. Finally, the 70S initiation complex (F) is formed by loss of IF1 from complex E and GTP hydrolysis. The 80S initiation complex is formed by similar reactions. Initially (A′), the 40S subparticle interacts with a capped mRNA molecule and initiation proteins, including a cap-binding protein and a methionyl tRNA molecule, to produce a 40S particle bound to message, proteins, and methionyl tRNA at a site upstream from the translation start site. (B′) With the aid of protein eIF4 and ATP hydrolysis, the 40S complex moves to the first AUG codon (C′). Addition of the 60S subparticle and release of many proteins produces the 80S initiation complex (D′).

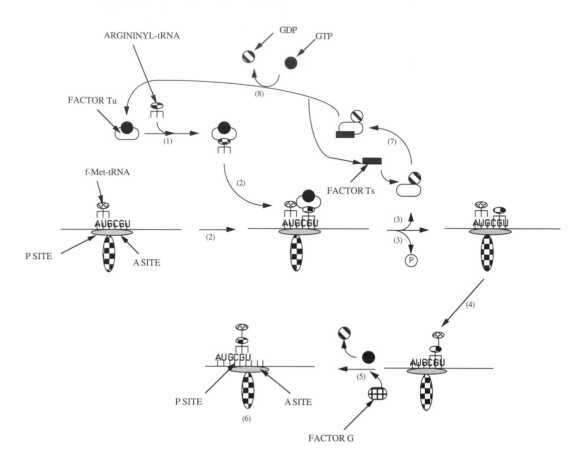

FIGURE 11.9 The essential aspects of procaryotic translation. (1) Formation of a tricomplex between elongation factor Tu, guanosine triphosphate, and a charged argininyl tRNA molecule. (2) Interaction of the tricomplex with a 70S ribosome whose peptide (P) site has been previously occupied by a formyl methionyl tRNA molecule, placing the arginine residue at the amino acid (A) site, adjacent to the P site. (3) Hydrolysis of the GTP in the A site complex and release of a dimer of factor Tu and guanosine diphosphate (GDP), leaving the arginine residue, bound by tRNA at the A site. This reaction only occurs when the appropriate aminoacyl tRNA molecule occupies the A site. (4) Transpeptidization, forming a peptide linkage between the formyl methionine residue and the arginine moiety, so that the formyl methionine carboxyl function interacts with the alpha amino group of arginine. For the code to be read, the reaction must occur in this manner. (5) Hydrolysis of GTP and movement of the ribosome along the message (translocation), in the 5' to 3' direction, so as to expose new A and P sites (6). This reaction requires translocation factor G. (7) Formation of a tricomplex between factor Ts, factor Tu, and GDP. (8) Hydrolysis of a free GTP molecule by the tricomplex formed in (7) and conversion of the factor Tu-GDP duplex to A factor Tu-GTP duplex, so that the latter may react with further amino acid-tRNA molecules.

When formation of functional ribosomal particles is complete, the elongation phase of protein synthesis may occur. Once again, differences are found in details, although analogous processes occur in elongation mediated by 70S and 80S ribosomes. The essential aspects of procaryotic translation (elongation) are shown in Figure 11.9.

Elongation is a carefully regulated and complex process. Initially, with a 70S ribosome, an aminoacyl tRNA molecule must be recognized at the A site (the aminoacyl site) of the ribosome, which is adjacent to the P site. Recognition requires formation of a tri-complex between a molecule of aminoacyl tRNA, a molecule of elongation factor Tu, and a molecule of GTP. The tri-complex reacts with a 70S ribosome to form a complex of ribosome, bound aminoacyl tRNA, GTP, and factor Tu. Hydrolysis of this complex results in formation of a ribosome, bound at its A site to a molecule of aminoacyl tRNA. In addition, a binary complex of GDP and factor Tu is formed, which reacts with elongation factor Ts to form a tertiary complex of GDP, factor Tu, and factor Ts. This complex then hydrolyzes a free molecule of GTP with the aid of factor Ts, yielding a factor Tu-GTP complex that may react with another molecule of aminoacyl tRNA. GTP hydrolysis appears to serve a "proofreading" function, allowing the ribosome to be occupied at the A site by the proper aminoacyl tRNA.

Once the appropriate aminoacyl tRNA has been positioned at the A site, adjacent to the P site, transpeptidization, or the formation of a peptide bond between the amino group of the A-site amino acid and the carboxyl group of the P-site amino acid, may occur. It is essential that peptide bond formation occurs in this manner. If it did not, the code would not be read since the triplicate code adjacent to the P site would remain constant. Peptide bond formation is *only* possible after elongation factor Tu is released from an aminoacyl tRNA molecule. The release is accompanied by hydrolysis of GTP to GDP, a reaction that leads to a conformational change in elongation factor Tu and allows release of a GDP-EFTu complex. Peptide bond formation is followed by translocation. This process requires GTP hydrolysis, which provides the energy for the translocation. The ribosome moves along the message in the 5' to 3' direction. During translocation, the uncharged tRNA molecule located at the P site is lost, the peptide is moved from the A to the P site and a new sequence of three bases is exposed adjacent to the P site so that the next amino acid may be added to the peptide chain.

Translation initiation in eucaryotes and procaryotes

The translation process in eucaryotes is similar to that in procaryotes. The transcription and initiation steps have been described above. The actual formation of peptide bonds in eucaryotes is mediated by protein initiation factors different from those of procaryotes, but functionally analogous. EFtu in procaryotes has a eucaryotic analogue, EF1α. Similarly, procaryotic factor Ts has a eucaryotic analogue, EFβg, and procaryotic factor G has a eucaryotic analogue, EF2. In both systems, presentation of an aminoacyl tRNA to the ribosome is mediated by a protein-GTP complex and peptide

bond formation requires GTP hydrolysis so that a GDP-protein complex may be formed that dissociates from the ribosome.

The initiation of translation in eucaryotes is regulated in a manner different from that in procaryotes. In eucaryotes, translation initiation is regulated by phosphorylation of eucaryotic elongation factor 2 (eIF2). When eIF2 is nonphosphorylated, it facilitates attachment of methionyl tRNA to the 40S subunit of the eucaryotic ribosome. However, when eIF2 is phosphorylated by a heme-containing protein kinase, it cannot participate in joining methionyl tRNA to the small subparticle of the 80S ribosome. The activity of the protein kinase under a given set of conditions thus regulates translation initiation in eucaryotes.

In contrast, translation initiation in procaryotes is controlled in large measure by the methylation state of initiating molecules of tRNA-bound methionine. In procaryotes, two kinds of tRNA bind to methionine. One type binds to the methionine that initiates translation and a second type binds to methionine molecules that are incorporated internally. A special formylating enzyme discriminates between methionine bound to initiation tRNA and methionine bound to tRNA destined for internal incorporation and formylates only the former.

Finally, translation initiation is regulated by the manner in which the initiating formyl methionine (procaryotes) or methionine (eucaryotes) is recognized. In procaryotes, recognition of the translation start site is primarily a function of compatibility between the purine rich regions on the message, the so-called Shine-Dalgarno sequences, and the 3' end of the RNA of the 30S ribosomal subunit. In eucaryotes, selection of the initiating methionine codon involves an ATP-mediated scanning process. A comparison of translation initiation in procaryotes and eucaryotes is shown in Figure 11.10.

Termination of translation

Translation termination in procaryotes and eucaryotes are similar processes. In both systems, release factor proteins interact with the stop codons UAA, UAG, and UGA. Various release factors bind specifically to the A site when the stop codons are encountered and also facilitate hydrolysis of the bond between the tRNA and the peptide at the P

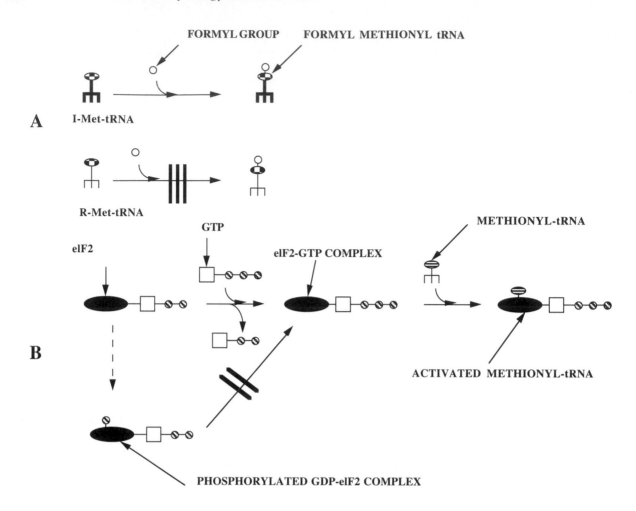

FIGURE 11.10 Regulation of translation initiation in procaryotes (A) and eucaryotes (B). (A) Methionine reacts with two kinds of tRNA, initiator RNA, and regular (R), tRNA. The initiator RNA is used for protein synthesis initiation, while the regular RNA is used to incorporate methionine residues at locations within the protein. The enzyme that converts methionine to formyl methionine discriminates between bound and free methionine and also between initiator-bound and regular-bound methionine and selectively formylates only initiator tRNA-bound methionine, allowing it to initiate translation. (B) Initiation of translation in eucaryotes requires formation of a complex between methionine, tRNA, guanosine triphosphate, and eucaryotic elongation factor eIF2. The latter complex interacts with the 40S subparticle of the 80S ribosome to initiate translation. At the completion of the addition of an amino acid residue to the protein, eIF2 is released in a GDP-bound form. Further participation of eIF2 in eucaryotic protein synthesis requires conversion of the eIF2-GDP complex to an eIF2-GTP complex. This process is accomplished by a guanyl nucleotide exchange protein. The protein cannot function when the eIF2-GDP complex is phosphorylated. A heme-dependent eIF2-GDP kinase regulates eucaryotic translation initiation. Under heme-deficient conditions, the kinase (indicated by the dashed arrow) phosphorylates the eIF2-GDP complex and inhibits translation.

site. Hydrolysis is followed by release of the protein, dissociation of polysomes from message molecules, and disaggregation of the large and small subunits of the ribosomal particles. In both systems, the smaller subunits interact with protein factors, IF3 for procaryotes and eIF3 for eucaryotes, so that nonspecific binding of the small and large ribosomal subunits does not occur.

The intimate role of proteins in their own formation

Proteins are intimately involved in all aspects of the protein synthesis process. They mediate the formation of the code, are required for, and facilitate, formation of functional ribosomes, and facilitate all of the aspects of peptide bond formation and its specificity. In addition, they are required for both initiation and termination of the protein synthesis process and they serve both catalytic and physiochemical functions. The intimate relationship between proteins and their formation is a cause for much reflection. Protein formation requires, above all, DNA that specifies the process. However, DNA replication cannot occur without protein. The two are inextricably intertwined.

Summary

Protein synthesis is an extremely complicated process that requires the integration and coordination of a variety of subprocesses and the cooperative functioning of many structures and locations within the cell. Although it occurs with the aid of slightly different structures in procaryotic and eucaryotic cells, the processes that must be accomplished in both systems are the same and the manner in which they occur are similar. However, in the midst of basic similarity, there are subtle differences between the processes in the two systems. Study of these small differences is helpful in understanding the comparative physiology and metabolism of procaryotic and eucaryotic cells and in exploiting those differences.

Selected References

Abelson, J. 1979. RNA processing and the intervening sequence problem. *Annual Review of Biochemistry.* **48:** 1035–1070.

Björk, G. R., J. U. Ericson, C. E. D. Gustafsson, T. G. Hagervall, Y. H. Jönsson, and P. M. Wikström. 1987. Transfer RNA modification. *Annual Review of Biochemistry.* **56:** 263–288.

Brenchley, J. E., and L. S. Williams. 1975. Transfer RNA involvement in the regulation of enzyme synthesis. *Annual Review of Microbiology.* **29:** 251–274.

Brimacombe, R., F. Stöffler, and H. G. Wittman. 1978. Ribosome structure. *Annual Review of Biochemistry.* **47:** 217–250.

Busch, H., R. Reddy, L. Rothblum, and Y. C. Choi. 1982. SnRNAs, SnRNPs, and RNA processing. *Annual Review of Biochemistry.* **51:** 617–654.

Cech, T. R., and B. L. Bass. 1986. Biological catalysis by RNA. *Annual Review of Biochemistry.* **55:** 599–630.

Chamber, P. 1975. Eukaryotic nuclear RNA polymerases. *Annual Review of Biochemistry.* **44:** 613–724.

Clark, B. 1980. The elongation step of protein biosynthesis. *Trends in Biochemical Sciences,* **5:** 207–210.

Doi, R. H. 1977. Role of ribonucleic acid in gene selection in procaryotes. *Bacteriological Reviews.* **41:** 568–594.

Doi, R. H., and L. F. Wang. 1986. Multiple procaryotic ribonucleic acid polymerase sigma factors. *Microbiological Reviews.* **50:** 227–243.

Eggertsson, G., and Dieter Söll. 1988. Transfer ribonucleic acid-mediated suppression of termination codons in *Escherichia coli. Microbiological Reviews.* **52:** 354–374.

Eguchi, Y. 1991. Antisense RNA. *Annual Review of Biochemistry.* **60:** 631–652.

Fournier, M. J., and H. Ozeki. 1985. Structure and organization of the transfer ribonucleic acid genes of *Escherichia coli* K-12. *Microbiological Reviews.* **49:** 379–397.

Friedman, S. M. 1968. Protein-synthesizing machinery of thermophilic bacteria. *Bacteriological Reviews.* **32:** 27–38.

Gegenheimer, P., and D. Apiron. 1981. Processing of procaryotic ribonucleic acid. *Microbiological Reviews.* **45:** 502–541.

Geiduschek, E. P., and G. P. Tocchini-Valentini. 1988. Transcription by RNA polymerase III. *Annual Review of Biochemistry.* **57:** 873–914.

Green, P. J., O. Pines, and M. Inouye. 1986. The role of antisense RNA in gene regulation. *Annual Review of Biochemistry.* **55:** 569–598.

Haselkorn, R., and L. B. Rothman-Denes. 1973. Protein synthesis. *Annual Review of Biochemistry.* **42:** 397–438.

King, T. C., R. Sirdeskmukh, and D. Schlessinger. 1986. Nucleolytic processing of ribonucleic acid transcripts in procaryotes. *Microbiological Reviews.* **50:** 428–451.

Kozak, M. 1983. Comparison of initiation of protein synthesis in prokaryotes, eukaryotes, and organelles. *Microbiological Reviews.* **47:** 1–45.

Kurland, C. G. 1972. Structure and function of the bacterial ribosome. *Annual Review of Biochemistry,* **41:** 377–408.

Kurland, C. G. 1977. Structure and function of the bacterial ribosome. *Annual Review of Biochemistry.* **46:** 173–200.

Lake, J. A. 1985. Evolving ribosome structure: Domains in archaebacteria, eubacteria, eocytes and eukaryotes. *Annual Review of Biochemistry.* **54:** 507–530.

Leff, S. E., M. G. Rosenfeld, and R. M. Evans. 1986. Complex transcriptional units: Diversity in gene expression by alternative RNA processing. *Annual Review of Biochemistry.* **55:** 1091–1118.

Lodish, H. F. 1976. Translational control of protein synthesis. *Annual Review of Biochemistry.* **45:** 39–72.

Maitra, U., E. A. Stringer, and A. Chaudhuri. 1982. Initiation factors in protein synthesis. *Annual Review of Biochemistry.* **51:** 869–900.

McClure, W. R. 1985. Mechanism and control of transcription initiation in prokaryotes. *Annual Review of Biochemistry.* **54:** 171–204.

Moldave, K. 1985. Eukaryotic protein synthesis. *Annual Review of Biochemistry.* **54:** 1109–1150.

Nierlich, D. P. 1978. Regulation of bacterial growth, RNA, and protein synthesis. *Annual Review of Microbiology.* **32:** 394–432.

Noller, H. F. 1991. Ribosomal RNA and translation. *Annual Review of Biochemistry.* **60:** 191–227.

Ochoa, S., and C. de Haro. 1979. Regulation of protein synthesis in eukaryotes. *Annual Review of Biochemistry.* **48:** 549–580.

Pabo, C. O., and R. T. Sauer. 1992. Transcription factors: Structural families and principles of DNA recognition. *Annual Review of Biochemistry.* **61:** 1053–1096.

Padgett, R. A., P. J. Grabowski, M. M. Konarska, S. Seiler, and Philip A. Sharp. 1986. Splicing of messenger RNA precursors. *Annual Review of Biochemistry.* **55:** 1119–1150.

Perry, R. P. 1976. Processing of RNA. *Annual Review of Biochemistry.* **45:** 605–630.

Reanney, D. C. 1982. The evolution of RNA viruses. *Annual Review of Microbiology.* **36:** 47–73.

Rich, A., and U. L. RajBhandary. 1976. Transfer RNA: Molecular structure, sequence, and properties. *Annual Review of Biochemistry.* **45:** 805–860.

Sawadogo, M., and A. Sentenac. 1990. RNA polymerase B (II) and general transcription factors. *Annual Review of Biochemistry.* **59:** 711–754.

Schimmel, P. R., and D. Söll. 1979. Aminoacyl-t-RNA synthetases: General features and recognition of transfer RNAs. *Annual Review of Biochemistry.* **48:** 601–648.

Srivastava, A. K., and D. Schlessinger. 1990. Mechanism and regulation of bacterial RNA processing. *Annual Review of Microbiology.* **44:** 105–129.

Symons, R. H. 1992. Small catalytic RNAs. *Annual Review of Biochemistry.* **61:** 641–672.

Taylor, John M. 1979. The isolation of eukaryotic messenger RNA. *Annual Review of Biochemistry.* **48:** 681–718.

Vold, B. S. 1985. Structure and organization of genes for transfer ribonucleic acid in *Bacillus subtilis. Microbiological Reviews.* **49:** 71–80.

Wahle, E., and W. Keller. 1992. The biochemistry of 3-end cleavage and polyadenylation of messenger RNA precursors. *Annual Review of Biochemistry.* **61:** 419–440.

Weissbach, H., and S. Ochoa. 1976. Soluble factors required for eukaryotic protein synthesis. *Annual Review of Biochemistry.* **45:** 191–216.

Wittmann, H. G. 1982. Components of bacterial ribosomes. *Annual Review of Biochemistry.* **51:** 155–184.

Wittmann, H. G. 1983. Architecture of prokaryotic ribosomes. *Annual Review of Biochemistry.* **52:** 35–66.

Wool, Ira G. 1979. The structure and function of eukaryotic ribosomes. *Annual Review of Biochemistry.* **48:** 719–754.

Young, R. A. 1991. RNA polymerase II. *Annual Review of Biochemistry.* **60:** 689–715.

CHAPTER 12

Nucleic Acid Metabolism

The intimate relationship between nucleic acids and the protein synthesis process was described in Chapter 11. Nucleic acids are essential for protein synthesis, but the converse is also true. Since nucleic acids serve as genetic material, in addition to their role in protein synthesis, nucleic acid synthesis and metabolism are topics that command our attention as separate subjects. It is particularly important that we understand the manner in which genetic material is replicated.

Although genetic material is universal in all living things, its nature is not. Double-stranded DNA is uniformly found in cellular life forms, but different forms of genetic material are found in other components of the natural world. The viruses, as a group, display a spectrum of genetic entities ranging from single-stranded DNA in the parvoviruses and double-stranded DNA in the pox viruses to double-stranded RNA in the reoviruses and single-stranded RNA in a diversity

of viruses. Understanding the ways in which replication and metabolism of nucleic acids occur in a variety of systems is useful in obtaining an understanding of the diversity of nucleic acid replication systems in nature and of the various ways in which nucleic acid metabolism may influence protein synthesis.

Contrasts in the functions of DNA and RNA

DNA serves *exclusively* as genetic material. When DNA is the genetic material, it manifests its action by allowing RNA formation so that protein may be formed. In contrast, RNA may serve either as genetic material or, in certain circumstances, directly as the message for protein formation. Such versatility is not possible when DNA is the genetic material.

Modes of RNA replication

In certain viruses for which RNA is the genetic material, several modes of RNA replication are possible. The consequences of RNA replication in these viruses depend upon the nature of the viral genome. When the virus is a single-stranded virus, and the strand is a *plus* strand (the strand that can serve directly as mRNA), the events of RNA replication and protein synthesis are different than when the single-stranded viral genome is a negative strand. In positive single-stranded RNA virus replication, a minus strand is formed, with the plus strand as template, by an RNA-dependent RNA polymerase, known as replicase, encoded by the plus strand. The double-stranded intermediate containing the minus strand produces additional plus strands that are translated into the proteins required for formation of the mature virus. The minus strands of the double-stranded intermediate also allow formation of additional plus strands that may serve as genetic material.

RNA replication in the negative single-stranded RNA viruses occurs in a different manner than that of the positively stranded viruses. In the negatively stranded viruses, the negative strand cannot serve directly as mRNA. Therefore, under the influence of an RNA-dependent RNA polymerase known as a transcriptase, positive mRNA strands are formed that serve as messages for protein. In addition, by a replicase enzyme, the negative strand produces a double-stranded intermediate. The positive strand of the double-stranded intermediate serves to allow formation of single, negative strands for new virus particles.

Replication of the double-stranded RNA viruses occurs in a different way than for single-stranded forms. The double-stranded virus enters the cell and a transcribed plus strand enters the cytoplasm. The plus strand serves as a template to allow formation of RNA-dependent RNA polymerase, which using the plus strand as template, allows formation of new minus strands. In addition, the plus strand serves as messenger to allow formation of viral proteins. When sufficient synthesis has occurred, double strands of RNA are assembled, along with virus proteins, to yield new virus particles. RNA replication in this manner is conservative, rather than semiconservative, because a double-stranded template gives rise to a single strand that serves as the template for a second, new single strand. The *two newly synthesized* strands are incorporated into the new virus, resulting in a conservative replication mode, as opposed to the semiconservative mode, in which newly replicated double nucleic strands would contain one parental strand and one newly synthesized strand.

The retroviruses are yet another class of RNA viruses whose replication is unique because replication of a single RNA strand depends upon formation of a double-stranded DNA intermediate. This is the reason that such viruses are referred to as retroviruses, since the literal meaning of *retro* is "reverse." Rather than forming RNA from a DNA template, the retroviruses form DNA from an RNA template by action of the unique enzyme, *reverse transcriptase*. This enzyme synthesizes a complementary minus strand of DNA from a single, plus-stranded RNA template. The enzyme also allows synthesis of a second strand of DNA, a plus strand, using the first DNA strand as template. Finally, reverse transcriptase digests the original, single RNA plus-strand. The total activities of the enzyme produce a double-stranded DNA intermediate from a single plus-strand of RNA. The double-stranded DNA molecule integrates into the chromosome of a susceptible cell. In the integrated state it allows transcription and formation of new viruses under specific conditions.

Retrovirus nucleic acid replication is, in many respects, similar to replication of lysogenic bacteriophage. In both cases, foreign DNA enters the host chromosome and changes the properties of the cell. In bacteria, the most frequent alteration is induction of toxin formation. In human cells, the analogue resulting from foreign DNA addition is cancer. Retrovirus replication and lysogeny share an additional common feature. In both cases, the integrated foreign DNA may replicate as a part of the host genome. With lysogenic phages, the integrated phage genome is a prophage, while in human cells, it is a provirus. The various modes of RNA replication are shown in Figure 12.1.

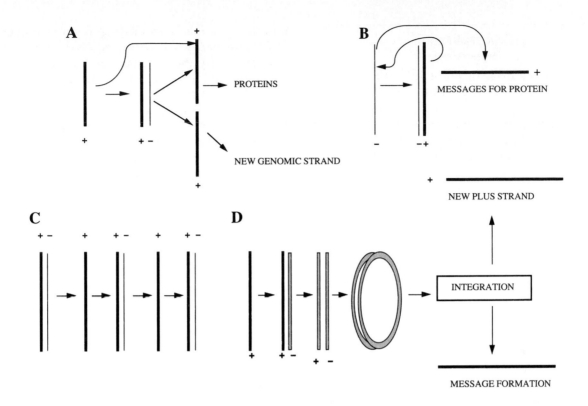

FIGURE 12.1 Various modes of viral RNA replication. Throughout the figure, plus RNA strands are shown as thick lines, minus strands are shown as thin strands, and DNA strands are stippled. (A) Replication of single-stranded RNA, in which the single strand is a plus strand. In this mode of replication, the positive strand codes directly for a replicase enzyme that allows formation of a double-stranded (+ and −) intermediate. The negative strand of the latter codes for the mRNA required to produce the proteins required for virus replication, and also serves as a template for formation of single plus strands for new viral genomes.
(B) Replication of a single strand in which the strand is a minus strand. In this mode, a preformed replicase enzyme allows formation of plus strands, which serve as mRNA for formation of viral structural and replication proteins. The synthesized plus strand also serves as a template for formation of new genomic minus strands. (C) Replication of double-stranded viral RNA. In this mode, the double-stranded genome allows formation of a single new plus strand. This serves both as mRNA and as a template for synthesis of a new double-stranded intermediate. Thus, each replication event is conservative, since the parental strands remain attached. (D) Retroviral RNA replication. In this mode, a single plus strand is replicated via a double-stranded DNA intermediate. Through reverse transcriptase, a negative DNA strand is formed on an RNA template. Reverse transcriptase or DNA polymerase then allows formation of a double-stranded DNA intermediate that circularizes and integrates into the chromosome of a susceptible eucaryotic cell. In the integrated state, which is obligatory in this mode, the double-stranded intermediate allows transcription of messages for retroviral replication and single positive RNA strands for the new virus.

DNA replication

The genetic material of all cellular life forms is double-stranded DNA. In addition, nongenomic double-stranded circular DNA is found in mitochondria and chloroplasts, as well as in certain plasmids and bacteriophage. Thus, throughout nature, various modes of DNA replication are found. Three modes of DNA replication are currently known: the theta mode, the sigma mode, and the linear mode. The theta mode occurs in replication of the procaryotic genome and in replication of both chloroplast and mitochondrial DNA. The sigma mode occurs in certain plasmids and viruses and the linear mode is used in replication of the eucaryotic genome.

The theta mode

DNA replication, by any mode, requires a template, ATP, a replication complex, and the deoxyribonucleotide triphosphates of adenine, guanine, thymine, and cytosine. The theta mode of DNA replication occurs in the following manner. A replication complex containing DNA polymerases III and I, plus a number of other proteins required to facilitate the replication process, is formed. The complex interacts with a molecule of DNA and allows unwinding of the supercoiled DNA molecule, producing a localized region known as the replication fork, where replication begins. The unwinding process is facilitated by a helicase enzyme (rep protein) and the unwound region of DNA is stabilized by interaction of the unwound strands with helix-destabilizing protein (HDP). The unwinding process requires ATP energy expenditure, but the amount of energy required is reduced with the aid of HDP and helicase. In addition, DNA gyrase periodically nicks one of the unwinding strands, thereby preventing the formation of positive supercoils.

When the unwinding of DNA at the replication fork is complete, replication can begin. The latter is semiconservative because each newly replicated DNA molecule contains a strand of the previous DNA molecule and a newly synthesized strand. Since DNA polymerases cannot, by themselves, attach to a strand of DNA, replication requires formation of a primer molecule to which the DNA polymerase can attach. Primer formation is accomplished by a special DNA-dependent RNA poly-

merase, primase, which forms short pieces of DNA-attached RNA with free 3'OH ends. DNA polymerase III, the major polymerizing enzyme of the replication complex, attaches to primer molecules. In the theta mode of replication, replication occurs in the 5' to 3' direction. The 5' phosphate end of the added deoxyribonucleotide reacts with the free 3'OH end of the primer, liberating a molecule of inorganic pyrophosphate and producing a bound deoxyribonucleotide residue with a free 3' end, to which the next ribonucleotide residue can be added, by its 5' moiety.

Replication, by the leading strand, the strand that replicates in the direction of the replication fork, is continuous, from a single primer molecule formed by primase, and always occurs in the 5' to 3' direction. The requirement for 5' to 3' addition of nucleotide residues to the growing nucleic acid dictates that the template strand for replication in the direction of the replication fork is in the 3' to 5' orientation. Because of the antiparallel nature of DNA, the strand of the original DNA molecule *complementary* to the strand that serves as template for replication in the direction of the replication fork is 5' to 3'. This direction is incompatible with continuous replication from a single primer attached to the end of the strand that replicates in the direction opposite to that of the replication fork. This difficulty is overcome by primase, through *discontinuous synthesis* of short RNA primer fragments attached to DNA molecules so that each primer molecule has a free 3' end. DNA polymerase III then attaches to the 3' end of the primer molecules and facilitates synthesis of short DNA fragments attached to molecules of RNA primer. After substantial synthesis by DNA polymerase III, DNA polymerase I removes the primer molecules of RNA and extends the previously synthesized pieces of DNA, producing a collection of short, discontinuous fragments known as Okazaki's fragments, in honor of their discoverer. Finally, DNA ligase joins the small pieces of DNA to form a continuous DNA strand. The result of these processes is formation of two new double-stranded DNA molecules, each of which contains a strand from the previous molecule and a newly synthesized strand. In most situations, replication by the theta mode is accomplished by two replication forks, which move in opposite directions around the circular chromosome. The theta mode of DNA replication is illustrated in Figure 12.2.

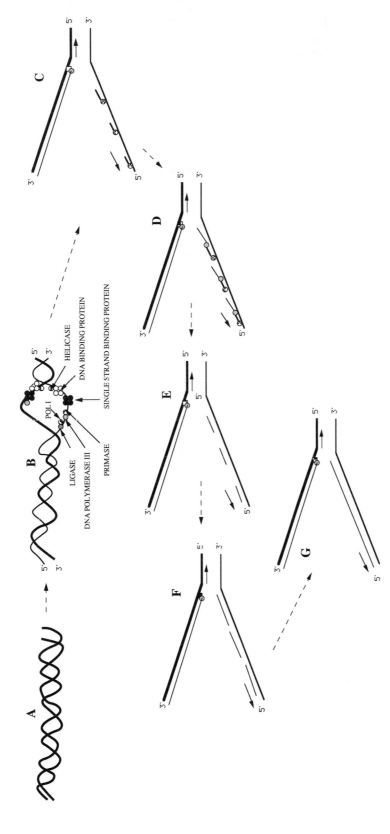

FIGURE 12.2 Replication of DNA by the theta mechanism. The dashed arrows show the progression of events. (A) The DNA is initially a highly coiled double-stranded helical molecule. (B) By interaction with the replication complex, which includes helicase, the DNA-binding protein, single-strand binding protein, DNA polymerases I and III, primase, DNA ligase, DNA gyrase, and many other proteins, the double-stranded molecule is unwound to form a replication fork. (C) After formation of a single primer and attachment of polymerase III, the leading strand that replicates in the direction of the replication fork, replicates continuously. Replication of the lagging strand, by contrast, is discontinuous and requires formation of small pieces of RNA, hybridized to the lagging template strand by a special DNA-dependent RNA polymerase, primase. (D) By the action of DNA polymerase III, which attaches to the 3'OH end of the short RNA pieces, small segments of RNA-bound DNA are produced. (E) DNA polymerase I now removes the RNA primer fragments to produce short DNA segments known as Okazaki's fragments. (F) DNA polymerase I extends the small fragments (F) and DNA ligase binds them to form a complete lagging strand (G).

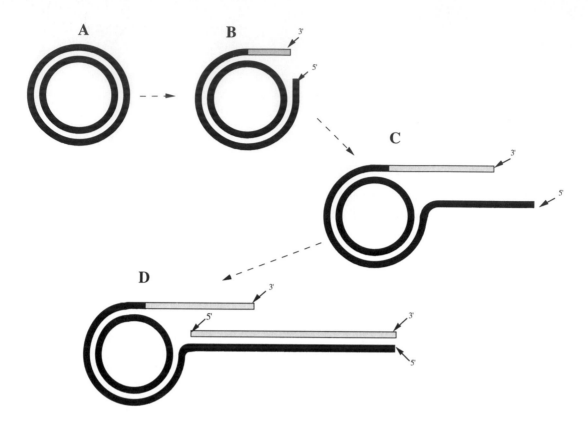

FIGURE 12.3 The rolling circle mode of DNA replication. The initially circular double-stranded DNA molecule (A) is converted to a double-stranded circular molecule in which the outer circle is nicked, producing a 3′OH end (B). By addition of nucleotides in the 5′ to 3′ direction, with the rotating nonnicked strand as template, a single strand of new DNA is synthesized and the 5′ end of the replicating strand is displaced (C). (D) As synthesis proceeds, a second new strand is formed using the displaced original replicating strand as template. This mode of DNA replication allows rapid and abundant formation of single DNA strands. Although not shown in the figure, it is often the case that the sigma mode of replication produces concatemers of single DNA strands and that the concatemers are cleaved by an endonuclease to produce the multiple copies.

The sigma mode

The sigma mode of DNA replication occurs in certain viruses and plasmids during conjugation. It is substantially less complicated than the theta mode. The sigma mode is often referred to as the rolling circle mechanism and is illustrated in Figure 12.3. In the sigma mode, a single DNA strand is nicked to produce a free 3′OH end. Using the other strand as a template, DNA polymerase III adds nucleotides to the 3′ end of the nicked molecule and displaces its 5′ end. Continuation of this process allows formation of many copies, called concatemers, of the single strand. Copies of the complementary strand are then synthesized, using the displaced strand as template. Finally, endonuclease enzymes convert the replicated concatemeric DNA into appropriately sized pieces.

Certain DNA viruses contain linear DNA, but replicate DNA by a mechanism involving a cyclic intermediate. In most cases, the linear molecules are double-stranded and formation of a cyclic intermediate is required before replication is possible. Circular DNA may be formed from linear DNA by one of two mechanisms. In one of the mechanisms, formation of bonds between cohesive ends, the ends that are complementary to each other and allow hybridization of adjacent linear strands, is followed by sealing of gaps in the resultant molecules by the action of DNA ligase.

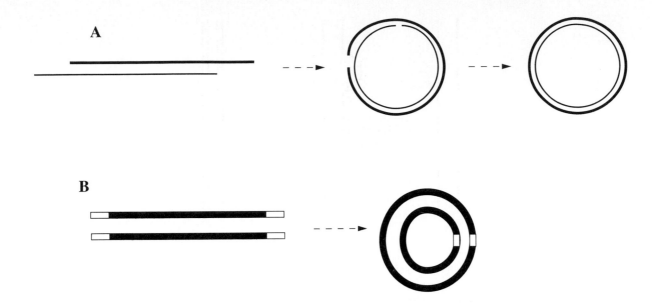

FIGURE 12.4 The major mechanisms for formation of circular DNA from double-stranded linear DNA. (A) Two linear strands possess cohesive ends. These ends are complementary to each other. The original doublet of single strands circularizes, producing, by hydrogen bonding, a circular double-stranded entity with gaps that correspond to the cohesive end region of each of the single strands. Polynucleotide ligase seals the gaps in each of the strands and yields a complete double-stranded molecule. (B) Two linear strands, denoted by the black lines, contain terminally repeated sequences at both ends, allowing formation of a circular doublet by recombination. In some situations, the linear strands are replicated prior to being circularized, leading to concatemeric circular doublets.

In an alternate mechanism, spontaneous formation of recombinants occurs between regions of terminal redundancy—terminal regions that contain completely complementary sequences. However, the cyclic molecule is formed, the cyclic molecule may then be replicated by a theta mechanism or by a combination of the theta and sigma mechanisms. The modes of formation of circular DNA from linear DNA are shown in Figure 12.4.

Eucaryotic DNA replication

Replication of chromosomal DNA in eucaryotic microbes differs substantially from the process in procaryotes, plasmids, and viruses. The genetic material in eucaryotes is both structurally and chemically more complex than in procaryotes. Each eucaryotic cell has more than one chromosome, and as a reflection of the inherently greater complexity of eucaryotic cells, the process of DNA replication in eucaryotes is substantially more complex than the analogous process in procaryotes.

In eucaryotes, DNA replication occurs in replication bubbles. Many bubbles are found on each chromosome and each replication bubble has its own replication origin. The replication apparatus of each bubble is known as a *replicon*. It appears that within a bubble, replication of one of the strands is discontinuous and that the replication of an entire chromosome is accomplished by the collective activities of adjacent bubbles, each of which replicates a portion of the chromosome. As replication proceeds, adjacent bubbles fuse and the entire chromosome is replicated. As many as 400 independent replication bubbles may occur on a eucaryotic chromosome. The process of DNA replication in eucaryotes is illustrated in Figure 12.5.

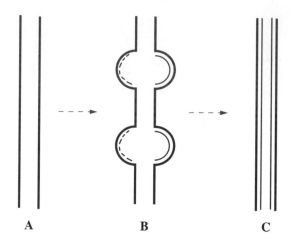

FIGURE 12.5 The general nature of eucaryotic DNA replication. The process is substantially more complex in eucaryotic organisms than in procaryotic. The eucaryotic chromosome is composed of a single double-stranded linear helical molecule (A). The size of the chromosome is such that, for each chromosome, multiple replication "bubbles" (B), functionally analogous to replication forks of the theta mode, are formed. Replication, at localized sites, is semiconservative and bidirectional. As is the case for procaryotes, replication is continuous in one direction and discontinuous in the other. The localized replication "bubbles" fuse to form the new strands. Because of the size of the eucaryotic chromosome, as many as 400 replication initiation sites that contain consensus recognition sequences, may occur on the same chromosome.

Summary

Microbial forms of all types participate in nucleic acid synthesis, whether the forms are cellular or acellular. Nucleic acid synthesis is required to allow perpetuation of the characteristics of an organism and also to allow the formation of proteins. In cellular forms, the relationship between nucleic acid synthesis, genome reproduction, and protein synthesis is distinct. DNA serves solely as genetic material and the synthesis of RNA on a DNA template is an obligate requirement for protein formation. By contrast, in the viruses, the relationship between nucleic acid synthesis and protein formation is less defined. In DNA viruses, and in chloroplasts, mitochondria, and plasmids as well, the relationship between DNA, RNA, and protein is the same as that in cellular forms. However, in the RNA viruses, the relationships are less clear. In some RNA viruses, the viral RNA may serve both as genetic material and as mRNA for protein synthesis. In other systems, such as the negative single-stranded RNA viruses, the message for protein synthesis is formed by transcription of an RNA message from the negative strand by an RNA-dependent RNA polymerase. In this case, the transcribed strand serves as both the message for protein formation and as template for formation of negative strands. The case of the retroviruses is even more confusing since they are single plus-stranded RNA viruses, but their replication is dependent on formation of a double-stranded DNA intermediate that functions in a manner analogous to a bacterial prophage. In the situations just described, the distinction between nucleic acids as informational molecules for protein synthesis and as genetic molecules becomes blurred. Irrespective of the system, it is only by careful study that the role of nucleic acids as genetic entities and as protein synthesis mediators, and the interrelationships between their functions in these two capacities, can be properly discerned.

Selected References

Alberts, B. 1980. *Mechanistic Studies of DNA Replication and Genetic Recombination.* Academic Press, Inc., New York.

Bremer, H., and G. Churchward, 1991. Control of cyclic chromosome replication in *Escherichia coli. Microbiological Reviews.* **55:** 459–475.

Cozzarelli, N. P. 1977. The mechanism of action of inhibitors of DNA synthesis. *Annual Review of Biochemistry.* **46:** 641–668.

Drlica, K. 1984. Biology of bacterial deoxyribonucleic acid topoisomerases. *Microbiological Reviews.* **48:** 273–289.

Erikson, R. L. 1968. Replication of RNA viruses. *Annual Review of Microbiology.* **22:** 305–322.

Geider, K. 1991. Proteins controlling the helical structure of DNA. *Annual Review of Biochemistry.* **50:** 233–260.

Harrison, S. C., and A. K. Aggarwal. 1990. DNA recognition by proteins with a helix-turn-helix motif. *Annual Review of Biochemistry.* **59:** 933–969.

Kornberg, A. 1980. *DNA Replication.* W. H. Freeman, San Francisco.

Lai, M. M. C. 1992. RNA recombination in animal and plant viruses. *Microbiological Reviews.* **56:** 61–79.

Lark, K. G. 1969. Initiation and Control of DNA Synthesis. *Annual Review of Biochemistry.* **38:** 569–604.

Lehman, I. R. 1974. DNA ligase: Structure, mechanism and function. *Science* **186:** 790.

Loeb, L. A., and T. A. Kunkel. 1982. Fidelity of DNA synthesis. *Annual Review of Biochemistry.* **51:** 429–458.

Nossal, N. G. 1983. Prokaryotic DNA replication systems. *Annual Review of Biochemistry.* **52:** 581–615.

Ogawa, T., and T. Okazaki. 1980. Discontinuous DNA replication. *Annual Review of Biochemistry.* **49:** 421–457.

Reanney, D. C. 1982. The evolution of RNA viruses. *Annual Review of Microbiology.* **36:** 47–73.

Reichard, P. 1988. Interactions between deoxyribonuclease and DNA synthesis. *Annual Review of Biochemistry.* **57:** 349–374.

Salas, M. 1991. Protein-priming of DNA replication. *Annual Review of Biochemistry.* **60:** 39–71.

Shapiro, L., and J. T. August. 1966. Ribonucleic acid virus replication. *Bacteriological Reviews.* **30:** 288–308.

Shatkin, A. J. 1974. Animal RNA viruses: Genome, structure and function. *Annual Review of Biochemistry.* **43:** 643–666.

Steinhauer, D. A., and J. J. Holland. 1987. Rapid evolution of RNA viruses. *Annual Review of Microbiology.* **41:** 409–434.

Temin, H. M. 1971. Mechanism of cell transformation by RNA tumor viruses. *Annual Review of Microbiology.* **25:** 609–648.

Tomizawa, J.-I., and G. Selzer. 1979. Initiation of DNA synthesis in *Escherichia coli. Annual Review of Biochemistry.* **48:** 999–1034.

Wang, T. S.-F. 1991. Eukaryotic DNA polymerases. *Annual Review of Biochemistry.* **60:** 513–552.

Wickner, R. B. 1992. Double-stranded and single-stranded RNA viruses of *Saccharomyces cerevisiae. Annual Review of Microbiology.* **46:** 347–376.

CHAPTER 13

Regulation

The complexity of regulation

Regulation of the physiological activities of microbes is an extremely complicated subject and understanding the nature of regulation is equally complex. Microbial cells have a diversity of mechanisms by which they regulate their activities in response to changes in their external environment and an equally diverse collection of mechanisms whereby they may regulate and coordinate intracellular events. In this chapter we will discuss selected mechanisms of cell regulation and indicate the complexity of the regulation process. Although we will not consider all of the possible regulation mechanisms in microbes, we will discuss a number sufficient to demonstrate the intricacies of regulation. It is the primary intent of this chapter to provide a general overview of regulation. Particular applications of the concepts discussed here are considered, in detail, in many other chapters.

Regulation may be regarded from a number of perspectives. We may, for example, ask how an organism regulates a *particular process* in response to a change in the external environment. Alternatively, we may consider the manner in which the cell regulates and coordinates its *intracellular* activities, irrespective of changes in the external environment. In addition to these perspectives, we may study the *level* at which regulation occurs. We may ask, for example, how the operation of regulatory mechanisms reflects alteration in the formation and function of *critical cell proteins*. Consideration of this question requires that we concurrently understand the nature and regulation of *nucleic acid metabolism*. In addition to these perspectives, we may ask how the formation and function of *structure* may serve as a regulatory mechanism. Finally, we may ask how changes in the physiological properties of microbes are reflective of changes in *genetic expression*. None of these viewpoints is mutually exclusive. It is usually not possible to fully understand the regulation of any cellular phenomenon without considering it from more than one perspective. In addition, because many physiological processes are interrelated, it is often impossible to fully understand the physiological consequences of a particular regulatory mechanism without considering the effects of a particular change on the total physiology of the organism.

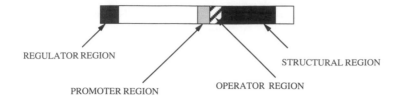

FIGURE 13.1 The general arrangement of the critical procaryotic genomic regions required for control of message formation for protein synthesis.

Regulation of protein formation

Control of protein formation is a major regulatory mechanism. This general mode of control may occur in several ways. We may regulate formation of the message for a particular protein. Alternatively, we may regulate formation of the protein itself. Within these two general possibilities, a number of mechanisms are known, some of which are primarily genetic and some of which are primarily physiological, in the sense that they involve aspects of the cell other than the genome.

For proteins to be formed, the genome and other cellular entities, particularly the ribosome, must operate cooperatively. If there is no message, there will be no protein but it is equally the case that if there is no ribosome, or if it does not function properly, protein formation is impossible. Since proteins are composed of amino acids, if the amino acids are in insufficient supply or are not appropriately activated, protein formation will not occur. Finally, if proteins are not present, new protein cannot be formed, for at least three reasons: 1) enzyme proteins are essential to the protein synthesis process; 2) proteins participate physically in protein formation; and 3) proteins are major components of the ribosomal particle, the particulate entity that joins amino acids into proteins.

Regulation of message formation

Message formation regulation is a major mechanism for protein synthesis control. To understand the complexity of message formation regulation, it is necessary to understand the structure of the genome and the manner in which the message is formed. The nature of the procaryotic genome as it

relates to message formation control is shown in Figure 13.1. Examination of the figure reveals that it contains at least four critical regions, the *regulator* or R region, the *promoter* or P region, the *operator* or O region and the *structural* or S region. The P, O, and S regions are normally adjacent to each and all three are physically separated from the R region.

The functions of the various regions in protein synthesis are diverse. The S region contains information associated with the actual formation of the proteins involved in a particular process. The O region is a regulatory region. When it is occupied by an active form of regulator protein, commonly known as a *repressor*, attachment of the enzyme required for code formation, DNA-dependent RNA polymerase, is prevented. Unless DNA-dependent RNA polymerase attaches to the DNA molecule, it cannot participate in formation of the code.

The P region, immediately adjacent to the O region is the actual region to which DNA-dependent RNA polymerase attaches. The sigma portion of the DNA-dependent RNA polymerase enzyme recognizes certain base sequences within the promoter region and allows attachment of the entire enzyme to the P region so that the process of code formation may be initiated. The R region codes for formation of regulator proteins, particularly repressor proteins. However, regulator proteins may be made in either an active or an inactive form. The form and activity of regulator proteins may be altered by interaction of the proteins with *effector* molecules, small nonprotein molecules that combine with certain proteins at allosteric sites, sites physically removed from active sites of proteins. Interaction of an effector molecule with a protein, at the allosteric site, changes the conformation of the protein, and thereby, its activity. Proteins whose conformation and activity may be altered by interaction with effector molecules are known as *allosteric proteins*.

The four major modes for control of message formation

Four major mechanisms of message formation control are currently well known, induction, catabolite activation, corepression, and attenuation. Induction and catabolite activation are mechanisms for control of catabolic sequences, whereas corepression and attenuation are mechanisms for control of biosynthetic metabolism.

Induction   In induction, a repressor protein initially made in an active form is inactivated by interaction with a carbon and an energy source. This phenomenon has been most thoroughly studied with regard to lactose use by *E. coli.* When the organism is supplied with alternate carbon sources, it uses the most readily usable one first. When supplied with both glucose and lactose, for example, it uses glucose first and uses lactose only when the glucose is exhausted. When glucose is exhausted, and only then, a lactose metabolite, *allolactose*, induces lactose utilization. Interaction of allolactose with an active repressor protein, inactivates the latter. Under these conditions, the repressor protein cannot occupy the operator locus on the genome and cannot prevent attachment of DNA-dependent RNA polymerase to the promoter region. DNA-dependent RNA polymerase attaches to the P region and the message for lactose utilization is formed. The message includes information for formation of both the transport protein for lactose, β-galactoside permease, and the enzyme galactosidase, which hydrolyzes lactose to glucose and galactose, both of which are utilizable by central metabolic sequences.

Catabolite activation   Catabolite activation is a second major mechanism for message formation control that affects degradative metabolism. In *E. coli* and many other organisms, this mechanism operates concomitantly with induction. However, mechanistically, the two modes of regulation are distinctly different. Induction may be regarded as a negative control mechanism because the presence of an active repressor molecule on the operator site inhibits attachment of DNA-dependent RNA polymerase to the P site. As a consequence, the code for lactose utilization is not formed. It is only if allolactose inactivates the active repressor molecule that attachment of DNA-dependent RNA polymerase, and message formation, are possible.

Catabolite activation operates to control lactose utilization in a manner distinctly different from the repression-induction system. In catabolite activation, the presence of an active molecule of a particular regulatory protein, *catabolite activator protein* (CAP), also known as catabolite repressor protein and cyclic adenosine monophosphate (cAMP) receptor protein, facilitates attachment of DNA-dependent RNA polymerase to the P site and allows formation of the message for lactose utilization. Catabolite activation is a positive control mechanism because an active protein facilitates DNA-dependent RNA polymerase attachment. In the repression-induction system, an active protein inhibits attachment of the enzyme.

The activity of catabolite activator protein is a function of the cellular concentration of cAMP. The catabolite activator protein (CAP) is normally synthesized in an inactive form. Interaction of CAP with cAMP converts the inactive CAP molecule to the active form. The cAMP-CAP complex binds to the P region of the genome and facilitates attachment of DNA-dependent RNA polymerase, allowing the code for lactose utilization to be transcribed.

The cAMP concentration is a function of the energy state of the cell. When presented with two energy sources, glucose and lactose, *E. coli* uses glucose in preference to lactose. Rapid utilization of glucose produces a relatively high concentration of ATP and a low concentration of AMP and ADP. As glucose is exhausted, the level of ATP falls, as energy is expended for life processes. The lowering of ATP concentration that accompanies exhaustion of glucose supplies is associated with an increase in the cAMP. The increase reflects alteration of the relative activities of the enzymes involved in cAMP formation and destruction, adenyl cyclase (production) and cyclic AMP phosphodiesterase (destruction). The level of cyclic AMP phosphodiesterase is relatively constant, irrespective of the cellular ATP concentration. Adenyl cyclase activity, by contrast, reflects ATP concentration. When the ATP concentration is high, adenyl cyclase activity is low and, conversely, as the ATP concentration falls, the level of adenyl cyclase increases, with the concomitant formation of increased levels of cAMP.

Corepression   Corepression is a mechanism for regulation of biosynthetic metabolism by message formation control. It is, in some respects, the *inverse* of the induction-repression mechanisms that regulate catabolism. In corepression, a repressor protein is synthesized in an inactive form and becomes

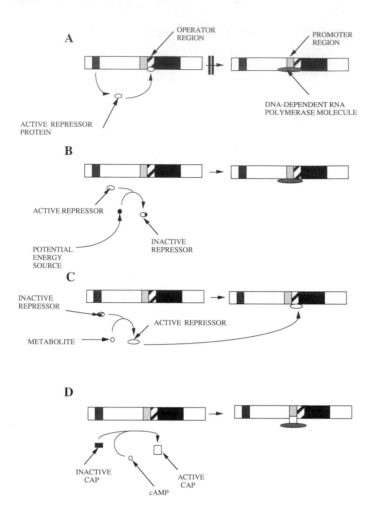

FIGURE 13.2 The four major mechanisms of message formation control. (A) Repression. In this mode, an active repressor protein occupies the operator region and prevents attachment of DNA-dependent RNA polymerase to the promoter region, thus preventing the message from being formed. (B) Induction. An active repressor protein is made inactive by interaction with a substance, normally a potential energy source. As a result of inactivation of the repressor, DNA polymerase attaches to the promoter region and the genes for utilization of the alternate energy source are activated. (C) Corepression. As a result of interaction with an anabolic metabolic product, a normally inactive repressor is made active. The active repressor occupies the operator locus and prevents formation of messages required for synthesis of the regulating substance. (D) Catabolite activation. Cyclic adenosine monophosphate (cAMP) interacts with inactive catabolite activator protein (CAP), making CAP active. The active CAP facilitates attachment of DNA-dependent RNA polymerase to the promoter region, allowing transcription to occur.

activated by interaction with a biosynthetic end product. The active repressor then attaches to the O site and inhibits formation of the message for formation of the corepressor, the biosynthetic product that activates the inactive regulator protein. Corepression serves to prevent excess formation of a biosynthetic metabolite that is present in sufficient quantities. Induction, repression, corepression, and catabolite activation are shown in Figure 13.2.

Attenuation   Attenuation is an additional means for regulation of biosynthetic metabolism by message formation control, but the mechanism by which message formation control occurs is substantially different from that for catabolite repression. The messages for proteins whose control is accomplished by attenuation are unusual, as is the location of the promoter region, relative to the structural region, on the genome. In situations in which attenuation controls message formation, the promoter region is situated substantially *upstream* from the S region. A series of about 160 bases intervenes between the P and the S regions. This region contains a transcription start site, a ribosomal-binding site, and a translation initiation site, and also codes for a leader sequence. The end of the leader sequence contains adjacent codons for the substance, normally an amino acid, that regulates the system.

In systems controlled by attenuation, translation and transcription are closely interconnected because a translating ribosome, attached to mRNA, closely follows the molecule of DNA-dependent RNA polymerase that participates in transcription. As transcription occurs, it is followed closely, both in time and in space, by translation. When the controlling substance is present in abundance, transcription and translation occur "normally"—through the region that requires addition of controlling substance, allowing the formation of an unusual transcript, a *pause loop,* by the polymerase enzyme. The formation of the pause loop slows down transcription for a period of time, but the translation product of the previously transcribed message alleviates the slowing of transcription and allows formation of a second, *terminator* loop. Under these conditions, both transcription and translation cease prior to formation of the structural message.

When the regulating amino acid is limited, translation of the leader sequence is also limited. Thus, the alleviation of the effects on pause loop formation does not occur and a terminator loop is not formed either. Rather, an alternative loop, an *antiterminator loop* is formed. The terminator portion of the genome is bypassed, and transcription and translation of the genes required for synthesis of the limiting amino acid occur. Although we have described attenuation as a phenomenon reflective of amino acid metabolism, the mechanism can, in theory, operate with other biosynthetic metabolites. Attenuation is shown diagrammatically in Figure 13.3.

Although there are distinct differences in the details of mechanisms by which transcription is controlled, there are also some distinct similarities. By whatever mode transcription regulation occurs, regulation is intimately associated with changes in the physical form of components of the system. From this viewpoint, induction, catabolite repression, and corepression may be considered as variations on a common theme, alteration of the function of a regulatory protein by change in physical form though interaction with a small nonprotein molecule. The consequences of the interaction may vary, but the mechanism is essentially the same. In induction, an active protein becomes inactive, while in both catabolite repression and corepression, an inactive protein is made active. In induction, the presence of a nutrient facilitates its utilization, while in corepression and attenuation, the presence of a nutrient inhibits its synthesis. Attenuation does not involve interaction between a protein and an effector molecule, but a translation product determines the manner in which a protein, DNA-dependent RNA polymerase, functions. In attenuation, interaction of DNA-dependent RNA polymerase with a product or its own action determines the manner in which the polymerase functions.

Control of translation

If one assumes that a message for protein is available, formation of the protein itself requires translation, the actual joining of amino acids to produce the protein. Translation, in turn, requires the availability of functioning ribosomes, a supply of amino acids in an appropriately activated form, and a collection of enzymes and auxiliary proteins so that activated amino acids may be recognized and joined to produce protein. As is the case with message formation, proteins are intimately involved both in the translation process and its regulation. Control of translation may have just as potentially profound an effect on protein synthesis as does control of transcription.

Translation may be regarded as a process that involves the reversible aggregation and deaggregation of particles. The nature of the particles involved in translation is different in procaryotic systems than in eucaryotic organisms, but in both cases, the process involves the formation and dissociation of particles. Factors that regulate the

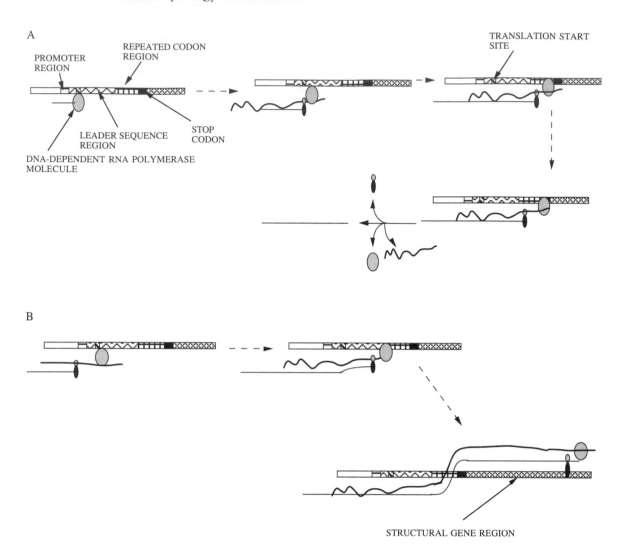

FIGURE 13.3 Control of message formation by attenuation. (A) Nonlimiting nutritional conditions. (B) Nutritionally limited conditions. Under nonlimited conditions of nutrition, a DNA-dependent RNA polymerase molecule attaches to the promoter region and begins transcription. At the same time, a ribosome attaches to the message and translation is initiated. Translation proceeds through the leader sequence until the region at the end of the leader sequence in which repeated codons for the amino acid that regulates the system occur is encountered. When the amino acid is present in abundance, tRNA molecules charged with that amino acid are also abundant. Translation continues through the entire leader sequence until the stop codon is encountered, at which point transcription and translation stop. Under nutritionally limited conditions, a different sequence of events occurs. In this situation, translation occurs through the leader sequence until the repeated codon region is reached. At this point, because of the sparsity of charged tRNA molecules, the relative rates of transcription and translation and the relative configurations of the ribosome and the messages are altered, allowing the stop codon to be bypassed and the structural genes to be transcribed and translated.

formation and physical aggregation of ribosomal subparticles and the formation and deaggregation of polysomes regulate translation by controlling the formation and function of the structural entities that accomplish protein formation.

Metabolic control of translation

Translation is regulated metabolically, as well as by control of ribosome formation and function. Amino acid availability is a potential means for metabolic control of translation. Amino acid supply reflects the proper functioning of synthetic pathways, but may sometimes reflect transport of exogenously supplied amino acids when synthetic pathways are inadequate to provide an amino acid pool that allows maximum translation efficiency. Since amino acids must be presented to the ribosome attached to tRNA molecules, RNA synthesis may regulate translation metabolically. In addition, the general state of cell energy metabolism influences translation. Aminoacyl tRNA molecules are formed from amino acid adenylates whose formation, in turn, requires ATP expenditure. The state of cell energy metabolism may influence translation not only by energy expenditure, but also because nucleotide phosphates are involved as *steric modifiers* that facilitate or inhibit ribosome-message and ribosome-aminoacyl tRNA interactions.

Regulation of protein activity

Cellular regulation may occur not only through control of transcription and translation, but also by control of the activity of pre-formed proteins. Translation and transcription are themselves controlled by such a mechanism, but many other proteins not involved in the protein synthesis process may be controlled in a similar way. In addition to regulatory molecules, the enzymes of both degradative and biosynthetic metabolism may be regulated by *allosteric interaction* with small molecules. The consequences of such a mechanism are far-reaching. In synthetic sequences, allosteric regulation of an enzyme at the beginning of a sequence may allow regulation of the entire sequence. The regulated enzyme is referred to as the *rate-limiting enzyme*. In branched metabolic sequences, the beginning enzyme of each branch is often controlled allosterically. The combined allosteric regulation of common reactions and the reactions unique to the branches, allows channeling of metabolism so that an adequate supply of common intermediates is available for both branches and, at the same time, physiologically appropriate proportions of common intermediates are directed to the branches.

Types of allosteric regulation

Allosteric regulation of enzymes may occur in several ways. In *simple feedback inhibition,* a single metabolite interacts with a regulating enzyme and concomitantly alters both its conformation and its activity. In *concerted feedback inhibition,* more than one metabolite interacts with the enzyme to alter its activity. In such situations, the regulated enzyme often is composed of subunits. In *sequential feedback inhibition,* which often occurs in branched metabolic sequences, two or more metabolites interact with the enzymes at the beginning of their branches. Because both enzymes that would normally metabolize the terminal intermediate of the intermediates common to both branches are inhibited, the terminal common intermediate accumulates, and in turn, interacts with an enzyme at the beginning of the entire sequence. As a result, the entire sequence, including the stem and branches, is regulated. The mechanisms of simple, concerted, and sequential inhibition, and their consequences, are shown in Figure 13.4.

Isoenzymes

In many metabolic sequences, particularly branched biosynthetic sequences, the flow of intermediates is regulated by *isoenzymes,* enzymes that, although mechanistically different, accomplish the same reaction. Such enzymes are typically found at the beginning of a reaction sequence that provides common intermediates for branched biosynthetic processes. Allosteric modification of the activity of isoenzymes by metabolites of a particular branch of a biosynthetic sequence, which involves common intermediates, allows regulation of intermediate flow to a particular branch of a branched metabolic sequence without intervention of intermediates from another branch. Isoenzymes provide an additional mode of regulation of the flow of intermediates in branched metabolic sequences. Regulation of metabolic sequences by isoenzymes is shown in Figure 13.5.

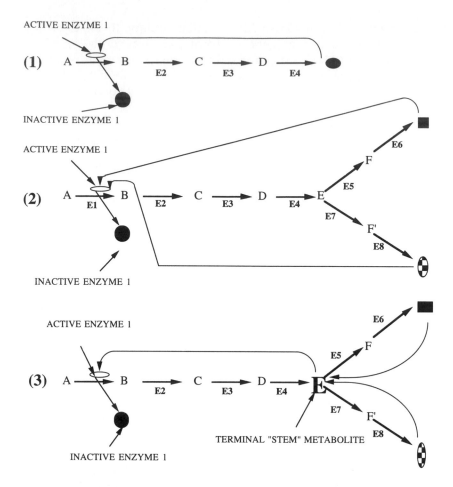

FIGURE 13.4 Regulation of preformed enzyme activity by simple feedback inhibition (1), concerted feedback inhibition (2), or sequential feedback inhibition (3). The capital letters represent metabolites and, except for the first enzyme, symbols represent biosynthetic end products. The various enzymes, other than the first one in each sequence, are represented by capital E and a number. In simple feedback inhibition, a single substance, normally a biosynthetic end product, interacts allosterically with the first enzyme in the sequence and changes both its conformation and its activity. In concerted inhibition, two products interact allosterically with an enzyme. In sequential feedback inhibition, metabolites or end products individually interact with the first enzyme of their respective branches, leading to accumulation of the first branch metabolite and also the terminal substance of the metabolites common to both branches. The terminal common metabolite, in turn, interacts with the first enzyme of the "stem" metabolic sequence.

Regulation by energy charge

In many cases, allosteric regulation of pre-formed enzymes is accomplished by interaction of particular metabolites with early enzymes in metabolic sequences. Energy charge (see Chapter 7) is an additional, more generalized mechanism for allosteric control of enzyme activity. In general, energy-yielding reactions are stimulated by AMP or ADP and inhibited by ATP. Conversely, energy-utilizing reactions are stimulated by ATP and inhibited by the products of ATP utilization, ADP and AMP.

Atkinson and colleagues suggested that the balance of catabolic and anabolic reactions in cells reflects the energy charge of the cell that is mathematically described below:

$$\text{Energy charge} = \frac{\Sigma\ \text{ATP} + 1/2\ \text{ADP}}{\Sigma\ \text{ATP} + \text{ADP} + \text{AMP}}$$

In the above equation, it is understood that the concentrations are in molar units. It appears that the concept of energy charge has much value as a general mechanism for allosteric regulation of

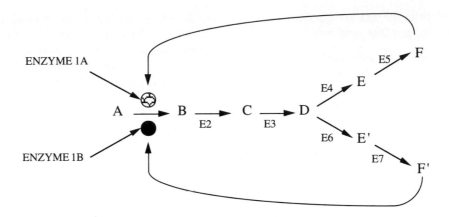

FIGURE 13.5 Regulation of isoenzymes. Terminal metabolites of the branches from a common metabolic sequence allosterically interact with two mechanistically and chemically distinct enzymes that catalyze the beginning chemical reaction of the reactions common to both branches.

metabolism, maintaining the appropriate interconnection between catabolism and anabolism.

Regulation by carrier molecules

Carrier molecules are present in only small amounts in cells but are critical to cellular processes and may serve as regulatory agents. We may regard carriers as one of two types; *metabolic carriers* and *structure synthesis carriers*. NAD, NADP, and FAD are examples of metabolic carriers, while undecaprenol, lipid A, and dolichol are examples of carriers that facilitate structure synthesis. The reduced forms of metabolic carriers may serve, either directly or indirectly, as sources of biosynthetic reducing power or as electron donors for ATP formation. Because of their dual role in energy metabolism and biosynthesis, metabolic carriers may regulate cellular activities. It is often the case that synthetic reactions are stimulated by NADH and NADPH, but inhibited by the NAD and NADP, while energy-yielding reactions exhibit the inverse pattern. The relative proportions of molecules in the reduced and oxidized states serve to regulate and coordinate metabolism in a manner analogous to the change in forms of nucleotides.

Structural carriers may also serve as regulatory agents. This is the case for two primary reasons. In the first case, different processes, such as murein and exopolysaccharide synthesis, require the same carrier molecule, and the same physical form of the carrier. In this situation, the amount of the carrier regulates the extent to which both processes may occur. In other cases, processes that occur in the same cell, that is, murein and teichoic acid synthesis, require alternate physical forms of the carrier, so the relative proportions of molecules in a particular form may serve to regulate the relative rates of synthetic processes.

Structural regulation

The structures of the cell exert regulatory effects on the total physiology of microbes. For example, neither the membrane or the wall operate in isolation. The membrane is intimately involved in several processes of the cell, both metabolic and structural. Particularly in procaryotes, the activities of the membrane influence all other aspects of the cell. The membrane, or invaginations of it, is the beginning point for cell wall synthesis. At the same time, the procaryotic membrane mediates electron transport, substrate transport, and DNA replication. The integrity of the membrane is, in turn, a function of the wall. Both the membrane and the wall participate in the process of sporulation. These examples illustrate the functionally and structurally interrelated nature of the structures of the cell. It is precisely because of these interrelationships that the structures may exert regulatory effects on the total physiology of the cell.

The role of the membrane in regulation of cell wall and DNA synthesis

The procaryotic membrane participates intimately in the portion of cell wall synthesis associated with septation. Mutational analysis has shown that in gram-negative bacteria, two types of murein synthesis occur. One type of synthesis occurs relatively early in the cell cycle, is inhibited by mecillinam, and is associated with elongation. The second type occurs late in the cell cycle and is associated specifically with septation. Septation is thus separate from elongation.

It is possible to disengage the septation and elongation with the antibiotic, *cephalexin*. This antibiotic inhibits septation without inhibiting elongation-associated murein synthesis. Study has shown that cephalexin binds to one of a class of proteins known as penicillin-binding proteins. Cephalexin binds specifically to penicillin-binding protein 3, a membrane-bound protein. Penicillin-binding proteins have been shown to participate in murein synthesis, so the finding that cephalexin binds specifically to penicillin-binding protein 3, a membrane-associated protein suggests that the protein is specifically involved in the murein synthesis associated with septum formation. The membrane regulates cell septum formation in at least two ways: by providing a surface, or in fact, a container, for newly synthesized murein and by the specific participation of one or more of its proteins in murein synthesis.

In addition to its role as a regulator of cell wall formation, the membrane is intimately involved in DNA replication. Much evidence indicates that DNA replication in procaryotes is a membrane-associated process. The membrane serves as an attachment site for initiation of DNA replication, and in consort with the wall, serves as the agent by which replicated DNA is partitioned into daughter cells. Because of these phenomena, both membrane and wall synthesis serve as regulators of DNA replication.

Endosporulation is yet another example of a process regulated, at least in part, by the membrane. At the completion of a round of DNA replication, a membrane-associated process, invagination of the membrane gives rise to the forespore. Engulfment of the forespore by the "mother cell," surrounds the forespore with a second membrane. The subsequent events of sporulation occur either within membranous boundaries (the cortex) or on membranous surfaces (the spore coats). Sporulation, from many perspectives, may be regarded as a membrane-mediated process and because of this, membrane synthesis influences it.

Multiple regulation

Thus far we have primarily discussed regulation of individual processes or the formation and functioning of the proteins that mediate them. Regulation may also occur in a manner such that a number of processes are simultaneously regulated by a common mechanism. The term *operon* describes a collection of physically adjacent genes involved in a common process, but controlled by a common regulatory protein, for example, an active repressor molecule. In some cases, a particular process is mediated by several operons that are physically separated but are controlled by a common regulatory protein. The term *regulon* describes such a situation. The term *modulon* has been used to describe a system of interconnected operons involved in different processes, all of which are controlled by individual regulatory proteins and, in addition, are collectively regulated by a "super" regulatory protein. Because the cell often responds coordinately to a particular extracellular stimulus, yet another term, *stimulon,* has been suggested to describe the coordinated response of many operons to a single environmental change. Examples of these regulatory mechanisms are discussed in Chapter 14.

Summary

Regulation of the physiological processes of microbes is extremely complex. In this chapter we have discussed a number of ways by which regulation may occur. Regulation, at a minimum, is simultaneously a genetic, metabolic, and structural phenomenon. Although we have described individual mechanisms of control, regulation is usually accomplished through the simultaneous operation of a variety of control systems, some of which are *metabolic*, some of which are *structural*, and some of which are *genetic*.

In practice, it is often difficult to decide precisely how we may best regard both particular regulatory systems and their collective action because we recognize regulation that involves processes that are not directly genetic phenomena, but it is DNA and its expression that allows everything that may occur in a cell to actually happen. Although we may study regulation at the structural, physiological, and metabolic levels, the fullest understanding of the implications of regulatory phenomena often requires genetic analysis.

Selected References

Atkinson, D. E. 1969. Regulation of enzyme of function. *Annual Review of Microbiology.* **23:** 47–68.

Bagg, A., and J. B. Neilands. 1987. Molecular mechanism of regulation of siderophore-mediated iron similation. *Microbiological Reviews.* **51:** 509–518.

Beckwith, J., J. Davies, and J. A. Gallant, (eds.) 1983. *Gene Function in Procaryotes.* Cold Spring Harbor Laboratory, New York.

Blau, H. M. 1992. Differentiation requires continuous active control. *Annual Review of Biochemistry.* **61:** 1213–1230.

Booth, Ian R. 1985. Regulation of cytoplasmic pH in bacteria. *Microbiological Reviews.* **49:** 359–378.

Botsford, J. L., and James G. Harman. 1992. Cyclic AMP in prokaryotes. *Microbiological Reviews.* **56:** 100–122.

Bremer, H., and G. Churchward. 1991. Control of cyclic chromosome replication in *Escherichia coli. Microbiological Reviews.* **55:** 459–475.

Brenchley, J. E., and L. S. Williams. 1975. Transfer RNA involvement in the regulation of enzyme synthesis. *Annual Review of Microbiology.* **29:** 251–274.

Cabib, E., R. Roberts, and B. Bowers. 1982. Synthesis of the yeast cell wall and its regulation. *Annual Review of Biochemistry.* **51:** 763–794.

Calendar, R. 1970. The regulation of phage development. *Annual Review of Microbiology.* **24:** 241–296.

Calvo, J. M., and G. R. Fink. 1971. Regulation of biosynthetic pathways in bacteria and fungi. *Annual Review of Biochemistry.* **40:** 943–968.

Cashel, M. 1975. Regulation of bacterial ppGpp and pppGpp. *Annual Review of Microbiology.* **29:** 301–318.

Chinkers, M. 1991. Signal transduction by guanylyl cyclases. *Annual Review of Biochemistry.* **60:** 553–575.

Cohen, P. 1989. The structure and regulation of protein phosphatases. *Annual Review of Biochemistry.* **58:** 453–508.

Crawford, I. P., and G. V. Stauffer. 1980. Regulation of tryptophan biosynthesis. *Annual Review of Biochemistry.* **49:** 163–196.

Davis, R. H. 1986. Compartmental and regulatory mechanisms in the arginine pathways of *Neurospora crassa* and *Saccharomyces cerevisiae.* Microbiological Reviews. **50:** 280–313.

Dawes, E. A., and P. J. Senior. 1973. The role and regulation of energy reserve polymers in microorganisms. *Advances in Microbial Physiology.* **10:** 136–266.

Doi, R. H., and L.-F. Wang. 1986. Multiple procaryotic ribonucleic acid polymerase sigma factors. *Microbiological Reviews.* **50:** 227–243.

Eggertsson, G., and D. Söll. 1988. Transfer ribonucleic acid-mediated suppression of termination codons in *Escherichia coli. Microbiological Reviews.* **52:** 354–374.

Gold, L. 1988. Posttranscriptional regulatory mechanisms in *Escherichia coli. Annual Review of Biochemistry.* **57:** 285–320.

Gottesman, S. 1984. Bacterial regulations: Global regulatory networks. *Annual Review of Genetics.* **18:** 415–551.

Green, P. J., O. Pines, and M. Inouye. 1986. The role of antisense RNA in gene regulation. *Annual Review of Biochemistry.* **55:** 569–598.

Harrison, D. E. F. 1976. The regulation of respiration rate in growing bacteria. *Advances in Microbial Physiology.* **14:** 243–313.

Hill, T. M. 1992. Arrest of bacterial DNA replication. *Annual Review of Microbiology.* **46:** 603–634.

Hinnebusch, A. G. 1988. Mechanisms of gene regulation in the general control of amino acid biosynthesis in *Saccharomyces cerevisiae. Microbiological Reviews.* **52:** 248–273.

Holzer, H., and W. Duntze. 1971. Metabolic regulation by chemical modification of enzymes. *Annual Review of Biochemistry.* **40:** 345–374.

Johnson, P. F., and S. L. McKnight. 1989. Eukaryotic transcriptional regulatory proteins. *Annual Review of Biochemistry.* **58:** 799–840.

Killick, K. A., and B. E. Wright. 1974. Regulation of enzyme activity during differentiation in *Dictyostelium discoideum. Annual Review of Microbiology.* **28:** 139–166.

Kim, S. K., D. Kaiser, and Adam Kuspa. 1992. Control of cell density and pattern by Intercellular signaling in *Myxococcus* development. *Annual Review of Microbiology.* **46:** 117–140.

Klier, A. F., and G. Rapoport. 1988. Genetics and regulation of carbohydrate catabolism in *Bacillus*. *Annual Review of Microbiology*. **42:** 65–96.

Lodish, H. F. 1976. Translational control of protein synthesis. *Annual Review of Biochemistry*. **45:** 39–72.

Lovett, J. S. 1975. Growth and differentiation of the water mold *Blastocladiella emersonii:* Cytodifferentiation and the role of ribonucleic acid and protein synthesis. *Bacteriological Reviews*. **39:** 345–404.

Merrick, W. C. 1992. Mechanism and regulation of eukaryotic protein synthesis. *Microbiological Reviews*. **56:** 291–315.

Miller, J., and W. S. Reznikoff, (eds.) 1978. *The Operon*. Cold Spring Harbor Laboratory, New York.

Niedhart, F. C., J. Parker, and W. G. McKeever. 1975. Function and regulation of aminoacyl-tRNA synthetases in prokaryotic and eukaryotic cells. *Annual Review of Microbiology*. **29:** 215–250.

Nierlich, D. P. 1978. Regulation of bacterial growth, RNA, and protein synthesis. *Annual Review of Microbiology*. **32:** 394–432.

Ochoa, S., and C. de Haro. 1979. Regulation of protein synthesis in eukaryotes. *Annual Review of Biochemistry*. **48:** 549–580.

Pato, M. L. 1972. Regulation of chromosome replication and the bacterial cell cycle. *Annual Review of Microbiology*. **26:** 347–368.

Platt, T. 1986. Transcription termination and the regulation of gene expression. *Annual Review of Biochemistry*. **55:** 339–372.

Poolman, B., A. J. M. Driessen, and W. N. Konings. 1987. Regulation of solute transport in *Streptococci* by external and internal pH values. *Microbiological Reviews*. **51:** 498–508.

Raetz, C. R. H. 1978. Enzymology, genetics, and regulation of membrane phospholipid synthesis in *Escherichia coli*. *Microbiological Reviews*. **42:** 614–659.

Razin, A., and H. Ceda. 1991. DNA methylation and gene expression. *Microbiological Reviews*. **55:** 451–458.

Rodriquz, R., and M. Chamberland, (eds.) 1982. *Promoters: Structure and Function*. Praeger, New York.

Silver, S., and M. Walderhaug. 1992. Gene regulation of plasmid- and chromosome-determined inorganic ion transport in bacteria. *Microbiological Reviews*. **56:** 195–228.

Srivastava, A. K., and D. Schlessinger. 1990. Mechanism and regulation of bacterial RNA processing. *Annual Review of Microbiology*. **44:** 105–129.

Stadtman, E. R. 1970. Mechanisms of enzyme regulation in metabolism. *In* P. D. Boyer, (ed.) *The Enzymes*, vol. 1, pp. 398–450. Academic Press, Inc., New York.

Struhl, K. 1989. Molecular mechanisms of transcriptional regulation in yeast. *Annual Review of Biochemistry*. **58:** 1051–1078.

Tabita, F. R. 1988. Molecular and cellular regulation of autotrophic carbon dioxide fixation in microorganisms. *Microbiological Reviews*. **52:** 155–189.

Umbarger, H. E. 1978. Amino acid biosynthesis and its regulation. *Annual Review of Biochemistry*. **47:** 533–606.

Villarreal, L. P. 1991. Relationship of eukaryotic DNA replication to committed gene expression: General theory for gene control. *Microbiological Reviews*. **55:** 512–542.

Volpe, J. J., and P. Roy Vagelos. 1973. Saturated fatty acid biosynthesis and its regulation. *Annual Review of Biochemistry*. **42:** 21–60.

Wakil, S. J., J. K. Stoops, and V. C. Joshi. 1983. Fatty acid synthesis and its regulation. *Annual Review of Biochemistry*. **52:** 537–580.

Watson, M. D. 1981. Attenuation: translational control of transcription termination. *Trends in Biochemical Sciences*. **6:** 180–182.

West, S. C. 1992. Enzymes and molecular mechanisms of genetic recombination. *Annual Review of Biochemistry*. **61:** 603–640.

Young, R. 1992. Bacteriophage lysis: Mechanism and regulation. *Microbiological Reviews*. **56:** 430–481.

CHAPTER 14

Genetics

The central importance of DNA

Genetics may be defined as the study of the nature and expression of, changes in, and inheritance of, genetic material. In organisms other than certain viruses, DNA is genetic material, so understanding of the nature, functioning, changes in, and inheritance of DNA is critical to understanding genetics.

All that an organism is, or may be, is determined by the nature of its DNA. Differences in the nature of the DNA of various organisms dictate whether an organism is a tree, a human, or a microbe. The term *genotype* describes the *nature* of the DNA of an organism and the limits of its genetic potential. The *phenotype*, in contrast, describes the *collection of characteristics* displayed by an organism at a particular time and place and reflects the precise way in which DNA is expressed under a given set of circumstances.

The bacteria as models for study of the structure and function of DNA

All cellular organisms must deal with the particular conditions in which they find themselves. For particular organisms, certain conditions are more amenable for growth and survival than are others.

Organisms differ in their ability to respond to, and survive in, various environments, but in spite of such differences, must accomplish a common set of tasks if life is to proceed. The adaptability of an organism is primarily a function of the diversity of ways in which it can alter its genetic expression in changing environments. Among all organisms, the bacteria collectively exhibit the greatest degree of adaptability and diversity of genetic expression found in nature. The bacteria survive in a greater variety of environments than is possible for other life forms. Their ability to persist in such a variety of environments reflects not only the diversity of bacterial DNA, but also the variety of mechanisms for its expression, and the plethora of ways in which its expression is controlled and regulated. We may, therefore, use bacteria as models for understanding the nature and consequences of changes in the function of DNA in all life forms. In addition, we may use bacteria to study the nature and consequences of genetic change through the processes of mutation and selection. The phenomenon of *mutation,* the random alteration of the nature of genetic material, when coupled with selection, allows bacteria to adapt to changing environments, not only by change in genetic expression, but by change in DNA itself.

The nature of the bacterial genome

The bacterial genome is a single, double-stranded circular DNA molecule that codes for all of the functions of the organism. Although functionally similar to DNA in other types of cells, the bacterial genome has certain unique properties. To begin with, it is *haploid,* containing only a single set of genes, in contrast to the *diploid* condition found in eucaryotic genomes. The haploid nature of the bacterial genome has substantial physiological consequences because a genotypic change is normally expressed rapidly, in some readily detectable manner. In eucaryotes, by contrast, genetic changes may be masked for extended periods of time because of the presence of multiple copies of the genome.

The absence of noncoding regions is a second major feature of the procaryotic genome. The eucaryotic genome has many noncoding regions between functional areas of the chromosome. The bacterial genome may be considered more efficient than its eucaryotic analogue since virtually all of the bacterial genome is used to code for physiological or regulatory functions.

The physical location of genes is another unique feature of the bacterial genome. It is frequently found that related genes are clustered in an adjacent linear fashion on the genome. Such groupings are known as *operons.* This feature of the bacterial genome often allows related genes to be coordinately controlled by a single regulatory molecule. In contrast, genes that code for similar functions in eucaryotes are often scattered around the chromosome or are on different chromosomes. Regulation of related genes in eucaryotes is therefore more complex than the usual case with procaryotes.

In addition to unique genomic features, procaryotic organisms often harbor nongenomic, double-stranded, circular, self-replicating, auxiliary genetic elements known as *plasmids.* Plasmids are most prevalent in, and have been most widely studied in, bacteria but are not restricted to procaryotes, since they are found in certain yeasts, and rarely, in other eucaryotic organisms. The discreteness of plasmids as genetic elements wherever they are found, is indicated by their lack of homology with genomic DNA. The plasmids of bacteria are relatively small (1–200 kilobase pairs) in comparison to the bacterial genome (1,000–5,000 kbp) and contain nonessential genetic information that may enhance the metabolic versatility and adaptability of the organisms that contain them. In addition to their role in providing auxiliary genetic information for procaryotic microbes, plasmids are of interest for at least four reasons. They may be used as models to study the nature of DNA replication and its regulation, as agents for study of procaryotic genome function, or they may be modified and used as tools for genetic engineering. Finally, they may sometimes cause practical difficulties if they contain genes that code for antibiotic resistance in pathogenic organisms. Certain plasmids contain information that allows their transfer from cell to cell, without cell division, besides the usual information regarding their replication. In such cases, plasmids may be vehicles of infectious multiple drug resistance, which is the simultaneous development of resistance to several antibiotics.

Genome organization

The genome of procaryotes has a number of structural features that are associated with its expression. In addition to being closely physically grouped, often genes that are frequently transcribed are physically located on the chromosome so that they may be transcribed without interference. Procaryotic genome replication proceeds bidirectionally from a common origin, using either the clockwise or the counterclockwise strand as the template. It is often found that DNA-dependent RNA polymerase enzymes transcribe in the same direction as the strand that serves as the sense strand.

Linkage studies have shown that, in general, the order and location of genes on the procaryotic genome reflects genetic relatedness. For example, the general organization of genes on the chromosomes of various enteric bacteria is similar within this group, but significantly different from the organization patterns found in a variety of other, taxonomically different, organisms.

Mutation

Mutation is the process by which, through random events, the nature of DNA is changed. Mutation is a process that occurs by chance, normally at a very low rate. The rate of spontaneous mutation may be expressed mathematically by the following formula:

$$a = \frac{m \ln 2}{n - n_0}$$

In the above equation, a = mutation rate, m is the number of mutations, and $\ln 2/n - n_0$, the natural logarithm of the number 2 divided by the difference in the number of cells at the beginning and at the end of a time period, is a measure of the number of generations in the time period. Extensive studies in the 1950s showed that mutation in the bacteria is a function of time and chance and is independent of the environment. However, the spontaneous rate of mutation can be increased to some extent by mutator genes that are believed to alter DNA polymerase in some way and thereby facilitate errors in DNA replication.

It is also possible for an original mutation to be countermanded by a *suppressor mutation,* a mutation that counteracts the effects of the first mutation. Two kinds of suppressor mutations are recognized: *intragenic suppression* and *extragenic suppression.* With intragenic suppression, the suppressor mutation occurs within the gene where the original mutation occurred and the extragenic suppressor mutation occurs in a different gene than the one in which the original mutation occurred, but is, none the less, able to reverse the phenotypic effects of the first mutation. Mechanistically, intragenic suppression may allow incorporation of a substitute amino acid that overcomes the effects of the original nonsense mutation, while extragenic suppression typically reflects alteration of tRNA molecules so that they may recognize nonsense codons and allow the insertion of an amino acid at a site that would otherwise be considered a nonsense code.

At the genome (DNA) level, mutation may occur in one of two ways—by *microlesions* or by *macrolesions.* As their names imply, microlesions involve only minor changes and macrolesions involve major changes in DNA and, in some cases, transposition, the movement of a large amount of DNA from one location to another.

Microlesions   Microlesions typically involve one, or at most two, base pairs and are typically transitions, transversions, or frame shift mutations. *Transitions* involve the homologous replacement of a base by another base, that is, replacement of a purine by a purine or a pyrimidine by a pyrimidine, while *transversions* involve heterologous base replacement, where replacement of a purine is by a pyrimidine or vice versa. The consequences of base pair substitutions are varied. In some cases, such changes have no effect because the base change does not change the amino acid coded by a particular triplicate code. For example, the amino acid glycine has four possible mRNA codes, GGU, GGC, GGA, or GGG. Variation in the third position of the mRNA code as a consequence of single change in the DNA base in position 3 of the glycine reading code could occur without any phenotypic consequences since any one of the four codes might allow incorporation of glycine into an incipient protein. Under these conditions, change in the third position of the reading frame (the collection of three DNA bases that determines a particular amino acid) would result in neutral mutation since the same amino acid would be added to the protein irrespective of change in the DNA code.

Missense mutations are a second phenotypic consequence of microlesions. In this situation, a single DNA base change leads to an altered mRNA code so that a different amino acid, rather than the usual one, is incorporated into a protein. The only DNA code for methionine, for example, is TAC (AUG on the message). In this example, change in position 3 of the reading frame would change the amino acid added to the protein from methionine to isoleucine. Finally, a single base change may fail to allow any amino acid to be added to a protein. This type of mutation is a *nonsense mutation,* which is normally the case when termination codes are encountered. In *frame shift mutation,* either a single base or an adjacent pair of bases is added or deleted. In this situation, the entire reading frame is changed. Such a genotypic change has major phenotypic consequences since all of the triplicate codes subsequent to the mutation location are

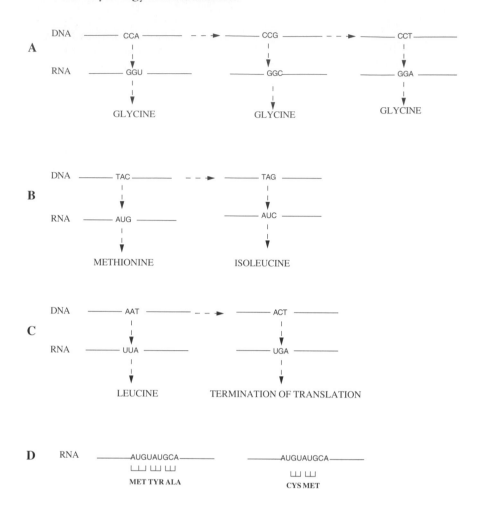

FIGURE 14.1 Various modes of microlesion mutation. (A) Neutral mutation. A transition or transversion change in the DNA gives rise to an mRNA code whose translation allows incorporation of the same amino acid as was added before the mutation. (B) Missense mutation. DNA change alters the mRNA transcript so that a different amino acid is added to the protein than would have occurred without the mutation. (C) Nonsense mutation. Mutation produces a transcript that fails to allow addition of any amino acid, terminating translation. (D) Frameshift mutation. Mutation occurs so that different triplicate codes are read than those read prior to mutation, resulting in major changes in the amino acids incorporated into the protein or termination.

affected. The nature and consequences of microlesion mutation are shown in Figure 14.1.

Macrolesions   Large amounts of DNA are affected by *macrolesions*. In *deletion mutations,* large amounts of DNA are removed. The phenotypic consequences of deletion depend upon the amount of DNA deleted and its location. Often deletion mutants completely abolish one or more phenotypic characteristics because of the removal of an entire gene. In many situations, such mutations are lethal and because of their nature, spontaneous reversion of deletion mutations is not possible.

Duplications are a second type of macrolesion. In this mutation, a gene is duplicated, and sometimes the duplicated gene is even further amplified. This type of process typically occurs during rapid bacterial growth in relatively rich culture media. Under these conditions, large numbers of ribosomes and large amounts of rRNA are required. As a result, duplication and amplification of genes that code for rRNA may occur. When growth conditions do not allow rapid growth and large numbers of ribosomes are no longer required, duplication and amplification typically stop.

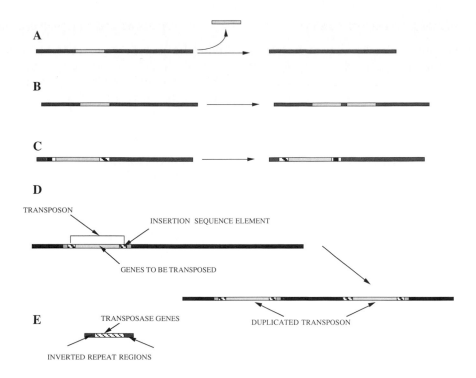

FIGURE 14.2 Various modes of macrolesion mutation. (A) Deletion. A large amount of genetic material is removed from the genome. (B) Duplication. An adjacent copy of a gene is made. (C) Inversion. By the action of an invertase enzyme, a substantial DNA piece is "flipped," while remaining at the same location. (D) Transposition. With the aid of insertion elements, which contain both transposase enzyme genes and inverted repeats, all of which flank the transposed genes, a copy of the transposed genes is inserted into a new chromosomal location. Normally, the transposed genes are retained at their original location. (E) The "fine structure" of an insertion sequence. The latter contains the genes required for transposition, surrounded by inverted repeat regions.

Inversions are yet another type of macrolesion, in which the sequence of bases in a substantial portion of the genome is reversed. Inversions apparently occur in nature as evidenced by the fact that the genomes of *E. coli* and *S. typhimurium* are extremely similar, except for inversion of a small portion of their chromosomes.

The final general mechanism by which macrolesions may occur is *transposition*. Transposable elements are of one of four types that differ in their size and nature. *Insertion sequences* (IS) typically are between 800 and 2,000 bp in size and serve only to allow the process of transposition to occur. A *transposon* (TP) is a relatively large translocatable element, normally greater than 2,000 bp long and it contains much more information than is found in insertion sequences. Transposons often have the property of duplication. When translocation is accompanied by duplication, an original copy is left at the old location and a new copy of the transposon is inserted at the site of translocation. Insertion of either an IS or a TP occurs by a staggered cut, necessitating synthesis of short segments of direct repeats (repeats of sequences in the same direction) of the target DNA. In addition, the transposon itself contains short repeated elements at its ends. In most cases, these repeated elements are inverted, in the opposite sequence at alternate ends of the transposon, although they may sometimes be direct, in the same direction as each other. Certain transposons contain antibiotic resistance markers, allowing them to be used in a deliberate fashion for mutagenesis, since the mutated cells can be selected for by their resistance to the antibiotic.

Invertable elements are the third type of transposable element. These entities code for an enzyme, DNA invertase, that allows inversion of a large piece of DNA without the necessity of its movement to a different location. Finally, the prophage of the Mu virus is a transposable element. The nature and genetic consequences of macrolesion mutation are shown diagrammatically in Figure 14.2.

Phenotypic consequences of mutation

Irrespective of the genetic mechanism of mutation, mutation results in a change in the physiological properties of the cell. It is often the case that mutation alters the pathway for synthesis of an essential nutrient. Such a mutation produces an *auxotroph*, an organism that requires the material whose synthesis is interrupted for growth. Because it can synthesize the required material, the parent or "wild type," also known as a *prototroph*, does not require the substance in question.

Mutations that result in a requirement for an altered carbon or nitrogen source are a second type of mutant. Whereas auxotrophs require a particular material because they cannot synthesize it, carbon and nitrogen source mutants result from the inability of the organism to use a previously utilizable carbon or nitrogen source. This inability may result from the loss of an enzyme, a transport system, or both. *Cryptic* mutants are those mutants that display an inability to use a nutrient even though such organisms contain, intracellularly, the appropriate enzymes.

Structural mutants are an important class of mutants. Their study, in comparison to normal organisms, has, among other things, provided information both about the nature and regulation of cell structure formation and about the intracellular locations at which components of structures are formed. Study of bacterial structural mutants has provided information about the formation and function of the cell wall, the membrane, and the flagellum, as well as other structures. Mutations of this nature have also revealed much about the coordination of cellular events with the division process (see Chapters 5 and 18).

Regulatory mutants also occur. For example, an organism that previously produced a protein only under certain conditions and was thereby subject to induction or repression, may develop the ability to produce a protein under all circumstances, that is, it becomes constitutive. Such a situation may result either from modification of a repressor protein so that it can no longer occupy the operator region or from modification of the operator region so that the repressor protein cannot attach to it. Many additional types of regulatory mutants are known.

Conditional and nonconditional mutation

Whatever their nature, mutations can be classified as those in which the altered phenotype is invariably expressed or those in which it is expressed only under certain conditions. Conditional mutants are useful, because the behavior of the same organism can be studied under two different conditions. Conditional mutants express the mutation under "permissive" conditions, and fail to do so under "nonpermissive" conditions. Temperature is often a conditional factor, one that may be used to alter mutation expression. Within those mutants whose expression is influenced by temperature, two classes can be distinguished, those that are heat sensitive and those that are cold sensitive. Heat sensitive mutants are of such a nature that elevated temperature inhibits their expression, while normal temperature allows it. Cold sensitive mutants display the inverse of this behavior. Cold temperatures inhibit expression and normal temperatures allow it. Mechanistically, the active repressor proteins are often formed at the nonpermissive temperatures and repression is relieved at permissive temperatures.

Mechanisms of mutagenesis

The normal spontaneous mutation rate may be increased by environmental factors. The genetic mechanisms by which exposure to environmental factors allows increased mutation are varied, ranging from formation of abnormal bonds between DNA bases to chemical alteration of bases to intercalation of molecules into the DNA helix. The cellular consequences of these effects reflect the manner in which genetic alteration affects the processes of DNA replication, transcription, and translation and the nature and extent of repair processes.

Ultraviolet light is a major mutagenic agent. It manifests its mutagenic effects primarily by the formation of pyrimidine dimers, often between adjacent thymine residues but also between thymine and cytosine and between adjacent cytosine residues. Ultraviolet radiation can also distort the helical structure of DNA. All of these effects may result in improper or altered DNA replication and transcription. Ionizing radiation, by the formation of highly reactive chemical entities such as free radicals, hydroxyperoxides, and the superoxide anion, may cause more extensive damage, such as single-stranded breaks in the DNA molecule and nongenomic damage to cells.

Chemical alteration of bases is a second mechanism of environmentally induced mutagenesis. Nitrous acid, for example, deaminates both adenine and cytosine, forming hypoxanthine and uracil, respectively, and altering the manner of base pairing. In addition, alkylation of bases on their nitrogen residues may lead to changes in base pairing ability. Incorporation of base analogues, such as halogenated bases, may result in the same phenomenon, the alteration of base pairing. Intercalation of small planar molecules into the DNA helix is also a major mechanism of mutagenesis. Ethidium bromide, proflavine, and the acridine dyes manifest their action in this manner. The net effect of all of these agents is alteration or abolition of normal processes of DNA replication and transcription, and the attendant phenotypic effects that result from altered or abolished translation.

Mutation repair

Microbes have a variety of mechanisms by which to mitigate the effects of mutation. Suppressor mutation is one such mechanism, but other, more generalized, mechanisms are known. Ultraviolet-induced mutation may be repaired by a single enzyme, *photoreactivation enzyme,* whose mode of action is to remove the linkages formed between adjacent pyrimidine molecules and to restore normal function of the genetic material. Photoreactivation enzyme is a product of the *phr* gene. In *E. coli,* the molecule has a molecular weight of 32,000 and its gene is located at 16 minutes on the chromosome. It binds to pyrimidine dimers in the dark, but will only photolyze the dimers in the presence of visible light between 340 nm and 400 nm. The general nature and consequences of photoreactivation is shown in Figure 14.3.

Excision repair is a more generalized mechanism for repair of DNA damaged more extensively than often occurs with UV light. Excision repair involves the participation of a complex endonuclease, which in *E. coli,* is coded for by the *uvrA,* the *uvrB,* and the *uvrC* genes located, respectively, at 92, 17, and 42 minutes on the chromosome. The

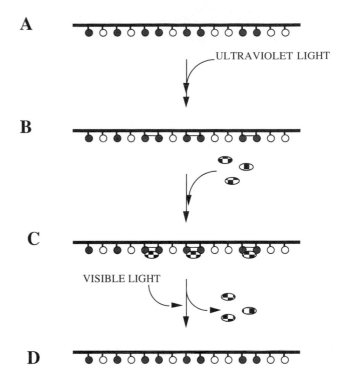

FIGURE 14.3 The mechanism of photoreactivation. Purine residues are shown as open circles and solid circles are pyrimidine residues. The checkered oblongs are photoreactivation enzyme molecules. (A,B) The initially functional DNA is exposed to ultraviolet light to produce dimers between adjacent pyrimidine residues (B). In the dark, the damaged DNA interacts with photoreactivation enzyme molecules to yield complexes (C). Exposure to visible light regenerates functional DNA and releases free molecules of photoreactivation enzyme (D).

molecular weight of the uvrA product is 100,000, while the molecular weights of the uvrB and uvrC gene products are 84,000 and 68,000, respectively. Collectively, the action of these three proteins allows recognition of the damaged region (apparently by the uvrA product), and double cleavage of the damaged strand by the combined action of the uvrA, uvrB, and uvrC products. Upon cleavage of the damaged area, it is removed. DNA polymerase I and DNA ligase then repair the damaged strand. Using the nondamaged strand as template, polymerase I synthesizes new material to replace the damaged portion and DNA ligase attaches the repaired fragment to the nicked strand. Excision repair is effective only for relatively short segments of damaged DNA. Repair of longer damaged sections is accomplished by other mechanisms, particularly the SOS system. Both photoreactivation and excision repair are regarded as *prereplication repair systems* because repair occurs prior to the replication process. Both photoreactivation and excision, by their nature, are error proof, photoreaction because it involves only bond breakage and excision because a template is available for repair of the damaged strand. The general nature of excision repair is shown in Figure 14.4.

In contrast to photoreactivation and excision repair, *mismatch repair* is a postreplication repair mechanism that as its name implies, removes improperly incorporated bases that have been missed by the 3′ to 5′ exonuclease activity of DNA polymerase III. Unless the inappropriate bases are removed prior to further replication, mutation will occur. For selective removal of inappropriate bases, some distinction must be made between the parental strand that is devoid of the inappropriate base, and the newly synthesized strand that contains it. Discrimination is accomplished by action of adenine methylase, the product of the *dam* locus. This enzyme recognizes the GATC sequence and methylates the N6 atom of its adenine residue. Since parental DNA contains more GATC residues than does newly synthesized DNA, the parental strand is more extensively methylated than is the new strand. Mismatch repair enzymes discriminate between the parental and newly synthesized strands and selectively remove the inappropriate base in the newly synthesized strand while retaining the appropriate base in the parental strand. In *Escherichia coli,* the products of the *mutH, mutL, mutS, mutT, mutD,* and *uvrD* loci are all involved in mismatch repair, but the precise modes of action of the various system components remain obscure.

The action of glycosylase enzymes is another postreplication DNA repair mechanism. Such enzymes attack the sugar-DNA base bonds in damaged DNA and selectively remove inappropriate bases while leaving the deoxysugar backbone intact. The result of such an action is to produce apurinic or apyrimidinic (AP) sites that can be removed by specific endonucleases that may cleave

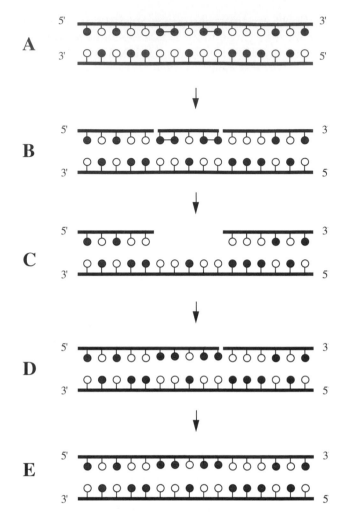

FIGURE 14.4 Excision repair. The solid spheres are pyrimidine residues, while the clear spheres are purine moieties. (A,B) Damaged DNA (A) is nicked by an endonuclease to produce double cuts on the 5′ and 3′ ends of the damaged area (B). The damaged area is then removed by an exonuclease (C). The removed piece is repaired with DNA polymerase I, using the undamaged strand as template (D). (E) The gap in the repaired DNA that results from polymerase I action is sealed by DNA ligase to produce functional DNA.

either at the 3′ (class I) or 5′ (class II) side of the AP site. In *E. coli,* exonuclease III, which also has class II endonuclease activity, endonuclease IV and endonuclease V participate generally in DNA glycosylase activity.

In addition to generalized glycosylase enzymes, specific enzymes are also found. Uracil-

DNA-glycosylase, a 24,500 molecular weight product of the *ung* locus, repairs inappropriately incorporated uridine residues formed from cytosine deamination and hypoxanthine-DNA-glycosylase, a 30,000 molecular weight enzyme, removes hypoxanthine residues inappropriately incorporated into DNA as a result of adenine deamination. Although many DNA-glycosylase enzymes are highly specific in their action, some are less so, and in certain cases, more than one enzyme accomplishes the same physiological effect. The 3-methyladenine-DNA-glycosylases exhibit these phenomena. Thus 3-methyladenine-DNA-glycosylase I, a constitutively formed, 20,000 molecular weight product of the *taq* locus, mitigates the effects of adenine methylation. However, a second enzyme, 3-methyladenine-DNA-glycosylase II, an induceable, 27,000 molecular weight product of the *alkA* locus accomplishes the same activity and, in addition, can remove 3-methylguanine and both 7-methylguanine and 7-methyladenine from methylated DNA. The action of 3-methyladenine-DNA-glycosylase II is different from that of 3-methyladenine-DNA-glycosylase I in regard to both its specificity and the conditions required for its production. It is of interest to note that the two functionally different enzymes are located at substantially different locations on the chromosome, the *tag* locus at 72 minutes and the *alkA* locus at 43 minutes.

Although many enzymes repair inappropriately incorporated nucleotides, including those nucleotides that contain an inappropriate methyl residue, by a glycosylase mechanism, in some cases only the methyl group is removed. Thus, O^6-alkylguanine-DNA-alkyltransferase removes the methyl group that may be attached by alkylation to the O^6 in guanine. The methyl residue is transferred to the cysteine component of the enzyme itself.

Postreplication gap repair of daughter strands is an additional postreplication repair mechanism. There are times when substantial DNA replication occurs prior to removal of altered bases. When this occurs, the replicated strand contains a gap because the damaged region of the parental template strand is not recognized as template. This situation can be alleviated in one of two ways, either by recombination of the incompletely replicated daughter strand with an appropriate portion of an undamaged parental strand or by the SOS system.

The general nature of postreplication repair systems (other than the SOS system) is shown in Figure 14.5.

The SOS system   This system is a mechanism of response to extensive DNA damage, normally the result of UV light exposure. The SOS response is a generalized response that involves the coordinated action of about 20 genes and, in addition, involves effects on the cell other than those strictly related to DNA repair. In addition to serving as a DNA repair system, the SOS response involves temporary inhibition of cell division and alteration of processes required for cell division to occur. From the viewpoint of DNA repair, a major physiological consequence of the SOS response is provision of sufficient time for the repair process. It also appears that the SOS response allows conservation of substances required for repair.

The SOS system is jointly regulated by operation of the recA and lexA gene products. The lexA gene product normally acts as a repressor for expression of many genes, including its own formation. The recA protein is multifunctional. It binds to single-stranded DNA and promotes repair of double-stranded DNA in a variety of ways. In addition, the recA gene product possesses protease activity. As a result of a signal formed in response to radiation exposure, particularly UV radiation, the protease activity of the recA protein is manifested and the lexA gene product is degraded, allowing relief from repression of gene product formation regulated by the lexA protein. As the repair of DNA is completed, the protease activity of the recA protein is reduced, allowing the lexA protein to accumulate and, by repression, to inhibit further DNA repair.

Destruction of the lexA protein under the influence of recA-mediated proteolysis has physiological consequences other than those associated with DNA repair. The sulA protein, an inhibitor of cell division initiation, is also under control of the lexA gene product. When the lexA protein is degraded, the product of the *sulfA* gene is formed. The latter reacts with the ftsZ gene product, a normal activator of the cell division process, and inactivates it, thereby inhibiting normal cell division. As the cell recovers from radiation damage, yet another gene product, the lon gene product, is activated. This protein has protease activity and destroys the sulA

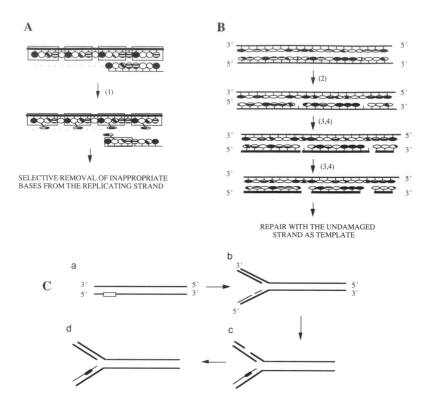

FIGURE 14.5 Various modes of postreplication DNA repair. Throughout the figure, back circles denote guanine residues, open circles denote adenine residues, circles with slanted lines are thymine moieties, horizontal-line circles are cytosine residues, and the elongated checkered ovals are methyl groups. (A) By selective methylation of adenine residues in GATC sequences by adenine methylase (1), endo- and exonucleases distinguish between parental and newly synthesized DNA and remove improperly incorporated bases from the new strand. (B) DNA glycosylase enzymes (2) remove inappropriate bases to produce apurinic and apyrimidinic sites, while leaving the deoxysugar backbone intact. (3,4) The AP sites are then selectively attacked by class I (**3′** side) and class II (**5′** side) endonucleases and exonucleases that collectively remove the AP sites. DNA polymerase I and DNA ligase then replace the incorrect base, using the undamaged strand as template. (C) Daughter strand gap repair. A DNA molecule with a damaged strand (a) begins replication, but the strand complementary to the damaged strand contains a gap because the damaged strand cannot properly serve as a template (b). By recombination (c), the gap is repaired, leaving a gap in the originally nondamaged parental strand. The new gap is repaired using the opposite strand as template (d).

protein, reversing inactivation of the ftsZ protein, and allowing normal cell division to occur. The general nature of the SOS response is shown in Figure 14.6.

Genetic exchange

The adaptability and survival potential of a species is, to a considerable extent, a function of genetic exchange. At the large organism level, we may manipulate the genetics of certain organisms so that we may combine desirable traits from one member of a species with the desirable traits of another member of the same species and obtain an organism that has the desirable traits of both. The hybrid

vigor phenomenon is an example of this idea. It is often possible, for example, in the plant world, to combine the traits of high productivity with disease resistance to produce an organism that displays both characters.

Compared to other life forms, genetic exchange in bacteria is a relatively rare event. However, because of the haploid nature of bacteria, genetic changes, including those that result from genetic exchange, can be readily recognized and studied. Although the mechanisms of genetic exchange in bacteria are different from those of other life forms, the consequences of exchange, such as mutation, are similar. The bacteria thus serve as excellent general models for study of the consequences of genetic change.

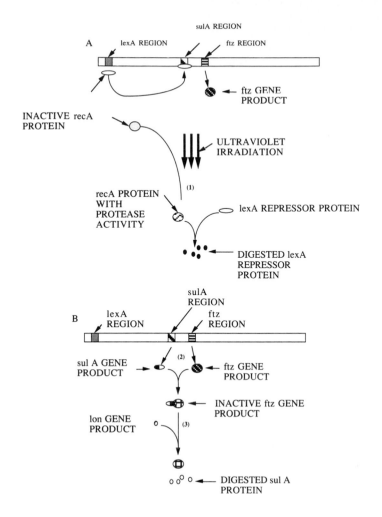

FIGURE 14.6 Essential features of the SOS DNA repair system. (A) In the absence of irradiation, the *lexA* gene codes for an active repressor of the sulA gene product and many other genes required for DNA repair. The ftz gene product, an activator of cell division, is synthesized in an active form. In addition, the recA gene product, an essential protein for DNA repair, is formed but does not display proteolytic activity. Under the influence of ultraviolet irradiation (1), the recA protein becomes proteolytic and digests the lexA-coded repressor protein. (B) Digestion of the lexA-coded repressor permits sulA gene product production. The sulA product interacts with an active ftz protein, inactivates it (2), and arrests cell division until DNA is repaired. As repair of DNA is completed, the lon gene product, a protease that digests abnormal proteins, digests the sulA component of the sulA-ftz protein complex (3), restoring ftz protein function, and allowing normal division to proceed.

Mechanisms of genetic exchange in the bacteria

Bacteria may exchange genetic material in one of three ways, *transformation, conjugation,* or *transduction*. Transformation, first discovered by Griffith in 1928, is mediated by "naked" DNA, DNA that is not associated with other molecules. Transduc-

tion is mediated by bacteriophages. In transduction, genetic exchange occurs because a small amount of foreign DNA is transferred to a recipient cell during the processes of bacteriophage replication. Conjugation is the third mechanism of bacterial genetic exchange. It is mediated by plasmids that have the ability to transfer their genetic material to cells other than those that they currently inhabit. Study of both plasmid- and phage-mediated genetic exchange has been useful in understanding not only of the nature of bacterial genetic exchange but also the ways in which virus-mediated, plasmid-mediated, and host cell-mediated genetic expressions are interconnected and regulated. In addition, studies of plasmid and phage replication are useful in understanding of the concept of *intracellular autonomy*, the extent to which the components of cells can determine their own "fate."

Transformation

Transformation was the earliest recognized mechanism of genetic exchange, and although in no way a simple process, it is perhaps simpler than the processes involved in transduction and conjugation. Transformation involves the uptake of DNA by a recipient cell, intracellular processing of the incorporated nucleic acid, recombination of the donor DNA with the host cell genome, and finally, expression of an altered trait as the result of the possession of new DNA. Although these general processes must be accomplished for transformation to occur, the manner in which transformation occurs in gram-positive bacteria differs substantially from the analogous process in gram-negative organisms, and within a particular gram reaction, differences are found between species in the details of the transformation process. Irrespective of these differences, certain mechanistic similarities are found in all organisms capable of transformation. Without regard to mechanistic differences, all cells that would receive naked DNA must be competent, or capable of the uptake of naked DNA. Although competence is a common feature of all transformation processes, the mechanisms by which the competence condition is achieved differ among organisms. In general, competence involves changes in the physiological state of the organism, changes in the nature of its outer layers, or both.

Streptococcal transformation   The process of transformation in the streptococci, as exemplified by *S. pneumoniae*, is mediated by a competence activator protein, produced by competent cells of *S. pneumoniae*. Attachment of the competence activator protein to surface receptors on the cell membrane allows attachment of donor DNA to the cell surface. However, simple attachment of DNA to the cell surface is insufficient, by itself, to allow transformation. Mutants that can bind DNA, but in which transformation is not possible, have been obtained, indicating that the DNA binding and transport of the bound DNA into the cell are separate processes. Following attachment at the cell surface, the donor DNA is extensively degraded by the activity of endonuclease I. The degradation process results in transport of polypeptide-bound single-stranded DNA into the cell interior. It appears that the transport of donor DNA into the cell interior is mediated by the competence factor. In addition to allowing DNA attachment to the cell surface, it appears that the competence activator protein facilitates partial autolysis of the recipient cell surface, enabling entry of the donor DNA.

In *S. pneumoniae*, and apparently other streptococci as well, differences are found in the efficiency of transformation. Although, to a certain extent, differences in transformation efficiency may reflect the nature of donor DNA, to a large degree, it appears that differences in transformation efficiency reflect differences in the genetic properties of the recipient cell. Certain low-efficiency (LE) genes are incorporated with only low efficiency, whereas others, high-efficiency (HE) genes, are incorporated with a high degree of efficiency. The ability to discriminate between HE and LE genes is known as the *hex function* and apparently reflects the ability of the host cell to discriminate between genes that are mismatched with recipient DNA and those that are not and, in some cases, to repair mismatches so that incorporation of the donor DNA may occur.

Transformation in *Bacillus subtilis*   Transformation in *B. subtilis* is, in many respects, similar to the process in streptococci, but differences can also be discerned in the precise details of the process in the two types of gram-positive bacteria. As is the case with the streptococci, transformation in *B. subtilis* requires binding of double-stranded DNA to the cell surface. Following binding, extensive degradation of donor DNA is accomplished by an active exonuclease that is characteristic of competent cells. In *B. subtilis*, competent cells appear to be substantially different from noncompetent cells since competent cells are less dense than noncompetent cells. Following degradation of one of the donor strands, the remaining strand is transported to the cell interior and almost immediately converted into a double-stranded intermediate. Shortly thereafter, the double-stranded forms are digested to single-stranded entities, which are used to accomplish the transformation process. A comparison of transformation in *S. pneumoniae* and *B. subtilis* is shown in Figure 14.7.

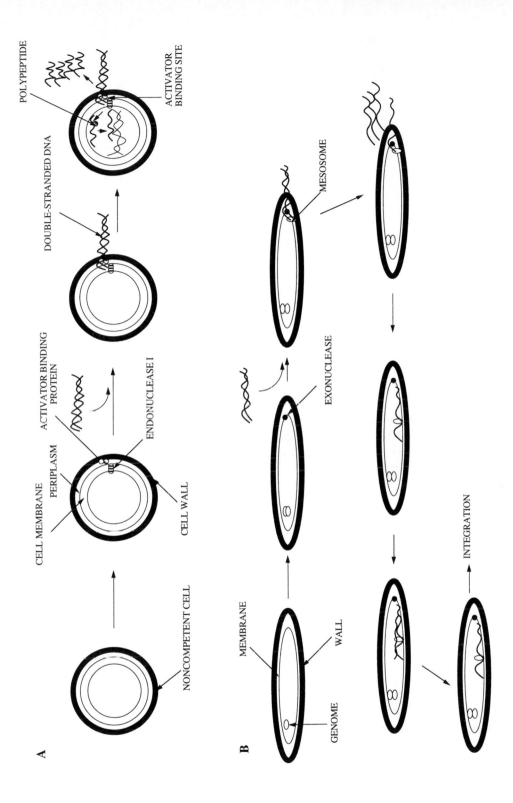

FIGURE 14.7 A schematic representation of the critical events of transformation in *Streptococcus pneumoniae* (A) and *Bacillus subtilis* (B). (A) *Streptococcus pneumoniae*. An initially noncompetent cell becomes competent. During this process, a competence activator protein, a competence activator binding site, and endonuclease I are induced. Double-stranded DNA then binds to the cell surface and is degraded to yield single DNA strands, some of which are extruded to the cell exterior. Others, attached to polypeptides, are transported to the interior where they are aligned with recipient DNA and integrated. (B) *Bacillus subtilis*. An initially noncompetent cell is induced to competence, during which time the genome is partially replicated, and exonuclease activity is induced. Double-stranded DNA then binds to the surface and is degraded to single-stranded material. Some of the single-stranded DNA, attached to a mesosome, enters the cell and is rapidly converted to a mesosome-attached double-stranded intracellular intermediate. The latter is converted to a mesosome-attached single-stranded intermediate that participates in the integration event.

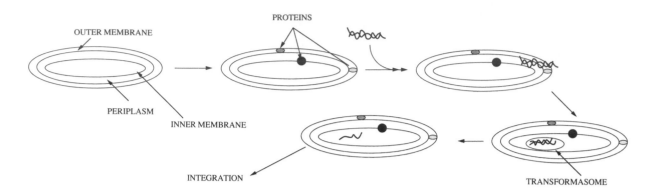

FIGURE 14.8 Transformation in *Haemophilus influenzae*. An initially noncompetent cell is induced to competence. During this time, substantial lipolysaccharide synthesis occurs, and three major proteins are formed, two in the outer membrane and one in the cytoplasmic membrane. Double-stranded DNA molecules then attach to the cell exterior and are transported, intact and encased in membrane vesicles known as transformasomes, to the cell interior. Intracellular nuclease activity then produces single strands that participate in integration.

Transformation in *Haemophilus influenzae*
Transformation in *Haemophilus influenzae*, and in gram-negative bacteria generally, differs substantially from the process in gram-positive organisms. In particular, soluble competence factors are not found in gram-negative organisms. In addition, extensive surface-associated nuclease activity is absent. As a result, in gram-negative bacteria, double-stranded DNA is transported to the cell interior, rather than the single-stranded molecules that are transported in gram-positive organisms. Although double-stranded DNA is absorbed at the surface in both gram-positive and gram-negative bacteria, the specificity of DNA adsorption in *Haemophilus* is extreme and dictated by a particular 11 base pair sequence, 5′-AAGTGCTGTCA-3′. This sequence dictates the ability of the *Haemophilus* DNA to interact with surface receptors. Such a specificity for adsorption is not found in gram-positive organisms studied thus far.

The mechanism of DNA transport in *Haemophilus* is different from that of gram-positive bacteria. In *Haemophilus*, double-stranded DNA is transported into the cell in the form of membrane-bound vesicles known as *transformasomes*. Although a double-stranded entity is transported into the cell in *Haemophilus*, only a single strand is involved in the transformation process. The other strand of the donor DNA and one strand of the recipient DNA are degraded, apparently at the same time, so that no single-stranded intracellular intermediates are formed. Transformation in *Haemophilus* is shown in Figure 14.8.

Artificial transformation Although transformation occurs naturally in some organisms, natural transformation is apparently absent, or of little significance, in others. However, certain organisms can be induced to receive DNA artificially. Such organisms, notably *E. coli*, are useful in genetic engineering because if artificially introduced, chemically modified, DNA can be incorporated into *E. coli* by extracellular chemical or biochemical modification of plasmid or phage DNA and subsequent transformation, it is possible to clone genes and to form proteins in microbial systems from heterologous microbial and nonmicrobial DNA. Plasmids and phages thus have extensive usefulness in biotechnology and genetic engineering, topics that are discussed in Chapter 20.

In order for plasmids and phages to serve in artificial transformation, their DNA must be made permeable to cells in such a manner that both the recipient cell and the transforming DNA are functional. A number of mechanisms have been devised to allow entry of DNA into cells, including treatment with Ca^{2+} at cold temperatures, freezing and thawing, formation of protoplasts followed by ethylene glycol treatment and electroporation, and disruption of cell envelope structure with pulses of high voltage electricity to an extent sufficient to allow penetration of DNA without significant recipient cell injury.

Transduction

Transfer of chromosomal DNA from one bacterium to another may be accomplished by transduction. Transduction allows only a small portion of the total chromosomal DNA to be transferred and may be accomplished in one of two ways, either by *generalized* transduction or by *specialized* transduction. In the former, any small portion of the chromosomal DNA may be transferred, while in specialized transduction, only genes closely associated with particular genetic loci are transferred.

Generalized transduction

The mechanism of generalized transduction has been particularly studied in *Salmonella typhimurium* phage P22. This phage is apparently representative of all generally transducing phages. Phage P22 is a double-stranded DNA phage that synthesizes its DNA in concatemers, repeated, complete segments of the phage genome. Typically, phage P22 concatemers contain 10 repeats of the phage genome. Certain sites along the concatemers are cleaved by a nuclease enzyme. Cleavage occurs so that each cleaved particle contains approximately 2% more DNA than is required for replication of the phage itself. The original cleavage event is rather nonspecific, but subsequent cleavages are dictated by the distance of a potential cleavage site from the original cleavage site. The general process thus produces a series of DNA pieces slightly larger than the phage genome. The ends of each piece contain identical regions and the pieces are thus considered terminally redundant.

In the normal process of phage replication, replicated phage genomes are incorporated into phage heads to form complete phages. Because of the lack of specificity of the initial cut by the nuclease, it is sometimes the case that the first cut of DNA occurs, not on the phage genome, but rather on the host cell chromosome. When this occurs, subsequent cuts are also made on the host chromosome. During the remaining phage replication events, small pieces of host genetic material are incorporated into phage particles in place of the normal phage genome. These particles have the ability to infect new host cells and transfer host genes by recombination. Such "phages," since they lack the phage genome, are incapable of replication. However, under certain circumstances, generally transducing phages, such as P22, may participate in a virulent infection. In such a situation, the phage heads contain the normal phage genome and the phage is capable of normal, lytic, replication. Generalized transduction may be mediated either by virulent or temperate phages. Virulent phages give rise only to lytic infections and temperate phages have the ability to either give rise to virulent infections or to exist in the state of lysogeny as prophages, that is, phage genomes incorporated into the chromosome of the host cell.

Specialized transduction

While generalized transduction may be accomplished either by virulent or temperate phages, specialized transduction is accomplished *only* by temperate phages. Specialized transduction, in contrast to generalized transduction, transfers only small portions of the host genome near the region where the prophage integrates into the host chromosome. Specialized transduction depends upon the phenomenon of "illegitimate" recombination between nonhomologous regions of the prophage (a phage genome integrated into an appropriate cell) and adjacent regions of the host cell genome. The specificity of gene transfer during specialized transduction is a function of the fact that the phage must integrate at a specific locus and only small portions of the host cell genome in the vicinity of the prophage may be transferred, since the total amount of DNA that may exist in a particular phage head is limited. Phage λ, the most studied of the specially transducing phages, integrates with the host genome only in the vicinity of the genes for galactose and biotin metabolism, that is, the *gal* and *bio* loci. Lambda transductants may be identified by their ability to convey the ability to catabolize galactose or to synthesize biotin upon recipient cells that previously lacked these abilities. Transduction by λ and other specialized transduction phages occurs when, as a result of abnormal removal of prophage DNA, a small portion of the host genome near the site of prophage integration is accidentally incorporated into the phage head. Occasionally, when host cell DNA is transferred, substantial portions of phage DNA are lost and the phage is therefore incapable of lytic infection. Such phages are known as *defective* phages and are often denoted by the letter *d*. In contrast, certain specialized transducing phages retain sufficient phage genes for replication and also contain certain host genes. Since such phages can sometimes undergo lytic infection and can, therefore, form plagues, they are designated as *p* phages. For phage λ, we may distinguish four kinds of phage particles, those that contain the galactose or the biotin gene, as well as certain host genes, and are also capable of phage replication, and those that contain the gal or biotin marker, but are incapable of replication.

Specialized transduction is an infrequent event because the improper removal of the λ prophage rarely occurs, so specialized transduction also occurs with a low frequency. However, by manipulation of the system, the frequency of transduction can be increased. If one transfers lysates of λ into a culture that is gal⁻ (unable to metabolize galactose) and selects for cells that use galactose, one may obtain cells that contain the *lambda gal* and associated genes in the prophage state. If the lysate contains dgal lambda genomes (genomes containing the *gal* locus and associated genes, but which cannot replicate), one may obtain cells which are double lysogens and contain, in the prophage state, DNA from both the defective lambda particle and from a *helper* phage particle, in this case, DNA from a normal lambda phage. When such a culture is induced, the DNA in the vicinity of the *gal* region is amplified, because it is replicated at both the dgal site and as a consequence of replication of the normal lambda phage. When λ is induced under these conditions, the proteins coded for by the genes associated with the lambda genome, including those associated with the transduction process, are frequently expressed.

By manipulation of the λ system, it is possible to obtain not only an increase in λ transduced genes, but also in their products. Success in amplification of both genes and their products normally associated with the lambda phage has led to efforts to allow such technology to be used in a more general fashion for amplification of genes not normally associated with λ. Two possibilities are reasonable,

movement of the λ attachment site to an unusual location, such as adjacent to genes of particular interest, or movement of genes not normally associated with λ to be near the λ integration site. Manipulations of this nature are becoming increasingly possible through the techniques of genetic engineering, and will undoubtably be increasingly applied to specialized transducing phages other than lambda. The mechanisms of generalized and specialized transduction are shown in Figure 14.9.

Conjugation

Conjugation is the third mechanism of bacterial genetic exchange. It is mediated by plasmids, and in certain cases, allows transfer of substantial amounts of host genetic material to a recipient cell. With the aid of plasmid-mediated conjugation, it is possible, in theory, to transfer the entire cell genome and to produce a diploid cell. However, in practice, such a situation seldom, if ever, occurs.

Plasmids are of interest for a variety of reasons. They may be involved in natural transfer of traits, such as occurs in development of multiple drug resistance. They may also be used as genetic tools or as vehicles for both gene cloning and gene expression in the form of useful proteins. In addition to their usefulness as agents of genetic manipulation, plasmids are of interest as models for understanding DNA replication and because many of them exert substantial physiological effects upon the cells that they inhabit. The relationship between many plasmids and their associated cells is an intimate one. To a substantial extent, plasmids and their host cells may be considered examples of "symbiosis," in the sense that their critical activities are interdependent.

To be useful as an agent of genetic exchange, a plasmid must, at a minimum, possess two properties, the ability to replicate and the ability to transfer itself to other cells. In other words, it must be a *conjugative* plasmid. The F factor is such a plasmid and its study has contributed to much of our understanding of both the general nature of the conjugation process and of the nature of plasmid-mediated genetic exchange.

With regard to the F (fertility) factor, cells may be one of two types, F⁺ and F⁻. F⁺ cells have the

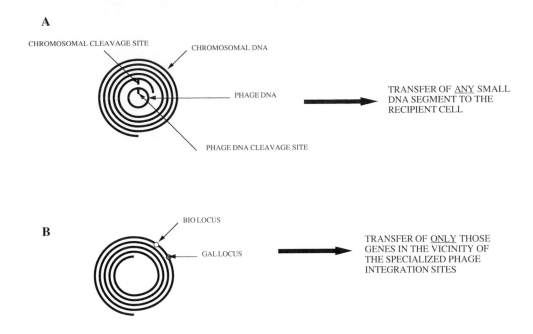

FIGURE 14.9 The mechanisms of generalized (A) and specialized (B) transduction. (A) In generalized transduction, a phage encoded nuclease "mistakenly" cleaves chromosomal DNA at a region analogous to the region it would normally attack on the phage DNA. Because of the nature of the nuclease, subsequent cuts are made at regions physically predictable in relationship to the first cut. The phage enzyme thus produces small pieces of chromosomal DNA that are randomly incorporated into intact phage particles. When the latter infect susceptible cells, small amounts of donor DNA are randomly transferred. (B) Specialized transduction. The phage genome integrates into the host cell chromosome at specific locations. In bacteriophage λ, integration occurs in the vicinity of the biotin (BIO) and galactose (GAL) loci. When the phage becomes virulent, it removes a few genes in the vicinity of the integration site and transfers only those genes to the recipient cell. It is likely that the mechanism of specialized transduction in other specialized transduction systems is similar or identical to that in phage λ.

ability to transfer genetic material to F⁻ cells, cells that are devoid of the F factor but are capable of receiving genetic material. As a self-replicating entity, the F factor can reproduce itself and facilitate its own transfer in the absence of host cell division.

The F particle is an *episome,* since it may exist either as a separate and discrete entity or may be integrated into the host chromosome. In the second state, it facilitates transfer of large amounts of host genetic material. Cells that contain integrated F factors are referred to as *Hfr cells,* cells that exhibit a high frequency of recombination. The ability of F particles to integrate into the host chromosome reflects the presence, in both the plasmid and in the recipient cell, of compatible insertion sequences.

The F particle may integrate into an appropriate cell at a variety of locations on the host cell chromosome. For this reason, Hfr cells are used extensively in genetic mapping, since a correlation exists between the appearance of recombinant traits in mutants and the distance between genes. By comparing the times of appearance of particular traits in mutant cells, using different Hfr cells, it has been possible to determine the distances between the major genes in the chromosomes of *E. coli* and *S. typhimurium* and several other bacteria. The structure of the F particle is depicted schematically in Figure 14.10. The particle contains genes for its own replication, genes for its transfer to other cells, and genes that promote its own perpetuation by regulating the relationship between its replication and that of the cell it inhabits (*fertility inhibition* genes) or by regulating the integrity of the host cell (*phage inhibition* genes).

F′ factors

Although the F factor may integrate into the host genome, it may also return to the independent state in one of two ways. In certain cases, the F factor leaves the chromosome, and leaves a portion of

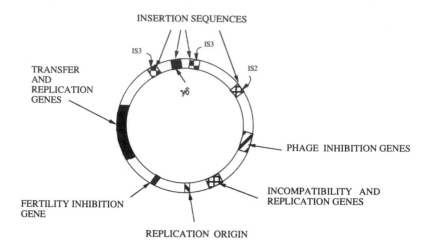

FIGURE 14.10 The structure of an F particle. The origin of replication region is to the left of the incompatibility region. To its right is the phage inhibition region. The fertility inhibition region is located to the left of the replication origin region. The tra, or transfer region, immediately to the left of the fertility inhibition region, codes for pilus formation and the transfer and replication of the F particle in the recipient cell. Insertion sequences facilitate the reversible integration of the F particle into the chromosome.

itself behind while acquiring a small portion of the host cell genome. Such a particle is known as a *type I* F′ particle. It is also possible that the F particle may leave the host chromosome so that it retains all of its original genetic material and also acquires certain host cell genes. Such particles are known as *type II* **F′** particles. In either situation, if mating occurs between an F⁺ and an F⁻ cell, small portions of host cell genetic material may be transferred to the recipient cell, producing a *merodiploid* cell, a cell that is only partially diploid, but which contains two copies of certain genes, those coded by material on the **F′** particle and their copies on the recipient cell chromosome. The possible states in which F factors may exist are shown in Figure 14.11.

F factor replication　Both for perpetuation of itself and for participation as a chromosome-mobilizing agent, F factor replication is essential. Both formation of the conjugation pilus and replication of the F particle are mediated by the *tra* genes that are distinguished from each other by capital letters between A and L. After attachment of an F⁺ and an F⁻ cell by a pilus formed under influence of *tra* genes, F factor replication begins. At a discrete location,

the *ori T* region, a single nick is made in one strand of the F particle. This process is followed by DNA replication, using the rolling circle mechanism (see Chapter 12), so that the 5′ end of a pre-exisiting F particle strand enters the F⁻ cell. Movement of the single strand into the F⁻ cell results after its displacement from the F⁺ cell because of new DNA synthesis. Within the F⁻ cell, a second strand is synthesized, using the transferred strand as template, and the resultant double-stranded molecule is recircularized. The elements of F factor replication are shown in Figure 14.12.

Genomic DNA transfer　The processes thus far outlined allow, at best, replication and transfer only of the components of the F particle and of small amounts of chromosomal material. More extensive transfer of chromosomal DNA requires integration of the F factor and the host cell chromosome. When this occurs, the properties of the F particle are retained. An Hfr cell interacts with an F⁻ cell, forms a pilus, and begins transfer of chromosomal DNA. Both the integrated F factor and a single strand of donor DNA are transferred. The position of insertion sequences on the F factor,

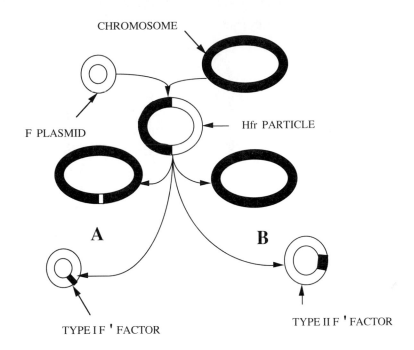

FIGURE 14.11 Possible consequences of F factor integration into the chromosome and the subsequent formation of an F′ factor. In process (A), integration of the F particle to form an Hfr cell is followed by F particle removal in a manner such that a portion of the F particle is left within the chromosome and a portion of the chromosome is imparted to the F particle. The process produces a type I F′ particle and results from the fact that recombination occurs between a region within the F particle DNA and a region of the chromosomal DNA. In process (B), Hfr formation is followed by removal of the F particle in a manner such that portions of chromosomal DNA on either side of the F particle recombine. Such an F′ particle is slightly larger than a type I F′ particle because the type II particle contains all of the original F particle DNA plus small portions of the chromosomal DNA that flanks it on both sides.

relative to the origin of transfer, dictates that the genes involved in the transfer process are transferred last. Conjugation in gram-negative bacteria is shown in Figure 14.13.

Conjugation in gram-positive organisms

Substantially less information regarding the conjugation processes of gram-positive organisms is available than for gram-negative species. For a long period of time, conjugation was unrecognized among gram-positive organisms. Relatively recent studies have revealed a variety of conjugative mechanisms in these organisms that simultaneously offer wonder and opportunity. It appears that a diversity of conjugation mechanisms exist among gram-positive bacteria and that few, if any,

of them involve pili. Conjugation in *Enterococcus faecalis* is mediated by heat resistant and protease-sensitive oligopeptides produced by certain cells devoid of plasmids. These substances have been designated *pheromones*. Under the influence of pheromones, certain plasmid-containing cells proliferate an aggregation substance that allows their aggregation with pheromone-producing cells and, in some as yet imperfectly understood manner, the transfer of DNA between them. Although conjugative plasmids are known in other gram-positive species, pheromones are recognized only in *E. faecalis* thus far.

In certain gram-positive bacteria, a completely different conjugation mechanism is found, the mechanism of *conjugative transposons*. Transposon Tn*916* is of particular interest, since it contains a

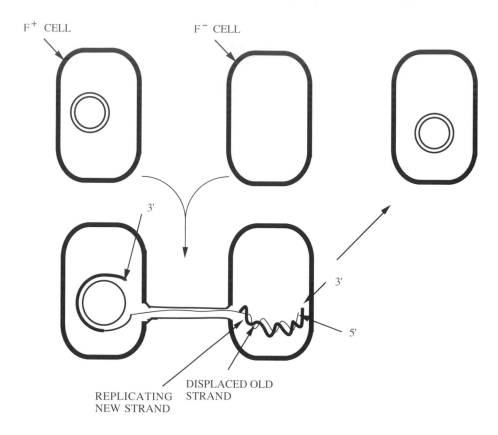

FIGURE 14.12 F factor replication. F plus and F minus cells come in close proximity and are connected by the F pilus coded for by the F factor. A nick is introduced into the outer strand of the F particle and the nicked strand is displaced into the recipient cell by replication in the 5′ to 3′ direction by a rolling circle mechanism, using the nonnicked strand of the F particle as template. The 5′ end of the displaced strand enters the recipient cell first. When the nicked strand is replicated, its complementary strand is replicated within the recipient cell, and the double-stranded entity is circularized.

gene for tetracycline resistance and therefore, can be readily detected. Transposon 916 and similar entities have the unique ability to mediate their own transfer. Transposons are apparently broad in their action and appear to be widely distributed in nature. Genetic manipulation of transposons offers substantial opportunities for their use as genetic engineering vehicles. At the moment, the precise ways in which transposons mediate genetic transfer are unknown.

The consequences of genetic transfer

From the viewpoint of the recipient, genetic transfer, by whatever mechanism, may convey new properties upon the cell. However, the extent to which new properties occur and are heritable, is a function of the posttransfer fate of the donated DNA. Three possibilities may occur. The donated DNA may be integrated into the host cell genome. The mechanism of integration of foreign DNA may involve a recA protein-dependent, *homology dependent* process. Alternatively, recombination may be recA protein independent and *homology independent*. RecA independent recombination may be further divided into two subcategories, site specific recombination that requires a minimum of homology and site independent, or illegitimate recombination.

If it is capable of self-replication, donated DNA may be converted to a form, normally circular, that allows it to exist as a discrete and heritable genetic entity apart from the genome. Such is the situation for plasmids. The persistence of such entities will reflect whatever selective pressures may be imposed upon them. Lastly, transferred

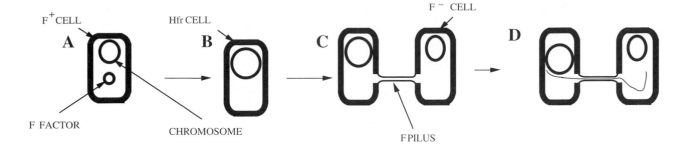

FIGURE 14.13 Conjugation in gram-negative bacteria. An F plus cell, in which the bacterial genome and the F factor are separate (A), is converted to an Hfr cell, in which the F factor is integrated into the bacteria chromosome with aid of compatible insertion sequences in both the F factor and the chromosome (B). The Hfr cell comes in contact with an F minus cell and a pilus is formed (C). A single strand of the F factor and the attached donor genome is then replicated via the rolling circle mechanism and transported into the recipient cell (D). The integrated F factor is transferred last. In theory, the entire chromosome may be transferred but it is unlikely that such is ever actually achieved. It was originally believed that the pilus served as both an attachment mechanism and as the vehicle for transfer of genetic material. More recent work suggests that, at least in some cases, the primary function of the pilus in conjugation is to allow pairing between Hfr and F minus cells and that the actual transfer of genetic material occurs at another location at which the two cells are in contact.

DNA may be destroyed. If this route is taken, destruction must be specific so that in the process of destroying the foreign DNA, the integrity of the original DNA may be retained. The necessity for selective DNA destruction requires that a discrimination mechanism exists between foreign and native DNA. As with other processes, such as motility, DNA repair mechanisms, and viral replication, selective methylation is the key. Although selective methylation is a generalized mechanism that allows for selective destruction of foreign DNA by restriction endonucleases, the basic mechanisms by which protection from DNA destruction occur are similar—differential methylation of adenine and cytosine residues. Furthermore, despite their wide distribution, restriction enzymes operate, mechanistically, in one of two general ways. *Type I* enzyme systems combine methylation and endonuclease activity in a single molecule and cleave unmethylated DNA randomly. In contrast, *type II* endonuclease systems involve separation of methylation from DNA degradation by the action of individual enzymes. Furthermore, type II endonuclease activity is site specific, so type II systems find broad use in molecular biology. The specificity of action of type II endonucleases is, in large measure, dictated by palindromes, characteristic regions of DNA in which the sequences of the DNA are identical in both directions.

Genetic expression

Expression is the ultimate consequence of the operation of the genetic apparatus of an organism. Selective expression allows an organism to use its genetic material in an effective manner under a particular set of circumstances. It makes no sense for an organism to devote its physiological energy toward tasks that, at a particular time, allow it no selective advantage. From the viewpoint of survival, it is useful for an organism to participate in *only* those processes that are appropriate to the conditions that it faces. At the same time, the environment with which a microbe must deal is usually changing. A microbe must therefore have the flexibility to adjust to different conditions. This ability is provided by changes in genetic expression. In this section, we will consider some of the major genetic mechanisms by which selective genetic expression is accomplished and their consequences. It is impossible to consider all of the alternatives, indeed, it is most unlikely that we are even fully aware of all of them at present. We will consider the regulation of selected operons as general examples and, in addition, will discuss regulatory mechanisms that apply to multiple operons. Finally, we will consider the manner in which genetic regulatory mechanisms operate in response to environmental change.

The *gal* and *ara* operons

In Chapter 13, we considered the *lac* operon, which we recognized was jointly controlled by induction and repression and by catabolite repression. The formation of cAMP was a central feature of this regulatory system. Cyclic AMP mediates catabolite-regulated operons other than the *lac* operon are known. The galactose (*gal*) and arabinose (*ara*) operons are additional examples of cAMP-mediated genetic regulation.

In *E. coli*, galactose utilization is mediated by 3 genes, the *galK*, the *galT*, and the *galE* genes that determine, respectively, galactose phosphorylation by a kinase enzyme that produces galactose-1-phosphate, transfer of galactose-1-phosphate to UDP to form UDP-galactose and epimerization of UDP-galactose to UDP-glucose so that the UDP-glucose may be used in Embden-Meyerhof reactions. The galactose operon is under the control of the *galR* regulatory gene and formation of messages is dependent on the action of two promoter regions that respond to cAMP differently. Gal promoter 1 (P_{G1}) is positively affected by cAMP, whereas the operation of gal promoter 2 (P_{G2}) is unaffected.

In addition to being a potential carbon and energy source, galactose is essential to the cell because UDP-galactose is an obligatory precursor to lipopolysaccharide, which must be formed with or without the presence of galactose. Regulation of the *gal* operon is thus intrinsically complex. The regulation mechanism reflects, mechanistically, the overlapping nature of the promoters. When cAMP levels are low, as when glucose is present and galactose is absent, transcription is mediated by the cAMP-insensitive promoter P_{G2}. However, when cAMP is high, as when glucose is exhausted and galactose is present, the P_{G1} promoter, which is facilitated by cAMP, is used. This mechanism allows discrimination between energy sources and, at the same time, the provision of an essential cell wall constituent when a precursor for its formation (i.e., galactose) is not abundantly available. Both promoters are under the control of the galR (regulator) product that appears to bind separately to independent operator regions in a dimeric form. Under certain conditions, it is believed that dimeric forms of the regulator protein associated with the individual operators unite to form a tetrameric form that protects both promoters from the action of DNA-dependent RNA polymerase. The operation of the *gal* operon is shown diagrammatically in Figure 14.14.

The *ara* operon for arabinose use is comprised of four genes, the *araA*, *B*, *C*, and *D* genes. Genes *A*, *B*, and *D* are involved as structural genes, while the *C* gene is involved in regulation. Functionally, the araA gene product converts L-arabinose to L-ribulose, an obligatory intermediate in L-arabinose utilization, the araB gene product converts L-ribulose to L-ribulose-5-phosphate, and the gene D product converts L-ribulose-5-phosphate to L-xylulose-5-phosphate. The general nature of the *ara* operon is shown in Figure 14.15.

Control of the *ara* operon is achieved by the differential action of two promoters, the P_C promoter that initiates transcription of the araC gene product and the P_{BAD} promoter. The promoters are located between the *araC* region and the *araBAD* region. Both promoters are stimulated by cAMP and catabolite activator protein (CAP, also known as catabolite repressor protein, CRP). When arabinose is present, the *araC* gene product activates transcription of the *A*, *B*, and *D* genes. In the absence of arabinose, the araC product changes its configuration so that both promoters are repressed. The same gene product thus regulates two activities, arabinose utilization and its own formation, by alteration of its configuration.

The central role of cAMP as a regulatory agent

Cyclic AMP serves as a regulatory agent for control of a number of operons that include the *ara* and *gal* operons, the *lac* operon, and many others. Substantial interest thus exists regarding the manner in which so many different operons might be regulated by a common mechanism since the critical regulatory products in the various operons are diverse. Recently, it has been suggested that the cAMP-sensitive systems may be regulated not by a particular metabolite but by the functioning of its transport system. According to this concept, catabolite activator (repressor) protein may exist in two forms. In one configuration, when bound to a transport protein, it is believed to bind to adenyl cyclase and inhibit the latter's activity. Under these conditions, little cAMP is formed and the catabolite repression phenomenon is observed. When a readily usable substrate is available, its transport protein is also available and use of an alternate substrate is not physiologically necessary. In contrast, when the readily used substrate is no

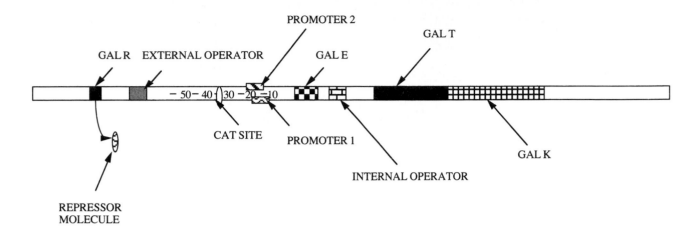

FIGURE 14.14 The structure of the gal operon. The operon contains two promoters, one of which, promoter 1, is activated by catabolite activator (repressor) protein and cyclic AMP. When the cyclic cAMP concentration is high, as occurs when glucose is absent and galactose is present, catabolite activator protein and cAMP interact with the cat site, 35 bases upstream from promoter 1, and promoter 1 is used to allow transcription of the enzymes required for use of the galactose. When galactose is absent and glucose is present, promoter 2 is used so that glucose metabolism may provide enzymes required to produce the galactose required for cell wall biosynthesis. Additional features of the operon are described in the text.

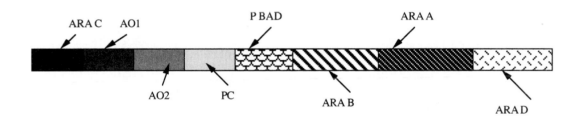

FIGURE 14.15 The essential features of the ara operon. The araC region codes for a repressor protein that when present in abundance, can attach to the operator regions associated with control of transcription of messages for the formation of the araC gene product (AO1) and the genes are arabinose utilization (AO2). In the absence of arabinose, the araC gene product binds to both operators and prevents transcription of the genes for both arabinose use and for its own formation. When cyclic AMP, catabolite activator protein, and arabinose are present in substantial amounts, the conformation of the araC protein is changed and it is removed from AO2. At the same time, the CAP-cAMP complex facilitates attachment of DNA-dependent RNA polymerase to the PBAD site so that messages for the araB, A, and D products may be formed. When the C protein concentration is low and the CAP-cAMP complex is abundant, DNA-dependent RNA polymerase attaches to the PC promoter, allowing messages for the C protein to be formed. The araC protein thus regulates the operon in two ways, controlling both the genes for arabinose utilization and for its own formation.

longer present, its transport protein is also decreased in abundance. Under these conditions, the catabolite activator protein cannot bind to the transport protein, so the complex may not inhibit adenyl cyclase. Adenyl cyclase activity thus increases, leading to an increase in cAMP, which binds to CAP, converting it to a form that allows it, by binding to a promoter, to facilitate attach-

ment of DNA-dependent RNA polymerase so that the message required for use of an alternate metabolite may be formed. Such a mechanism would allow a variety of systems to be regulated by a common mechanism, independent of a particular metabolite as such. Although the mechanism has been proposed as a general mechanism for regulation of use of sugars mobilized by the PTS

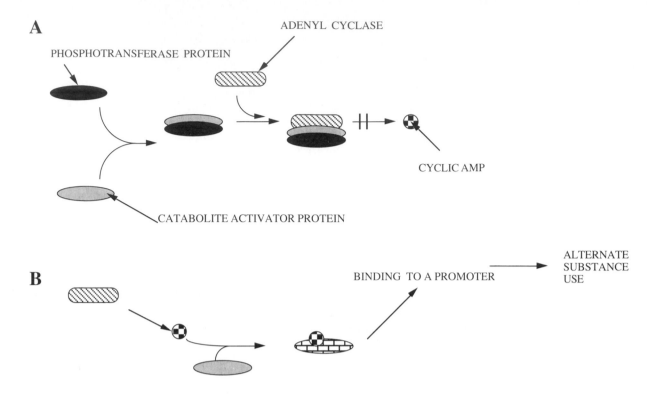

FIGURE 14.16 A possible general mechanism for catabolite regulation of metabolism. The theory proposes that regulation involves an interaction between catabolite activator protein and a transport protein and also suggests that catabolite activator protein has the ability to exist in two forms. In situation (A), when phosphotransferase activity is high, catabolite activator protein interacts with a phosphotransferase protein to form a complex, which, in turn, interacts with active adenyl cyclase enzyme and inactivates it. As a result, cyclic AMP does not accumulate. When phosphotransferase activity is absent or diminished (B), adenyl cyclase is active and produces cyclic AMP in abundance. The latter interacts with catabolite activator protein, converting it to a form in which it can bind to the promoter region associated with use of an altered energy source, allowing formation of proteins required for its transport and metabolism.

transport system, it may well be a concept that applies generally. Ideas regarding generalized regulation of a variety of operons by a CAP-dependent mechanism are shown in Figure 14.16.

The arginine regulon

Corepression is apparently the major mechanism that regulates use of arginine. Arginine utilization regulation is under control of the argR gene product that by binding with arginine, becomes activated and turns off arginine synthesis when arginine is in abundant supply. When arginine is limited, the regulatory protein becomes inactive, allowing arginine synthesis. Although the enzymes for arginine synthesis are coordinately controlled by the argR gene product, the arginine synthesis system is regarded as a *regulon*, rather than an operon, since the majority of enzymes involved in arginine synthesis are scattered around the chromosome. However, the *argE, C, B,* and *H* genes are clustered.

The ability of a single regulator to coordinately regulate a collection of scattered genes on one hand, and, on the other, a gene cluster, all of which participate in a common process, requires explanation. The solution appears to be differential binding of the argR gene product by "arg boxes," sequences of approximately 18 base pairs found upstream from the start of the coding portions of the arginine genes. It is believed that differential binding by different arg boxes as a consequence of differences in their base sequences, combined with differences among arg boxes in their number and distance from coding regions, may account for differences in the extent of repression or induction, by

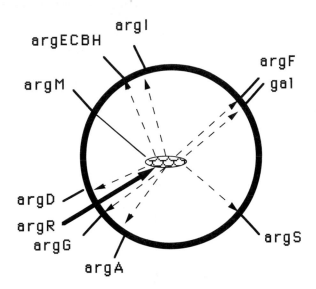

FIGURE 14.17 The arginine regulon. The various genes concerned with arginine synthesis are, for the most part, scattered around the chromosome, but arginine genes E, C, B, and H are clustered. The product of the argR locus regulates all of the arginine genes by corepression. In the absence of arginine, the product is produced in an inactive form, allowing arginine synthesis. When arginine is present in abundance, it interacts with the inactive argR gene product to make it active, thus shutting off arginine synthesis. Additional details are described in the text.

a common gene product, of the various enzymes involved in arginine synthesis. The elements of the *arg* regulon are shown in Figure 14.17.

Global regulation

The complexity of regulatory processes at the genetic level is astounding. The systems previously discussed are, comparatively speaking, simple in that they are concerned, for the most part, with regulation of a particular process or system. Other, more generalized regulatory systems are also known and, as time proceeds, it is most likely that additional systems will be discovered.

The stringent response

When organisms are grown in the presence of limiting amounts of a nutrient, particularly amino acids, guanosine 5-diphosphate-3'diphosphate (ppGpp) accumulates in the cells. At the same time, a deficiency of charged aminoacyl tRNA molecules

is observed. Accumulation of the latter is apparently the triggering mechanism that gives rise to the elevated ppGpp concentration. Binding of uncharged tRNA molecule at the ribosome surface leads to formation of the relA gene product, or *stringent factor*, a ribosome-bound pyrophosphatase enzyme that allows formation of ppGpp from either GDP or the combination of GTP and ATP. These processes lead to a variety of physiological responses, including temporary cessation of growth and decreases in both RNA and DNA synthesis. In addition, a variety of other metabolic changes are observed.

The physiological mechanisms underlying the stringent response have been the subject of much study. Some of the suggested modes by which the stringent response might be manifested are shown in Figure 14.18. Originally, it was believed that the major effect of ppGpp accumulation was inhibition of protein synthesis initiation. It now appears that the effects of ppGpp accumulation are more broadly based than originally thought. According to some studies, ppGpp influences the action, and perhaps the formation, of DNA-dependent RNA polymerase. It is thought by some that the effects of ppGpp on RNA polymerase are multifaceted. It has been suggested that ppGpp may affect the configuration, and, therefore, the activity of the enzyme and may also promote formation of terminator loops. It has further been suggested that ppGpp manifests its action by specifically inhibiting transcription of rRNA. Inhibition of rRNA would, in turn, diminish protein synthesis. In addition to these effects, ppGpp formation has been suggested as a means of diminishing the cellular pool of GTP, thus diminishing the occurrence of reactions, particularly those involved in protein synthesis, for which GTP is required. Whatever the precise mechanisms by which ppGpp may exert its effect, it is increasingly clear that the stringent response is much more complex than originally recognized and that it exerts a variety of regulatory effects on cells.

Phosphate starvation

Limitation of phosphate has major effects on the cell. In certain gram-positive bacteria, phosphate starvation changes the teichoic acids to teichuronic acids. It is believed that such a change facilitates Mg^{2+} accumulation so that, among other things, kinase enzymes may function more efficiently in a

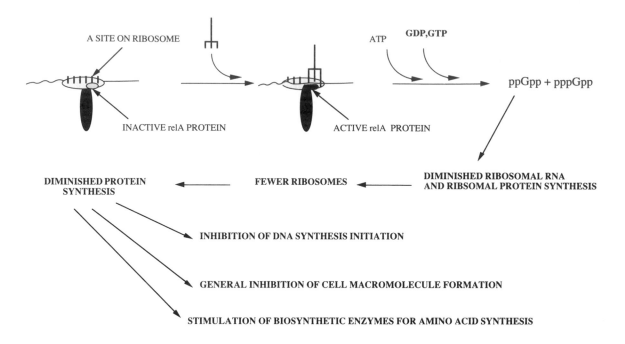

FIGURE 14.18 Some manifestations of the stringent response.

phosphate-deficient environment. Phosphate limitation serves as a *stimulon*, a condition that induces generalized physiological and genetic responses to a single stress.

Phosphate limitation exerts a variety of effects on microbial cells, particularly changes in the nature of the components of the gram-negative outer membrane and the periplasmic space. Phosphate limitation leads, among other effects, to activation of the *phoA* and *phoE* genes. The former codes for an alkaline phosphatase and the latter for an outer membrane protein that facilitates phosphate transport. In addition to *phoA* and *phoE*, studies with *E. coli* have shown that as many as 20 promoters respond to phosphate limitation, indicating that the effects of phosphate limitation on cells are extremely complex.

The aerobic to anaerobic shift

Change in the gaseous phase can exert major effects on microbes in the manner of a stimulon. Thus it has been shown that as many as 19 proteins in *E. coli* may be specifically induced by transition from anaerobic to aerobic conditions and that, conversely, the shift from aerobiosis to anaerobiosis may specifically induce as many as 18 enzymes. In general, the induced enzymes are those that might

be expected in response to the environmental change. Thus, presumably as a function of oxygen, aerobically induced organisms contain high levels of TCA cycle enzymes and the pyruvate dehydrogenase complex, as well as superoxide dismutase. It is presumed that the increase in TCA cycle enzymes reflects the more extensive metabolism possible under aerobic conditions and that the elevated superoxide dismutase reflects an increased requirement for protection against toxic oxygen metabolites. Similarly, the increased glycolytic enzymes and pyruvate-formate lyase activity under anaerobic conditions undoubtably reflect involvement of these enzymes in energy metabolism under anaerobiosis.

The heat shock response

The heat shock response is a generalized response to temperature change that occurs throughout nature. In *E. coli*, heat shock induces formation of a collection of at least 17 proteins. One of the proteins formed is an unusual sigma factor, a 32,000 molecular weight molecule. It is believed that, under conditions of elevated temperature, it is this unusual sigma factor that allows expression of unique heat shock proteins. It is of interest that not only heat, but other stimulae, also elicit cellular responses similar or identical to those of the heat shock response. Ultraviolet light and ethanol exposure result in similar responses. Although heat shock gives rise to a characteristic response, it is not clear that the response to heat is unique to heat itself. Understanding of the heat shock response requires further study.

Summary

The topics discussed in this chapter concern the nature of the bacterial genome, the ways in which it may be modified, the manner in which microbial cells respond to change in genetic material with regard to repair mechanisms, genetic exchange, mechanisms of change in genetic expression, and the physiological consequences of all of these phenomena. It is abundantly clear that although the physiological consequences of genome function are many, amidst the diversity of responses, there are patterns in regard to both genotypic and phenotypic mechanisms, and that genetic expression reflects a cooperative relationship between the genome, per se, and the remaining structures of the cell through whose operation genomic changes are expressed. As knowledge progresses, our understanding of that relationship will also increase, as will our ability to exploit that understanding.

Selected References

Anderson, S. G. E., and C. G. Kurland. 1990. Codon preferences in free-living microorganisms. *Microbiological Reviews.* **54:** 198–210.

Balbinder, E., D. Kerry, and C. I. Reich. 1983. Deletion induction in bacteria. I. The role of mutagens and cellular error-prone repair. *Mutation Research.* **112:** 147–168.

Bremer, H., and G. Churchward. 1991. Control of cyclic chromosome replication in *Escherichia coli. Microbiological Reviews.* **55:** 459–475.

Cech, T. T. 1990. Self-splicing of group I introns. *Annual Review of Biochemistry.* **59:** 543–568.

Clark, A. J. 1971. Toward a metabolic interpretation of genetic recombination. *Annual Review of Microbiology.* **25:** 437–464.

Clewell, D. B. 1981. Plasmids, drug resistance and gene transfer in the genus *Streptococcus. Microbiological Reviews.* **45:** 409–436.

Collado-Vides, J., B. Magasanik, and J. D. Gralla. 1991. Control site location and transcriptional regulation in *Escherichia coli. Microbiological Reviews.* **55:** 371–394.

Csonka, L. N. 1989. Physiological and genetic responses of bacteria to osmotic stress. *Microbiological Reviews.* **53:** 121–147.

Doolittle, W. Ford. 1979. The cyanobacterial genome, its expression, and the control of that expression. *Advances in Microbial Physiology.* **20:** 2–102.

Downie, A. W. 1972. Pneumococcal transformation—a backward view. *Journal of General Microbiology.* **73:** 1–12.

Drake, J. W., B. W. Glickman, and L. S. Ripley. 1983. Updating the theory of mutation. *American Scientist.* **71:** 621–630.

Dressler, D., and H. Potter. 1982. Molecular mechanisms in genetic recombination. *Annual Review of Biochemistry.* **51:** 727–762.

Goodgal, S. H. 1982. DNA uptake in *Haemophilus* transformation. *Annual Review of Genetics.* **16:** 167–192.

Grindley, N. D. F., and R. R. Reed. 1985. Transpositional recombination in prokaryotes. *Annual Review of Biochemistry.* **54:** 863–896.

Hamkalo, B. A., and O. L. Miller, Jr. 1973. Electron microscopy of genetic activity. *Annual Review of Biochemistry.* **42:** 379–398.

Hannawalt, P. C., P. K. Cooper, A. K. Granesan, and C. A. Smith. 1979. DNA repair in bacteria and mammalian cells. *Annual Review of Biochemistry.* **48:** 783–836.

Harayama, S., M. Kok, and E. L. Neidle. 1992. Functional and evolutionary relationships among diverse oxygenases. *Annual Review of Microbiology.* **46:** 565–602.

Haseltine, W. A. 1983. Ultraviolet light repair and mutagenesis revisited. *Cell.* **33:** 13–17.

Hegeman, G. D., and S. L. Rosenberg. 1970. The evolution of bacterial enzyme systems. *Annual Review of Microbiology.* **24:** 429–462.

Hiraga, S. 1992. Chromosome and plasmid partition in *Escherchia coli. Annual Review of Biochemistry.* **61:** 283–306.

Holloway, B. W., and A. F. Morgan. 1986. Genome organization in *Pseudomonas. Annual Review of Microbiology.* **40:** 79–106.

Hopwood, D. A. 1981. Genetic studies with bacterial protoplasts. *Annual Review of Microbiology.* **35:** 237–272.

Hotchkiss, R. D. 1974. Models of genetic recombination. *Annual Review of Microbiology.* **28:** 445–468.

Ingraham, J. L., O. Maaloe, and F. C. Neidhardt. 1983. *Growth of the Bacterial Cell.* Sinauer Associates, Inc., Sunderland, Massachusetts.

Joklik, W. K. 1981. Structure and function of the reovirus genome. *Microbiological Reviews.* **45:** 483–501.

Kaiser, D., C. Manoil, and M. Dworkin. 1979. Myxobacteria: Cell interactions, genetics, and development. *Annual Review of Microbiology.* **33:** 595–640.

Kessin, R. H. 1988. Genetics of early *Dictyostelium discoideium* development. *Microbiological Reviews.* **52:** 29–49.

Krawiec, S. and M. Riley. 1990. Organization of the bacterial chromosome. *Microbiological Reviews.* **54:** 502–539.

Levinthal, M. 1974. Bacterial genetics excluding *E. coli. Annual Review of Microbiology.* **28:** 219–230.

Lindahl, T. 1982. DNA repair enzymes. *Annual Review of Biochemistry.* **51:** 61–87.

Matin, A., E. A. Auger, P. H. Blum, and J. E. Schultz. 1989. Genetic basis of starvation survival in nondifferentiating bacteria. *Annual Review of Microbiology.* **43:** 293–316.

McGeoch, D. J. 1989. The genomes of the human herpes viruses: Contents, relationships, and evolution. *Annual Review of Microbiology.* **43:** 235–266.

Miller, R. V., and T. A. Kokjohn. 1990. General microbiology of recA: Environmental and evolutionary significance. *Annual Review of Microbiology.* **44:** 365–394.

Mizuuchi, K. 1992. Transpositional recombination: Mechanistic insights from studies of Mu and other elements. *Annual Review of Biochemistry.* **61:** 1011–1052.

Moat, A. G., and J. W. Foster. 1988. *Microbial Physiology,* 2nd edition. John Wiley and Sons, New York.

Nagley, P., K. S. Sriprakash, and A. W. Linnane. 1977. Structure, synthesis and genetics of yeast mitochondrial DNA. *Advances in Microbial Physiology.* **16:** 158–277.

Neidhardt, F. C., J. L. Ingraham, and M. Schaechter. 1990. *Physiology of the Bacterial Cell.* Sinauer Associates, Inc., Sunderland, Massachusetts.

Novick, R. P. 1969. Extrachromosomal inheritance in bacteria. *Bacteriological Reviews.* **33:** 210–263.

Parker, J. 1989. Errors and alternatives in reading the universal genetic code. *Microbiological Reviews.* **53:** 273–298.

Petes, T. D. 1980. Molecular genetics of yeast. *Annual Review of Biochemistry.* **49:** 845–876.

Radding, C. M. 1978. Genetic recombination: Strand transfer and mismatch repair. *Annual Review of Biochemistry.* **47:** 847–880.

Razin, A., and H. Cedar. 1991. DNA methylation and gene expression. *Microbiological Reviews.* **55:** 451–458.

Reeve, John N. 1992. Molecular biology of methanogens. *Annual Review of Microbiology.* **46:** 165–192.

Riley, M., and A. Anilionis. 1978. Evolution of the bacterial genome. *Annual Review of Microbiology.* **32:** 519–560.

Sakaguchi, K. 1990. Invertons, a class of structurally and functionally related genetic elements that includes linear DNA plasmids, transposable elements, and genomes of Adeno-type viruses. *Microbiological Reviews.* **54:** 66–74.

Scolnik, P. A., and B. L. Marrs. 1987. Genetic research with photosynthetic bacteria. *Annual Review of Microbiology.* **41:** 703–726.

Shatkin, A. J. 1971. Viruses with segmented ribonucleic acid genomes: Multiplication of influenza versus reovirus. *Bacteriological Reviews.* **35:** 250–266.

Simpson, L. 1987. The mitochondrial genome of kinetoplastid protozoa: Genomic organization, transcription, replication, and evolution. *Annual Review of Microbiology.* **41:** 363–382.

Smith, H. O., B. Donner, and R. A. Deich. 1981. Genetic transformation. *Annual Review of Biochemistry.* **50:** 41–68.

Tzagoloff, A., and A. M. Myers. 1986. Genetics of mitochrondial biogenesis. *Annual Review of Biochemistry.* **55:** 249–286.

Villarreal, L. P. 1991. Relationship of eukaryotic DNA replication to committed gene expression: General theory for gene control. *Microbiological Reviews.* **55:** 512–542.

Walker, G. C. 1984. Mutagenesis and inducible responses top deoxyribonucleic acid damage in *Escherichia coli. Microbiological Reviews.* **48:** 60–93.

Wallace, D. C. 1982. Structure and evolution of organelle genomes. *Microbiological Reviews.* **46:** 208–240.

Willetts, N., and R. Skurray. 1980. The conjugation system of F-like plasmids. *Annual Review of Genetics.* **14:** 41–76.

Willetts, N., and B. Wilkins. 1984. Processing of plasmid DNA during bacterial conjugation. *Microbiological Reviews.* **48:** 24–41.

Wood, W. B., and H. R. Revel. 1976. The genome of bacteriophage T4. *Bacteriological Reviews.* **40:** 847–868.

CHAPTER 15

The Effects of Environmental Factors on Microbes

Microbes exist in a diversity of physical and chemical environments. In order to survive, they must adjust to a diversity of environmental factors. Study of the ways in which microbes adapt to a variety of external factors allows us to understand the reasons why particular organisms may persist in a particular environment. Study of the effects of environmental factors on microbes also allows us to understand the limits of their physiological abilities. In addition, study of the effects of environmental factors on microbes allows understanding of the genetic mechanisms underlying physiological responses.

Temperature

Temperature is a major criterion by which microbes are classified. We recognize, for example, that the phenomenon of thermophily is a property unique to procaryotes. We further understand that many disease-producing microbes exhibit optimal growth at temperatures similar or identical to the body temperatures of their hosts, while certain microbes, most notably the fungi, fail to grow readily under these conditions. Within the microbial world, we find organisms known as *thermophiles*, organisms that require elevated temperatures, normally at or above 50° C. In addition we find a large group of *mesophiles*, those organisms that normally grow at temperatures between 25° C and 45° C. Finally, we find the *psychrophiles*, organisms capable of growth at or below 20° C.

Sometimes, an organism may normally grow in one temperature range but can, under certain conditions, grow within a different temperature range. We distinguish between psychrophiles, organisms that "love" a temperature range between approximately 5° C and 20° C and will not initiate growth at temperatures above 20° C, and *psychrotrophs*, organisms that normally grow within the mesophilic range but that are capable of growth at psychrophilic temperatures. We also recognize organisms that normally grow within the thermophilic range, but under certain conditions, may grow within the mesophilic range. Finally, we recognize organisms that grow only within the thermophilic range, typically found in hot springs. Within these various temperature groupings of organisms, further distinctions can be made. For a given organism, we recognize a minimum temperature below which growth is not possible, an optimum temperature, the temperature of maximum and most rapid growth, and a maximum temperature, the temperature above which growth does not occur. These latter three temperatures, taken together, constitute the *cardinal temperatures* for the organism. Finally, within a given temperature range, differences are found among organisms in the nature of their cardinal temperatures.

Physiological consequences of temperature change

The diversity of responses to temperature change displayed by various microbes raises many physiological questions. What determines the lower and upper temperature limits for growth? Why are some organisms restricted to a particular temperature range, yet others may grow in more than one range? How does temperature change affect the cell? Does the cell respond structurally, biochemically, genetically, or in a variety of ways? Although all of the above are interesting and challenging questions, there is much yet to be learned about their answers. In most cases, the answers are not simple and we are just beginning to fully understand them.

The master reaction concept

Many years ago, Arrhenius studied the relationship between the rate of a chemical reaction and temperature and found an interesting relationship between reaction rate and temperature change that is shown in the formula below:

$$V = Se^{-\Delta E^*/RT} \tag{1}$$

Conversion of equation (1) to a logarithmic form yields:

$$\ln V = -(\Delta E^*/R \times 1/T) + S \tag{2}$$

In the above equations, S and R are constants, E^* is the activation energy, V is the reaction velocity, ln is the natural logarithm, and T is the absolute temperature. This equation, originally applied to chemical reactions in the test tube, gave rise to the recognition that in many situations, a chemical re-

action increases in speed by a factor of 2 for every 10° C change in temperature. Since bacterial growth results, in large measure, from a combination of enzyme-mediated chemical reactions, it seemed reasonable to plot bacterial growth rates in a logarithmic representation as a function of the reciprocal of the temperature. Although variations are found between organisms in the precise nature of the relationship between temperature and growth rate, the general nature of the relationship is shown in Figure 15.1, drawn so that it includes the cardinal temperatures.

Consideration of Figure 15.1 reveals that the relationship between growth rate and temperature is not a constant one. Over the optimal temperature range for growth, a linear relationship with a negative slope is found between the reciprocal of the temperature and the logarithm of the growth rate. However, the slope of the relationship between growth rate and temperature becomes effectively vertical when the temperature is adjusted either below or above the optimal range. Many have suggested that this phenomenon reflects the control of growth by a *master reaction* or process. According to this thinking, either above or below a certain temperature range, a critical reaction or process is inactivated, thereby abolishing growth. The vertical nature of the slope of the relationship indicates that cessation of growth occurs rapidly at either extreme. Extension of this logic would suggest that at the lower temperature limit, insufficient activation energy is available to allow the critical process and that at the upper limit, the increased rate of the process, as a function of increased temperature, is offset by the changes in protein structure that also occur during temperature elevation resulting, at some point, in inactivation or denaturation of a critical cell protein, with an accompanying cessation of growth.

Whatever the mechanisms that may be involved in growth limitation by temperature, the effects of temperature change on microbial growth cannot be explained entirely by consideration of the effects of temperature on protein-mediated processes such as enzyme activity or transport. The upper and lower temperature limits for growth of many organisms may also reflect the nature of the nutritional environment. It is sometimes possible to extend the normal upper or lower temperature limit for growth of an organism by nutritional supplementation, but no general

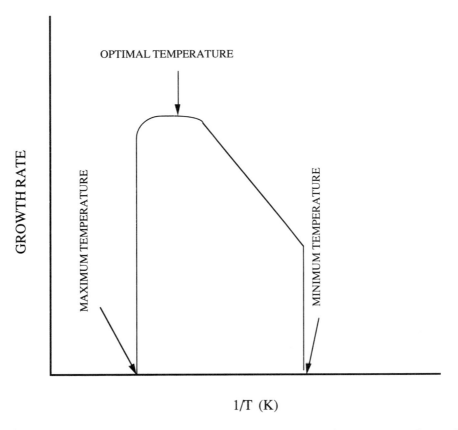

FIGURE 15.1 The general nature of the relationship between temperature and the growth rate of microbes. The temperatures are expressed as degrees Kelvin.

rule has yet been found in regard to the effects of particular nutrients on the temperature range for growth. Although variation is found between species, and strains within a species, in the effects of particular nutrients on the temperature characteristics of various organisms, for each particular organism, nutrient effects are specific. The specificity of nutrients in alteration of both the upper and lower limits of growth has been interpreted as indicating that for particular organisms, particular nutrients alter different processes at the upper and lower temperature extremes, thereby exerting different types of metabolic control at the different temperature limits.

The effects of temperature on cell structure and function

The effects of heat on microbes are diverse and are not solely reflective of the manner in which heat affects the rates of protein-mediated processes. Structural effects are also observed both in regard to

proteins and to cell structures, such as the genome and the membrane. The effects of heat on cellular entities reflect the chemical composition of the particular structure or macromolecule in question and the collective effects of heat on the physiology of particular microbes result from the summation of various effects. The stability and activity of proteins, for example, reflects the primary structure of the protein. Primary structure is required to maintain structural integrity, irrespective of the protein's other properties. Studies of temperature-sensitive mutants have shown substantial differences in the effects of mutation on the temperature ranges that allow growth. Detection of these differences suggests that the various effects of changes in temperature range for growth in different mutants reflect genetically determined, diverse effects on the stability and functions of both structural and process-mediating proteins. A variety of specific temperature-associated regulatory effects on proteins have been observed. Some reflect changes in the structure and the availability of water, while other regulatory effects are manifested as changes in protein structure that alter the ability of the protein to respond to regulatory molecules.

Temperature-associated effects on microbial cell physiology are not restricted to proteins. It is frequently found that DNA of thermophilic microbes contains a relatively large proportion of guanosine (G) and cytosine (C) residues. Such a finding is unlikely to be an evolutionary accident. It is probable that the relatively high GC content of the DNA of many thermophiles reflects greater stability of DNA, because guanosine and cytosine are interconnected by three bonds, where adenine and thymine are interconnected by only two bonds.

In addition to affecting critical cell proteins and nucleic acid, heat affects the nature and function of the cytoplasmic membrane. It is often the case that exposure to elevated temperature is accompanied by an increase in the amounts and proportion of saturated, as opposed to unsaturated, lipids. It is well established that the temperature required for melting saturated lipids is higher than that for unsaturated lipids. It is believed that the increase in saturated membrane lipid content that attends temperature elevation is an adaptive response by microbes in an effort to maintain the membrane in an appropriate degree of fluidity. Compared to the effects of temperature elevation, an inverse effect on membrane lipid composition is often found when microbes are exposed to temperatures at the lower end of their temperature range. Under these conditions, unsaturated lipids may predominate and saturated lipids are normally a decreased proportion of total membrane lipids. The physiological effect of this change is to mitigate freezing, since unsaturated lipids resist freezing to a greater extent than do saturated lipids. The increase in unsaturated lipids that accompanies low-temperature exposure is believed, once again, to be an adaptive response in an effort to maintain membrane fluidity.

The heat shock response

When an organism is exposed to elevated temperatures, a collection of proteins is formed that are not present during growth under normal growth temperatures. Formation of *heat shock proteins* is a widespread phenomenon in nature and occurs in both microbial and nonmicrobial forms. Although the precise function of heat shock proteins is, at the moment, unclear, and may differ some-

what from organism to organism, the manner of control of their formation has been established. Formation of heat shock proteins is mediated by formation of a particular sigma factor that allows DNA-dependent RNA polymerase to recognize particular promoters and to allow formation of heat shock protein mRNA molecules.

pH

Although many microbes grow optimally at pH values that approximate neutrality, a significant number of microbes grow at pH values substantially different from a pH of 7. *Thiobacillus* species, for example, may grow at a pH as low as 2.0, and *Bacillus akalophilus* may grow at a pH value as high as 10.5. The ability of these, and other organisms, to grow at pH values incompatible with the growth of the majority of microbes raises many interesting physiological questions. How do the proteins in such organisms remain functional? How are the activities of protein-mediated processes accomplished? How does a microbe that persists in an unusual pH environment maintain its internal pH at a physiologically acceptable level? How does such an organism accomplish both energy generation and transport?

The effects of acidity on protein form and function

In addition to development of equations that describe the relationship between the rates of protein-mediated processes and the concentration of the materials upon which proteins act, Michaelis proposed equations relating acidity to protein form and function. Michaelis suggested that proteins should be regarded as analogous to diprotic acids. According to such thinking, the initial reaction of the protein in response to pH change was single ionization, yielding a singly ionized protein and a proton. With further pH change, a second ionization was possible, yielding a doubly ionized protein and a second proton. At a particular pH value, the total cell protein was envisioned to be comprised of a portion that was un-ionized, a portion that was singly ionized, and a portion that was doubly ionized. The proportions of the total protein associated with a particular ionic form would reflect the properties of the protein and the pH of its environment. These ideas are displayed in the equation below:

$$E_t = EH_2 + EH^- + E^{2-} \qquad (3)$$

In this equation, E_t is the total enzyme and EH_2, and EH^-, and E^{2-} denote its un-ionized, singly ionized, and doubly ionized forms, respectively.

The various ionic forms of the enzyme result from the operation of two dissociation constants, K_1 that relates EH_2 and EH^- and K_2 that relates EH^- and E^{2-}. Michaelis further proposed that the relationships between the forms of an enzyme and pH could be described by a series of equations that allowed calculation of the amounts of all forms of an enzyme from any one of them plus dissociation constants that related the various forms of the enzyme to each other and to pH. The equation that relates all forms of the enzyme to the un-ionized form and pH is shown below:

$$E_t = EH_2 \left(1 + K_1/H^+ + K_1K_2/(H^+)^2\right) \qquad (4)$$

In this equation, the 1 is the factor that allows calculation of the amount of total enzyme in the un-ionized form, the K_1/H^+ term allows calculation of the concentration of EH^- from EH_2 and $K_1K_2/(H^+)^2$ allows calculation of the concentration of the doubly ionized form, E^{2-} from EH_2. By an analogous series of equations, all forms of the enzyme may be calculated from EH^-, the singly ionized form:

$$E_t = EH \ \left(1 + H^1/K_1 + K_2/H^+\right) \qquad (5)$$

Finally, all forms of the enzyme may be calculated from the doubly ionized form, E^{2-}:

$$E_t = E^{2-} \left(1 + H^+/K_2 + (H^+)^2/K_1K_2 \right. \qquad (6)$$

A plot of the theoretical values of the three functions in relationship to pH is shown in Figure 15.2. The equation that relates EH_2 to all of the other forms is denoted f_0 whereas f' and f'' denote the equations that relate the singly and doubly ionized enzyme forms to all forms of the enzyme. Examination of Figure 15.2 shows that the values of f_0, are large near the Y axis and asymptotically approach the X axis as the pH becomes more alkaline. Conversely, values of f'' begin close to the X axis and increase in value as the pH number increases. The values of f' as a function of pH change are numerically of the same form as those typically observed with enzymes and transport molecules. For enzyme activity or transport, the shape of the activity curve as a function of pH approximates a bell shape, with suboptimal activity on either side of an optimum. Although various enzymes and transport molecules exhibit optimal activity at different pH values, the general shape of the activity curve as a function of pH is identical for most of them. Collectively, the correspondence between the observed activity and pH and that expected if the functional protein was in the singly ionized condition, strongly support the suggestion that the active form of many critical proteins is the singly ionized form. Such a suggestion is consistent with the fact that active sites of many proteins contain sulfhydryl groups. It is likely that organisms whose optimal pH values for growth are other than pH 7 have unusual mechanisms for maintaining many of their proteins in the singly ionized form. The ability of some alkalophilic and acidophilic organisms to withstand pH extremes undoubtably also reflects the intrinsic stability of some or all of their proteins in unusual pH environments.

pH and the PMF

Organisms that grow at pH extremes must find unusual ways to accomplish functions such as transport, motility, and energy generation. This is particularly the case for the alkalophiles, whose extracellular pH is substantially less than that within the cell. Since, in many organisms, a proton gradient is the major component of the proton motive force required for ATP formation, active transport, and other critical physiological processes, organisms for which the extracellular pH concentration is less than that of the interior must devise alternate ways to provide sufficient PMF for critical life processes to occur. Recent research with *Bacillus alkalolphilus* suggests that the alkalophiles may generate PMF by alteration of their metabolism so that the *membrane potential*, rather than the proton gradient, becomes the major component of the PMF. In this way, the alkalophiles carry out PMF-requiring processes in the absence of a substantial external pH concentration. It appears that a significant portion of the ability of alkalophiles to grow at alkaline pH values reflects net outward movement of Na^+ ions with the aid of unusual transport proteins.

In contrast to the alkalophiles, the obligate acidophiles require a pH of at least 4.0 for growth. It appears that in many cases, the obligate nature

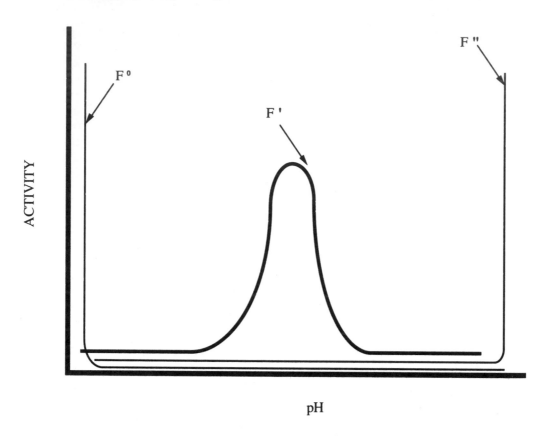

FIGURE 15.2 The theoretical relationship between enzyme activity and pH when all forms of the enzyme are considered as functions of the unionized (F^0), the singly ionized (**F'**) or the doubly ionized (**F''**) forms of the enzyme molecule. The curve expected for the singly ionized form is that which is actually observed, suggesting that the active forms of enzymes are singly ionized entities.

of the acidic growth requirements of these organisms reflects the inability of the organisms to generate a natural proton gradient. Thus, all of the PMF-dependent activities of these organisms depend upon the presence of an artificial gradient.

Maintenance of internal pH

Although a significant number of microbes grow in external pH environments, substantially different from neutrality, the internal pH of most microbial cells is close to pH 7. Maintenance of the internal pH at neutrality is a significant physiology problem for organisms whose external environment differs substantially from neutrality. In this regard, the problems that confront many acidophiles are the inverse of those that face the alkalophiles. Many acidophiles require continual proton extrusion so that the external environment remains, relatively speaking, acidic, while the alkalophiles require continual proton influx. Hydrogen-oxidizing bacteria, particularly members of the genus *Desulfovibrio*, face special problems in regard to intracellular pH regulation. The strategies by which evolution has allowed solution of these problems are diverse and range from unusual transport mechanisms to the action of hydrogenase enzymes.

Oxygen

Oxygen is a critical substance for the vitality of many microbes, but many bacteria thrive in its absence. Because of the centrality of oxygen to the survival of many microbes, considerable attention has been devoted to the effects of oxygen upon them. The ability to grow in the presence or absence of oxygen is a major taxonomic characteristic, which was initially used to classify microbes.

Those that grew only in its presence were *aerobes* and those that grew both with or without oxygen were *facultative* organisms. Those that required the absence of oxygen for growth were *obligate anaerobes.*

Our increasing understanding of the physiological consequences of oxygen exposure and removal indicates that our original appreciation of oxygen's role in the physiology of microbes was simplistic. We currently recognize degrees of requirements for oxygen. Thus some organisms require substantial amounts and others, the *microaerophiles,* require smaller amounts, relative to the amounts found in air. In addition, some organisms, although capable of growth in the presence or absence of oxygen, fail to use oxygen in their metabolism. Such organisms are *anoxybiontic,* in contrast to the oxygen-utilizing, *oxybiontic* organisms. In addition to these distinctions, we recognize differences in the physiological effects that attend oxygen removal. Certain organisms exhibit varying degrees of aerotolerance, while oxygen, or its metabolites, is toxic to other organisms. Finally, we recognize that the effects of oxygen removal are, to a substantial extent, a function of the gas that replaces it. It is often the case that oxygen removal, manipulated by an investigator or as a natural con sequence, is accompanied by an increase in carbon dioxide concentration. Studies have shown that CO_2 exposure affects various organisms differently. We recognize that the growth of many heterotrophic organisms requires, or is stimulated by, CO_2. It is equally apparent that differences exist among heterotrophs with regard to their quantitative requirements for carbon dioxide. Those that require relatively small amounts of CO_2 are *capneic* and other organisms require relatively large CO_2 concentrations. The latter organisms use substantial amounts of CO_2 in their energy-yielding and biosynthetic metabolism.

Full understanding of the consequences of oxygen's effects on microbes requires intense study, both of the effects of oxygen on the organism and of the consequences of its removal. The effects of oxygen on *Streptococcus pyogenes,* an anoxybiontic facultative organism, for example, are substantially different from those observed for a wide variety of obligately anaerobic gastrointestinal bacteria. In *S. pyogenes,* growth of the organism is impervious to oxygen and the major effect of oxygen removal is enhancement of hemolysis. By contrast, oxygen exerts a variety of effects in many obligate anaerobes.

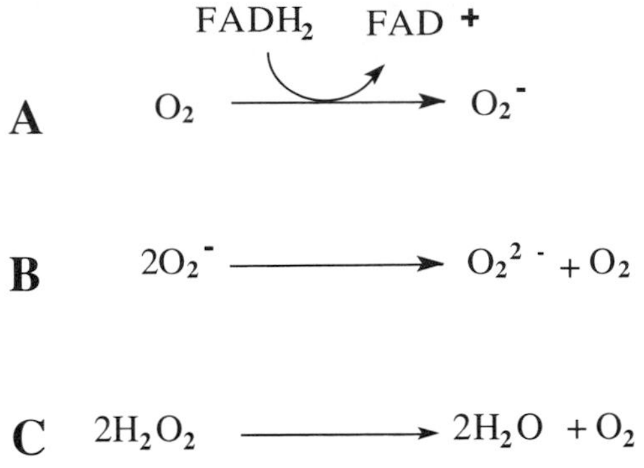

$$\text{A} \quad O_2 \xrightarrow{\text{FADH}_2 \quad \text{FAD}^+} O_2^-$$

$$\text{B} \quad 2O_2^- \longrightarrow O_2^{2-} + O_2$$

$$\text{C} \quad 2H_2O_2 \longrightarrow 2H_2O + O_2$$

FIGURE 15.3 Reactions of oxygen metabolism and detoxification in microbes. (A) An oxidase. (B) Superoxide disumutase. (C) Catalase.

A substantial portion of obligate anaerobes withstand oxygen exposure for substantial periods of time, whereas other anaerobic organisms exhibit much less ability to resist oxygen, and certain anaerobic species (e.g., the methanogens) are inhibited by miniscule exposure to oxygen. Differences in the ability of various anaerobes to resist oxygen exposure reflect, to a large extent, the presence and amounts of two enzymes, superoxide dismutase and catalase. The former enzyme converts the superoxide anion, O_2^-, to the peroxide anion, O_2^{2-}, and oxygen, while catalase converts the peroxide anion to water and oxygen. Aerotolerant anaerobes normally contain superoxide dismutase, but not catalase, although obligately aerobic forms typically contain both. The actions of superoxide dismutase and catalase, as well as oxidases, that produce the superoxide anion from univalent reduction of oxygen with flavoprotein-derived reducing power, are shown in Figure 15.3. Superoxide anion formation is an obligatory consequence of oxybiontic metabolism, and the superoxide anion, in turn, gives rise to other toxic entities, including the peroxide anion. Protection from exposure to oxygen and its toxic metabolites thus depends upon enzymes that destroy toxic forms of oxygen or its metabolites.

In addition to understanding the effects of oxygen on the physiology of microbes, understanding the nature of anaerobiosis often requires study of the manner in which anaerobes use CO_2. Many anaerobes use carbon dioxide extensively in their metabolism. The details of carbon dioxide utilization and its consequences are discussed in Chapters 8 and 17.

Osmolality

Microbes exist in a variety of osmotic environments. Because of their rigid walls, the plantlike forms have the ability to withstand tremendous variation in osmotic pressure, while forms that lack walls are extremely sensitive to osmotic change. Differences are even found among the organisms that withstand elevated osmotic pressure. Salt, for example, is used as a selective agent for isolation of *Staphylococcus aureus* because the organism resists salt exposure. The marine bacteria exhibit the phenomenon of *halophilism*, the specific requirement for salt in their physiology.

The physiological mechanisms allowing halotolerant organisms to survive salt exposure are substantially different from the mechanisms of halophilism. Most *halotolerant* organisms withstand salt exposure either because their membranes resist salt or because they possess specific transport mechanisms for ion removal. By these mechanisms, they maintain their internal osmotic environment in a condition compatible with life in a hostile external environment. Truly halophilic organisms, in contrast, obligatorily require salt for structural integrity or for accomplishment of critical life processes. Although the true halophiles specifically require salt, the ways in which salt affects halophilic physiology vary among organisms. For some organisms, Na^+ or K^+ is required for enzyme activity and in others, the same ions mediate transport processes. Still other organisms require Na^+ or K^+ to maintain structural integrity. The stability of the proteinaceous walls of some halophiles specifically requires Na^+ and the ribosomes of others are stabilized by K^+. All halophilic organisms must control the nature of their internal water environment. The mechanisms of halophilism are diverse and require careful study if their consequences are to be fully understood.

Physical pressure

Certain microbes, and other life forms, exhibit the phenomenon of *barophilism*, the seeking of elevated atmospheric pressure. Microbes may serve as an example of the mechanisms and consequences of this phenomenon, although in the natural world, various effects of pressure on essential physiological functions have been found. It is often the case that the particular effects of pressure reflect the environment occupied by the organism. Pressure has been shown to adversely affect enzymatic proteins found in both fishes and microbes that inhabit shallow aqueous environments and to facilitate the enzymes of organisms that inhabit deeper aqueous regions. The effects of pressure on various microbes appear to reflect the molecular volume of the active state of the protein. The stimulatory or inhibitory effects of pressure are thus dependent upon whether the particular pressure change facilitates or inhibits the optimum reaction volume.

The effects of temperature and pressure on critical cell proteins are interrelated. Classical studies with luciferase, a barophilic enzyme found in certain marine bacteria, showed that the activity of the enzyme could be maintained at a higher temperature during exposure to elevated pressure than could be maintained at atmospheric pressure. It is believed that elevated pressure facilitates maintenance of the enzyme in a physiologically active state. Recent work with members of the microbial community that exists in the vicinity of marine thermal vents supports such a hypothesis. Thermal vent organisms were originally thought to be capable of growth at temperatures as high as 250° C, a temperature incompatible with activity of proteins commonly found in organisms that exist at normal atmospheric pressure. Although it appears that original estimates of the temperature that allows growth of these organisms were excessive, it is likely that such organisms may grow at temperatures as high as 110° C. It is reasonable to suggest that such an ability reflects, at least in part, the mitigating effects of pressure on the detrimental effects that temperature elevation exerts on critical cell proteins. Pressure apparently exerts little effect upon the enzymes of microbes that normally exist at atmospheric pressure. Studies with *E. coli* and other organisms, have shown that the pressure can be elevated by a factor as large as 20 with no appreciable effect on the organisms.

Radiation

Radiation exerts a variety of effects on microbes. Its effects are largely reflective of two factors, its energy and the extent to which a particular microbe can withstand radiation. Radiation resistance, in turn, is a function of the genetics of the organism, whether or not the organism contains materials that may absorb the radiation and what mechanisms of radiation damage repair that the organism may possess. The effects of radiation on a particular microbe are a collective function of all of these factors.

Irrespective of the above considerations, radiation's effects upon microbes and other life forms are a function of its energy, which is itself a function of the wavelength. An inverse relationship exists between wavelength and energy. Thus small wavelength radiation is substantially more energetic, penetrating, and damaging than large wavelength radiation. Cosmic radiation and high-energy X-rays, whose dimensions are in fractions of nanometers have intrinsically greater effects than do radio waves, whose dimensions are in meters. Our current concern with ozone layer destruction reflects this fact.

Mechanisms of cell interaction with radiation

The ways in which radiation exposure affect microbes are diverse. Radio waves and infrared rays generally have little or no deleterious effects since they are insufficiently energetic to affect the physiology of the organisms. The effects of radiations within the visible energy range are variable, depending on the organism. In some cases, deleterious effects have been found as a function of visible light exposure at wavelengths greater than 400 nm. However, in many cases, little if any harmful effects result from visible light exposure. Pigmented organisms may absorb particular wavelengths and both infrared and visible light may be used in energy generation in phototrophic species. Carotenoids and similar pigments may protect these species from excessive radiation exposure.

Because of their relatively high energy, ultraviolet (UV) and X-radiation may exert substantially more detrimental effects on microbes. The primary effects of UV radiation are fourfold: 1) damage to DNA by the formation of pyrimidine dimers; 2) excitation of orbital electrons in pigments; 3) fluorescence, in which energy originally absorbed in the UV region is emitted as visible light; and 4) photosensitization in which energy absorbed by one molecule is transferred to a second molecule, activating it.

As a function of their intense energy and penetrating power, X-rays and cosmic rays exert more disastrous effects on microbes than any radiation form discussed thus far. In addition to DNA damage, these radiations induce formation of a collection of detrimental entities that range from singlet oxygen to free radicals, superoxide and peroxide anions, and other highly reactive molecular and ionic species. Collectively, these elements are highly detrimental to cells via destruction and alteration of DNA and generalized oxidative damage to essential cell components.

Radiation resistance

Microbes display intrinsic, genetically determined differences in their ability to resist radiation exposure. The general effects of radiation exposure are reflective, among other things, of the overall properties of the organism, the environment of exposure, and the intensity and length of exposure. Certain organisms, notably *Deinococcus radiodurans,* are exceptionally radiation resistant, yet other species are highly radiation affected. Uncertainty persists, with regard to both microbes and other life forms, concerning the precise effect of radiation exposure, particularly when exposure is extended and of relatively low-energy.

DNA repair mechanisms

The mutagenic effects of radiation on microbes have long been recognized. Indeed, in many situations, UV irradiation is routinely used for mutagenic purposes. Because of their long exposure to radiation, and because of its damaging effects on genetic material, evolution has allowed microbes to develop a number of mechanisms by which radiation-damaged DNA may be repaired. The major mechanisms of DNA repair are discussed in Chapter 14.

Nutrient limitation

Although it is not normally regarded as an environmental factor in the same sense as the factors previously described, nutrient limitation is an environmental factor that profoundly affects the growth of microbes. It is a major factor that may limit growth and also may exert a variety of additional effects. Phosphate limitation may induce change in the nature of the cell wall polymers of gram-positive bacteria, allowing teichuronic acid formation in place of the normal teichoic acid. Phosphate limitation, in addition, may induce proteins in gram-negative bacteria that facilitate its own transport. Siderophore formation occurs during the limitation of iron. Nutrient limitation may also induce both capsule formation and endosporulation. Finally, tactic responses are, in certain cases, a reflection of nutritional deprivation or, conversely, may reflect a physiological requirement for a critical nutrient.

Summary

Microbes exist in a tremendous diversity of environments and must adapt to a variety of environmental factors in order to survive. The precise ways in which particular microbes deal with environmental factors are a function of their genetic potential, but within genetic limits, a high degree of versatility in adaptive responses is possible. Microbes respond simultaneously to whatever conditions pertain to their particular environment. It is frequently, if not invariably, the case that a number of adaptations must be simultaneously accomplished. It is also usually the case that a particular adaption in response to a particular environmental factor affects the manner of its response to another factor. Adaptations may be metabolic, structural, or both, and often reflect significant differences in the manner of genetic expression. The diversity of microbes, and the environments that harbor them, is such that through study of microbial systems, we may learn much, not only about microbes, but about the adaptive responses of nonmicrobial forms. In the environmental realm, as in all other areas, microbes may serve as models for understanding a diversity of general biological phenomena.

Selected References

Aliabadi, Z. F., W. Mya, and J. W. Foster. 1986. Oxygen-regulated stimulons of *Salmonella typhimurium* identified by Mud (Aplac) operon fusions. *Journal of Bacteriology.* **165**: 780–786.

Booth, I. R. 1985. Regulation of cytoplasmic pH in bacteria. *Microbiological Reviews.* **49**: 359–378.

Dunlap, V. J., and L. N. Csonka. 1985. Osmotic regulation of L-proline transport in *Salmonella typhimurium. Journal of Bacteriology.* **163**: 296–304.

Ingraham, J. L., O. Maaloe, and F. C. Neidhardt. 1983. *Growth of the Bacterial Cell.* Sinauer Associates, Inc., Sunderland, Massachusetts.

Lanyi, J. K. 1974. Salt-dependent properties of proteins from extremely halophilic bacteria. *Bacteriological Reviews.* **38**: 272–290.

Moat, A. G., and J. W. Foster. 1988. *Microbial Physiology,* 2nd edition. John Wiley and Sons, New York.

Morita, R. 1975. Psychrophilic bacteria. *Bacteriological Reviews.* **39**: 144–167.

Neidhardt, F. C., J. L. Ingraham, and M. Schaechter. 1990. *Physiology of the Bacterial Cell.* Sinauer Associates, Inc., Sunderland, Massachusetts.

Neidhardt, F. C., R. A. VanBogelen, and V. Vaughn. 1984. Genetics and regulation of heat-shock proteins. *Annual Review of Genetics.* **18**: 295–329.

Neidhardt, F. C., V. Vaughn, T. A. Phillips, and P. A. Bloch. 1983. Gene protein index of *Escherichia coli* K12. *Microbiological Reviews.* **47**: 231–284.

Poolman, B. A., J. M. Driessen, and W. N. Konings. 1987. Regulation of solute transport in streptococci by external and internal pH values. *Microbiological Reviews.* **51**: 498–508.

Stanier, R. Y., J. L. Ingraham, M. L. Wheelis, and P. R. Painter. 1986. *The Microbial World,* 5th edition. Prentice-Hall, Englewood Cliffs, New Jersey.

CHAPTER 16

The Physiology of Antimicrobial Chemicals

The history of antimicrobial chemotherapy

Disease treatment was a major concern for ancient man. Long before humankind recognized the existence or nature of microbes, it was forced to deal with disease. Smallpox, typhus fever, the plague, leprosy, typhoid fever, dysentery, cholera, yellow fever, and malaria, whose presence throughout history is well documented, are only some of the microbes that by their effects have changed human history. The roots of microbiology as a discipline are grounded, to a great extent, in the phenomenon of disease. With the aid of agar, we have isolated the major bacterial agents of disease and through the skillful use of antimicrobial chemicals, we are able to control *bacterial* diseases. However, at present, most *viral* infections remain beyond our control. By diligent effort, we have apparently eliminated the threat of smallpox, but most viral infections continue to haunt us. The human immunodeficiency virus is a horrifying and civilization-threatening contemporary reality.

Our ability to control bacterial infections, and our increasing ability to deal with diseases caused by eucaryotic microbes reflects, in large measure, selectively active *antimicrobial chemicals* and our increasingly sophisticated understanding of their modes of action. In contrast, our current limited ability to treat viral infections chemotherapeutically reflects our relative lack of understanding of the manner in which antimicrobial agents may affect viruses and the subtly different ways in which such chemicals may affect the processes mediated by viruses and those accomplished by the cells that viruses invade.

Ancient man believed that disease was the result of spirit possession. Early attempts to treat disease employed ingestion of body fluids, animal excrement, and other materials designed to make the body inhospitable to the evil spirits. Over time, interest in the use of organic materials in disease treatment increased. Theophrastus (4th century B.C.) developed a compendium of medicinal plants. In the early 17th century, the bark of the chincona tree was recognized as useful for treatment of malaria.

The search for specific chemicals

At the beginning of the 20th century, an earnest search was begun for chemicals that would specifically inhibit the microbial agents of disease without, at the same time, damaging their hosts. Motivated largely by a concern with syphilis, Ehrlich discovered compound 606, the "magic bullet," now known as arsphenamine. Ehrlich delineated a critical principle, as valid today as when first proposed, that a chemotherapeutic antimicrobial chemical must be *selectively toxic*. Out of the 606 compounds he tested, only the 606th compound fulfilled this requirement. The remaining 605 compounds were therapeutically useless! During a similar time period, Fleming discovered penicillin, largely by accident. Fleming's work was expanded by Florey and Chain, allowing penicillin's development into the tremendously successful chemotherapeutic agent it remains to the present.

Additional chemotherapeutic agents were recognized in the early 20th century. Domagk identified protonsil red as an agent that selectively inhibited the growth of pathogenic streptococci and staphylococci. It was shortly thereafter shown by Jacques and Therese Trefouel, that the active component of protonsil red was its sulfanilamide moiety. The work of D. D. Woods in 1946 focused attention on the mode of action of sulfanilamide. Woods found that its action against staphylococci could be overcome by *para*-aminobenzoic acid (PABA). By comparing the structures of the sulfanilamide and PABA, he recognized a similarity between them. The structures of sulfanilamide and PABA are shown in Figure 16.1. Woods suggested that sulfanilamide inhibited growth by inhibiting PABA metabolism through competition for the active site of the enzyme required for the incorporation of PABA into folic acid. We now understand that the mode of action of all sulfa drugs results from this phenomenon. We also understand the basis for the selective toxicity of the sulfa drugs. Sulfa sensitive microbes synthesize folic acid, while sulfa insensitive organisms, including humans do not! Woods, in the course of elucidating the interrelationships between sulfanilamide and PABA, discovered the physiological basis for the selective action of all the sulfa drugs. In addition, he discovered the principle of *competitive inhibition*. This principle has since been shown to be generally ap-

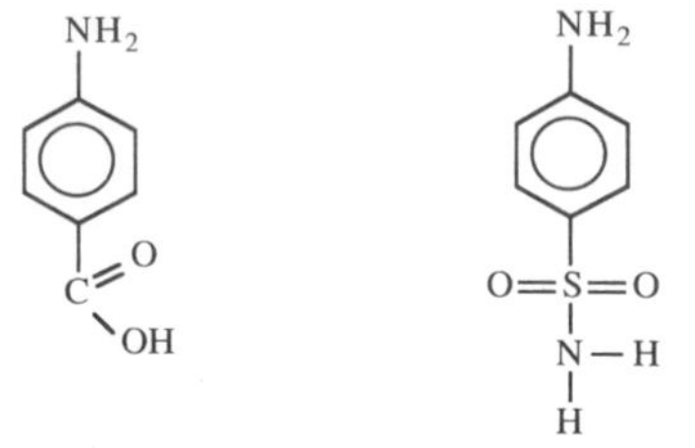

FIGURE 16.1 The structures of para-aminobenzoic acid (PABA) and sulfanilamide.

plicable to a host of enzyme reactions, many of which have no direct relevance to antimicrobial chemotherapy. In retrospect, it is difficult to decide which of Wood's several discoveries was of greatest importance.

The discovery of streptomycin and related substances by Selman Waksman was the final "major" discovery regarding antimicrobial chemotherapy in the first half of the 20th century. Streptomycin's discovery was important for several reasons. It identified a new class of naturally produced antimicrobial agents. In addition, study of streptomycin's action in comparison to the actions of sulfa and penicillin led to the recognition of important concepts. Waksman recognized the distinction between chemically synthesized antimicrobial agents and those of natural origin. In the process, he codified the definition of the term *antibiotic*, an antimicrobial chemical produced naturally by one microbe that was inhibitory to others. Waksman contrasted such substances with chemically synthesized agents.

The distinction between *broad spectrum* and *narrow spectrum* antibiotics was a second major theoretical concept developed by Waksman. In the process of studying the diversity of organisms affected by streptomycin and other substances, he recognized that some antimicrobial chemicals were active against many organisms while others were active against only a few. Those substances active against many organisms were considered broad spectrum and those that affected only a few organisms were regarded as narrow spectrum. While we continue to debate the limits of the terms "broad" and "narrow," the concepts are still valid. In a similar manner, although we may now synthesize chemical substances that were once produced naturally, the original source of most antimicrobial chemicals is natural. Although our ability to chemically synthesize materials that were originally obtained from biosynthetic metabolism has, to a certain extent, changed our understanding of what is natural and what is not, the distinction proposed by Waksman is still valid in many circumstances.

Selectivity

Intelligent and skillful use of antimicrobial chemicals requires that the action of the material used is indeed selective and that we fully understand the basis by which the selective action is possible. It makes no sense to treat a bacterial infection with an agent that selectively inhibits yeasts. It is equally inappropriate to treat a protozoal infection with an agent designed for bacteria. It is inappropriate to treat a gram-negative infection with an agent that preferentially inhibits gram-positive organisms. Finally, it is most inappropriate to employ any substance for treatment of microbially-mediated disease that simultaneously exerts a substantial effect upon the host.

The precise modes of action of representative antimicrobial agents will be discussed shortly. However, prior to such a discussion it is useful to generally consider the various ways in which antimicrobial chemicals exert their actions and the potential reasons for their selectivity. It will readily become apparent that in the midst of the myriad of agents and inhibition mechanisms currently recognized, there is a unity in their general modes of action, a unity that reflects the fact that in basic outline, life requires the accomplishment of common physiological tasks. In addition, except for the viruses, the tasks are accomplished by structures, which though different in some respects, are functionally analogous. By understanding the basis of action of antimicrobial agents we may, on the one hand, devise intelligent strategies for disease treatment and, on the other hand, more intimately understand subtle differences in the relationships between structure and function that are found throughout nature. When considered from the biological perspective, the relative importance of antimicrobial chemicals as agents for disease treatment and as agents for study of physiological processes is a question that requires reflection.

General modes of action of antimicrobial chemicals

Inhibition of enzyme function is a major way in which antimicrobial action is manifested. That this mode of action is widespread is a reflection of the fact that the critical biochemical activities of all life forms are enzyme-mediated processes. *Inhibition of nucleic acid function* is a second mechanism for selectivity that is generally applicable in principle, but is difficult to use practically. Although nucleic acids are critical to all life, their modes of formation and functions are so similar in all living systems that it is difficult, at present, to selectively inhibit the nucleic acid metabolism.

Disruption of cell wall formation or function is the third major way mode of antimicrobial action. This mode of action is more generally useful from the chemotherapeutic viewpoint, since not all organisms have walls. In addition, both structural and biochemical differences are found among cell wall-containing organisms.

Inhibition of protein synthesis is the fourth major mode of antimicrobial action. This method is widely used in chemotherapy. Although protein synthesis is essential to life, the manner in which it is accomplished in various organisms is not identical. In many cases, we may inhibit the process in one organism without, at the same time, disturbing another. *Alteration of membrane function* is the final major way to selectively inhibit microbes. Although membranes are a universal characteristic of all cellular life forms, and of some viruses, their nature is sufficiently different among living things, so that membrane function in one form of life may often be disrupted without simultaneous disruption of membrane function in another.

Agents that affect enzymes

A diversity of substances selectively affect the formation and function of enzymes. Within those agents that affect enzymes, many modes of action are known. The arsenicals, such as arsphenamine, exert their inhibitory effect in a general way by interaction with sulfhydryl groups. The physiological consequences of the latter reaction may be varied and range from inactivation of the enzymes themselves to inactivation of other sulfhydryl-containing molecules required for their action, such as lipoic acid and glutathione. The selective action of various arsenicals appears primarily to reflect differential solubility in the

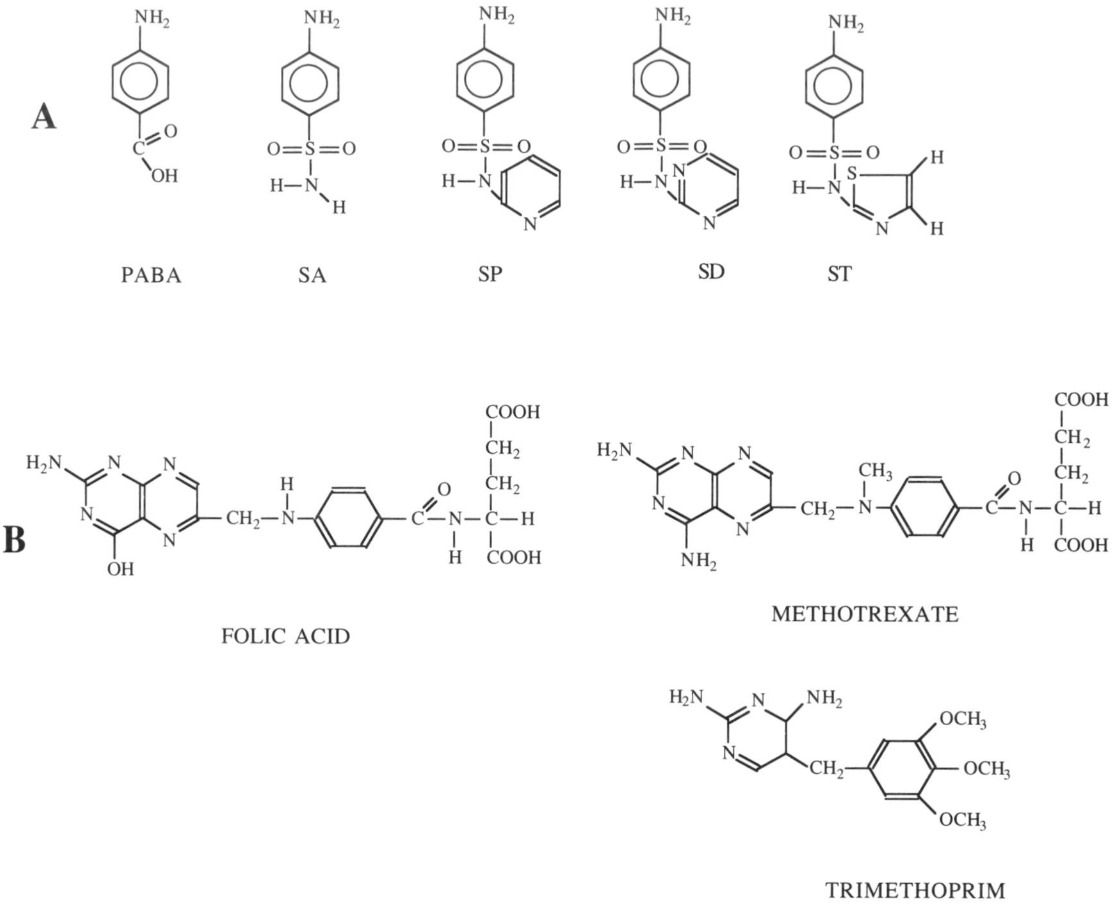

FIGURE 16.2 The formulae of para-aminobenzoic acid (PABA), folic acid, and of structurally similar substances that interfere with their metabolism. (A) sulfanilamide (SA), sulfapyridine (SP), sulfadiazine (SD), and sulfathiazole (ST) All of the above compounds are derivatives of sulfanilic acid and manifest their actions by interfering with the incorporation of PABA into a functional form of folic acid. The physiological consequence of this metabolic disruption is, primarily, interference with DNA synthesis. (B) Folic acid and selected structural analogues. The analogues interfere with the normal functioning of folic acid, particularly in its role as a methylating agent.

membranes of the organism that they affect in comparison to those of the host.

Many enzyme-affecting agents manifest their action in a manner more specific than that found for arsenicals. Competitive inhibition, as exemplified by the sulfa drugs, is a widely distributed mechanism for selective interference with critical life processes. Although it is perhaps most thoroughly documented with regard to the sulfa drugs, it is a general selective mechanism. Methotrexate and trimethoprim, folic acid antagonists, exert their action, at least in part, by this mechanism. In all situations in which competitive inhibition operates, a structural similarity exists between the normal metabolite and the inhibitory material. The struc-

tures of PABA, selected sulfa drugs, folic acid, trimethoprim, and methotrexate shown in Figure 16.2 illustrate this phenomenon.

In addition to competitive inhibition, substances structurally similar to normal metabolites may exert their actions in other ways. Formation of an altered product is such a mechanism. The incorporation of halogenated derivatives of uracil into RNA exemplifies this mechanism. As a result of incorporation of altered bases into the RNA molecule, the molecule functions improperly, or not at all. In addition, iodinated forms of uracil disrupt thymidine synthesis. In a manner analogous to that of pyrimidines, structural analogues of purines may disrupt normal purine formation and

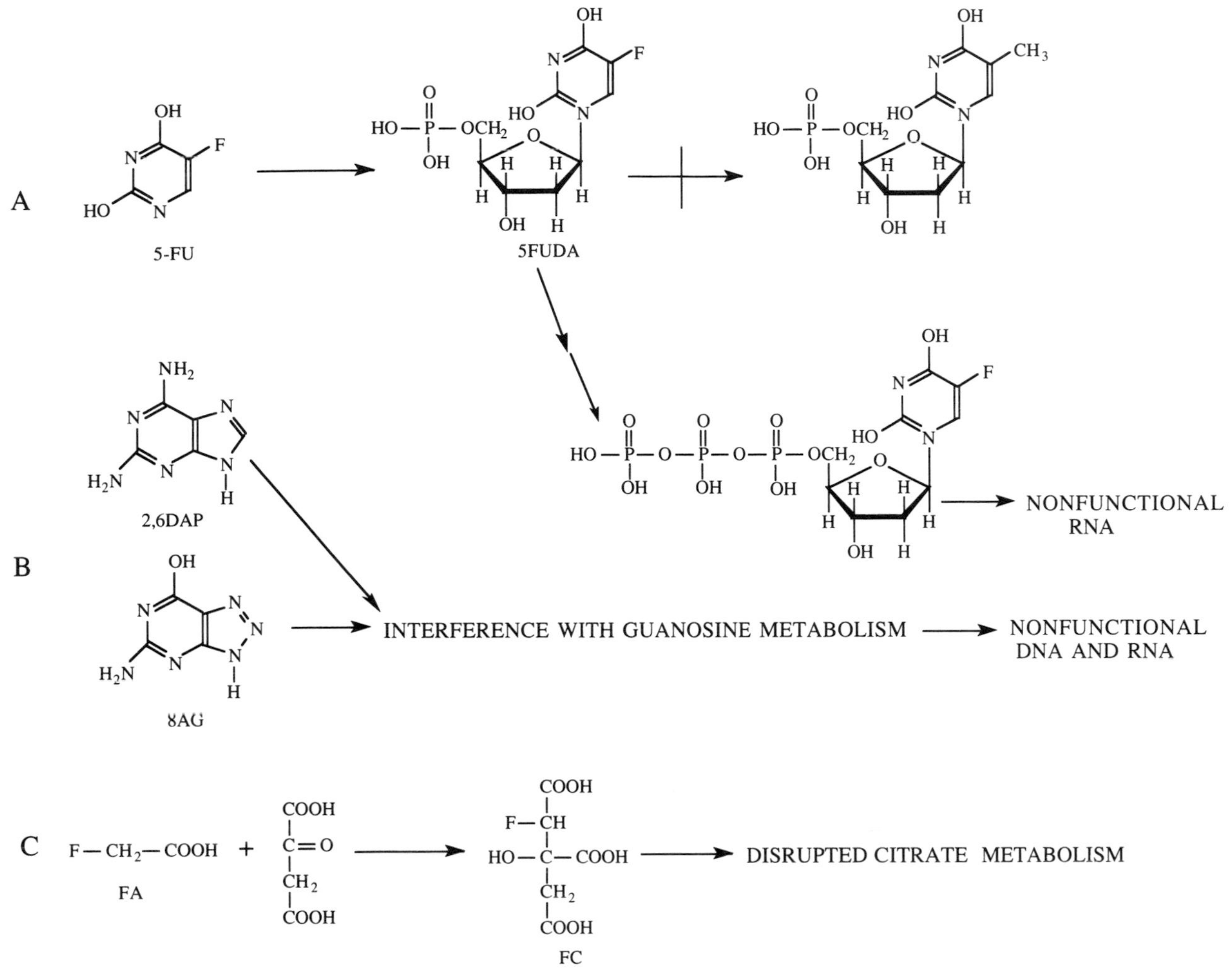

FIGURE 16.3 Modes of action, and the physiological consequences, of representative structural analogues of normal metabolites. (A) Analogues of pyrimidines. 5-fluorouracil (5FU) is converted to 5-fluorouridylic acid (5FUDA), which may inhibit by interfering with thymidine synthesis or by incorporation as a fluoro-5-phosphate into RNA, leading to production of nonfunctional forms of the latter. (B) Purine analogues. 8-azaguanine (8AG) and 2,6 diaminopurine (2,6DAP) serve as guanine antagonists and lead to the formation of nonfunctional forms of both DNA and RNA. (C) Fluoroacetate (FA) leads to formation of fluorocitrate (FC). The latter fails to serve as a substrate for aconitase and thus interferes with normal citrate metabolism.

function. For example, nonfunctional purines are formed when 2,6-diaminopurine or 8-azaguanine are substituted for normal metabolites.

The use of fluoroacetate in place of acetate for citrate synthesis is an additional example of altered product formation as a consequence of a structural analogue use. Such a procedure results in formation of fluorocitrate, a nonmetabolizable substrate for aconitase. As a result, normal operation of the TCA cycle is disrupted or abolished. Mechanisms that involve incorporation of a nonfunctional structural analogue into a product are known as *lethal syntheses*.

Structural analogues may manifest their action in additional ways. They may serve as *false repressors, false corepresssors,* or *false inducers*. In this manner, structural analogues manifest their action not only by altering protein function, but its formation as well. Thiohistidine regulates the histidine operon in such a fashion. The consequences of altered product function, false repression, false corepression, and false induction are shown in Figure 16.3. Enzyme formation and function may be altered by structural analogues in a diversity of ways.

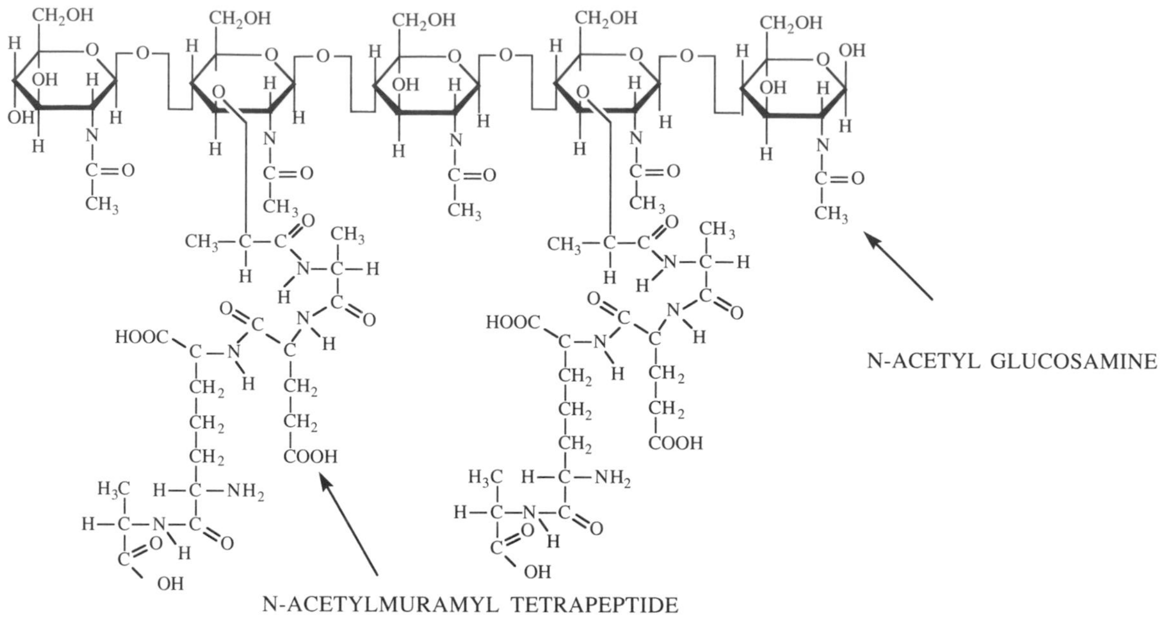

CH_2OH CH_2OH CH_2OH CH_2OH CH_2OH

N-ACETYL GLUCOSAMINE

N-ACETYLMURAMYL TETRAPEPTIDE

FIGURE 16.4 The fundamental structure of murein. The molecule is composed of alternating molecules of *N*-acetylglucosamine and *N*-acetylmuramyl tetrapeptide. Precursors to the finished product contain a pentapeptide attached to *N*-acetylmuramyl residues. The pentapeptide contains, beginning with the amino acid proximal to the carbohydrate chain, L-alanine, D-glutamic acid, meso-diaminopimelic acid and two molecules of D-alanine. During formation of the completed murein molecule, the terminal D-alanine residue is removed. Interconnections between murein strands are accomplished by peptide linkages that typically occur between the amino group of a diaminopimelic acid residue and the carboxyl function of the terminal D-alanine residue from a side chain of another strand. Lysine sometimes replaces diaminopimelic acid. Linkages between murein strands more complicated than simple peptide linkages occur in many organisms.

Wall-affecting agents

Many antimicrobial agents exert their inhibitory action on wall synthesis. Since rigid walls are found in both eucaryotic and procaryotic microbes, but are absent in animal cells, wall-inhibiting agents have widespread clinical use. The majority of wall-affecting agents are directed toward disruption of synthesis of the rigidity polymer of the wall—murein in the bacteria and chitin in the fungi. We will begin our discussion by consideration of agents that affect murein synthesis. The general structure of murein is shown in Figure 16.4. The details of murein synthesis are discussed in Chapter 3. Reflection on Figure 16.4 reveals the basic structure of murein. It is composed of a repeating polymer of *N*-acetylglucosamine (NAG) and *N*-acetylmuramic acid (NAM) that are connected by β-1,4 linkages. Murein chains are interconnected by peptide bonds between the amino acids attached by a lactic acid moiety to carbon 3 of NAM.

You will recall from Chapter 3 that murein synthesis is a compartmentalized phenomenon. Murein precursors are formed within the cytoplasm or at the cytoplasm-cytoplasmic membrane junction, transported through the membrane, and incorporated into the growing polymer at an extra-cytoplasmic membrane location. Both the structural complexity of murein and the complexity of its synthesis make the process of murein synthesis readily amenable to disruption.

It is possible to disrupt murein synthesis in a variety of ways including: 1) disruption of polymer formation; 2) disruption of formation of the carbohydrate backbone; 3) disruption of formation of the lactic acid moiety; 4) disruption of amino acid addition to the side chains; 5) interference with fine structure formation by disruption of peptide bond synthesis; and 6) disruption of the

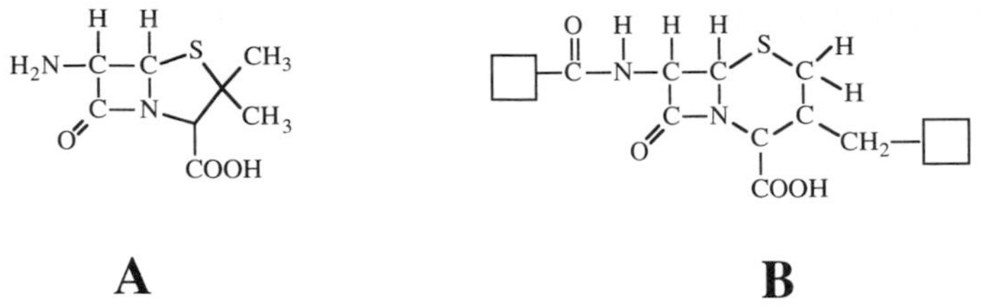

FIGURE 16.5 The general structures of penicillin (A) and cephalosporin (B) antibiotics. The various penicillins are produced by replacement of a hydrogen of the free amino group of the basic penicillin structure with a variety of organic moieties. The various cephalosporins are derived from the basic cephalosporin molecule by the addition of organic residues to the positions indicated by the boxes. Collectively, these modifications of basic molecules convey different specificities of action on particular molecules and enhance their resistance to enzymatic destruction.

function of the lipid carrier molecule, undecaprenol, required for transport of murein precursors to their addition site on the polymer. Examples of all of the above mechanisms are known. Marcarbomycin treatment of *S. aureus* interferes with formation of the carbohydrate backbone and leads to accumulation of peptidoglycan precursors. Fosphonomycin inhibits the addition of a phosphoenol pyruvate (lactyl side chain) residue to an incipient NAM molecule. Cycloserine manifests its action in two ways: it inhibits alanine racemase, the enzyme required for formation of D-alanine residues from L-alanine and also inhibits peptide bond formation between D-alanine molecules. Cycloserine primarily functions by competitive inhibition.

Penicillins and cephalosporins

Clinically, the penicillins and cephalosporins are widely used and a great variety of both compounds are available. Although many penicillins and cephalosporins are known, they may all be regarded as derivatives of two compounds, 6-aminopenicillic acid and 7-aminocephalosporanic acid. The structures of these substances are shown in Figure 16.5. The derivatives of the basic molecules result from alterations of the R groups. The primary clinical advantages of these alterations are to facilitate solubility in a selective fashion, making them differentially available to various organisms and to make the compounds less amenable to the action of the β-lactamase enzymes that may inactivate them.

In spite of structural differences, the basic mode of action of penicillins and cephalosporins is similar, the disruption of murein fine structure by inhibition of peptide linkages between adjacent murein chains. Both penicillins and cephalosporins inhibit the action of a variety of peptidase enzymes. Penicillin is apparently an analogue of the di-D-alanyl residues of the amino acid side chain.

Penicillin-binding proteins Subtle differences are found in the antimicrobial action of various penicillins. The manner of action of various penicillins has been elucidated, in large measure, by study of their selective reactivity with penicillin-binding proteins (PBPs). The latter are a collection of proteins that react, selectively and preferentially, with various penicillins. This property has allowed isolation of specific PBPs by a combination of penicillin-mediated affinity chromatography on agarose and electrophoresis. Study of the selective binding of various penicillins by particular PBPs and the effects of binding on the physiological and morphological properties of penicillin-sensitive cells has allowed understanding of the nature of the differences in action of particular penicillins, and in the functions of particular PBPs. These studies have shown that synthesis of the murein associated with cell septum formation differs from that which occurs during elongation and that various penicillin-binding proteins mediate the two processes. Septum formation is inhibited by cephalexin, which selectively binds PBP3, while mecillinam that preferentially binds PBP2, inhibits murein synthesis associated with elongation, but does not affect the one required for septation.

Additional wall-affecting agents

Many antibiotics affect procaryotic wall synthesis in ways substantially different from those displayed by the penicillins and the cephalosporins. Vancomycin, ristocetin, and similar substances inhibit murein formation by inhibition of amino acid incorporation into the side chains of N-acetyl muramic acid (NAM). Hybridization studies have shown that vancomycin specifically interacts with di-D-alanine residues at the C terminus.

Synthesis of intact murein requires that precursors synthesized within the cytoplasm or at the cytoplasm-cytoplasmic membrane junction are transported to the addition site on the growing polymer. The latter process is mediated by the 55-carbon lipid carrier undecaprenol (see Chapter 3). To be active in murein precursor transport, undecaprenol must be in a monophosphorylated form. However, during addition of a murein precursor to the growing rigidity polymer, a diphosphorylated form of undecaprenol is released. Further participation of undecaprenol in murein synthesis requires that the monophosphorylated form of undecaprenol be regenerated from the diphosphorylated form produced during addition of a building block to the growing murein polymer. Bacitracin interferes with murein synthesis by inhibiting formation of undecaprenol monophosphate from undecaprenol diphosphate, thus inhibiting the ability of undecaprenol to serve as a carrier molecule.

Fungal cell wall inhibitors

Synthesis of the rigidity polymer of the fungal cell wall is similar, in certain respects, to the analogous process in procaryotes. Fungal cell walls contain chitin, a β-1,4-linked polymer of N-acetylglucosamine. In chitin synthesis, N-acetylglucosamine is converted to a uridine diphosphate-activated form, in a process similar to that required for murein formation, and this form is transferred to a lipid carrier, dolichol phosphate, a material functionally analogous to the undecaprenol carrier involved in synthesis of the murein rigidity polymer. From the viewpoints of both building block synthesis and building block incorporation into the rigidity polymer, similarities exist in the manner of rigidity polymer synthesis in procaryotic and eucaryotic microbes. Polyoxin D inhibits chitin synthetase, the enzyme involved in building block formation and tunicamycin inhibits transfer of the building block to its lipoidal carrier molecule, dolichol phosphate.

Membrane-affecting substances

The cytoplasmic membrane is a critical structure for cells. It serves in all cases as the major agent of selective permeability and, in procaryotes, serves a variety of additional functions. The membrane can be adversely affected in many ways. Its composition may be altered, its fluidity may be changed, its organization may be destroyed, or its charge may be abolished. Any of these phenomena exert devastating effects on microbes, ranging from leakage of critical intracellular materials to the external environment, to the disruption or destruction of enzyme and transport activities, to the abolition of the ability to generate energy. In some cases, membrane disruption destroys the morphological integrity of the cell.

A variety of substances affect the membranes of microbial cells. Some are generalized in their activities, while others are specific. Detergents and phenols are generally broad in their actions. Detergents manifest their action primarily by disruption of membrane protein-lipid interactions, leading to severe membrane structure changes and leakage of small molecules from the cell interior. Additional effects of detergents include protein degradation and cell lysis by the action of autolytic enzymes. Phenols also manifest relatively nonselective action by generalized membrane disruption and enzyme inhibition. The effectiveness of phenols is frequently affected by the nature of the materials attached to the basic molecule. In general, the more lipid soluble the phenol, the more effective its antimicrobial activity. The relative solubility of phenols may be measured by study of their partitioning between water and oil.

Membrane-affecting antibiotics

A variety of antibiotics affect membrane function. The polymyxins, cyclic polypeptides produced by *Bacillus polymyxa*, preferentially affect the membranes of gram-negative bacteria. The colistins and circulins exert similar effects. Although all these substances preferentially affect gram-negative bacteria, their precise effects are imper-

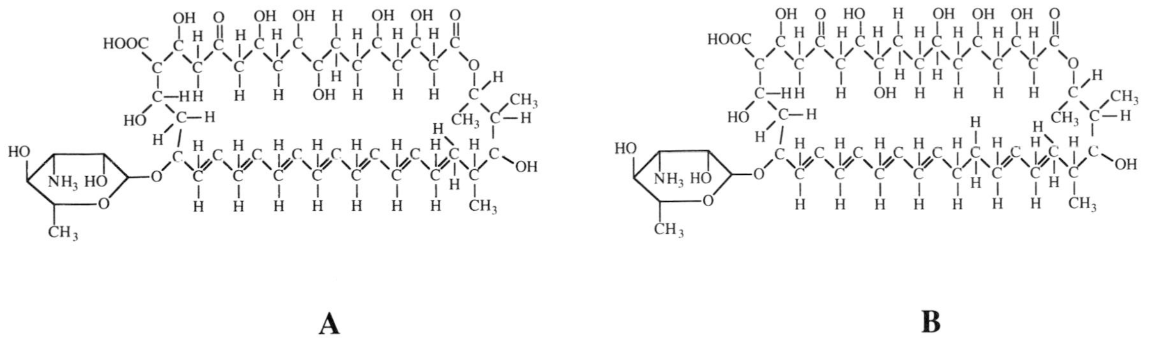

FIGURE 16.6 The structures of amphotericin B (A) and nystatin (B). The polyene antibiotics differ in the number and location of the double bonds and the number and location of various functional groups attached to the basic molecule.

fectly understood. Exposure to these compounds leads to leakage of 260 nm-absorbing material and other intracellular substances. It is believed that these phenomena reflect generalized interaction with phospholipids, particularly phosphatidyl ethanolamine, with resultant generalized membrane disruption and loss of selective permeability. The gramicidins, linear peptides that preferentially affect gram-positive bacteria, also appear to exert their effects by generalized membrane disruption.

The polyene antibiotics, exemplified by nystatin and amphotericin, derive their name from the fact that irrespective of structural differences, they contain numerous unsaturated bonds. These agents affect the membranes of eucaryotic microbes and members of the genus *Mycobacterium*, but are ineffective against the eubacteria. Generally, the polyene antibiotics contain a lactone ring connected to a linear structure that contains many double-bonded carbons and some carbons that are attached to hydroxyl or carbonyl moieties. The structures of representative polyenes, amphotericin and nystatin, are shown in Figure 16.6. The selectivity of the polyene antibiotics is a function of their interaction with various sterols in eucaryotic membranes, although the precise consequences of those interactions are imperfectly understood.

Agents affecting membrane charge

Although any agent that allows substantial membrane disruption may affect membrane charge, a number of agents specifically exert this effect. Thus, 2,4-dinitrophenol, the salicylanilides and re-

lated compounds such as tetrachlorosalicylanilide (TCS), tetramethyldipicrylamine (TMPA), and carbonyl-cyanide chlorophenylhydrazone (CCCP) manifest their action specifically by destroying the proton gradient required for electron transport-mediated ATP synthesis. Valinomycin operates in a similar manner. Other substances affect ATP synthesis in different ways. The aurovertins are inhibitors of soluble ATPase activity, oligomycin binds to the membrane component (but not the soluble component) of F_1F_0 ATPase, and dicyclohexylcarbodiimide (DCCD) binds to the proton translocase protein of the F_1F_0 ATPase complex. These examples illustrate the diversity of ways in which antimicrobial chemicals may inhibit membrane structure and function. In addition to these materials, a variety of substances are known that affect synthesis of the components of the membrane.

Protein synthesis agents

Many substances selectively affect protein synthesis. Selective protein synthesis-inhibiting substances find wide clinical use in treatment of microbially-mediated disease, primarily for two reasons: 1) protein synthesis is a critical activity for all life forms and 2) differences in the organelles involved in protein synthesis by procaryotic and eucaryotic forms allow the same process to be selectively inhibited. Although it is difficult to consider the protein synthesis process without reference to activities of both the genome and the ribosome and their interactions, this section will

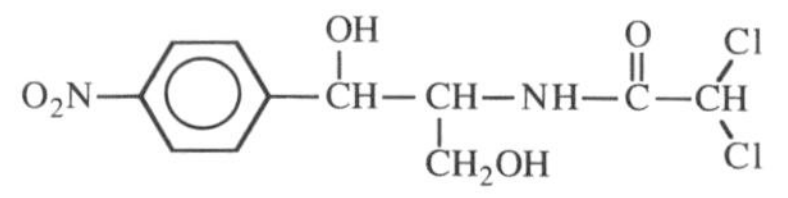

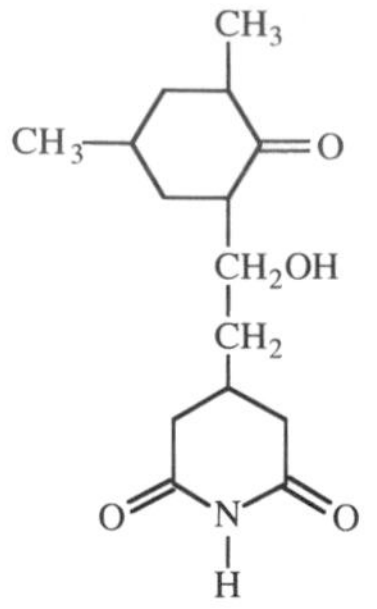

FIGURE 16.7 The structures of chloramphenicol (A) and cycloheximide (B).

focus on inhibitors of ribosome function, that is, inhibitors of translation. Agents affecting transcription and DNA replication will be considered separately.

Whether it occurs in procaryotes or eucaryotes, translation is a multifaceted process that is mediated by the ribosomal particle. This entity is, in turn, made up of subparticles that must aggregate into a functioning, complete ribosome so that the translation process may be accomplished. Any substance that impairs any aspect of normal ribosome function (see Chapter 11) will therefore drastically affect the entire translation process. The selective action of particular substances on protein synthesis reflects, to a large extent, intrinsic differences between the formation and function of 70S and 80S ribosomes and differences in the ability of particular substances to attach to 70S and 80S ribosomal subparticles. The examples discussed here are illustrative of the diversity of ways in which selectively inhibitory substances react with, and inhibit the function of, ribosomes or their subparticles.

Chloramphenicol and cycloheximide

Chloramphenicol and cycloheximide are particularly interesting inhibitors of translation. They are highly specific in their action, since chloramphenicol exclusively affects 70S ribosome-mediated protein synthesis and cycloheximide exclusively affects the 80S ribosome-mediated process. Chloramphenicol inhibits transpeptidization by binding to the 23S subunit of the 50S subparticle of the 70S ribosome. Cycloheximide is unable to bind to any component of the 50S subunit of the 70S ribosome and thus does not affect 70S-mediated translation. Similarly, as a consequence of its inability to bind

to the 80S ribosome, choramphenicol does not affect 80S ribosome-mediated translation. It does, however, inhibit translation by the 70S ribosomes in mitochondria and chloroplasts found within the eucaryotic cell. Cycloheximide manifests its effect by interaction with the intact 80S ribosome and by the prevention of the ribosome's movement along the mRNA molecule. As a result of this interaction, transpeptidization cannot occur. Although mechanistically different, the consequences of the actions of cycloheximide and chloramphenicol on the protein synthesis process are similar. The structures of these two substances are shown in Figure 16.7.

Streptomycins and tetracyclines

The streptomycins and tetracyclines are produced by members of the genus *Streptomyces* and manifest their antibacterial activity in similar, but not identical, ways. They are similar, in that both of them interact with the small ribosomal subparticle, but differ in the consequences of that interaction. Ribosomal interaction with streptomycin leads to inhibition of transpeptidization and to misreading of the mRNA code, whereas tetracycline interaction inhibits amino acid tRNA binding to the 30S ribosomal subunit. The comparative structures of streptomycin and tetracycline are shown in Figure 16.8.

The macrolide antibiotics

The macrolide antibiotics include a variety of substances that share the common structure shown in Figure 16.9. The overriding characteristics of the macrolide antibiotics are the macrolide ring structure, containing a lactone moiety with 12 to

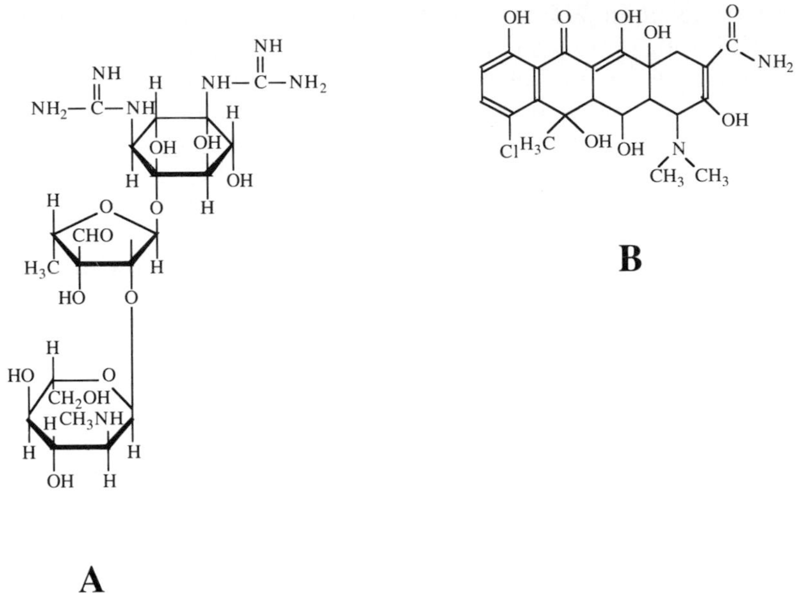

FIGURE 16.8 The structures of streptomycin (A) and tetracycline (B). Both antibiotics interact with the small subunit of the 70S ribosome and inhibit translation, but the mechanisms by which the two substances inhibit translation are different.

22 carbon atoms, but very few, if any, double bonds. Erythromycin is the most widely used member of this group. The mechanism of action of the macrolide antibiotics is the subject of considerable discussion. Although it involves interaction with the 50S subunit of the 70S ribosome, the precise consequences of that interaction are only incompletely known. It appears that macrolide antibiotics inhibit transpeptidization, and in some cases, particularly erythromycin, translocation as well.

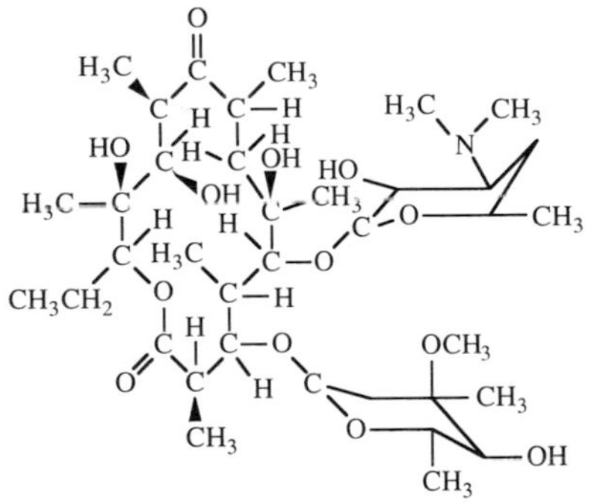

Nucleic acid inhibitors

Many inhibitors of nucleic acid synthesis and function are known, but agents of this type are used less frequently in antimicrobial chemotherapy than many other agents, since it is difficult to selectively inhibit the nucleotide metabolism of microbe and host. However, a number of mechanisms are available for interference with normal nucleic acid formation or function. Interference with normal purine and pyrimidine formation, and interconversions among purine and pyrimidine precursors, is normally manifested as alteration of the function of critical synthetic enzymes. Analogues of glutamine, for example, such as L-azaserine and 6-diazo-5-oxo-L-norleucine (DON), inhibit phosphoribosyl-

FIGURE 16.9 The general nature of the macrolide antibiotics as exemplified by erythromycin A. All macrolide antibiotics contain lactone rings attached to two sugar moieties, but the precise nature of the rings and sugars varies among the various compounds.

formylglycineamide synthetase and hadadin, an aspartic acid analogue, inhibits adenyl succinate synthetase.

Interference with nucleic acid function may be physiologically manifested in a variety of ways. Direct incorporation of nucleotide base analogues such as halogenated or 2 deoxy-derivatives may result in abnormal nucleic acid replication or in the formation of inaccurate transcripts. Rifampin and the actinomycins interfere with transcription with disastrous phenotypic consequences and they manifest these effects by inhibition of DNA-dependent RNA polymerase. The effects of rifampin are concentration independent. In contrast, the effects of actinomycins are concentration dependent. At low concentrations, actinomycins interact with single-stranded RNA molecules and inhibit transcription, while at higher concentrations, the same substances act as intercalating agents and inhibit replication of DNA.

Interference with DNA replication is the most physiologically devastating effect of antimicrobial chemicals. Physically, interference may be accomplished in one of three ways: 1) interference with strand separation; 2) false residue incorporation; and 3) interference with the action of DNA polymerases, particularly DNA polymerase III (Pol III). Intercalating agents—small, polycyclic, planar molecules—may intersperse themselves between adjacent bases in a DNA molecule without disruption of hydrogen bonds, but in a manner such that the DNA helix is distorted, allowing errors in the function of pol III. Proflavine, actinomycin D, and ethidium bromide are all highly potent intercalating agents, whose structures are shown in Figure 16.10.

In general, frame shift mutation, the changing of the code by one or possibly two nucleotides so that a different triplicate group of bases than the normal would be read, is the major genetic consequence of intercalation. In addition to intercalation, cross-link formation between opposite strands is a mechanism for inhibition of DNA function. It is believed that mitomycin and similar substances manifest antimicrobial activity in this manner. Finally, DNA gyrase inhibitors, such as nalidixic acid, are highly antimicrobial because of their inhibition of strand unwinding.

Difficulties with anti-nucleic acid-directed chemotherapy

Because of universality in the mechanism of synthesis and because of the disastrous consequences of disruption of nucleic acid function, antimicrobial chemotherapy with nucleic acid analogues is extremely difficult. In most cases, it is impracticable because, quite literally, the cure is worse than the disease. As our knowledge of microbial physiology progresses, it may become increasingly possible to use nucleic analogues for chemotherapeutic purposes. At present, nucleic acid analogue chemotherapy is primarily used in situations in which there is no other reasonable choice, for example, in the treatment of the AIDS virus and certain types of cancer, although amantidine and acyclovir are sometimes used for less serious conditions. Amantidine, which blocks the uncoating of the influenza virus, is used to treat influenza and acyclovir and idoxuridine (5-iodo-2-deoxyuridine) are employed in *Herpes* infections. Acyclovir, in a phosphorylated form, serves as a structural analogue of deoxyguanosine triphosphate and inhibits the action of DNA polymerase, while the mode of idoxuridine's action is not currently known. In addition to these substances, cytosine and adeninyl derivatives of arabinose are used as antiviral and antitumor agents. The mechanism of action of these substances is, apparently, selective inhibition of DNA polymerase, although the precise reasons for their selectivity are not yet known.

Selectivity

By whatever mechanism antimicrobial chemicals function, understanding of the reasons for the selective action of antimicrobial chemicals is a matter of both substantial interest and practical concern. It is useful, and often essential, to understand the reasons why particular agents may be used chemotherapeutically, but it is often of equal or greater importance to understand why a particular substance may *not* be used for chemotherapeutic purposes. When considering the basis for the selectivity of antimicrobial chemical action, two interrelated, but not identical, questions become important. First we may ask why a particular agent is effective against one microbe and secondly, why is it inactive against another? As we have seen throughout this chapter, the answers to these questions are as useful to our understanding of the physiological differences among microbes as they are for the application of antimicrobial chemotherapy. Study of the physiological bases for the differential actions of antimicrobial chemicals contributes most significantly to our understanding of subtle differences in the nature of extremely similar life processes and in the ways in which slightly different, but extremely similar, structures function. Understanding of the differences and, at the same

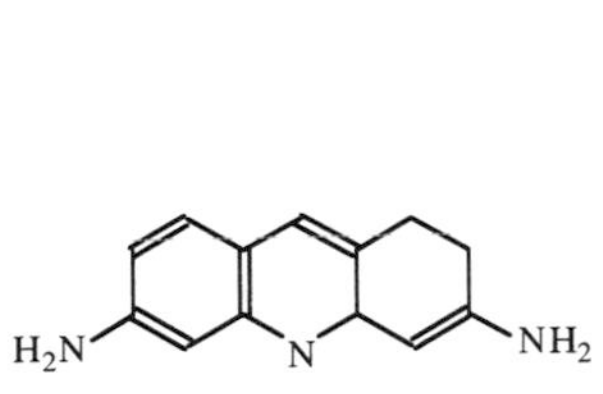

PROFLAVINE

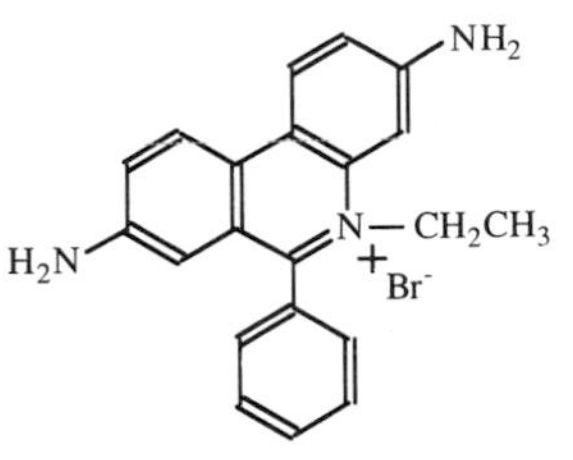

ETHIDIUM BROMIDE

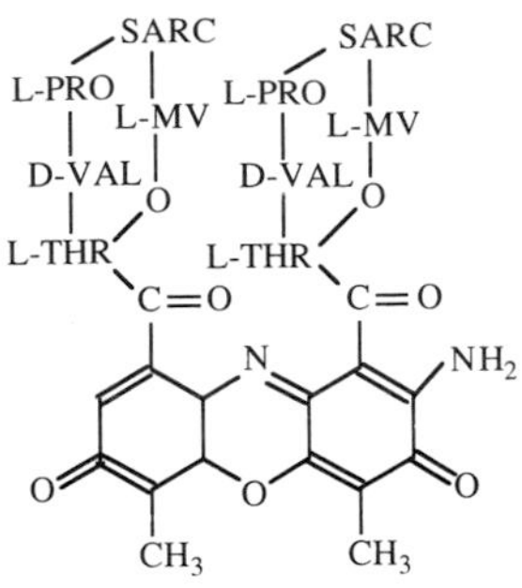

ACTINOMYCIN D

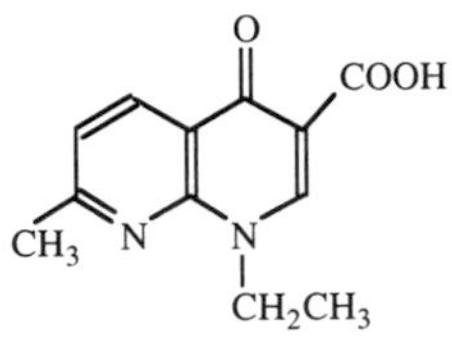

NALADIXIC ACID

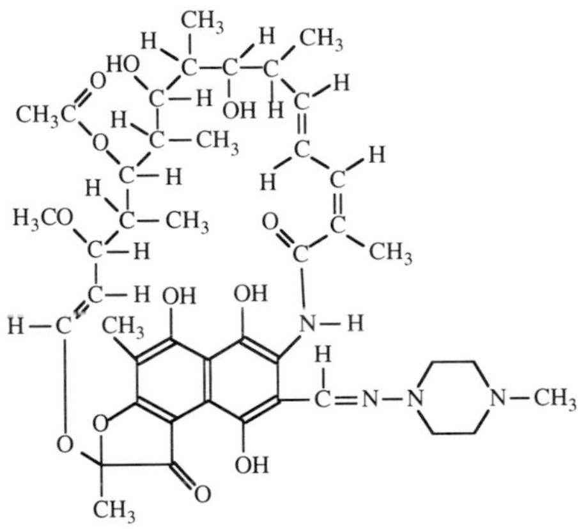

RIFAMPIN

FIGURE 16.10 The structures of agents that interfere with DNA and RNA formation and function. The various modes of action of these and other substances are discussed in the text. The components of the nitrogenous ring of actinomycin D are L-threonine (L-THR), D-valine (D-VAL), L-proline (L-PRO), sarcosine (SARC), and L-*N*-methylvaline (LMV).

time, the similarities between the functioning of similar, yet different, life forms helps us understand, simultaneously, the unity and diversity of cellular life. In addition, the areas about which we require more information, such as the chemotherapy of viral diseases and cancer are brought more clearly into focus.

Resistance

The discovery of antibiotics and other antimicrobial chemicals, and studies of their modes of action, has allowed us to control a great variety of diseases. It is truly the case that when skillfully and carefully used, antibiotics and other antimicrobial

drugs are "wonders." We must, however, use chemotherapeutic antimicrobial chemicals with skill and caution so that their selectivity is maintained. At the present time, largely because of human carelessness and the abuse of previously effective substances, diseases such as tuberculosis that we once considered conquered, are resurging. Resistance to antimicrobial agents is a matter of constant concern, not only from the viewpoint of human health but also for veterinary medicine and agriculture.

Resistance, in the general sense of the word, is both a "boon and a bane." Innate resistance to the effects of particular substances is the basis for all of chemotherapy. The subtle differences, at the cellular level, between host and microbe are precisely the factors that allow us to employ chemotherapy. Problems with antimicrobial chemotherapy arise when one of two situations occur. On the one hand, as is currently the case with viruses and cancer, we may not be able to be sufficiently selective in our use of chemicals because of the host's lack of "resistance" to substances that inhibit the microbe. On the other hand, in the more traditional situation, a microbe may become resistant to the agent we are using to selectively inhibit it. Both situations are difficult, but for converse reasons.

Mechanisms of resistance to antimicrobial chemicals

Microbes display a diversity of physiological mechanisms by which they may become resistant to antimicrobial chemicals. They may become impermeable to the material. This mode of response involves a change in the nature of the cytoplasmic membrane so that the organism is no longer susceptible to the inhibitory material. Alternatively, the organism may proliferate an enzyme, such as a β-lactamase that destroys the antimicrobial substance. Change in the nature of the intracellular target is yet another mechanism of resistance. In addition to these mechanisms, an organism may alter an existing metabolic sequence, or develop an alternate one, so that the once sensitive process is no longer affected by the antimicrobial agent. All of the above responses involve metabolic and physiological adaptations upon the part of the organism and are reflections of metabolic changes, genetic changes, or both. We may, therefore, use study of resistance to antimicrobial chemicals as a vehicle for more complete understanding of both physiological change and change in genetic expression. Studies of penicillin resistance have, for example, provided a wealth of understanding regarding the nature of enzyme induction. Even multiple drug resistance is helpful at the same time that it is a difficulty. Although the phenomenon has caused substantial difficulty in disease treatment, it has stimulated study of subjects as profound as the manner of plasmid replication, the nature of conjugation, and the nature of the symbiotic relationship between plasmid and host.

Synthetic antimicrobial substances

Although most antimicrobial agents are originally derived from natural synthesis, we are increasingly able to synthesize them in the laboratory. In addition, we are able to synthesize modified forms of originally natural materials to enhance their selectivity, effectiveness, and resistance to destruction. The development of new chemotherapeutic agents is a challenging and complex endeavor that requires the simultaneous use of a variety of technologies. The subtlety with which we may devise new agents is a direct reflection of the state of our understanding, both of their modes of action and of their differential action on host and pathogen. As our knowledge of microbial physiology, in relation to the action of potential antimicrobial agents, increases, our strategies for development of new substances will become ever more sophisticated and it is likely that we will, some day, through the use of selectively active chemicals, become adept at controlling diseases, which at present, elude us. It is virtually certain that the development of new substances will involve the combined use of techniques as diverse as organic and biochemistry, computer modelling, molecular biology, and biotechnology.

Summary

From primitive beginnings, humankind has advanced to the point at which, by selective use of chemicals, we may control a variety of microbially mediated diseases. Our ability to chemically control microbial disease agents requires that we understand, in great detail, the nature of the processes in microbe and host so that we may devise a strategy for selective inhibition of a process that occurs in the microbe but not in the host. Alternatively, and more frequently, we may inhibit a process that does occur in both microbe and host, but does not occur identically in both. Through study of the precise modes of action of antimicrobial chemicals on critical life processes, we may learn of subtle differences in the manner in which they occur in different life forms and, at the same time, the fundamental unity of life.

Selected References

Bycroft, B. W. 1987. *Dictionary of Antibiotics and Related Substances.* Chapman and Hall, New York.

Franklin, T. J., and G. A. Snow. 1981. *Biochemistry of Antimicrobial Action.* Methuen, New York.

Hahn, F. E. 1979. *The Mechanism of Action of Anti Bacterial Agents.* Springer-Verlag, New York.

Hahn, F. E. 1983. *The Modes of Action of Microbial Growth Inhibitors.* Springer-Verlag, New York.

Kucers, A. 1987. *The Use of Antibiotics: A Comprehensive Review With Clinical Emphasis.* Heinemann Medical Press, New York.

Lamber, H. P., and F. W. O'Grady. 1992. *Antibiotic and Chemotherapy.* Churchill Livingston, New York.

Lancini, G. 1982. *Antibiotics, An Integrated View.* Springer-Verlag, New York.

Vandamme, E. J. 1984. *Biotechnology of Industrial Antibiotics.* Marcel Dekker, New York.

Vinins, L. C. 1983. *Biochemistry and Genetic Regulation of Commercially Important Antibiotics.* Addison-Wesley, Reading, Massachusetts.

Weinstein, M. J., and G. H. Wagman. 1978. *Antibiotics, Isolation, Separation and Purification.* American Elsevier, New York.

CHAPTER 17

The Autotrophs

The diversity of the autotrophs

Autotrophic microbes display a tremendous degree of both metabolic and physiological diversity. Within those organisms that exhibit the autotrophic life-style, we find those that obtain energy from chemotrophic or phototrophic processes. Within the *chemolithotrophs*, organisms that obtain energy and reducing power from the oxidation of inorganic materials, we find organisms that oxidize substances as diverse as the reduced forms of sulfur, various forms of nitrogen, the ferrous ion, carbon monoxide, and hydrogen gas. An equally diverse spectrum of organisms is found within the *phototrophic* microbes. Although phototrophs share the common property of obtaining biologically useful energy from light, a diversity of photosystems are found within this group, a diversity that reflects the various environments within which phototrophs are found.

In addition to metabolic and physiological diversity, autotrophic microbes are both morphologically and taxonomically diverse. They are found as rods, cocci, or filaments; they may be motile or nonmotile; they may grow in fresh or salt water; and they may grow over a temperature range from relatively cold temperatures to temperatures that approach or exceed the boiling point of water. Within phototrophic autotrophs, we find members of the cyanobacteria, the eubacteria, and the archaebacteria. Finally, we find organisms for which the autotrophic life-style is obligatory as well as those for whom it is an option but who may grow, under certain conditions, heterotrophically. Such organisms are known as *mixotrophs.*

The common properties of autotrophs

Amidst the diversity found within the autotrophs, there is a high degree of unity. During growth in the autotrophic mode, autotrophs, wherever they are found, must accomplish the same common tasks of life that heterotrophs must also do. They must generate energy, form organic carbon compounds, assimilate nitrogen and sulfur, and synthesize their essential components. They normally accomplish these tasks in a nutritionally sparse environment that includes CO_2 as the sole carbon source, plus a few minerals. The simplicity of their nutrition is accompanied by a high degree of both metabolic and physiological complexity, but in the midst of diversity and complexity, the processes that the autotrophs must accomplish are achieved, both in principle and often in detail, in similar, or identical, ways. Although all of the autotrophs must fix carbon, the mechanisms by which they do so are few. Although differences exist in details, electron transport processes are normally their major mode of energy generation. In addition, similarities are often found in the manner in which the autotrophs assimilate nitrogen and sulfur.

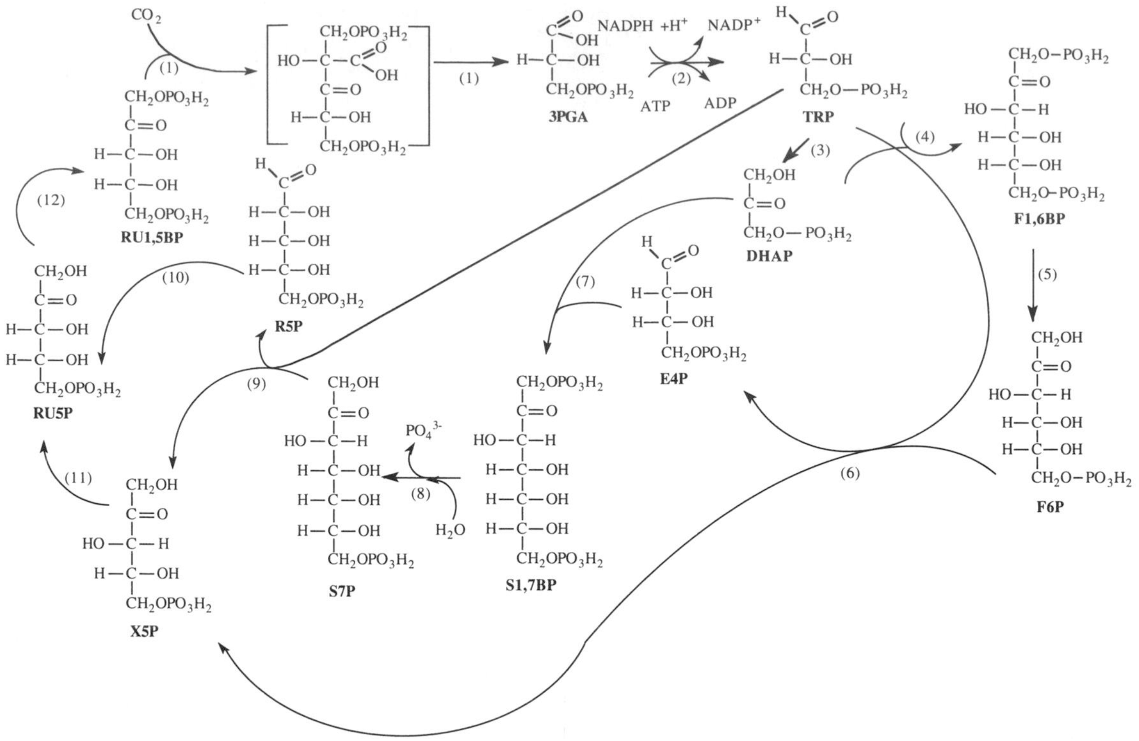

FIGURE 17.1 The Calvin cycle for carbon dioxide fixation. (1) The formation of 3-phosphoglyceric acid (3PGA) from ribulose-1,5-bisphosphate (RU1,5BP) and carbon dioxide by RU1,5BP carboxylase. (2) reduction of 3PGA to triose phosphate (TRP) by TRP dehydrogenase with NADPH and ATP expenditure. (3) Conversion of TRP to dihydroxyacetone phosphate (DHAP) by TRP isomerase. (4) Formation of fructose 1,6-bisphosphate (F1,6BP) from TRP and DHAP by F1,6BP aldolase. (5) Dephosphorylation of F,16BP to form fructose-6-phosphate (F6P) by F1,6BPase. (6) Transketolase-mediated formation of erythrose-4-phosphate (E4P) and xylulose-5-phosphate (X5P) by reaction of F6P and TRP. (7) Conversion of DHAP and E4P to sedoheptulose-1,7-bisphosphate (S1,7BP) by S1,7BP aldolase. (8) Formation of sedoheptulose-7-phosphate (S7P) from S1,7BP by S1,7BPase. In some systems it appears that the dephosphorylation of S1,7BP and F1,6BP is accomplished by a single, multifunctional enzyme. (9) Transketolase-mediated formation of ribose-5-phosphate (R5P) and X5P by the interaction of S7P and TRP. (10) Isomerization of R5 to ribulose-5-phosphate (RU5P) by phosphopentose isomerase. (11) Epimerization of X5P to RU5P by phosphopentose isomerase. (12) Regeneration of RU1,5BP from RU5P by phosphoribulokinase with ATP expenditure.

Modes of carbon dioxide fixation

Fixation of CO_2 is an obligate requirement for autotrophy. Within autotrophic organisms, at least three modes of carbon dioxide fixation are currently known, the Calvin cycle, the reductive tricarboxylic acid (TCA) cycle, and the acetyl CoA pathway. For a long time, the Calvin cycle was believed to be the sole means of CO_2 fixation in nature. Recently, additional modes of carbon dioxide fixation have been discovered. The *reductive TCA cycle*, first proposed by Evans, is known in *Chloro-bium limicola, Hydrogenobacter thermophilus*, and certain members of the sulfate-reducing bacteria and the *acetyl CoA pathway* is widely distributed among the anaerobes. It is useful to compare the nature of these sequences.

The Calvin cycle

The Calvin cycle is found in aerobic and facultative eubacteria and is shown in Figure 17.1. The pathway displays similarity to the hexose monophosphate shunt (HMS) pathways of heterotrophs, but also has unique enzymes that allow its detection. The key enzymes of the sequence are phos-

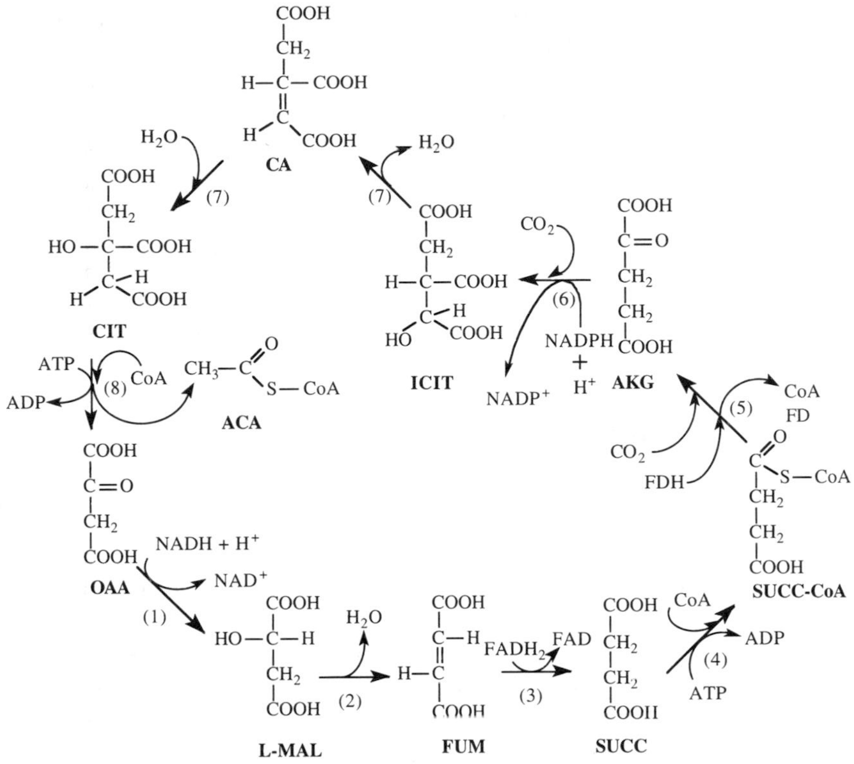

FIGURE 17.2 The reductive TCA cycle. (1) Conversion of oxaloacetate (OAA) to L-malate (L-MAL) by malic dehydrogenase with NADH as reductant. (2) Dehydration of L-MAL to fumarate (FUM) by fumarase. (3) Reduction of FUM to succinate (SUCC) with $FADH_2$ as reductant, by the action FUM reductase. (4) Conversion of SUCC to succinyl CoA (SUCC-CoA) by SUCC-CoA synthetase, an ATP requiring enzyme. (5) Fixation of CO_2 into SUCC-CoA to form α-ketoglutarate (AKG) by the action of AKG synthetase. Reduced ferredoxin (FDH) overcomes a thermodynamic barrier in this reaction. (6) Fixation of CO_2 into AKG to form isocitrate (ICIT) by ICIT dehydrogenase with NADPH as reductant. (7) Formation of citric acid (CIT) from ICIT, with the intermediate formation of *cis*-aconitate (CA) by aconitase. (8) Cleavage of CIT to form acetyl CoA, by an ATP-dependent CIT lyase. The action of this enzyme also regenerates OAA.

phoribulokinase (PRK), ribulose-1, 5-bisphosphate carboxylase (RUBisCO), sedoheptulose-1,7-bisphosphatase (SBP), fructose-1, 6-bisphosphatase (FBP), and sedoheptulose-1, 7-aldolase (SA). The remaining enzymes of the cycle are similar or identical to those of the Embden Meyerhof (EM) and HMS pathways (see Chapter 7). Operation of the Calvin cycle allows formation of many of the central intermediates for anabolic processes, but the remaining critical intermediates are derived from further metabolism. *Prima facie* evidence for the presence of the Calvin cycle in unknown organisms is demonstration of the action of its critical enzymes and rapid detection of radioactivity from $^{14}CO_2$ in 3-phosphoglycerate, ribulose-5-phosphate, and other sugar metabolism intermediates.

The reductive TCA cycle

Failure to detect activity of Calvin cycle enzymes or to demonstrate $^{14}CO_2$ radioactivity in sugar intermediates led to discovery of the reductive TCA cycle in *Chlorobium limicola*. Subsequent studies have shown its presence in other organisms, but determination of the ubiquity of the cycle in nature awaits further study. The reductive TCA cycle is shown in Figure 17.2. The key enzymes of the cycle are α-ketoglutarate (AKG) synthetase, a ferredoxin-requiring enzyme that allows synthesis of AKG from CO_2 and succinyl CoA, and an ATP-dependent isocitrate lyase that allows acetyl CoA and oxaloacetate formation from isocitrate, with ATP expenditure. The ferredoxin-dependent AKG synthetase overcomes a thermodynamic barrier

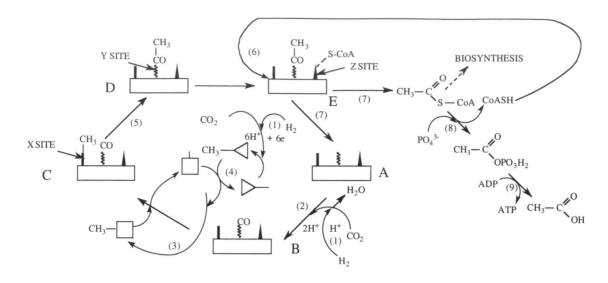

FIGURE 17.3 The acetyl-CoA pathway of autotrophy. The rectangle represents the enzyme carbon monoxide dehydrogenase that has three active centers: the X site that binds to a methyl residue, the Y site that binds to carbon monoxide, and the Z site that binds to coenzyme A. Various states of the enzyme are denoted by the letters A–E. Separate carbon dioxide molecules are reduced to carbon monoxide and methyl residues, with hydrogenase activity (1) as the source of reducing power. A molecule of carbon monoxide reacts with a free enzyme molecule at the Y site (A) to produce enzyme state B (2). The CO-bound enzyme then reacts with a methyl-bound corrinoid enzyme, denoted by the methyl-bound square, to produce a molecule of carbon monoxide dehydrogenase bound to a CO molecule at the Y site and a methyl residue at the X site (C). This process is mediated by methyl transferase (3) and results in a free molecule of corrinoid enzyme. The latter is remethylated by reaction with a molecule of methyl-bound tetrahydrofolate (THF), denoted by the methylated triangle (4). Methyl-bound THF is formed by CO_2 reduction, with formate as an intermediate. CO dehydrogenase enzyme state D is produced from state C by transfer of the methyl group to the bound CO residue at site Y (5). The resultant of this process reacts with coenzyme A (CoA) via carbon monoxide disulfide reductase (6) to yield enzyme state E, with CoA bound to the Z site and the methyl-CO moiety bound to site Y. Transfer of the 2-carbon moiety to CoA produces acetyl-CoA and generates free CO dehydrogenase (7). The acetyl-CoA molecule may be used for biosynthesis or converted to free acetate and ATP with the intermediate formation of acetyl phosphate (8,9). The reduced CoA produced is used as the substrate for reaction 6.

that would prevent net AKG formation if the normal TCA cycle enzyme, AKG dehydrogenase, were present. Formation of central intermediates other than those produced by the reductive TCA cycle results from further metabolism, for example, carboxylation, of acetyl CoA.

Under certain circumstances, the reductive TCA cycle may operate in both directions. This situation may be possible when the reductase enzyme used for succinate production from fumarate is membrane-bound and hydrogen or formate is the electron donor. Under these conditions, an expenditure of only 1 ATP molecule is required for succinate production, allowing the reductive TCA cycle to function in both directions, that is, for CO_2 fixation during autotrophic growth and for acetate oxidation during growth under heterotrophic con-

ditions. When the fumarate reductase enzyme is soluble, the cycle appears to function in a single direction only.

The acetyl-CoA pathway

It is becoming increasingly clear that the acetyl-CoA pathway is the major CO_2-fixing system in chemolithotrophic anaerobes. The system has been found in the majority of sulfate-reducing and methanogenic organisms thus far studied. The pathway is shown in Figure 17.3. Operation of this sequence depends upon the multifunctional enzyme carbon monoxide (COD) dehydrogenase. Demonstration of the action of this enzyme, as well as rapid labelling from $^{14}CO_2$ in acetyl-CoA, is compelling evidence for the presence of the sequence.

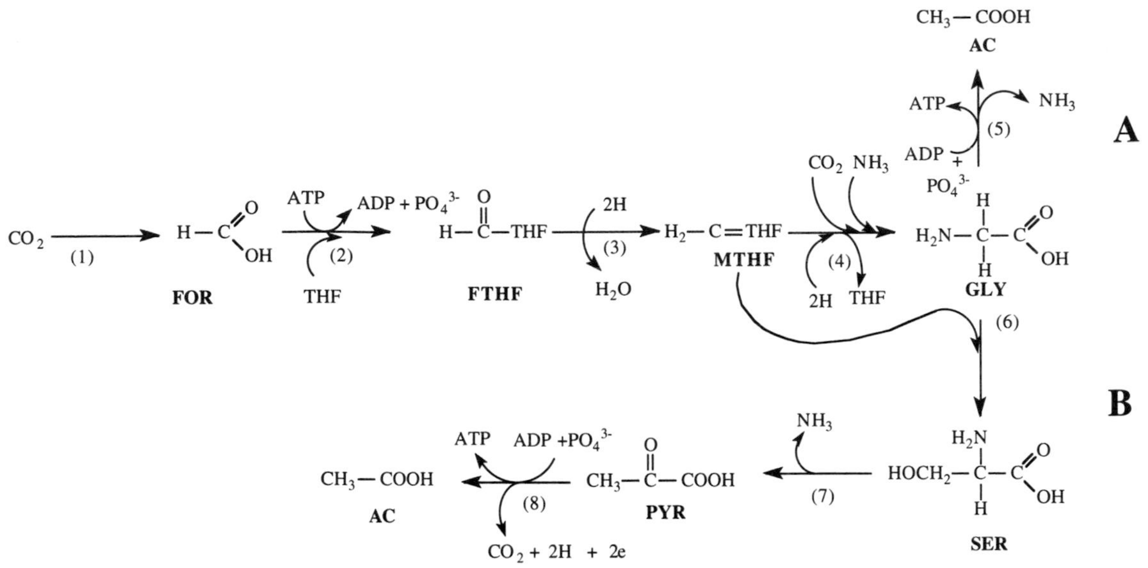

CH₃—COOH → CH_3-COOH
AC

$CO_2 \xrightarrow{(1)}$ H—C(=O)(OH) FOR $\xrightarrow[(2)]{ATP \quad ADP + PO_4^{3-}}$ H—C(=O)—THF FTHF $\xrightarrow[(3)]{2H \quad H_2O}$ H₂—C=THF MTHF $\xrightarrow[(4)]{CO_2 \; NH_3 \quad 2H \; THF}$ H₂N—CH₂—C(=O)(OH) GLY

(5) ATP ← AC (CH₃—COOH) + NH₃, ADP + PO₄³⁻   A

(6) → SER   B

CH_3-COOH AC $\xleftarrow[(8)]{ATP \quad ADP +PO_4^{3-}}$ CH₃—C(=O)—COOH PYR $\xleftarrow[(7)]{NH_3}$ HOCH₂—CH(NH₂)—C(=O)(OH) SER

CO₂ + 2H + 2e

FIGURE 17.4 The glycine synthetase pathway for acetate synthesis from carbon dioxide in *Clostridum acidiurici* and similar organisms. Although not shown in the diagram, all of the reactions except for glycine reductase are reversible. In the presence of selenium, branch A occurs, while branch B is operative in its absence. (1) Conversion of carbon dioxide to formate (FOR) by formate dehydrogenase. (2) Synthesis of formyl tetrahydrofolate (FTHF) from FOR and tetrahydrofolate (THF) by formyl tetrahydrofolate synthetase. (3) Conversion of FTHF to 5,10 methenyl tetrahydrofolate (MTHF) by methenyl-tetrahydrofolate cyclohydrolase. (4) Formation of glycine (GLY) from MTHF, CO_2 and NH_3 by glycine synthetase. (5) Formation of acetate (AC), NH_3 and ATP from glycine by glycine reductase. (6) In the absence of selenium, conversion of GLY to serine (SER) by SER hydroxymethylase. (7) Formation of pyruvate (PYR) and NH_3 from SER by SER dehydratase. (8) Oxidation of PYR to AC and CO_2 by PYR oxidase with ATP formation.

The initial reactions of the acetyl-CoA pathway involve reduction of two carbon dioxide molecules by the operation of COD dehydrogenase and formate dehydrogenase. COD dehydrogenase activity provides the carboxyl moiety of the incipient acetyl-CoA, while formate dehydrogenase activity initiates formation of the methyl carbon. The formate formed from carbon dioxide reduction is further reduced with the aid of formyl-pterin synthetase to methenyl-H₄-pterin, with the removal of water. Hydrogenation of methenyl-H₄-pterin by the action of methenyl-H4 pterin reductase yields methyl-H₄-pterin. This binds to COD in conjunction with acetyl CoA synthetase and methyl transferase, yielding, in the homoacetogens, acetyl CoA. In homoacetogens, a portion of the acetyl-CoA may give rise to free acetate and ATP, and the remaining acetyl-CoA may be used for synthesis of intermediates for anabolism by fur-

ther carboxylation reactions. Carbon assimilation and the formation of anabolic intermediates in methanogens occurs by a similar set of reactions.

The glycine synthetase pathway

Although it has not yet been shown to be used by autotrophic organisms, an additional carbon dioxide fixing system that allows total synthesis of acetate from CO_2, has been found in *Clostridium acidiurici, Clostridium cylindrosporum* and *Diplococcus (Streptococcus) glycinophilus*. Glycine may arise from this mechanism in the last organism. The pathway is shown in Figure 17.4. A molecule of carbon dioxide is reduced to formate by the action of formate dehydrogenase and the formate is further converted by reactions analogous to those of the acetyl-CoA pathway to methyl-tetrahydofolate. This last substance combines with CO_2 and NH_3, under the influence of glycine synthetase (decarboxylase), to form glycine. Methylation of glycine, with serine hydroxymethylase, can yield serine, and the action of glycine reductase on glycine may yield acetate. It is possible that as work progresses, a system of this nature will be found within the autotrophs.

Obligate and facultative autotrophy

Although the capacity for autotrophy is widespread among microbes, relatively few organisms are restricted to it. Many organisms capable of autotrophic growth may grow heterotrophically under certain conditions, and substantial amounts of organic carbon may be incorporated by organisms growing under conditions in which CO_2 is the major source of carbon. On the other hand, obligately autotrophic growth is a characteristic of the green phototrophic bacteria and, apparently, the nitrifying organisms. The finding of organisms restricted to the autotrophic life-style makes us ponder the causes that may result in obligate autotrophy. In spite of extensive study, definitive reasons for obligate autotrophy remain elusive. Currently popular suggestions include toxicity of organic substances, inability to use organic substances for energy, and the absence of an NADH-dependent AKG dehydrogenase.

Energy generation

A variety of mechanisms are known for energy generation among the autotrophs. Phototrophic forms generate energy by mechanisms involving the PMF generated by electron transport, considered in Chapter 9. Most of the Fe^{2+}-oxidizing, NH_3 and NO_2^- oxidizing, sulfur oxidizing, carbon monoxide oxidizing, sulfate reducing, and methanogenic bacteria also obtain energy from the action of PMF-driven ATPase enzymes. With the exception of certain iron bacteria, the PMF arises from electron transport. Instability of the Fe^{2+} ion at neutral pH values prevents iron oxidation under

these conditions. However, the obligately acidophilic iron bacteria appear capable of iron oxidation at a pH near 4.0. At this external pH, the intracellular pH approximates 6.0, allowing ATP formation by an artificial ion gradient. The physiological function of Fe^{2+} oxidation by organisms at pH 4.0 appears associated with reduction of protons entering the cell during ATP synthesis and with formation of reduced pyridine nucleotides for anabolism.

Chemolithotrophic electron transport systems

A diversity of electron transport systems is found among chemolithoautotrophic microbes, but substantial differences are found in both the nature of these systems and their consequences. In addition to providing the PMF and ATP, it is often, but not invariably, the case that chemolithotrophic electron transport results in formation of the reduced NAD(P) required for synthetic processes. However, in many cases, although electron transport allows PMF and ATP formation, NAD(P)H is not formed. The inability to form anabolic reducing power as a function of electron transport reflects, primarily, two factors: 1) the relative oxidation potentials of the oxidized material and NAD(P) and 2) the structure of the electron transport system. It is useful to compare the physiology of electron transport in chemolithoautotrophic organisms that oxidize different materials.

Electron transport in the aerobic chemolithotrophic hydrogen-oxidizing bacteria

Hydrogen utilization in electron transport systems requires the presence of hydrogenase enzymes, which are, throughout nature, Ni-containing, structurally similar enzymes that may be one of two types, membrane-bound or soluble. Membrane-bound hydrogenase enzymes are incapable of direct NAD reduction, while the soluble enzymes readily form NADH. To date, among the chemolithotrophic hydrogen utilizers, only *Alcaligenes eutrophus* has been shown to contain both the soluble and membrane-bound enzymes, although *Nocardia autotrophica* and *Pseudomonas denitrificans* contain a soluble hydrogenase. Most other aerobic hydrogen utilizers form NAD(P)H by reverse electron transport in which a portion of the PMF generated during electron transport is used to form anabolic reducing equivalents. In the aerobic hydrogen bacteria, electrons derived from hydrogen oxidation are channeled, normally through the iron-sulfur protein ferredoxin, into "normal" electron transport chains and are ultimately used for O_2 reduction to water.

The sulfate-reducing hydrogen bacteria

Hydrogen utilization by the methanogens and the sulfate-reducing bacteria is similar in principle to that of aerobic forms, although the carriers in electron transport and the hydrogenase enzymes differ from those found in aerobic hydrogen oxidizers. In the sulfate-reducing bacteria substantial evidence indicates that sulfate reduction is coupled to ATP formation by a PMF derived from electron transport. Sulfate is reduced, in a stepwise fashion, to H_2S by intracellular enzymes, necessitating transport of sulfate from the external environment. This is apparently accomplished by active transport, which may involve ATP expenditure. Once inside the cell, sulfate is reduced to sulfite and H_2S by a collection of enzymes. Initially, sulfate is reduced to sulfite by the sequential actions of the enzymes ATP sulfurylase and adenosine phosphosulfate (APS) reductase. ATP sulfurylase converts sulfate to APS. The sulfate moiety of APS is then reduced to sulfite by APS reductase. Sulfide is formed from the action of sulfite reductase, an enzyme whose action releases sufficient energy for formation of 2 to 3 ATP moles per mole of sulfite reduced. Hydrogen serves as the source of electrons for sulfite reduction by apparently "normal" electron transport chains, which include one or more quinones, low electron carriers such as ferredoxin, rubredoxin, and flavodoxin, and cytochromes. Substantial diversity appears to exist among the hydrogen-utilizing, sulfate-reducing bacteria in the precise nature of their electron chains and in the physiological connections between hydrogen oxidation, sulfate reduction, and ATP formation.

The precise mechanism of ATP formation in sulfate-reducing bacteria remains to be elucidated, although both ATP formation and sulfate reduction to sulfite can be abolished by both protonophores and dicyclohexylcarbodimide (DCCD), a substance known to uncouple proton translocating ATPases. An amount of ATP is formed by proton gradient to allow not only sufficient energy for sulfate transport, but for growth as well.

Methanogenic hydrogen utilization

The methanogenic bacteria may use hydrogen in electron transport-facilitated energy generation. Energy generation and methanogenesis are intimately interconnected processes in these organisms. The scheme for carbon dioxide reduction to methane is shown in Figure 17.5 and displays substantial similarity to the mechanism of acetyl CoA formation by the organisms. Methane formation from hydrogen by methanogens occurs through a series of reductive processes that involve unique electron carriers whose structures are shown in Figure 17.6. The reductive steps are associated with development of a proton gradient that may be abolished by protonophores. Furthermore, ATP formation is eliminated, or greatly reduced, by DCCD exposure. In addition, an F_1F_0 ATPase is found in several methanogenic bacteria. Collectively, these results indicate that electron transport mediates most, if not all, of the ATP formed by methanogenic bacteria. Determination of precisely how PMF formation is accomplished is of substantial interest, but is complicated by the presence of a highly active Na^+/H^+ antiporting system in the organisms. The presence of this system makes it difficult to determine if a proton gradient is formed during hydrogen oxidation. Studies with *Methanosarcina barkeri*, an organism that forms methane from methanol and H_2, in a process that does not require Na^+, are instructional. When cells of this organism were incubated under hydrogen in the presence of a very low sodium concentration and a high K^+ concentration, generation of a proton gradient was possible. The proton gradient was then eliminated by addition of protonophores. These experiments suggest that at least in *M. barkeri*, electron transport does, indeed, generate a PMF

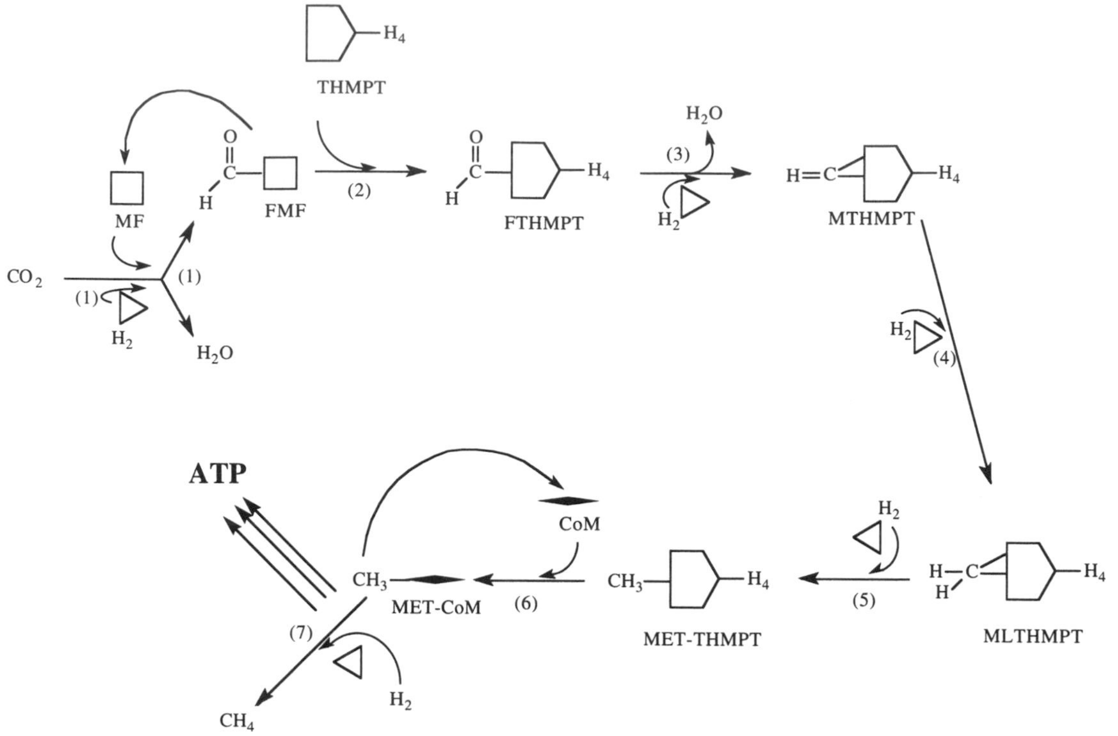

FIGURE 17.5 The reduction of CO_2 to methane. Throughout the diagram, the square represents methanofuran (MF), the pentagon represents tetrahydromethanopterin (THMPT), and the triangle represents factor 420 (F420). In addition, the black rod represents coenzyme M (CoM). (1) Interaction of methanofuran, carbon dioxide, and hydrogen, with the mediation of F420, to produce formyl methanofuran (FMF) and water. (2) Transfer of the formyl group to THMPT to yield formyl tetrahydromethanopterin (FTHMPT). (3–5) Sequential H_2 and F420-mediated reduction of the THMPT-attached 1-carbon moiety to produce methenyl tetrahydromethanopterein (MTHMPT), methylenyltetrahydromethanopterin (MLTHMPT), and methyl tetrahydromethanopterin (MET-THMPT). (6) Transfer of the methyl residue to CoM to produce methyl CoM (MET-CoM). (7) Reduction of MET-CoM to produce free CoM and methane with the concomitant formation of ATP.

and that the connection between methanogenesis, PMF generation, and ATP formation is the reduction of the immediate precursor of methane, methyl-Coenzyme M. It is likely that a similar connection between electron transport, PMF generation, and ATP formation is found in most methanogenic bacteria.

The nitrifying bacteria

Electron transport is the obligatory means of energy generation by the nitrifying bacteria, who generate a PMF either by NH_3 oxidation to NO_2^- or by oxidation of NO_2^- to NO_3^-. Ammonia oxidation to NO_2^- is accomplished in NH_3 oxidiz-ers by two enzymes, ammonia monooxidase (AMO) and hydroxyl amine oxioreductase (HAO). Either NH_3 or NH_2OH may serve as nitrite precursors. The connection between ammonia oxidation and PMF formation in *Nitrosomonas* is shown in Figure 17.7. Examination of Figure 17.7 reveals that HA formation from NH_3 occurs at the cytoplasm-cytoplasmic membrane junction and that HA traverses the membrane to the periplasm, where HAO then oxidizes HA to NO_2^-. Both processes result in electron donation to a cytochrome-containing electron transport system that allows proton extrusion to the periplasm and ATP formation by a membrane-bound proton translocating ATPase.

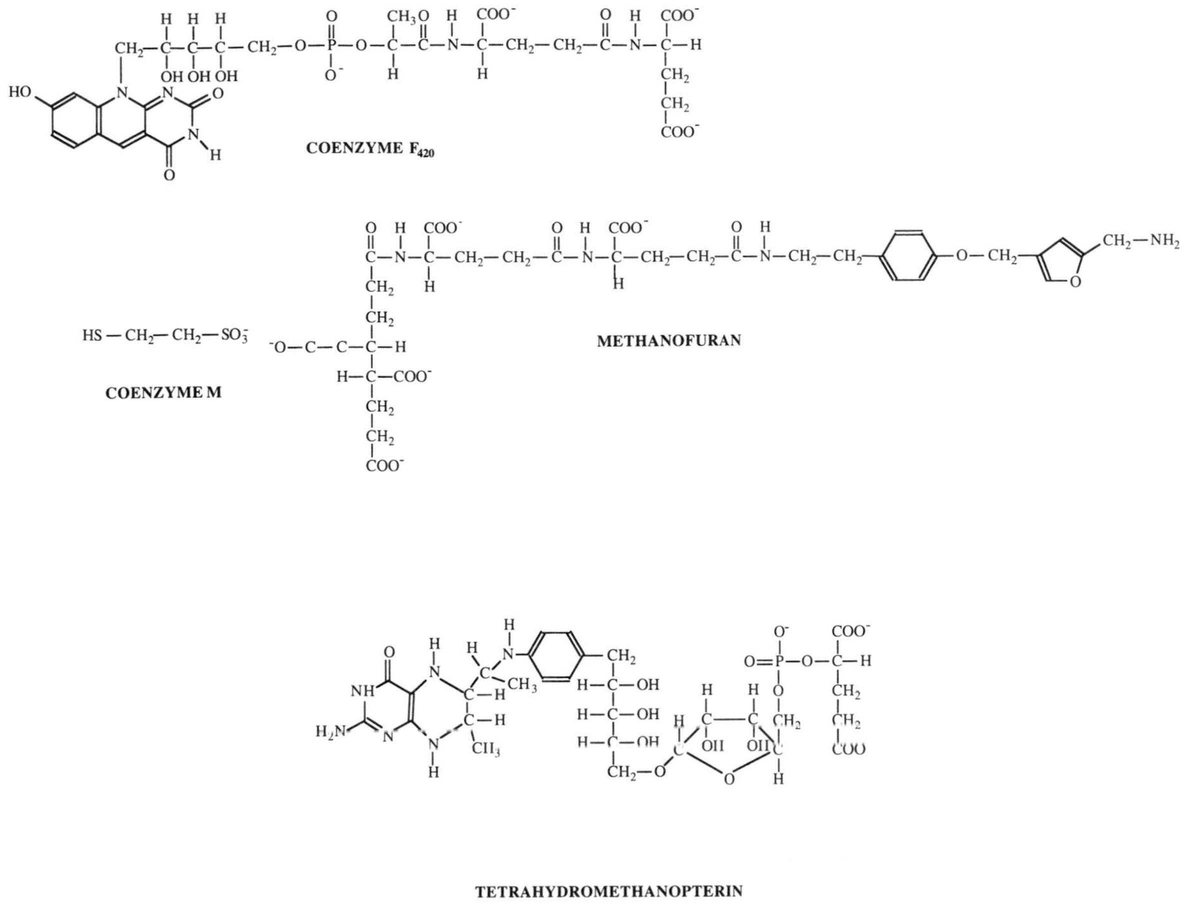

FIGURE 17.6 The structures of coenzyme F420, coenzyme M, methanofuran, and tetrahydromethanopterin, the unique electron carriers in methanogenic electron transport.

Energy generation by NO_2^--oxidizing organisms is a function of nitrite dehydrogenase. Development of information about nitrate dehydrogenase is complicated by the fact that members of the genus *Nitrobacter* display assimilatory nitrate reductase activity in addition to nitrite oxidase. Whether or not these activities are mediated by the same, or separate, proteins is not entirely clear. Similarly, although it is generally believed that ATP is generated by a proton pumping mechanism, the nature of the substance(s) involved in PMF formation is not yet completely determined. Further study is required to delineate the nature of electron transport-associated ATP formation and nitrite oxidation.

Sulfur-oxidizing bacteria

A diversity of forms of sulfur are oxidized by a variety of chemoautotrophic bacteria. The most definitive information regarding sulfur-oxidizing organisms is available from studies of the unicellular sulfur-oxidizing species. A substantial variety of the latter are known, and an equal diversity exists in their ability to oxidize sulfur. Species are known that degrade the sulfide anion, elemental sulfur, polythionates, thiosulfate, and sulfite. The general ways in which these processes occur is shown in Figure 17.8.

Although the details of oxidation of sulfur compounds are not entirely established in many cases, the general outlines of the process are

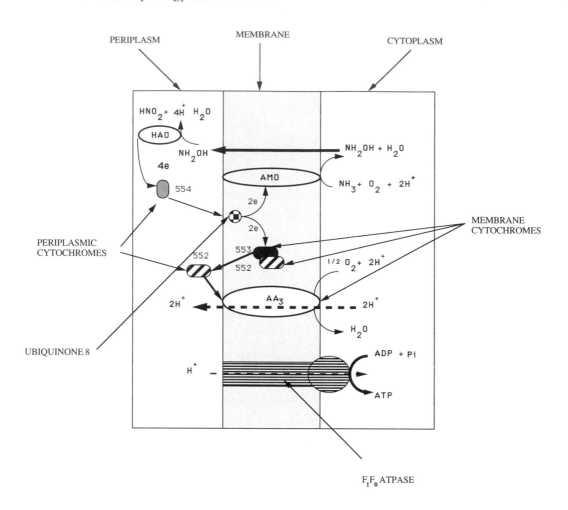

PERIPLASM   MEMBRANE   CYTOPLASM

$HNO_2 + 4H^+ H_2O$

HAO

NH_2OH

4e

554

AMO $NH_2OH + H_2O$

$NH_3 + O_2 + 2H^+$

2e

MEMBRANE
CYTOCHROMES

2e

553

PERIPLASMIC
CYTOCHROMES   552   552 $1/2 O_2 + 2H^+$

$2H^+$ AA_3 $2H^+$

H_2O

UBIQUINONE 8

ADP + Pi

H^+ —

ATP

F_1F_0 ATPASE

FIGURE 17.7 The oxidation of ammonia and hydroxylamine by *Nitrosomonas.* The process is highly compartmentalized and involves three major enzymes: ammonia monoxidiase (AMO) that spans the membrane, hydroxylamine oxidase (HAO) in the periplasm, and a cytochrome oxidase (AA_3) complex that spans the membrane. In addition, both membrane and periplasmic cytochromes, distinguished by their characteristic absorption peaks (552, 553, and 554 nm), are involved. At the cytoplasm-membrane junction, AMO oxidizes ammonia to hydroxyl amine. The latter passes through the membrane to the periplasm, where HAO converts it to the nitrite anion. The electrons derived from nitrite formation are passed through a series of electron transport compounds and are used to reduce oxygen to water. The periplasmic proton gradient is used to allow cytoplasmic ATP formation from ATP and inorganic phosphate, by the action of membrane spanning F_1F_0 ATPase.

similar. A relatively reduced form of sulfur is oxidized and electrons from the oxidative process are transferred to an electron transport system, allowing PMF formation, ATP generation, NAD(P)H formation, and CO_2 reduction, as well as the remaining processes required for growth.

Coupling of sulfur compound oxidation to electron transport is achieved in a diversity of ways. Representative schemes are shown in Figures 17.9 and 17.10. *Thiobacillus versutus* oxidizes $S_2O_3^{2-}$ to SO_4^{2-} as shown in Figure 17.9. A multicomponent periplasmic enzyme system oxidizes $S_2O_3^{2-}$ to SO_4^{2-} and transfers electrons to membrane-associated cytochromes with the concomitant formation of periplasmic protons. Within the cytoplasmic membrane, electron transport occurs in such a manner that electrons are transferred via a terminal cytochrome oxidase to cytoplasmic protons, reducing oxygen to water. The periplasmic protons drive a membrane-bound F_1F_0 ATPase in the direction of

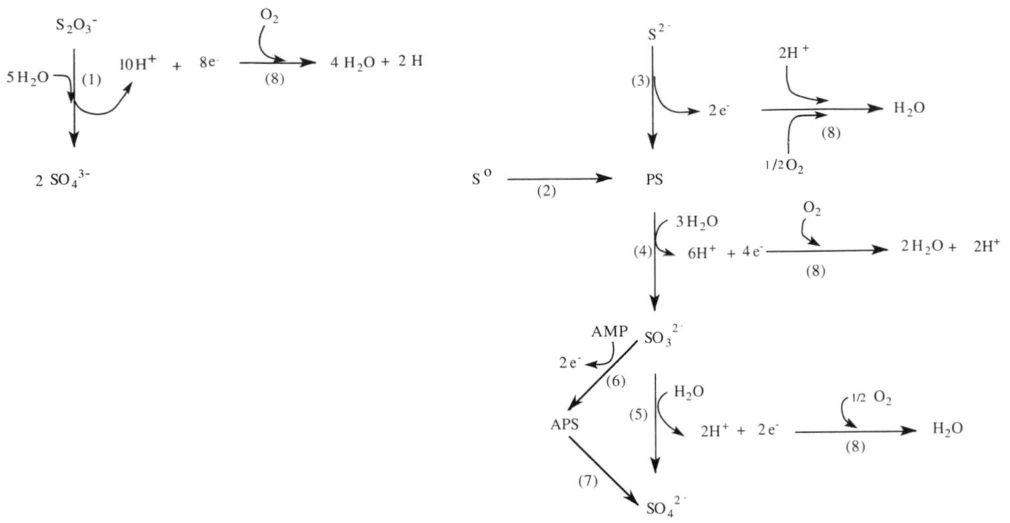

$$S_2O_3^- \quad 5H_2O \xrightarrow{(1)} 10H^+ + 8e^- \xrightarrow[(8)]{O_2} 4H_2O + 2H$$

$$2\,SO_4^{3-}$$

$$S^{2-} \xrightarrow{(3)} 2e^- \xrightarrow[(8)]{2H^+} H_2O \quad 1/2\,O_2$$

$$S^\circ \xrightarrow{(2)} PS$$

$$\xrightarrow{(4)} 3H_2O \quad 6H^+ + 4e^- \xrightarrow[(8)]{O_2} 2H_2O + 2H^+$$

$$AMP \quad SO_3^{2-}$$

$$2e^- \xrightarrow{(6)} APS \xrightarrow{(5)} H_2O \quad 2H^+ + 2e^- \xrightarrow[(8)]{1/2\,O_2} H_2O$$

$$\xrightarrow{(7)} SO_4^{2-}$$

FIGURE 17.8 A generalized representation of the various paths for sulfur compound oxidation in chemolithoautotrophs. Various sulfur-oxidizing species possess different capacities for sulfur compound oxidation and the enzymes found in particular species reflect the ability of a particular species to degrade particular spectra of sulfur compounds. Ultimately, sulfate is produced and the electrons derived from its formation are channeled into electron transport so that a proton motive force may be generated and used for ATP synthesis and other critical cell activities. (1) Oxidation of thiosulfate by a multienzyme complex. (2) Conversion of elemental sulfur (S°) to polysulfide sulfur (PS) by sulfur oxidase. This reaction, by itself does not yield ATP. (3) Formation of polysulfide sulfur from the sulfide anion. This oxidative process allows generation of a proton gradient and ATP synthesis. Glutathione is required for this reaction. (4) Conversion of polysulfide sulfur to sulfite by sulfur oxidase. (5) Formation of sulfate from sulfite by sulfite oxidase. (6,7) Conversion of sulfite to sulfate by the joint action of adenosine phosphosulfate (APS) reductase (6) and ADP sulfurylase (7). The electrons released by reaction 6 are donated to an electron transport system and reaction 7 yields ADP from APS by a substrate-level mechanism. The majority of the ATP formed from sulfur oxidation, however, results from electron transport-mediated processes reaction (8).

ATP synthesis. ATP hydrolysis, the PMF, or a combination of the two provide the energy for NAD(P) reduction and carbon dioxide fixation.

Tetrathionate oxidation by *Thiobacillus tepidarius* (Figure 17.10) occurs in a somewhat different manner. Periplasmic tetrathionate is transported to the cytoplasm by a symportic, proton-dependent uptake system. Within the cytoplasm, $S_4O_6^{2-}$ is oxidized to SO_4^{2-} with the extrusion of protons to the periplasm. Electrons derived from SO_4^{2-} formation are transferred to a terminal oxidase by a membrane-bound electron transport system and are subsequently used in the cytoplasm to reduce oxygen to water. Periplasmic protons allow ATP formation by a membrane-bound ATPase. Carbon dioxide reduction is PMF, and possibly also ATP-dependent.

Although the enzymes and mechanisms by which sulfur oxidation occurs are as yet largely unknown in the filamentous sulfur bacteria, evidence largely suggests that these organisms, as exemplified by members of the genus *Beggiatoa*, may degrade sulfur compounds under lithotrophic conditions by a series of reactions similar to those of the unicellular forms.

Carbon monoxide-oxidizing organisms

Although it is toxic to many life forms, carbon monoxide (CO) may be used as the sole energy source by some of the lithoautotrophic bacteria. CO oxidation is accomplished by a membrane-associated carbon monoxide dehydrogenase (COD) enzyme. Lithotrophic growth is not obligatory for CO-utilizing organisms since they may grow

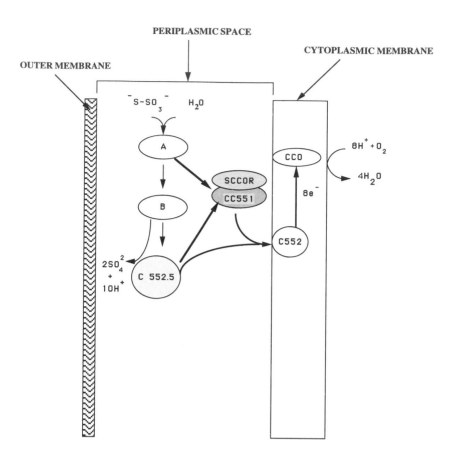

FIGURE 17.9 The coupling of thiosulfate oxidation and electron transport in *Thiobacillus versutus*. Within the periplasm, thiosulfate and water interact with enzymes A and B to produce sulfate and protons. Electrons derived from thiosulfate oxidation are transferred to periplasmic cytochromes (C552.5 and C551). Reduction of C551 is mediated by an associated sulfite: cytochrome c oxidoreductase (SCCOR). Electrons from both C552.5 and C551 are donated to cytoplasmic membrane-bound cytochrome 552 (C552) and then, via cytochrome oxidase (CCO), to oxygen to yield water. The periplasmic proton gradient is used for ATP formation and other cell functions.

heterotrophically as well. However, many CO-oxidizing organisms can grow lithoautotrophically with H_2 and CO_2 as energy and carbon sources, respectively. Operation of the Calvin cycle has been shown in all CO-using organisms thus far studied.

Carboxydotrophic bacteria contain unusual electron transport systems, one of which operates primarily during heterotrophic growth and the other, a carbon monoxide resistant enzyme, is involved during autotrophic growth. The comparative nature of the systems in *Pseudomonas carboxydovorus* is shown in Figure 17.11. During CO-H_2 growth of the organism, the CO-insensitive system is functional and during heterotrophic growth, the CO-sensitive system operates. CO inactivates the CO-sensitive system by interaction with

its cytochrome a_1. In contrast, the functionality of the CO-sensitive system is maintained by preferential binding of O_2 by its cytochrome c component. This property allows the system to function in the presence of substantial amounts of CO. ATP formation and NAD(P) reduction are both a function of the PMF.

Sulfur and nitrogen assimilation

Although some of the autotrophic bacteria use sulfate as an electron acceptor for energy generation, all bacteria, including the autotrophs, must synthesize essential sulfurous substances. Methionine and cysteine, for example, are essential components of

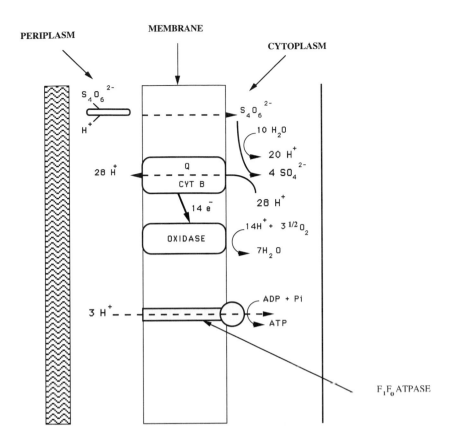

FIGURE 17.10 Tetrathionate oxidation and ATP formation in *Thiobacillus tepidarius*. Both structurally and biochemically, the process is substantially different from the manner of thiosulfate oxidation in *T. versutus* (Figure 17.9). In *T. tepidarius*, tetrathionate is transported to the cytoplasm by a proton transporting symport protein. Oxidation of tetrathionate occurs in the cytoplasm, while in *T. versutus*, thiosulfate oxidation is accomplished by a multienzyme complex in the periplasm. Although electron transport occurs in both systems, it is accomplished by C-type cytochromes in *T. versutus* and by B-type cytochromes in *T. tepidarius*. In both systems ATP is synthesized by a proton driven, membrane-bound ATPase.

proteins, while coenzyme A and lipoic acid are integral components of a variety of metabolic processes. For both structural and metabolic reasons, microbes must be capable of sulfur assimilation.

Substantial variations in the manner in which sulfur assimilation occurs are found among the various autotrophic microbes. Some use H_2S to synthesize cysteine, from which the remaining sulfur compounds are formed. Others use exogenously supplied cysteine. Still others synthesize reduced sulfur compounds by sulfate reduction, using enzymes similar or identical to those found in nonautotrophic forms, such as ATP sulfurylase, APS reductase, and sulfite reductase. Equal diversity is found among the autotrophs in regard to

assimilation of nitrogen. Nitrogen may be assimilated by fixation of nitrogen gas, use of preformed nitrogenous compounds, or ammonia uptake.

Ammonia is the preferred nitrogen source for many microbes including, but not restricted to, the autotrophs. Although it is widely used by microbes, ammonia assimilation is normally accomplished by one of two mechanisms, operation of the glutamine synthesis/glutamic acid synthesis (GLS/GLUT) pathway or the glutamic dehydrogenase (GD) pathway. Although these pathways are found within the autotrophs, it is difficult to discern a pattern of their occurrence in autotrophic bacteria, and the general mechanisms for ammonia use by autotrophs appear similar or identical to

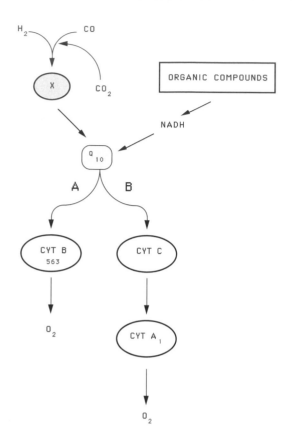

FIGURE 17.11 Electron transport in the carbon monoxide oxidizing bacteria as exemplified by *Pseudomonas carboxydovorans*. The organism has two electron transport systems: system A, which is carbon monoxide insensitive and system B, which is carbon monoxide sensitive. The systems share a common quinone, but differ in their cytochromes. When carbon monoxide serves as energy source, it is oxidized by carbon monoxide dehydrogenase and carbon dioxide serves as carbon source. Hydrogen may also be used as an energy source during autotrophic growth. In addition to autotrophic growth, the organism may also grow heterotrophically. The CO sensitivity of system B reflects interaction of CO with its cytochrome a_1 oxidase. The relative insensitivity of system A to CO reflects the strong affinity for oxygen of cytochrome b563, which serves as an oxidase in system A.

those found in heterotrophic organisms (see Chapter 10). Additional forms of nitrogen that may be used by some autotrophs include hydroxyl amine, urea, and, in some cases, NO_2^- and NO_3^-. The ability to use NO_3^- depends upon the presence of an assimilatory nitrate reductase.

Summary

The autotrophic bacteria are a diverse collection of organisms that inhabit a tremendous variety of habitats. Although they are metabolically, morphologically, and taxonomically diverse, and exhibit a remarkable adaptability in the habitats they may occupy, autotrophs must accomplish the same physiological tasks that are accomplished by heterotrophs. In order to accomplish those tasks, autotrophs have developed a high degree of both morphological and biochemical complexity. Evolution has allowed a degree of specialization and complexity in autotrophs that exceeds that found in most heterotrophic forms. Although tremendous diversity is found in the autotrophic world, at the same time, a commonality exists in the manner in which essential tasks are accomplished. Just as for heterotrophs, essential energetic and biosynthetic activities are accomplished in relatively few ways. Among the autotrophic microbes there are, at most, only four major ways by which carbon is fixed, and only two ways for ammonia fixation. In addition, a single enzyme, such as CO dehydrogenase, may serve similar, or only slightly different, functions in various microbes. With regard to energy generation, a similar unity in the midst of diversity is found. Although they differ in the nature of the components of their electron transport schemes, electron transport is the major mode of energy generation for most autotrophs, the same as for heterotrophs. Within the autotrophs, both phototrophs and chemotrophs generate energy by electron transport-mediated processes. In principle, in regard to energy, phototrophs and chemotrophs differ from each other only in the manner of reducing power generation—trapping radiant energy on the one hand and by chemical oxidation on the other.

Selected References

Bowien, B. 1989. Molecular biology of carbon dioxide assimilation in aerobic chemolithotrophs. *In* H. G. Schlegel and B. Bowien (eds.) *Autotrophic Bacteria.* pp. 437–460. Springer-Verlag, New York, Berlin, Heidelberg, Tokyo.

Fuchs, G. 1989. Alternative pathways of autotrophic CO_2 fixation. *In* H. G. Schlegel and B. Bowien (eds.) *Autotrophic Bacteria.* pp. 365–382. Springer-Verlag, New York, Berlin, Heidelberg, Tokyo.

Gottschalk, G. 1989. Bioenergetics of methangenic and acetogenic bacteria. *In* H. G. Schlegel and B. Bowien (eds.) *Autotrophic Bacteria.* pp. 383–396. Springer-Verlag, New York, Berlin, Heidelberg, Tokyo.

Hooper, A. B. 1989. Biochemistry of the nitrifying lithoautotrophic bacteria. *In* H. G. Schlegel and B. Bowien (eds.) *Autotrophic Bacteria.* pp. 239–266. Springer-Verlag, New York, Berlin, Heidelberg, Tokyo.

Kelley, D. P. 1989. Physiology and biochemistry of the unicellular sulfur bacteria. *In* H. G. Schlegel and B. Bowien (eds.) *Autotrophic Bacteria.* pp. 193–218. Springer-Verlag, New York, Heidelberg, Tokyo.

Kondratieva, E. N. 1989. Chemolithotrophy of phototrophic bacteria. *In* H. G. Schlegel and B. Bowien (eds.) *Autotrophic Bacteria.* pp. 283–288. Springer-Verlag, New York, Berlin, Heidelberg, Tokyo.

Kuenen, J. G., and P. Bos. 1989. Habitats and ecological niches of chemolitho (auto) trophic bacteria. *In* H. G. Schlegel and B. Bowien (eds.) *Autotrophic Bacteria.* pp. 53–80. Springer-Verlag, New York, Berlin, Heidelberg, Tokyo.

McFadden, B. A. 1989. The ribulose bisphosphate pathway of CO_2 fixation. *In* H. G. Schlegel and B. Bowien (eds.) *Autotrophic Bacteria.* pp. 351–364. Springer-Verlag, New York, Berlin, Heidelberg, Tokyo.

Nelson, D. C. 1989. Physiology and biochemistry of the filamentous sulfur bacteria. *In* H. G. Schlegel and B. Bowien (eds.) *Autotrophic Bacteria.* pp. 219–238. Springer-Verlag, New York, Heidelberg, Tokyo.

Pfennig, N. 1989. Ecology of phototrophic purple and green sulfur bacteria. *In* H. G. Schlegel and B. Bowien (eds.) *Autotrophic Bacteria.* pp. 97–116. Springer-Verlag, New York, Berlin, Heidelberg, Tokyo.

Schlegel, H. G., and B. Bowien. (eds.) 1989. *Autotrophic Bacteria.* Springer-Verlag, New York, Berlin, Heidelberg, Tokyo.

Stanier, R. Y., J. L. Ingraham, M. L. Wheelis, and P. R. Painter. 1986. *The Microbial World,* 5th edition. Prentice Hall, Englewood Cliffs, New Jersey.

Stetter, K. O. 1989. Extremely thermophilic chemolithotrophic archaebacteria. *In* H. G. Schlegel and B. Bowien (eds.) *Autotrophic Bacteria.* pp. 167–176. Springer-Verlag, New York, Berlin, Heidelberg, Tokyo.

Tabita, F. R. 1989. Molecular biology of carbon dioxide fixation in photosynthetic bacteria. *In* H. G. Schlegel and B. Bowien (eds.) *Autotrophic Bacteria.* pp. 481–498. Springer-Verlag, New York, Berlin, Heidelberg, Tokyo.

Thauer, R. 1989. Energy metabolism of sulfate reducing bacteria. *In* H. G. Schlegel and B. Bowien (eds.) *Autotrophic Bacteria.* pp. 397–414. Springer-Verlag, New York, Berlin, Heidelberg, Tokyo.

Truper, H. 1989. Physiology and biochemistry of phototrophic bacteria. *In* H. G. Schlegel and B. Bowien (eds.) *Autotrophic Bacteria.* pp. 267–282. Springer-Verlag, New York, Berlin, Heidelberg, Tokyo.

Wood, Harlan G. 1989. Past and present of CO_2 utilization. *In* H. G. Schlegel and B. Bowien (eds.) *Autotrophic Bacteria.* pp. 33–52. Springer-Verlag, New York, Berlin, Heidelberg, Tokyo.

CHAPTER 18

Differentiation

The nature of differentiation

Differentiation is a widespread phenomenon in the microbial world and elsewhere. It may be defined as a change in the genetic expression of an organism. Differentiation manifests itself in a variety of ways such as the formation, or lack of formation, of a critical cell protein, motility or its absence, the formation of a particular structure, or a series of morphological and biochemical changes. In each of these situations, when faced with a new environment, the organism adapts so that it may survive under the altered conditions by altering the manner in which its genetic material is expressed. The nature of the adaptation, and the manner of genetic expression, vary from situation to situation, but the underlying mechanism does not.

Physiological questions raised by differentiation

The phenomenon of differentiation gives rise to a number of physiological questions. What induces the change? When the change occurs over an extended time period, as is often the case, why do the components of the change occur in a time-dependent manner? What determines the time range over which particular events occur? What

are the mechanisms by which particular events occur? In what manner are they controlled? These are only some of the questions that might be asked. In this chapter, we will consider some examples of the ways in which differentiation occurs in the microbial world.

Bacterial endosporulation as a model of differentiation

Differentiation processes in morphologically complicated organisms are extremely complex. Their complexity currently defies elucidation. Because of our inability to study differentiation critically in higher life forms, differentiation has been studied in simpler organisms, the bacteria. It is believed that at the cellular level, the processes of differentiation are similar, irrespective of cell type, and that study of microbial systems will help in understanding analogous processes in more morphologically complex forms. Over an extended time, endosporulation has been used as a model system for the study of differentiation. The focus on endospore formation as a tool for understanding differentiation has resulted from many factors. To begin with, although sporulation is a relatively infrequent occurrence in bacteria, it is of great practical importance. *Bacillus anthracis,* the causative agent of anthrax, owes its pathogenicity, in large part, to its ability to persist as a spore in the soil. *Clostridium tetani,* the causative agent of tetanus is also an extremely dangerous spore-forming soil inhabitant, while *Clostridium botulinum* is a serious potential hazard to the food industry. For very practical reasons, it is important to understand both the properties of spores and the conditions that allow their formation.

Endosporulation is an excellent model for study of differentiation for several reasons: 1) It is a time ordered process that occurs over a sufficient time period (7–10 hours), so that its subprocesses can be readily identified; 2) it exhibits distinct stages; 3) it is manifested in both biochemical and morphological changes; and 4) relatively speaking, it is amenable to genetic analysis.

The events of sporulation

The process of sporulation in a typical *Bacillus* organism is shown diagrammatically in Figure 18.1. The stages of sporulation are denoted by the Roman numerals 0 to VII, indicating the approximate time, in hours, at which particular events occur. Mutations in the sporulation process are, for each time period, denoted by capital letters in alphabetical order. Thus we may distinguish spo 0A from spo 0B and both spo 0A and spo 0B from spo IIIC. Both spo 0A and spo 0B occur during the initial stage of sporulation, whereas a spo III mutation of any kind occurs at a later point in time. In addition, within a time period, an *A* mutant affects a process earlier than does a *B* mutant, which manifests its action prior to a *C* mutant. In this way, correlation can be made between a particular mutant and its time of occurrence in the sporulation process.

Sporulation is initiated at stage 0. Stage 0 is followed by stage I, during which replicated DNA forms an axial filament in the center of a vegetative cell. Stage I is followed by stage II, during which time the spore septum is formed and unequal cell division occurs, resulting in a relatively large *mother cell* and a much smaller *forespore.* During stage III, the forespore is surrounded by the mother cell, giving rise to an incipient spore containing the spore genome and cytoplasm surrounded by a membrane. The entity just described is further surrounded by the membrane and wall of the engulfing mother cell to produce a double

membrane-bound incipient spore surrounded by the mother cell. During stage IV, a *cortex* of carbohydrate, composed of a modified form of murein, is deposited between the spore membranes. Stage IV is followed by stage V, during which calcium dipicolinate, a characteristic spore component, and heat resistant proteins, the *coat* proteins, are added. During stage VI spore maturation occurs by the further addition of proteins, particularly at the surface of the exterior spore membrane. Finally, during stage VII, the complete, mature spore is released to the external environment by lysis of the mother cell.

The morphological events of sporulation just described are accompanied by an equally elaborate sequence of biochemical events. During the period of stages 0 to II, a variety of antibiotics and protease enzymes are formed. A spore-specific alkaline phosphatase appears during stage III and calcium dipicolinate and heat-resistant proteins are added during stage V.

It is believed that endosporulation is accomplished by at least 50 genetic locations around the chromosome which are expressed as operons, that is, genes involved in similar parts of the process. That the subprocesses of sporulation are ordered and time-dependent is confirmed by genetic analysis. Spo 0 mutations do not affect vegetative cell growth, but prevent spore septum formation. Mutations in the spo II region prevent engulfment and spo III mutants fail to allow both cortex and spore coat deposition.

Although the morphological events of sporulation have been established and a substantial number of biochemical changes associated with the process have been demonstrated, many questions remain about the precise factors that influence and control it. Although sporulation initiation appears correlated with nutritional limitation, the role that nutrient limitation may exert on initiation is less clear. The nature of the limitation seems not to be a critical factor since nitrogen, carbon, and phosphate limitation can all allow sporulation. Decreases in GTP and GDP concentrations are frequently found during initiation and it is believed that the decrease is associated with binding of proteins to guanosine nucleotides, but the consequences of this protein-nucleotide binding are unclear. Uncertainty exists regarding whether nucleotide-protein binding serves to directly activate events required for initiation or whether the

VEGETATIVE CELL   AXIAL FILAMENT   FORESPORE FORMATION

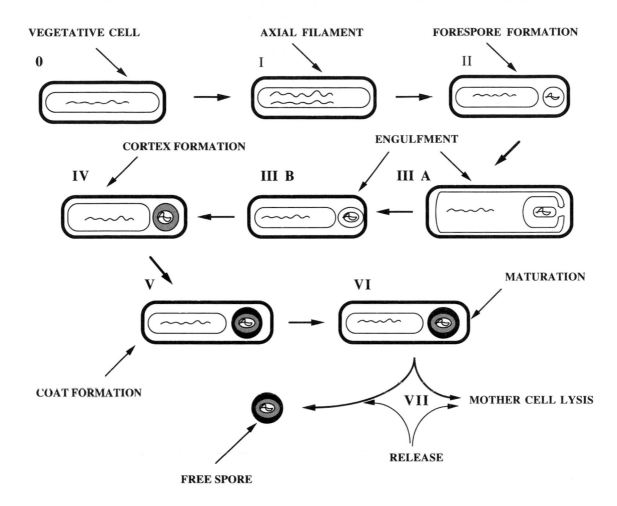

FIGURE 18.1 The morphological events of bacterial sporulation. The vegetative cell (0) completes a round of DNA replication, giving rise to replicated DNA aligned in an axial filament in the center of the cell (I). By unequal division, the replicated DNA is partitioned into the forespore and the mother cell (II). The mother cell then surrounds the forespore (IIIA, IIIB) to produce the spore genome, encased by a double membrane. A cortex, composed primarily of carbohydrate, is then deposited between the membranes that surround the spore genome (IV). This process is followed by the addition of protein coats at the exterior of the outer spore membrane (V), maturation of the incipient spore (VI), and release (VII) of the completed spore to the external environment, with concomitant lysis of the mother cell. The biochemical events that accompany morphological changes are best described in the text.

process relieves inhibition of sporulation. In addition, it is not clear whether the initiation signals for sporulation are metabolic, transcriptional, or a combination of both. It is generally conceded, however, that the major events of endosporulation are regulated to a substantial degree by transcriptional control mechanisms. Evidence for transcriptional control is obtained particularly from studies of the consequences of time-dependent mutants and from hybridization studies that show the reproduceable and time-dependent appearance of specific mRNA molecules at particular points of the sporulation cycle. In addition, multiple forms of DNA-dependent RNA polymerase and multiple sigma factors, with affinities for different promoter regions have been found. These findings indicate that the messages for specific proteins are formed in time-dependent manner. It is virtually certain that translational control is also a major regulatory mechanism for endosporulation, but direct evidence for this control is more difficult to obtain.

Old versus new

The extent to which the spore is a newly synthesized entity is a question of substantial importance. Studies of sporulation have been complicated by the fact that most of the mRNA molecules formed during sporulation are formed in the mother cell. Particularly in the early stages of the process, it is difficult to determine the extent to which the processes observed are a function of the mother cell or the incipient spore. It is equally difficult to determine the extent to which spore components are derived, directly or indirectly, from the mother cell or arise from *de-novo* synthesis. Studies with radiolabelled materials have shown that during the initial sporulation stages, extensive digestion of mother cell wall components occurs and that the digestion products, or their metabolites, appear in the spore. Studies have also shown that extensive synthesis of new material from exogenous nutrients accompanies sporulation. At a particular point, however, sporulation can occur in the absence of exogenous nutrients. Beyond a certain point, when commitment has occurred, sporulation can occur *endotrophically* in distilled water. This finding implies that although the early stages of sporulation involve use of mother cell materials or mother cell-synthesized substances and although sporulation involves substantial *de-novo* synthesis, at some point, completion of the process can occur from within. At some point, the developing spore has *within itself* all that is needed to complete its own formation. The completed spore is a composite of the old and the new.

Activation, germination, and outgrowth

Just as the events leading to the formation of endospores are under stringent genetic and biochemical control, spore germination, the formation of vegetative cells from spores, is also a complex and highly regulated process. Activation, the commitment of the sporulated cell to return to the vegetative state, is a prerequisite to germination. Activated cells retain many of the physical properties of spores, such as resistance to heat and chemicals, but the activation process is, in some cases, reversible. Germination, in contrast, is an irreversible process, signalled by the release of calcium dipicolinate. At least five genetic locations have been associated specifically with germination. In addition, specific acid-soluble spore proteins, (SASSPs) have been associated with germination and mutational analysis has shown that the process of outgrowth is distinct from the germination process. Mutants are known that are specific for the two processes.

Return to the vegetative state requires that the spore, a dehydrated and relatively metabolically inactive structure, reverts to active metabolism so that a new vegetative cell may be formed. Germination requires a considerable amount of energy that is, at least initially, derived from energy-rich storage compounds within the spore, particularly 3-phosphoglycerate. The latter is found in high concentrations within the spore and is rapidly converted to ATP and pyruvate by the enzymes 3-phosphoglycerate mutase, enolase, and pyruvate kinase, all of which are found in abundance within the spore. A variety of additional spore storage compounds are also mobilized. SASSPs are degraded to provide the building blocks for new protein formation. In addition, both RNA and protein synthesis are rapidly initiated at the beginning of germination. However, DNA synthesis and chromosome replication occur toward the end of the process.

The process of germination can be subdivided, on the basis of time, into two periods, the first 70 minutes, commonly known as stage I, and the time beyond the first 70 minutes, referred to as stage II. During stage I, storage compounds found within the spore are degraded and mobilized for RNA and protein synthesis so that the components of the incipient vegetative cell may be formed. By the end of stage II, the germinated cell is capable of synthesizing all of its small molecules, generating its own energy, and, in general, of accomplishing all of the critical processes for life.

Differentiation in *Caulobacter crescentus*

Bacterial endosporulation displays the general nature of differentiation, a time ordered occurrence of a series of genetically regulated events. However, differentiation manifests itself differently in other systems. *Caulobacter crescentus* displays a differentiation process substantially different from that of endosporulation. Endosporulation is sub-

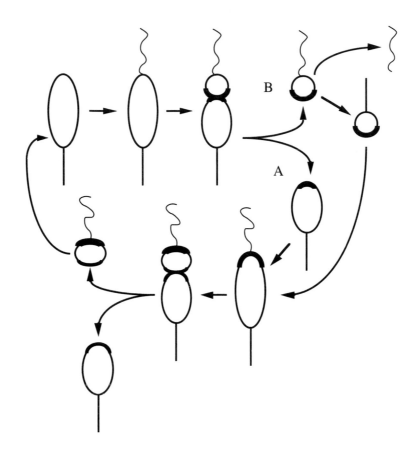

FIGURE 18.2 Differentiation in *Caulobacter crescentus*. Beginning at the top left, a stalked cell gives rise to a stalked and flagellated cell. The flagellum is formed at a "pole" formed during a previous division event. Further development gives rise to a small incipient "swarmer" cell, a component of a larger "predivision cell," containing the incipient swarmer cell attached to a larger stalked cell. During this time period, pole material (denoted by the heavy lines) is synthesized. Division of the predivision cell yields a stalked cell (A) and a swarmer cell (B). The stalked cell immediately initiates DNA replication and the other events required for differentiation and reproduction. The swarmer cell, however, undergoes a different sequence of events. DNA replication in the swarmer cells occurs only after a "resting period," during which, the swarmer cell loses its flagellum and a stalk is formed at the exact site previously occupied by the flagellum. In the same time period, the cell enlarges in size, and pole development occurs to an extent sufficient to allow new flagellum formation. In addition, DNA replication is initiated and from this point, the remaining events of division and differentiation of swarmer cells occur as for a stalked cell. However, the delay in initiation of DNA replication and flagellum formation in the swarmer cells is such that the overall rates of production of swarmer and stalked cells are unequal.

stantially affected by external conditions and mediates interconversion between a vegetative and a dormant state. By contrast, differentiation of *C. crescentus* is relatively uninfluenced by external conditions and involves conversion between two cell types, neither of which is dormant. The elements of differentiation in *C. crescentus* are shown in Figure 18.2. The process begins by the conversion of a motile form, a swarmer cell, into a nonmotile cell by the loss of a flagellum. The flagellum

is subsequently replaced by a stalk at the *exact* location previously occupied by the flagellum. Differentiation in *C. crescentus* thus displays a degree of polarity. The stalk cell is further converted into an incipient "bicelled," stalked and flagellated entity that is itself converted into a second bicelled entity, the *predivision cell*. Pili are associated with the flagellum-containing component of the predivision cell. The predivision cell then divides, producing a new swarm cell and a stalk cell. Only the stalk cell is immediately capable of DNA synthesis. The events of the *C. crescentus* differentiation process are, however, cyclic because a stalked cell gives rise to swarm cell. The order of events in the *C. crescentus* differentiation cycle is independent of the growth rate of the organism and occurs sequentially, irrespective of the absolute rate of growth, providing evidence that the process is, indeed, a process of differentiation.

A series of ordered biochemical events occurs during differentiation of *C. crescentus*. The swarmer stage is associated with synthesis of flagellin A, a 25,000 molecular weight protein, which is one of two flagellins. Flagellin A is also synthesized during the formation of a flagellated bicell. Hook protein and flagellin B, a 27,500 molecular weight, immunologically distinct, filament protein, are also formed at this time. In addition, methyl-accepting chemotaxis proteins, as well as carboxymethyl transferase and methylesterase, are formed. Phospholipid synthesis occurs during conversion of the swarmer cell to a stalked cell. Finally, DNA synthesis occurs during the conversion of the bicelled stalked entity to the flagellated, piliated, and stalked entity. Phospholipid synthesis is an obligatory prerequisite for both DNA synthesis and stalk formation; and DNA synthesis is also required before both flagellin and hook protein synthesis.

The morphological and biochemical events of the *C. crescentus* differentiation cycle are a carefully controlled sequence of functionally interrelated phenomena. *C. crescentus* is a chemotactic organism. It is, therefore, reasonable to suggest that the elements of the chemotactic system would be formed during the motile stages of the developmental cycle. Such is indeed the case. Methyl-accepting chemotactic proteins, methyl esterase, and carboxymethyl transferase are all formed during stages of the cycle associated with motility. In a similar manner, phospholipid synthesis precedes both stalk protein and pilin synthesis and DNA replication. It is believed that the temporal relationship between phospholipid synthesis and the other cellular activities reflects a functional relationship, that is, the requirement for membrane synthesis for stalk and pilin insertion into the outer membrane of *C. crescentus* and for separation of replicated DNA.

Differentiation in *Dictyostelium discoideum*

The life cycle of the eucaryotic cellular slime mold, *Dictyostelium discoideum*, displays a complicated pattern of both morphological and biochemical differentiation. The morphological aspects of the cycle are shown in Figure 18.3. The cycle begins with the germination of a spore to form a vegetative amoeba. After about 7 hours, the amoebas begin an aggregation process and form a pseudoplasmodium, which by further differentiation, forms a fruiting body. The fruiting body, in turn, yields spores that by germination, produce amoebas and reinitiate the cycle.

Differentiation in *D. discoideum* exhibits unusual features. It requires chemotaxis and an aggregation of homogeneous cells to form a structure, the fruiting body, composed of differentiated cells, the stalk cells and spore cells. Formation of stalk and spore cells occurs in an orderly manner. The anterior third of the aggregated amoeboid mass becomes stalk cells and the remainder of the amoebas become spore cells.

The morphological changes of the *D. discoideum* developmental cycle are associated with an array of biochemical and metabolic changes. During the free-living stage of the cycle, six types of mRNA are produced. In contrast, an abundance of mRNA types are formed during the aggregation stage. Aggregation is mediated by a gradient of cAMP. Cyclic AMP is formed in waves of about two minutes duration, followed by a five minute rest period. The wave, initially produced by a small subset of the cells in the population is amplified as cAMP-exposed cells produce additional cAMP. Fluctuations in the cAMP are apparently accomplished by changes in the degree to which cAMP occupies receptors on the surface of aggregating cells. Changes in occupancy of cell surface receptors appear to change the extent of synthesis of adenyl cyclase, altering the relationship between cAMP formation by adenyl cyclase and its destruction by cAMP phosphodiesterase. These changes result in periodic changes in cAMP concentrations, with accompanying changes in cell migration.

The mechanisms of the various aspects of the differentiation cycle of *D. discoideum* are distinctly different. The differentiation processes for prestalk

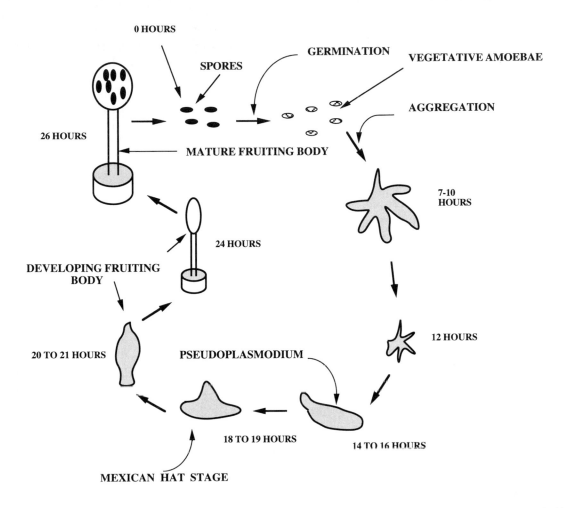

FIGURE 18.3 The essential events of differentiation of *Dictyostelium discoideum.* Spores are released from a mature fruiting body and germinate to form free-living amoeboae. At 7–10 hours after germination, the amoeboae aggregate under the influence of cyclic AMP pulses. Aggregation becomes more compact at 12 hours. By 16 hours postgermination, a "slug," or pseudoplasmodium, is formed. By 19 hours, it has been converted to a "mexican hat." The mexican hat form, in turn, gives rise to increasingly more morphologically complex structures formed from differentiated stalk and spore cells. Ultimately, a mature fruiting body is formed, so that the cycle may be repeated.

and prespore cells, for example, are accomplished in different ways. Within each cell type, two classes of mRNA are recognized by hybridization analysis, prestalk I and II and prespore I and II. Hybridization probes designed for particular RNAs do not cross react. The messages for the various aspects of stalk and spore cell differentiation are discrete. In addition, the various aspects of differentiation respond differently to environmental factors. Cell aggregation and formation of prestalk I, prestalk II,

and prespore I mRNAs are all affected by cAMP pulses but prespore II mRNA formation is not. Finally, messages involved in various parts of *C. crescentus* differentiation are processed at different rates. Messages formed *throughout* the entire process react, at all stages, with a common set of hybridization probes while messages produced transiently, at specific times, do not. The latter react only with probes specific for messages produced at a specific time.

Differentiation during bacterial cell division

The cell division process in bacteria may be regarded as a process of differentiation. Sequential expression of genes is required so that the process may occur in an orderly manner. Disruption of that order has major physiological consequences. The process of bacterial cell division has been studied in some detail. It has been shown that chromosome replication occurs at a constant speed at a given temperature and that cell division occurs at a constant time after the completion of DNA replication. It is further known that the D time, the time between completion of a round of DNA replication and the division event, is independent of growth rate.

Within the temperature range that allows growth of an organism, substantial differences are found in the growth rate of an organism. At rapid growth rates, different amounts of DNA and other substances are found in cells than during periods of less rapid growth. When the growth temperature is such that the rates at which the other growth-associated processes are compatible with the rate of DNA replication, "normal" division occurs. Each cell contains a single copy of the genome and a constant chemical composition. At other temperatures, however, the relationship between DNA replication and the other components of cell division differs as a function of particular conditions.

Cell division requires the partitioning of protoplasmic components into daughter cells. This process requires three subprocesses, septum formation, cell envelope synthesis, and cleavage. Each of these processes, as well as DNA replication, is under intense genetic control.

The relationships among DNA synthesis initiation, septation, and cell division

Although many factors influence the microbial cell division process, evidence indicates that the interconnections between initiation of DNA synthesis, septation, and the division event are the critical factors that influence microbial cell growth. Most of the remaining subprocesses of cell growth occur quite randomly throughout the cell cycle, but cell division, partitioning, and the initiation of DNA synthesis occur with regularity. It is for this reason that these factors are considered of prime importance.

The interconnections between the three processes are complex. For example, it is generally established that membrane or wall attachment is required for initiation of DNA synthesis and for partitioning of replicated DNA to occur, but simple attachment is insufficient for these processes to occur. It is also well established that the initiation of DNA synthesis and the synthetic process are distinctly separated, but intimately connected, phenomena. Inhibitors of protein synthesis inhibit the initiation process that requires *de novo* protein synthesis, but do not affect completion of previously initiated synthetic events that can occur by the operation of preformed proteins. In addition to these factors, partitioning of replicated DNA is mediated, at least to a certain extent, by membrane growth.

The septum

The interconnections between septum formation and function and other critical cell events are of equal complexity to those of DNA. Although septum formation is a requirement for growth, septum formation *per se* is insufficient to allow the process of cell division required for growth. Although septum formation is required for cell division, unless cell envelope synthesis also occurs, normal cell division is not possible. The interrelationships between the factors critical to cell growth are complex.

Genetic aspects of cell division

Each of the critical aspects of cell division is under complex genetic control. DNA synthesis requires the presence of enzymes, each of which must be synthesized through the use of genetically controlled messages. In addition, substantial evidence indicates that initiation of synthesis is regulated by the DNA A gene product, although the precise manner in which the gene product operates in control of initiation remains obscure. It is as yet unclear whether the product must accumulate to a certain concentration, or must be diluted below a critical concentration. Whatever its mode of action, evidence indicates that the DNA A gene product regulates its own formation, presumably by controlling formation of the code for its own synthesis.

Both envelope and septum formation are under genetic control of a complexity equal to that for control of DNA. Studies of mutants have shown that although murein synthesis is required for both morphogenesis and septum formation, different genes regulate the murein required for elongation than regulate septum-associated murein.

Septum formation is also a carefully regulated process. Mutants have been obtained that form murein, but do not form septa. The antibiotic cephalexin selectively inhibits septum formation but does not affect elongation, and mecillinam affects murein formation involved in elongation, but does not affect septation, suggesting that particular and discrete proteins are involved in elongation and septum formation. The complexity of regulation of septum formation and function is illustrated by the isolation of mutants that regulate septum location, initiation, formation, separation, and inactivation.

Growth under normal conditions

Although the critical processes for microbial growth are each precisely regulated in a variety of ways, under normal growth conditions, they are coordinated so that the cell may grow and compete in the environment in which it must exist. By as yet imperfectly understood control mechanisms, the synthesis of wall material, the initiation of DNA synthesis, the synthesis process itself, the formation of septa, the replication of ribosomes, and all of the events critical to new cell formation occur in a time ordered and interconnected manner so that the organism may survive. Although the most intense studies of the cell cycle to date have employed *E. coli*, it is undoubtably the case that the concepts derived from studies of this organism, in regard to control of the interconnected processes of cell growth and division, will apply to most, if not all, cells.

Other examples of differentiation

The examples of this chapter illustrate the diversity of differentiation and its consequences, but many additional examples may be found. Additional examples of differentiation include the growth cycle of the myxobacteriales and the formation of spores by both the streptomyces and by yeasts. Less complicated examples include the induction of enzymes and transport systems, modifications in function of the phosphotransferase system, and the phenomenon of diauxic growth.

Summary

In both the microbial and nonmicrobial world, differentiation manifests itself in a diversity of ways. Irrespective of these differences in expression, the fundamental properties of differentiation are the same, the sequentially and carefully regulated occurrence of interconnected events. Although the consequences of differentiation processes are different in various organisms and environments, all of the processes, in whatever system studied, exemplify the diversity of genetic expression and its control. Although at the present time, we are restricted, primarily, to the study of differentiation in microbial systems, it is certain that information derived from such studies will be generally instructional and useful.

Selected References

American Society for Microbiology. 1983. Molecular Biology of Microbial Differentiation. International Spore Conference, American Society for Microbiology, Washington, D.C.

Bonner, John T. 1971. Aggregation and differentiation in the cellular slime molds. *Annual Review of Microbiology.* **25:** 75–92.

Cole, G. T. 1986. Models of cell differentiation in conidial fungi. *Microbiological Reviews.* **50:** 95–132.

DeMello, W. C. 1987. *Cell to Cell Communication.* Plenum Press, New York.

Dring, G. J., D. J. Ellar, and G. W. Gould. 1985. Fundamental and Applied Aspects of Bacterial Spores. FEMS Symposium. Academic Press, New York.

Gerisch, G. 1987. Cyclic AMP and other signals controlling cell development and differentiation in *Dictyostelium. Annual Review of Biochemistry.* **56:** 853–880.

Killick, K. A., and B. E. Wright. 1974. Regulation of enzyme activity during differentiation in *Dictyostelium discoideum. Annual Review of Microbiology.* **28:** 139–166.

Losick, R., and L. Shapiro. 1984. Microbial Development. Cold Spring Harbor Symposium, Cold Spring Harbor Press.

Moir, A., and D. A. Smith. 1990. The genetics of bacterial spore germination. *Annual Review of Microbiology.* **44:** 531–553.

Newton, A., and N. Ohta. 1990. Regulation of the cell division cycle and differentiation in bacteria. *Annual Review of Microbiology.* **44:** 689–719.

Shapiro, L. 1976. Differentiation in the *Caulobacter* cell cycle. *Annual Review of Microbiology.* **30:** 377–407.

Smith, J. E., and J. C. Galbraith. 1971. Biochemical and physiological aspects of differentiation in the fungi. *Advances in Microbial Physiology.* **5:** 45–134.

CHAPTER 19

The Physiology of Microbial Ecology

Microbial ecology as applied microbial physiology

We study the properties of microbes in pure cultures, but pure cultures seldom occur in nature. Except in extreme environments, microbes are normally found in nature as mixed cultures, composed of many different species, each of which is present as a different part of the whole. The members of a microbial community collectively participate in critical biological processes, but seldom are the collective activities of microbes in natural environments identical to those of the individuals that comprise the community. Although we may study, individually, the features of each member of a microbial community *in vitro*, rarely is the spectrum of activities displayed by an individual organism in pure culture identical to activities of the same organism that are expressed in mixed culture. Although the limits of the possibilities for microbial activity are genetically determined, the expressed characteristics reflect the nature of the environment in which the organism is found.

The test tube or the flask are seldom appropriate simulations of nature. In nature, microbes coexist with each other and with other life forms. Whereas in the test tube, a particular microbe responds to the chemical and physical conditions it encounters, in nature, a collection of organisms are jointly influenced by the environment. By their joint interaction, microbes modify the nature of their environment. This phenomenon often leads to changes in the expressed characteristics of a particular organism, as a function of time, in response to changing environmental conditions. All of these factors make it difficult to understand the role of particular microbes in the environment. We are caught in a dilemma. We cannot understand the functioning of particular microbes without studying their individual properties. On the other hand, it is difficult to determine whether the properties exhibited in the test tube are those expressed in the natural environment. In addition, study of the action of a community of microbes is intrinsically more difficult than understanding a single biological entity. Study of microbial ecology is a challenging and difficult task.

The essentiality of the discipline

Although microbial ecology and its physiological basis are intrinsically difficult subjects, understanding the role of microbes in nature is critical. The activities of microbes are essential to the welfare and survival of all of life. Microbes participate intimately in trapping radiant energy, in fixation of nitrogen, in fixation of carbon dioxide and replenishment of oxygen, in degradation of organic materials, and in detoxification of hazardous materials. In addition, they are essential or integral components of geochemical cycles and are major agents of disease. Understanding microbial interactions is essential for our survival. Such is particularly important in light of the increasingly numerous *xenobiotic* (manmade) chemicals that enter the environment. We increasingly rely on microbes as agents of bioremediation in addition to their normal ecological roles. Study of microbial interactions allows us to understand the role of microbial activity in particular environments. Study of particular environments allows us to understand general biological processes and principles that apply to all environments. Through study of microbial ecology at the physiological level, we may understand not only the manner in which microbes interact, but may also understand more clearly, the interactive processes that occur among more morphologically complicated life forms.

Types of ecological interactions

Microbes are physiologically interrelated to each other, and to other life forms, in a variety of ways. We may classify these interactions in accordance with their consequences. In certain interactions, organisms may coexist with little or no effect upon each other. This condition was, until recently, described by the term *commensalism*, literally, "eating at the same table." More recently, we distinguish two subcategories of the condition in which two organisms may live together without injury. In *neutralism*, the organisms live together without affecting each other. Commensalism is restricted to the condition in which two organisms live together in a manner so that one is benefited and the other is unaffected.

Mutualism is a second mode of symbiotic interaction. In this situation, organisms obligatorily exist together in a way so that both benefit. In addi-

tion to these associations, there are times when organisms live together so that together, they accomplish something that neither organism could achieve alone. We describe such a condition as *synergism.*

Not all symbiotic associations are beneficial or of little consequence. In *antagonistic* interactions, organisms interact in a manner so that one or more of the interacting organisms is harmed. Finally, we recognize *parasitism*, a term about whose meaning there is no uniform agreement. To some parasitism implies a physical condition, the living of one organism within or upon the surface of another living organism. When considered from this viewpoint, *saprophytism* is the condition that pertains when an organism lives on dead or decaying material. It is perhaps more "ecological" to consider parasitism as the condition that exists when an organism lives in association with another that serves as host in a relationship that harms the host. From this perspective, pathogenic microbes may be perceived as displaying a particular type of a more generalized phenomenon, *symbiosis,* the living together of two or more organisms, irrespective of consequences.

The ecological niche

The concept of an ecological niche is central to all of ecology, microbial and otherwise. From the ecological perspective, the term *niche* implies the role that a particular organism displays in its environment. Understanding the nature of the niche that a particular organism plays in the totality of the community is critical to understanding the activities in which an organism may participate, the impact of those activities on the environment, and the factors that influence the activities of the organism under study.

To understand the niche occupied by a particular microbe, it is essential to determine precisely where, physically, the organism may exist. This is determined, in large measure, by the nature of nutritional, chemical, and physical conditions found at a particular location. By identifying conditions essential for the growth of a microbe, we may understand the physiological reasons for its occurrence at a particular location within an ecosystem.

To further assess the nature of the ecological niche of an organism, we must assess the nature of the total population within the environment and the proportion of the total population attributable to the organism under study. It is equally important to identify both the nature of the activities in which the organism participates and the extent to which a particular organism is essential to the functioning of the entire community. Essentiality is largely a function of factors, such as whether or not other members of the community participate in a similar activity, whether the activity is a *macro*activity or a *micro*activity, and the rate at which the process may occur. For a particular organism, occurrence rate often reflects growth rate. The growth of an organism is, in turn, largely determined by its physiological environment. A keen understanding of the physiological properties of a microbe is vital to delineation of the nature and significance of its activities in the environment of which it is a part.

Is the organism significant?

Recognizing that many factors must be considered in determining the significance of a particular microbe in a microbial community and that we must understand the physiology of the organism in intimate detail to assess the organism's importance, questions still remain regarding the criteria by which we may evaluate the consequences of the presence of a particular organism. Is the organism a transient or is it an integral part of the community and how do we decide which is the case? In the early 1950s, Elsden and Phillipson proposed criteria by which to evaluate the ecological significance of an organism. Although their criteria were originally proposed to evaluate gut microbes, particularly those of the rumen, the criteria are generally applicable. Elsden and Phillipson's criteria are the following: 1) The organism must be present in the environment; 2) the organism must be capable of growth at a rate allowing it to remain in the environment; 3) the organism must conduct an activity that occurs in the environment; and 4) the organism must conduct that activity at a rate and extent compatible with its contributing quantitatively to the activity in the environment.

Assessment of importance

Although the criteria of Elsden and Phillipson appear straightforward, definitive answers to the questions raised by the criteria are often difficult to obtain. These difficulties reflect the complexity of environmental factors that influence particular microbes.

The continued existence of a microbe in an environment requires that the conditions within the environment be such that the organism may grow at a rate compatible with its rate of removal from the environment or its death. The precise conditions that allow growth of the organism are determined by compatibility between the physiology of the organism and the nature of the environment. Within a particular environment, *microhabitats* may exist that are compatible with some organisms but incompatible with others. For example, in the mouth, the gingival crevice allows growth of anaerobes and facultative organisms but precludes growth of aerobes and in aqueous environments, the presence of particular phototrophic organisms at particular locations reflects the nature of the light to which they are exposed, the depth of water, and the nature of the pigments in particular species. As an additional example of the effects of environmental factors on the presence of particular organisms at particular locations, the sulfate-reducing bacteria typically occupy locations at which oxygen is absent or minimal, light cannot penetrate, and sulfate is available. Rarely does a single factor determine the presence and persistence of a microbe.

If we can determine the precise conditions compatible with growth and persistence of an organism, questions still remain about the nature and extent of its activities. In this regard two questions are critical: 1) How much activity does the organism exhibit; and 2) how fast does the activity occur? These questions cannot be considered separately. It is only if the organism contributes in a significant manner to the activities of the habitat that it may be considered a significant member of the community, but assessment of the extent of an organism's involvement is difficult, even if quantitatively large amounts of the organism's activity are required for its significance.

The extent of an organism's participation in an activity cannot be determined until we assess not only how much and how fast the organism grows and metabolizes, but also the rate at which the product of action of a particular microbe is used by other microbes. An organism may appear, quantitatively, to exert only minimal influence on the environment when, in fact, such is not the case because as fast as a product is produced by a particular organism, it is used by others. This situation is well illustrated by consideration of situations found in the microbial community of the rumen. In pure culture, hydrogen gas is produced by many rumen microbes, but hydrogen is an exceedingly small portion of the rumen gaseous phase.

The rumen gaseous phase is normally comprised of approximately 60% carbon dioxide and 40% methane. The hydrogen produced by some organisms is used by others to reduce carbon dioxide to methane. In a similar manner, formate and succinate, although products of microbial activity in pure culture, are not normally found as major ruminal metabolites. The synergistic action of the mixed microbial community is such that formate is converted to CO_2 and H_2 that are, in turn, converted to methane, while succinate is converted to propionate. In both of the previously described examples, the combined action of interacting components of the ecosystem is such that substances produced abundantly in *pure* culture do not normally accumulate in the mixed environment.

In contrast to the situations just described, an organism may be ecologically significant even though its activities in the habitat are, in fact, quantitatively small. Production of a vitamin, for example, may require only minimum amounts before the material is physiologically significant. Gut microbial vitamin synthesis frequently provides an abundance of materials sufficient not only for other gut microbes, but for their warm-blooded animal hosts as well.

Microbial interactions as environmental modifiers

Interconnected microbial activities are not always physiologically compatible. The soil harbors an abundance of microbes, particularly in the rhizosphere, where nutrients are plentiful. Soil microbes display an abundance of microbial interactions, many of which are antagonistic. *Antibiosis*, the inhibition of one organism by the metabolic product(s) of another, is a frequently observed phenomenon. Although many antibiotics exert their action on a wide variety of organisms, other substances such as the bacteriocins, antimicrobial substances pro-

duced by certain bacteria that inhibit other taxonomically similar bacteria, are more restricted in their action. The producing organism forms proteins that react with and inactivate the potentially toxic substance, allowing its production by the organism without injury to itself. The material, however, inactivates close taxonomical relatives of the producing organism. Although most studies of such substances thus far done have concerned primarily bacteria-produced materials, the phenomenon is general and understanding its significance in microbial ecosystems is presently imperfect.

Metabolism as an ecological modifier

In addition to formation of intrinsically toxic materials, normal microbial metabolism may serve as a selective mechanism within the ecosystem. Acid production is often a selective factor. *E. coli*, for example, although an invariant gut inhabitant, is normally a quantitatively insignificant fraction of the normal human gut microflora. The organism is inhibited by the propionic, acetic, and butyric acids that result from carbohydrate fermentation by other species. Acid production is also an environmental determinant in the rumen ecosystem. Although it is frequently the case that the microbial population of the rumen contains both bacterial and protozoal species, the protozoal population normally is absent if the rumen pH is lower than 5.0. At pH values below 5.0, the protozoa are killed or inactivated, but the bacterial population remains large and is sufficient, by itself, to provide materials essential for the ruminant animal. If the rumen pH decreases to 4.0, another acid-related effect is observed. In this situation, normally induced by ingestion of readily soluble nutrients, rapidly growing lactic acid-producing bacteria lower the pH and inactivate the normal microflora.

Competition for nutrients

Competition for nutrients is a determinant in microbial ecosystems. In the human gut, for instance, bacteria are normally the predominant microbes and yeasts constitute only a small portion of the total microbial population. Because of their rapid metabolism, large surface-to-volume ratio, and the relative ease with which they grow at body temperatures, bacteria normally outcompete the fungi for nutrients in the gut environment. However, during normal flora disruption, such as may occur as the result of antimicrobial chemotherapy, the diminished bacteria population may allow development of a substantial gut yeast population. The bacteria no longer compete with the yeasts for nutrients and the concentration of bacterially produced inhibitory metabolic products is diminished, allowing the yeasts that would normally be inhibited, to thrive.

Although, in the previously described situation, the presence of an inhibitory substance, an antibiotic, facilitates a nutritionally influenced microbial change, nutrient competition may serve as a selective environmental factor in the absence of inhibitory chemicals. Intrinsic genetic differences among organisms in growth rates, transport systems, enzymatic sequences, and regulatory mechanisms may facilitate the growth of one organism at the expense of another. Nutrient availability is also a selective factor. Readily soluble nutrients are utilized more rapidly than are those produced from degradation of polymeric molecules, and species that may use readily available nutrients have a selective advantage over those whose growth requires polymer digestion.

Growth factors

The functioning of microbes in an ecosystem is highly influenced by the presence or absence of major nutrients, such as carbon and energy sources, but equally reflects the availability of micronutrients such as vitamins and growth factors. If the latter are absent in the environment, they may sometimes be supplied by microbial activity. Cellulolytic gut bacteria, for example, require, or are stimulated by, branched-chain organic acid growth factors, which are intimately involved in both amino acid and lipid synthesis by the organisms. Branched-chain organic acids normally result from proteolysis by noncellulolytic organisms. Thus, in the gut ecosystem, proteolysis and cellulose digestion are intimately connected phenomena. The rate of cellulose digestion is a function of the availability of growth factors whose formation results, primarily or exclusively, from protein digestion.

Strategies for study of microbial ecosystems

The complexity of microbial ecosystems makes their study difficult. Assessment of the nature of the habitat is, by itself, a challenging task. It is intrinsically easier to study physically homogeneous environments such as the gut, rather than habitats such as the soil, which is both physically and chemically heterogeneous and contains a diversity of microhabitats. For critical evaluation of a habitat, it is essential that all significant microhabitats and their associated microbes are identified and studied. Although complete identification of the habits and microbes in an environment is essential, it is, practically speaking, an ideal. Even with relatively simple habitats, identification of all of the pertinent microhabitats and their associated microbes is difficult, if not impossible.

In addition to microhabitat identification and study, efforts must be made to identify, both qualitatively and quantitatively, the critical processes in which the microbial community participates, and to identify and cultivate the critical organisms involved. Attempts to cultivate organisms from little studied ecosystems are intrinsically difficult and frustrating. Study of the nature of the physical, chemical, and biochemical aspects of the habitat is essential, but is often insufficient to delineate the critical factors required to cultivate important members of the microbial community. Careful attention must be paid to correlation between microscopic and cultural results. Although it is indeed possible to miss a quantitatively small microbial component of the ecosystem, failure to cultivate a morphotype detectable microscopically is a source of major concern since it usually indicates substantial inability to simulate the natural environment. At the same time, cultivation of an organism that *cannot* be detected microscopically is of concern since such a situation may reflect the use of conditions so selective that the isolated organism may not be representative of the community under study. At least initially, it is wise, if not essential, to use cultural conditions that simulate the environment to the greatest extent possible, if we would recover *representative* organisms. With all of their difficulties, the criteria of Elsden and Phillipson are helpful in discerning the truly significant organisms and may allow distinction between organisms that are truly important and those which are not.

To the greatest extent practicable, efforts should be made to assess the degree to which cultural and microscopic results agree, that is, to determine in a quantitative way the relationship between the microbes *observed* in a habitat and those that are cultured by the methods at hand. Assessment of information in this regard is difficult, because even with the best habitat simulation, discrepancies exist between the number and nature of organisms observed and those that can be cultured under a given set of conditions. It is virtually impossible to devise cultural conditions that simultaneously allow growth of all the important microbial components of the habitat. Indeed, even with detailed knowledge about the habitat under study, it is usually impossible to devise cultural conditions of *any* type that allow cultivation of *all* the pertinent organisms.

When discrepancies are found between the number and nature of the microbes observed in a habitat and those that can be cultured, attempts should be made to determine the reasons for the discrepancy. While it is virtually certain that a portion of the nonrecovered organisms are dead, the possibility that some of the organisms, although not dead, may be incapable of growth under the established conditions should be studied. This possibility can be studied with radiochemicals, vital stains, and selective metabolic inhibitors.

The physical and biological complexity of natural environments is such that complete understanding of most, if not all, ecological systems is currently impossible, particularly at the microbial level. The diversity of activities in which microbes participate and the complexity of their interactions are tremendous. Although we may understand the coordinated action of pairs of microbes in certain circumstances, it is increasingly understood that many biological processes are mediated by *consortia*, collections of several interacting microbial species. Understanding the functioning of interacting microbes becomes increasingly difficult as the number of microbial components of the system increases, because the number and nature of the physiological interactions of a particular consortium, and the consequences of consortial action on the environment, become increasingly complex as the diversity of the consortium increases. Although

it is highly desirable that such systems be understood, understanding the operation of most microbial consortia requires more intimate knowledge of the nature and physiological functions of its members than is presently possible.

Study of incompletely characterized ecosystems

In many situations, it is impossible or impracticable to assess a natural environment by culturing the microbial species within it. Even when such is the case, it is possible to determine, to a certain extent, the nature and activities of a microbial population. We may extract and characterize environmental DNA by sequence analysis of cloned DNA fragments from an uncharacterized ecosystem. Comparison of environmentally derived sequences with sequences of known, previously characterized, organisms can provide insight into the general nature of an unknown microbial population. However, the use of noncultural methods must be done with caution and often reflects an inability to adequately simulate an unknown environment. Genetic analysis of unknown habitats does not allow identification of the components of the system with precision and fails to allow study of their interaction.

Summary

Understanding the collective activities of microbes in natural environments is a complex and challenging task. Although we may delineate the nature and properties of the microbial components of an environment, determine the environmental factors that allow the organisms to grow, and identify the nature and rate of processes in which the organisms participate, assessment of the functioning of organisms in nature is difficult because we cannot be assured that the properties displayed by the organisms in the laboratory will be displayed in the natural environment. Even when this is the case, it is often found that the manner in which particular microbial activities are manifested in nature differs substantially from that displayed in the laboratory. The discrepancy between the laboratory and the environment reflects not only differences between the *in vitro* and *in vivo* environments, but also a variety of interactions between the components of a mixed microbial population. Understanding those interactions and their consequences is essential if we would understand the functioning of a microbial community. In most, if not all, cases, complete understanding of the functioning of a microbial ecosystem is an ideal rather than a reality.

Selected References

Andrews, J. H. 1991. *Comparative Ecology of Microorganisms and Macroorganisms.* Springer-Verlag, New York.

Atlas, R. M. 1987. *Microbial Ecology: Fundamentals and Applications.* Addison Wesley, Reading, Massachusetts.

Bauchop, T., and R. T. J. Clarke. 1977. *Microbial Ecology of the Gut.* Academic Press, Inc., New York.

Hentges, D. 1983. (ed.) *Human Intestinal Microflora in Health and Disease.* Academic Press, Inc., New York.

Horikoshi, K., and W. D. Grant. 1991. (eds.) *Superbugs: Microorganisms in Extreme Environments.* Japan Scientific Societies Press, Berlin.

Hungate, R. E. 1966. *The Rumen and its Microbes.* Academic Press, Inc., New York.

Mitchell, R. 1992. (ed.) *Environmental Microbiology.* Wiley-Liss, New York.

Seidler, R. J., M. A. Lewin, and M. Rogul. (eds.) 1992. *Microbial Ecology: Principles, Methods and Applications.* McGraw Hill, New York.

CHAPTER 20

Molecular Microbial Physiology

Why should we be concerned with molecular physiology?

A chapter entitled "Molecular Microbial Physiology" may seem to some, at first glance, simplistic, confusing, or unnecessary, depending on their viewpoint. "Why should we have such a chapter?" some might ask. Is not all of microbial physiology and metabolism molecular? Others, who perceive microbes as genetic tools and consider only DNA, RNA, and protein important, might say, "Why bother to have such a chapter? All of biology can be explained in terms of DNA, RNA and protein. The other molecules are unimportant!" The truth is somewhere in between such thoughts. It is most assuredly true that all of microbial physiology is molecular, but it is also true that DNA, RNA, and protein are critical and special molecules because they determine the physiological limitations of an organism. However, under a given set of conditions, an organism's manner of genetic expression determines its physiological response. To fully understand microbes and how we may best exploit them by genetic manipulation, we must understand both the manner and the functioning of their genetic aspects and the way in which changes in genetic expression are reflected as phenotypic change. In our concern with their genetic aspects, we must not forget that microbes are more than simply storage containers for enzymes and other macromolecules. They are living, viable entities with discrete and functional regions and structures that collectively accomplish critical life processes. If we would understand the nature of those processes, and the ways in which they are accomplished, we must understand the functions of all of the parts of the cell. Our understanding must include the intimate details of protein and nucleic acid metabolism and function, but cannot be restricted to such matters. To fully understand and exploit microbes, we must understand the molecular aspects of all cell processes and all cell molecules, genetic and otherwise. We must become molecular microbial physiologists.

Applied molecular microbial physiology

From the advent of "molecular" biology, some have regarded microbes simply as genetic tools, vehicles for genetic manipulation rather than as organisms in their own right. Although we may regard microbes as genetic tools, to regard them *only* from this perspective is inappropriate for at least two reasons. By considering microbes simply as "gene and protein factories" we: 1) circumscribe their usefulness and 2) neglect the fact that microbes are physiological entities that function in natural environments, in addition to whatever use they have in the laboratory. To fully exploit the genetic potential of microbes, we must understand them as completely as possible so that we may use them not only in the laboratory to produce genes and proteins, but also in nonlaboratory environments as ecological agents. If we would use genetically altered microbes for nonlaboratory purposes, we must understand the extent to which the physiological responses to genetic change in the laboratory approximate those that occur in nature.

The use of microbes to produce nucleic acids and proteins

Although our ultimate goal is to understand the totality of the physiology of microbes at the molecular and genetic levels and to practically exploit that knowledge in a variety of ways, at the present time, protein and nucleic acid production are the physiological aspects of microbes most extensively studied and exploited by genetic manipulation. Manipulation of microbial physiology for protein and nucleic acid synthesis is most useful and has allowed the production of a variety of nonmicrobial proteins and large amounts of nonmicrobial DNA in microbial systems. The details by which both processes are accomplished are described in many manuals that the student who desires more detailed information is invited to consult. It is intended here to describe general strategies for protein and nucleic acid production by microbes and to indicate their physiological implications and consequences.

Strategies for DNA and protein production by genetic engineering

DNA and protein production by the techniques of genetic engineering are intimately interrelated, because without the gene, we can have no protein, and without an expression system, the gene is, practically speaking, useless. To produce protein by genetic engineering we must accomplish a collection of tasks. First, we must identify and isolate the gene for the desired protein. Next, we must clone the gene. After cloning the gene, we must insert it into an expression system—a carrier system commonly known as a *vector*, so that we may induce the vector to form protein. Finally, we must purify the product.

Gene isolation

Gene isolation is undoubtedly the most difficult aspect of genetically engineered protein production. A number of strategies are available by which genes may be isolated. Precisely how we proceed depends upon our purposes and the cell type we are using. Irrespective of cell type, however, two basic approaches are available. We may isolate the gene *directly* or we may isolate the *message* for the protein for which the gene codes and with reverse transcriptase, synthesize the gene.

Often we seek to isolate the gene directly and proceed in the following way. We obtain a purified DNA preparation, digest it with restriction enzymes into a collection of fragments, separate these fragments from each other by electrophoresis on the basis of size and charge, denature the resultants, and transfer single strands of the isolated fragments to a membrane filter. We may then "probe" the fragments with a detection agent designed specifically to identify the gene. Normally, the detection probe is a chemically synthesized, radiolabelled nucleic acid sequence complementary to a portion of the gene of interest, although methods are now available to identify genes with nonradioactive probes.

Great care must be taken in preparation of a probe, since if it is not specific, it may identify a piece of DNA that does not contain the gene. Probe preparation is complicated by the fact that in most cases, because of code degeneracy, more than one DNA code describes a particular amino acid. The probe must be designed so that it will "cover all the bases." Unless we have specific information regarding which of the various codons for an amino acid are used preferentially (the codon usage frequency), it is necessary that the probe include all of the possibilities. This necessity diminishes the specificity and sensitivity of the probe but is unavoidable in the vast majority of cases. In devising a suitable probe, it is preferable, if not essential, to obtain a pure sample of the protein whose gene we seek to isolate and to determine its sequence. It is only in this way that we may determine the most appropriate probe, which should be, *to the greatest extent possible*, specific for the protein in question, and whenever possible, representative of unique sequences within it. It is often useful or essential to use more than one probe to identify a gene, usually a probe for its carboxy-terminal end and a second one for its amino-terminal. With care, and preferably with the guidance of experienced workers, we may identify and isolate a gene with the previously described techniques. The general procedures required for gene isolation by this approach are shown in Figure 20.1.

An alternative strategy for gene isolation is to synthesize it from its message. Such an approach is often used in eucaryotic systems. The most critical aspect of this approach is messenger RNA isolation. If possible, we begin with a system actively synthesizing the protein of interest, thereby enhancing the prospects of isolating its message.

Several approaches are possible to isolate mRNA. One approach exploits the fact that eucaryotic mRNA molecules typically contain a series of adenine residues at their 3' ends. A column containing immobilized thymidine residues (a poly T column) allows us, by complementarity, to selectively isolate mRNA molecules, but does not, by itself, identify any particular mRNA. Identification may be accomplished by use of specific labelled probes, in this situation, normally a radiolabelled portion of the gene to which the message is complementary. Prior to probing, the mRNA is chemically eluted from the poly T column, purified by selective enzyme digestion and by electrophoresis, and

attached to a membrane filter. The techniques just described are shown schematically in Figure 20.2.

Precipitation of a rapidly synthesizing mRNA-ribosome-protein complex by an antibody specific for the translation product of a particular message is also a useful method for obtaining particular RNA messages. This method takes advantage of the fact that during active protein synthesis, a significant portion of the transcribed messages are, at the same time, attached to ribosomes that simultaneously translate them, forming an incipient polypeptide. It is thus possible to use an antibody to the translation product to precipitate the entire complex that is synthesizing a particular protein, including the portion of the complex that contains its specific mRNA molecule. After precipitation of the complex, the mRNA is purified by selective enzyme digestion, electrophoresis, and other procedures. The general procedures used for mRNA purification with specific antibodies are shown in Figure 20.3. Confirmation of the nature of the isolated mRNA can be obtained by probing it with appropriate DNA or RNA probes, as described earlier.

However a particular species of mRNA is identified, the next step is to synthesize the gene. One accomplishes this process by the use of reverse transcriptase, also known as RNA-dependent DNA polymerase. As its name implies, RNA-dependent DNA polymerase forms a single strand of DNA from an RNA template. The primer for action of this DNA-polymerizing enzyme is normally a poly T sequence complementary to the poly A region of the 3' end of the mRNA molecule. Attachment of a poly T primer to the message allows the first DNA strand to be synthesized in the 5' to 3' direction. By hairpin formation, because it is a multifunctional enzyme, RNA-dependent DNA polymerase can also synthesize the second DNA strand. More often, however, DNA-dependent DNA polymerase I, in combination with an enzyme that nicks the mRNA strand (RNAase H) is used to form the second DNA strand. The nicking enzyme provides the RNA primers from fragments of the original mRNA molecule that are required for operation of the DNA-dependent DNA polymerase I in the synthesis of the second strand.

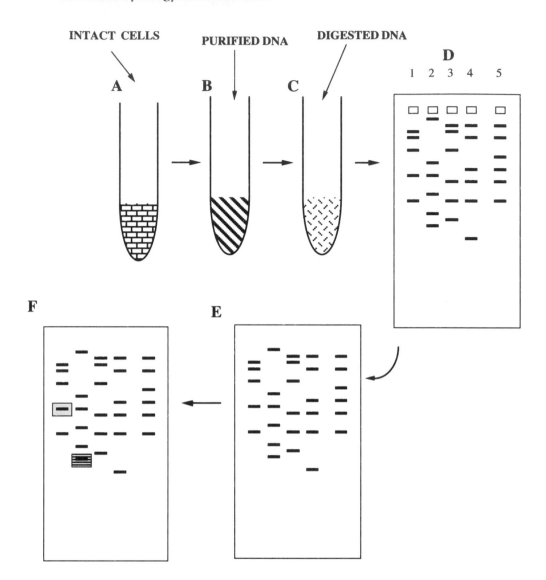

FIGURE 20.1 The general procedures for direct isolation of a gene. (A–C) Intact microbial cells (A) are extracted to produce purified DNA (B). The latter is digested by restriction enzymes (C) to produce fragments of DNA of varying sizes, some of which contain the gene of interest and some of which do not. (D) The fragments are separated according to size by electrophoresis, along with known molecular weight markers. In (D), the lanes numbered 1 through 4 are electrophoretically separated digests from the actions of different restriction endonucleases, while lane 5 is a separation of DNA digested to produce known molecular weight markers. From left to right, the lanes in the remaining diagrams correspond to those of (D). (E) The digested and separated DNA is denatured and transferred, by the techniques developed by F. M. Southern, to a membrane filter. (F) The filter is "probed" with oligonucleotides oligomers complementary to the gene of interest to identify digestion bands that contain the gene. In the present example, fragments identified by different probes are highlighted. When genes are identified, they may be cloned.

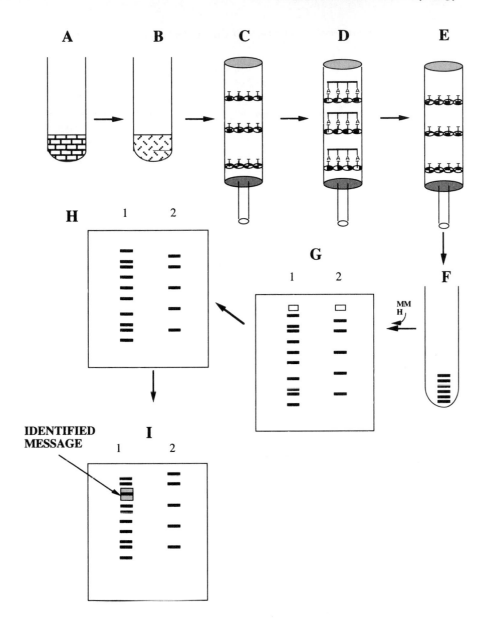

FIGURE 20.2 Isolation of messenger RNA with a poly T column. (A,B) The cells (A) are extracted to produce a crude RNA preparation (B). The crude RNA preparation is applied to a column containing particles that have thymidine (T) residues attached (C). Because of the fact that eucaryotic mRNA molecules have a series of adenine (A) residues at their 3′ ends, eucaryotic mRNA molecules adhere to the poly T column (D). The adhered message molecules are eluted from the column (E,F) and the message molecules (F) are denatured with methyl mercury hydroxide (MMH) and electrophoresed to separate mRNA molecules of various sizes (G). The electrophoresed RNA is transferred to a membrane (H) and examined with a probe complementary to the message desired (I). The identified message is then produced in preparative amounts and used to synthesize the gene for which it codes.

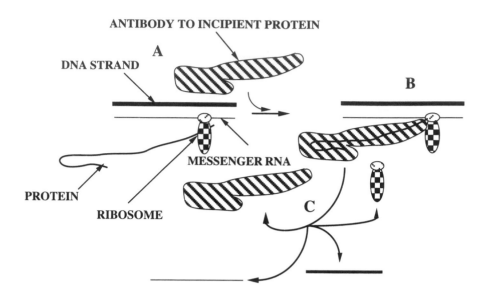

FIGURE 20.3 Messenger RNA isolation with antibody. (A) A rapidly operating, protein synthesizing system is treated with an antibody specific to the protein of interest, producing a complex between the protein synthesizing system and the antibody (B). (C) Centrifugation of the complex leads to specific precipitation of a particular message molecule and its associated protein synthesizing system. By a variety of methods, the components of the entire system are separated and the specific mRNA molecule is used to synthesize its gene.

Cloning the gene

Once a gene has been obtained, it must be cloned. A variety of procedures are available for this purpose. The precise procedure selected will depend on many factors, particularly: 1) the purposes of the investigator; 2) the size of the DNA to be cloned; and 3) the nature of the cloning vector. When relatively small segments (1 to 10 kbp) of DNA are to be cloned, plasmids are excellent cloning vehicles. A great variety of cloning plasmids are available that differ in their properties. The major ways in which cloning plasmids differ are in their size, the nature and number of their restriction sites, the ways in which insertion of foreign DNA into the plasmid may be detected, and the manner in which plasmid insertion into cells may be detected.

A schematic drawing of a typical cloning plasmid is shown in Figure 20.4. To be useful as a cloning vector, the plasmid must contain, at a minimum, an origin of replication, a cloning region, and a selection region, the region that codes for a property that allows selection of plasmid-containing cells. The selection region is normally a region that codes for antibiotic resistance. Ampicillin is the antibiotic most frequently used to select for cells containing inserted plasmids.

In addition to the properties indicated earlier, the plasmid must have a "recognition" region, a region that allows recognition of the insertion of foreign DNA into the plasmid. In many plasmids, recognition of foreign DNA insertion into a plasmid takes advantage of the *lacZ* gene, the gene that codes for the enzyme β-galactosidease. The plasmid is constructed so that under nonrecombinant conditions, it codes for formation of the amino-terminal region of the enzyme. The *host strain*, in which the plasmid is grown, is arranged so that it codes for the carboxy-terminal region of the gene. In the presence of isopropyl thiogalactoside (IPTG) both the host and the plasmid synthesize a portion of the enzyme, allowing formation of functional β-galactosidease by α-complementation. In the presence of the artificial substrate *X-gal* (5-bromo-4-chloro-3-indole-β-D-galactoside), under the influence of functional β-galactosidease, blue-colored colonies are formed. When foreign DNA is introduced into the plasmid in the general region that

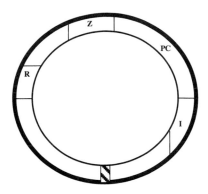

FIGURE 20.4 The general nature of a "typical" cloning plasmid. Except for the origin of replication, which is depicted by the striped region, the various regions of the plasmid are delineated by the lines that surround the bold letters. The I and Z regions are two of the genes in the lac operon. The PC region is the polycloning region into which foreign DNA may be inserted, with the aid of specific restriction enzymes. The R region is the region that codes for antibiotic resistance, normally to ampicillin. This region allows selection for cells that have received a plasmid. Cells that have received a recombinant plasmid appear white in color because the normal functioning of β-galactosidease is disrupted. Nonrecombinant plasmid-containing colonies appear blue because complementation between host cell and plasmid DNA allows normal functioning of β-galactosidease in such cells.

codes for the α-region of β-galactosidease, the formation of functional enzyme is abolished. Colonies produced by cells of this nature appear white. Nonrecombinant plasmid-containing colonies are blue and ampicillin-resistant, while recombinant colonies are white and ampicillin-resistant. Picking of white colonies to fresh, ampicillin-containing agar plates, incubation, and screening of plasmid DNA prepared from the white colonies grown in ampicillin broth media with appropriate DNA probes, allows identification of the recombinant plasmid that contains the gene of interest. Growth of the identified colony in ampicillin-containing broth allows production of large amounts of the desired plasmid DNA, from which the gene of interest may be extracted.

Insertional inactivation is an alternate method for identifying recombinant plasmids. In this procedure, foreign DNA is inserted into a plasmid that contains genes for resistance to two antibiotics, frequently tetracycline and ampicillin. The

foreign DNA is inserted into the region that codes for tetracycline resistance. Recombinant plasmid-containing colonies are distinguished from non-recombinant plasmid-containing colonies by the fact that as a result of foreign DNA insertion, recombinant plasmid DNA-containing cells are sensitive to tetracycline, and nonrecombinant colonies contain plasmids that retain tetracycline resistance. Both recombinant and nonrecombinant plasmid-containing cells display ampicillin resistance, since it is retained during foreign DNA insertion, even though the insertion process abolishes tetracycline resistance. Insertional inactivation is illustrated in Figure 20.5.

The general procedures for obtaining a recombinant plasmid for cloning are shown in Figure 20.6. The plasmid and the foreign DNA are digested by one or more restriction endonucleases. When possible, it is desirable that two enzymes are used. Use of two enzymes allows insertion of the foreign DNA into the plasmid in a particular orientation and also minimizes spontaneous recylization of the plasmid, that can be further minimized by treatment of the digested plasmid DNA with alkaline phosphatase. The digested plasmid and foreign DNA are recombined with the aid of DNA ligase. The recombinant plasmid is then used to artificially transform appropriate host cells made permeable by treatment with divalent cations, notably calcium.

Cloning in phage λ   Bacteriophage λ is a temperate, double-stranded DNA phage that may replicate either by lysogeny (i.e., integration of phage DNA into the host cell chromosome) or by the lytic mechanism. During the lytic mechanism, the DNA in the phage particle is linear. When a susceptible cell is invaded, λ DNA becomes circular, and for a period of time, replicates by the theta mode. As replication proceeds, the mode of viral DNA replication changes from the theta mode to the rolling circle mode, producing linear molecules that are incorporated into mature phage particles. The general nature of replication of λ phage is shown in Figure 20.7.

Phage λ may be used as a vehicle by which to clone a gene. The ability to use phage lambda as a cloning agent depends upon a number of unusual features of the virus. To begin with, its DNA contains repetitive sequences at its ends, which during replication, are cleaved so that the ends are

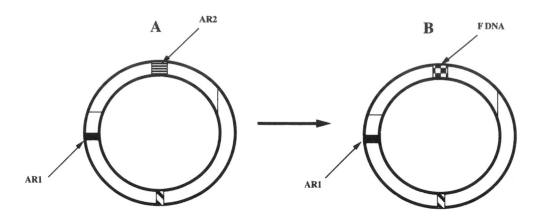

FIGURE 20.5 Insertional inactivation as a means of detecting incorporation of foreign DNA into a plasmid. (A) A plasmid is constructed so that it has two regions that code for antibiotic resistance, AR1 and AR2. (B) By the use of restriction enzymes, a region that contains the resistance gene for AR2 is removed and replaced with foreign DNA (F DNA). The plasmid-containing organism is then grown in the presence of the antibiotic whose resistance is coded for by AR1. Recombination is recognized by the detection of clones that are resistant to the AR1 antibiotic, but have lost their resistance to the antibiotic associated with AR2. Insertion of foreign DNA (F DNA) into the region that coded for resistance to AR2 and removal of the AR2 gene has caused the organism to lose its ability to resist AR2. Although the example shown illustrates the use of two antibiotic markers, the loss of almost any characteristic might be used to assess recombination as long as the plasmid retained a mechanism to allow selection, e.g., a gene for resistance to a different antibiotic.

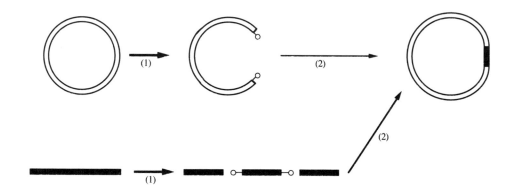

FIGURE 20.6 The procedures for obtaining a recombinant plasmid. (1) The plasmid and the foreign DNA are cut with the same restriction enzyme, producing fragments of both that have compatible ends. Whenever possible, it is desirable to cut with two different enzymes. Such a procedure enhances the specificity of cleavage and, in some cases, allows directional insertion of the foreign DNA. (2) The cleaved pieces of plasmid and foreign DNA are joined by DNA ligase, and the resultant recombinant plasmid is recovered with a selection marker on the plasmid.

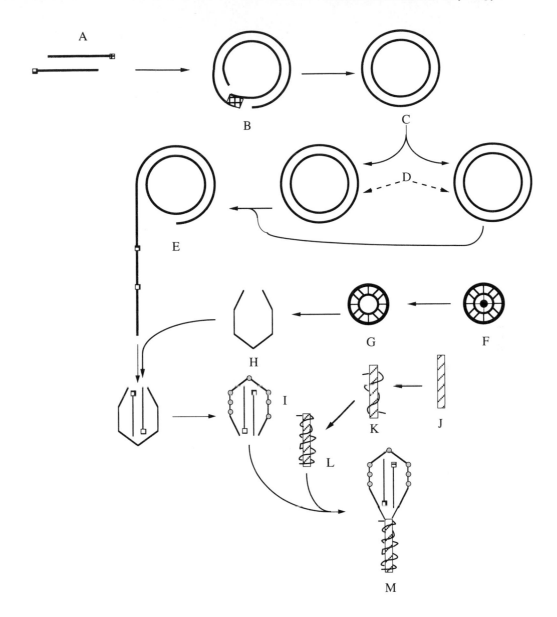

FIGURE 20.7 The general processes involved in the lytic mode of λ phage replication. (A–C) The linear phage DNA becomes circular when the phage invades the cell. The cohesive ends of the strands cause the extracellular linear DNA to become circular and the initial gaps in the circular intracellular entity are closed by host cell polymerase activity. (D) The circular DNA is replicated initially by the theta mechanism. (E) Further DNA replication occurs by the sigma mode, producing concatemers that are delineated by *cos* sites (the small squares). The phage head (H) is formed from a scaffold prehead (F) that is converted into a prehead (G) by removal of the scaffold. The head is then filled with DNA that is cut at the *cos* sites, by the *ter* function of the A replication protein. The filled head is coated with decoration protein and combined with tails (L) formed from precursors (J,K) to produce the complete phage particle (M).

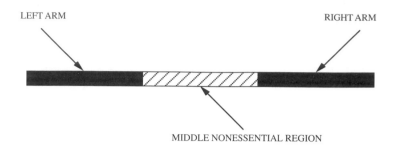

FIGURE 20.8 The structure of the λ phage genome. The genome is composed of a right arm, a left arm, and a middle section that is not required for phage replication. Engineered forms of the phage have been developed that contain restriction enzymes unique to the middle zone. Such vectors may contain one or two unique restriction sites within the middle region. Vectors that contain only one site are known as "insertion" vectors and vectors that contain two unique restriction sites in the middle zone are known as "replacement" vectors. When phage λ is used as a cloning or expression vector, the middle portion of the genome is removed and replaced with a piece of foreign DNA of a size similar or identical to it. When skillfully done, the resultant recombinant DNA piece is incorporated into mature λ phage particles and the foreign DNA is replicated when the modified phage particles participate in the lytic cycle of reproduction. Some genetically modified phage particles contain promoters and other features that allow the phage to be used, not only to clone foreign DNA but also for production of foreign proteins. For λ to be used in either capacity, care must be taken to ensure that the modified phage genome is of an appropriate size so that it will be incorporated into the phage head. Modified λ genomes containing less than 78% or greater than 105% of the wild type λ genome will not be highly viable. The use of phage λ allows manipulation of substantially larger amounts of foreign DNA than is possible with many other cloning agents.

cohesive. The cohesive regions of the lambda genome are known as *cos sites.* Possession of cos sites allows lambda DNA, upon host cell invasion, to become circular and to be replicated intracellularly for a substantial period of time by the theta mechanism, allowing production of very substantial amounts of DNA.

A second factor that facilitates use of phage λ as a cloning vehicle is the fact that approximately 1/3 of the λ genome (Figure 20.8) is not required for replication of the virus. Insertion of foreign DNA into this region allows its amplification during the normal processes of phage replication. The final aspect of phage lambda that makes it a useful cloning vector is the complexity of its replication processes. The lambda virus is composed of a head and a tail. Each of these entities is formed from precursors, and after their formation, heads and tails are assembled to form the intact virus. The intact virus is released from the cell. All of the subprocesses of phage lambda replication are under genetic control. The A gene product, for example, cuts concatemeric forms of lambda DNA produced from the rolling circle mode replication at the cos sites, into pieces of sufficient length for incorpora-

tion into phage heads and, in addition, facilitates insertion of DNA into the head. The D and E gene products also participate in head formation. By combining a series of mutants of phage λ together physically, a mixture of mutant forms can be obtained that allows the packaging of DNA to which foreign DNA has been added into heads and the formation of functional, genetically modified λ phage particles, which are used to amplify foreign DNA. Packaging mixtures contain mutants of phage lambda, all of which participate in portions of the replication cycle but none of which can, separately, carry out the replication process. The general procedures used to obtain functional, genetically altered lambda particles are shown in Figure 20.9. The use of phage λ as a cloning vector permits cloning of substantially larger fragments of DNA than is normally possible with plasmids.

The versatility of phage λ Phage lambda is extremely versatile as a genetic engineering tool. It may be used both as a cloning vector for DNA production and as an expression vector that allows the formation of proteins. The ability to use phage λ in a variety of ways is the result of detailed studies of

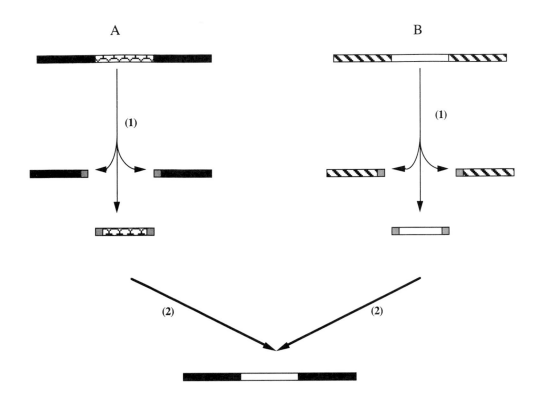

FIGURE 20.9 Procedures required to obtain a piece of recombinant DNA suitable for cloning in bacteriophage λ. The λ DNA (A) and the foreign DNA (B) are digested with the same restriction enzyme (1). When large amounts of DNA are involved, it is necessary to use a replacement vector rather than an insertion vector. Digestion with a common enzyme produces fragments that have ends (denoted by stipling) compatible in subsequent ligation reactions. The compatible fragments are treated with a ligase enzyme to produce recombinant pieces that contain the genes for phage replication plus a piece of foreign DNA in place of a portion of the middle section of the λ genome. The recombinant piece is used to produce recombinant λ phage and, in the process, produces foreign DNA or protein.

the virus. Analysis of its DNA has allowed identification of its restriction sites, the genetic locations that are attacked by specific endonuclease enzymes and lambda mutants have been obtained in which particular restriction sites have been removed from the portions of the genome required for phage replication. These sites have, however, been retained at one or two sites within the nonessential portion of the genome. Mutants of this nature are used to incorporate foreign DNA specifically into the nonessential region of the lambda genome.

The distinction between cloning vectors that amplify DNA and expression vectors that produce foreign proteins, is not absolute. Mutants of phage lambda (e.g., λgt10) may be specifically used for DNA cloning, whereas other mutants (i.e., λgt11)

may be used as expression vectors. Lambda expression vectors contain promoters that allow transcription of both lambda and foreign proteins during virus replication. The nature of the processes required to use phage lambda for cloning or expression are shown in Figure 20.10.

Cosmid cloning

Cosmids are plasmid-cloning vectors that contain, at a minimum, an origin of replication, an antibiotic resistance marker, a number of restriction sites, and a copy of the cos sites required to package DNA into phage lambda particles. The presence of these elements allows the use of cosmids as cloning vectors. They are particularly useful in cloning

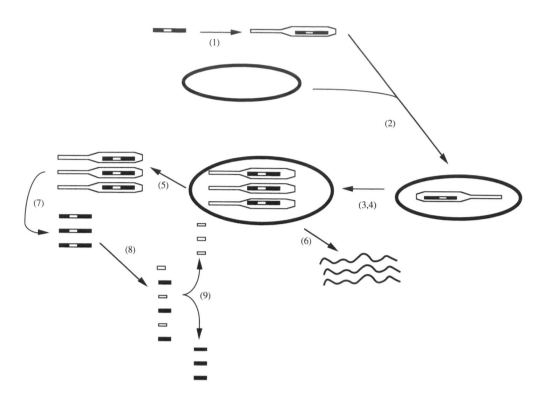

FIGURE 20.10 The processes required to use λ phage DNA for cloning or expression. (1) Genetically altered phage DNA, prepared as described in Figure 20.9, is converted to a complete phage particle with a "packaging mixture." (2) The complete phage is used to infect *E. coli* cells so that the phage may replicate itself (3–5) by the lytic mode and during the replication process, amplify the foreign DNA as well. (7) If the phage is a cloning vector, amplified foreign DNA is produced, but no foreign proteins are obtained. If the phage is an expression vector, and is induced, foreign proteins may be produced (6) during phage replication. Induction may occur because the phage DNA contains artificially added or naturally occurring inducible promoter regions if the inserted DNA is transcribed under the influence of phage promoters. If DNA is amplified, it is purified (8) and digested (9) with the same restriction system used to produce it so that, after digestion, the foreign DNA may be separated from phage DNA. The foreign DNA is then probed to identify the gene of interest. In many cases, plaques resulting from recombination are screened directly, without digestion and extraction of their DNA.

very large segments of DNA—fragments typically between 35 and 45 kb in size. Cloning is achieved by digesting both the plasmid and the foreign DNA segment with a restriction enzyme that restricts in such a way as to yield compatible ends between the cosmid and the foreign DNA piece. The digested pieces are then ligated, normally with phage T_4 DNA ligase, to produce pieces of foreign DNA surrounded by cos sites. Use of this linear DNA to produce infectious phage lambda particles and use of the particles to infect susceptible cells produces cosmid plasmids (cosmids) that contain the ability to replicate, an antibiotic resistance marker, cos sites, and a large fragment of foreign DNA. In this situation, the lambda particles, al-

though infective, do not replicate since they lack the genetic elements required for replication and contain only the cos sites required to package DNA into lambda particles. Recombinant, cosmid-containing cells may be selected for by growth in agar media containing the substance to which the cosmid conveys resistance and the clones containing the gene of interest can be identified by probing colonies or the cosmid DNA preparations obtained from them. When the clone of interest has been identified, it can be cultivated in an antibiotic-containing broth medium and the recombinant DNA can be purified by the normal techniques for plasmid DNA isolation.

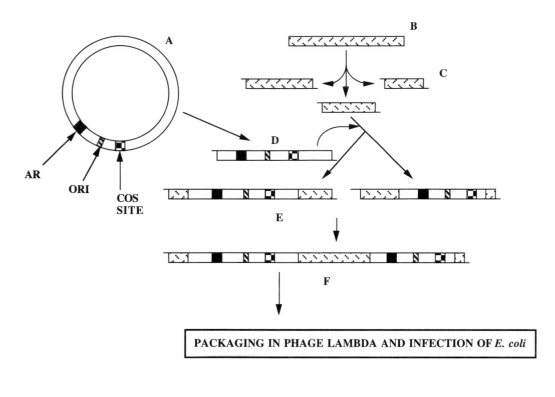

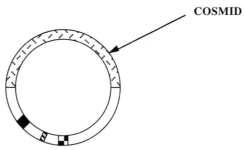

FIGURE 20.11 General procedures for preparation of cosmid cloning vectors. A cosmid, containing a replication origin (ORI), an antibiotic resistance gene (AR), and a *cos* site from bacteriophage λ (A) is cleaved with a restriction enzyme to produce fragments containing the origin, the resistance marker, and the a *cos* site (D). Foreign DNA (B) is digested with the same enzyme to produce fragments (C) compatible with the digested cosmid (D). Ligation of these entities (E) produces concatemeric fragments in which foreign DNA is accompanied by replication origins and resistance genes and is surrounded by *cos* sites (F). When these entities are used to produce λ phages and the latter are used to infect *E. coli*, recombinant cosmids are produced that allow foreign DNA amplification.

Cosmids can be used either to amplify DNA in a particular cell or to transfer DNA from one recombinant cell to another cell by transduction. When the transfer technique is done, the cosmid-containing organism is usually grown in the presence of a phage λ strain capable of replication. During phage replication, the cosmid DNA is packaged into phage heads because of cleavage specifically at the cos sites. The phages thus obtained may be used for transduction. The techniques commonly used for cosmid cloning and for cosmid transduction are shown in Figure 20.11. Because they allow cloning of relatively large DNA fragments, cosmids are particularly useful for study of eucaryotic genes.

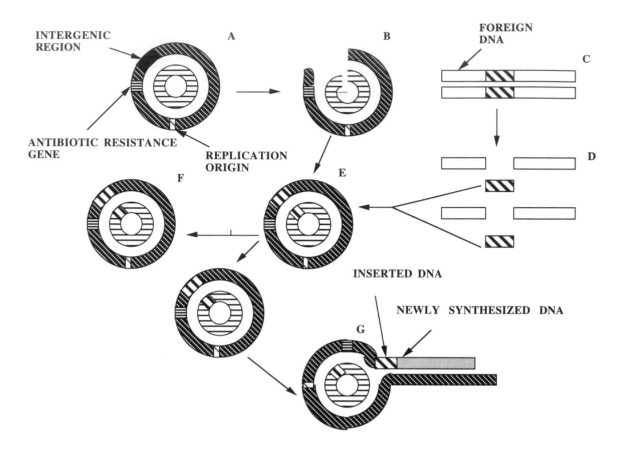

FIGURE 20.12 The use of phagemids as cloning vectors. The double-stranded replicative form of the phagemid (A) is cleaved within the polycloning site in the intergenic region of the phagemid to produce a cleavage product (B). The foreign DNA (C) is cleaved with the same restriction enzyme, producing digestion products (D) with ends compatible with the digested replicative form. With the aid of ligase enzymes, the compatible digested foreign DNA fragments and the cleaved replicative form are joined. The resultant phagemid, with foreign DNA inserted, may be propagated as a plasmid in an appropriate bacterium with antibiotic resistance as the selection mechanism (F). This process yields double-stranded molecules of amplified DNA, including the foreign insert. If a phagemid-containing cell population is exposed to a single-stranded filamentous phage, the phagemid DNA replication mode is changed. The plus strand is nicked and new DNA is formed by a rolling circle mechanism with addition to the 3′ end of the nicked strand (G). This process produces single-stranded copies of the DNA, including the foreign insert.

Phagemids

Phagemids are plasmids that contain an origin of replication, an antibiotic resistance marker, and a copy of the intergenic region of a filamentous phage. This region codes for all of the processes required for filamentous phage replication. In the absence of an appropriate filamentous phage, these entities replicate as plasmids, allowing replication of foreign DNA in the process of normal plasmid replication. When cells containing plasmids of this type are exposed to an appropriate filamentous phage, the mode of replication of their DNA

changes. Under the influence of the gene II product of the phage, a nick is introduced in the double-stranded phagemid DNA molecule. Subsequent DNA replication produces single-stranded copies of the phagemid DNA, which may be isolated and are particularly useful for sequencing. These processes are outlined in Figure 20.12.

Single-stranded DNA may also be obtained by use of a filamentous phage directly. The unique properties of filamentous phage DNA replication are shown in Figure 20.13. The initially single-stranded phage containing the plus (coding) strand

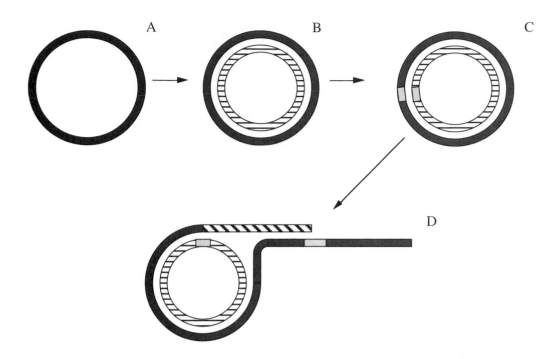

FIGURE 20.13 Direct single-stranded DNA production with filamentous phage. The single-stranded form of the phage (A) becomes the double-stranded replicative form (B) during normal phage replication. The replicative form is isolated and digested as depicted in Figure 20.12. Foreign DNA is also digested, allowing introduction of foreign DNA into the replicative form (C). The latter is used to transform appropriate cells so that viral replication may continue. In the process of phage replication, the foreign DNA is also amplified. When this strategy is employed, only single DNA strands are obtained (D), whereas the procedures outlined in Figure 20.12 allow formation of either single- or double-stranded DNA, depending upon the conditions used.

invades a susceptible cell. Under the influence of host enzymes, the plus strand is converted to a double-stranded replicative form containing both a plus and minus form. After a period of time, the viral gene II product nicks the plus strand of the replicative form, providing a 3′ end upon which new plus strands are synthesized by the rolling circle mechanism with the minus strand as template. The single-stranded plus strand is subsequently incorporated into new virus particles. The minus strand of the replicative form is transcribed to give mRNA molecules for the virus proteins.

To exploit the replication system for single-stranded phage replication to form single-stranded copies of foreign DNA, a problem must be circumvented. Single-stranded DNA is resistant to the action of restriction endonucleases that are required to digest foreign DNA and are normally used, with the aid of ligases, to insert foreign DNA specifically into vectors. Since single-stranded DNA molecules are resistant to endonucleases, an unusual method must be found to insert foreign DNA into single-stranded phages. The difficulty is overcome by isolation of double-stranded molecules of the replicative form (RF) during replication of phage-infected cells, digestion of the double-stranded molecules, and insertion of foreign DNA. Transformation of cells with the modified DNA produces phagemids that contain double-stranded foreign DNA. When these phagemids are exposed to an appropriate single-stranded filamentous phage and virus replication is allowed to proceed, phage particles will eventually be formed that contain only the plus strands of foreign DNA.

The polymerase chain reaction

Recently, DNA amplification has been achieved by the polymerase chain reaction (PCR). Because of its simplicity and rapidity, it is likely that PCR technology will be increasingly used. The elements of PCR technology are shown in Figure 20.14. DNA

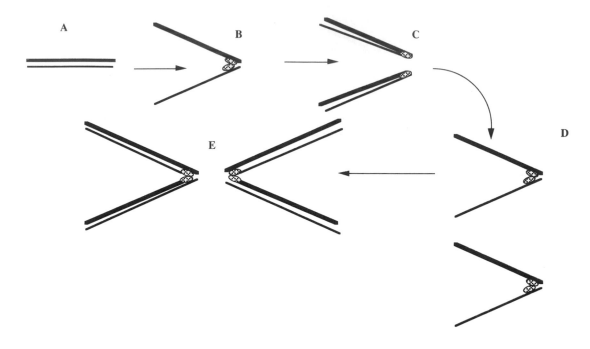

FIGURE 20.14 DNA amplification with polymerase chain reaction (PCR) technology. Strands of DNA (A) are separated by heating to allow primers to attach and then cooled to a temperature between 37° C and 55° C, so that primers may anneal. (B) The annealed strands are then reheated to 72° C, so that replication with a heat stable polymerase may occur (C). The replicated strands are cooled (D) to allow primer attachment and annealing a second time, and reheated to allow replication of the separated strands a second time (E). These processes are repeated over and over again, doubling the amount of DNA at each step, because the replicated strands serve as surfaces for new primer attachment and as templates for further replication. Normally, the primers for PCR are known oligomeric sequences that flank the DNA fragment whose amplification is desired.

obtained from a particular source is repeatedly heated and cooled in the presence of a heat-stable, DNA-dependent DNA polymerase and an excess of oligoribonucleotide primers, that can be made specific. The heating allows periodic separation of double-stranded DNA molecules so that primers may attach and the subsequent cooling allows the attached primers to anneal. The polymerase then attaches to primers and synthesizes DNA. Alternation between heating and cooling allows alternate primer attachment and annealing and DNA synthesis. Such a system is extremely simple in principle, and in skilled hands, can yield large amounts of DNA from miniscule amounts of starting material. Although it is simple in principle, PCR technology is not always practicable. Unless one can be assured that the initial DNA is extremely pure, it is inadvertently possible to produce substantial

amounts of DNA other than the desired one. In spite of this potential difficulty, the rapidity and simplicity of the technique are such that it will undoubtedly be increasingly used.

Expression

Expression, the production of the protein coded for by a gene, is the final major aspect of genetic engineering. Expression may be achieved in many systems. As we have previously seen, phage lambda may serve as a vehicle for DNA production, protein production, or both, depending upon how the particular lambda strain is constructed. To serve as an expression vector, the vector must contain the gene for the desired protein, a suitable promoter so that the gene may be transcribed, a sequence allowing the transcribed gene to attach

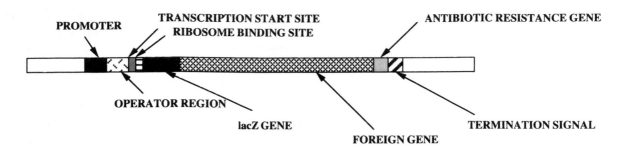

FIGURE 20.15 A model expression vector. The vector contains a promoter region and an operator region, a transcription start site, and a ribosome binding site. The initial structural region contains a portion of a gene for a protein that is under the control of the promoter. Often, the gene is the β-galactosidease gene of *E. coli (LacZ)*. Such an arrangement allows relatively easy control of the system. The vector is normally arranged so that the foreign gene can be inserted in any of the three reading frames. It is also inserted so that it is in harmony with the *lacZ* gene, allowing the entire system to be activated by an inducer for the lac operon. By genetic engineering techniques, the structure shown in the figure is incorporated into either a plasmid or a phage which is used to transfect or transform an appropriate cell system so that protein may be formed. The antibiotic resistance gene is used as a selection agent. Detection of transformation or transfection in this system normally depends on complementation between portions of the *lacZ* gene carried partly by the vector and partly by the host cell. Cells containing the recombinant vector appear white because β-galactosidease is inactivated by insertion of foreign DNA into the vector. Cells devoid of the recombinant vector appear blue. Protein formation is induced by exposure of the cells to an inducer of the *lacZ* gene, causing formation of a fusion protein containing a portion of the β-galactosidease protein, in addition to the protein of interest. Protein production is normally detected by antibodies to the foreign protein. When the protein is identified, it may be purified.

to ribosomes, and a mechanism that allows induction of the system. In addition, the physical organization of the vector must be such that it is functional. The cloned gene must be physically close enough to the promoter so that it can be expressed and the gene must also be in proper orientation. It must be situated in the proper direction relative to its promoter and must also be arranged so that it is in the proper reading frame. If this is not the case, the protein will be made improperly or not at all. Finally, in a good expression vector, a mechanism must exist for termination of transcription at the appropriate point so that the desired gene product, and only that product, is obtained. If all of the above features are present, the precise vector used is of relatively little consequence. The vector is typically a phage or a plasmid. It is seldom, if ever, the case that *perfect* expression vectors are constructed. It is often necessary to manipulate a system so that optimum expression may be obtained or to process the original product in some way so that the desired material is obtained. A model expression vector is shown in Figure 20.15.

Detection of protein formation

Detection of protein production is critical to assessment of the success of genetic engineering. Detection is normally accomplished in one of two ways, either by reaction with an antibody to the protein of interest or by staining procedures. Antibodies may be obtained either with purified protein or, in some cases, with a *fusion protein* composed of the desired protein plus an attached portion of an additional protein produced by the expression system.

In many cases, detection of an unusual protein is accomplished with staining, normally with Coomassie brilliant blue or with Ag^+ ions. Whatever detection system is used, care must be taken to ensure that the protein observed is truly produced by the expression system. Comparisons between the expression system and the system that most closely approximates it are instructional. For example, study of the behavior of the cells containing a genetically engineered plasmid may be compared to the behavior of cells containing the same plasmid but devoid of the added DNA. The novel protein should be formed in the expression system and should not be formed, or at least be formed much less abundantly, in the nongenetically engineered system. In addition, conditions required for induction should preferentially affect the genetically engineered system, and with the genetically engineered organism, the amount of protein formed should be a function of time. If all these criteria are met, it can be reasonably assured that formation of the protein in question is a function of the expression system.

Purification

Purification is the final step in production of a genetically engineered protein. Purification is an intrinsically difficult process. Precisely how we proceed depends on many factors. Normally, a variety of techniques are required. It is often the case that substantial purification may be achieved by selective precipitation, but it is extremely rare that this process is sufficient to allow enough purification so that the product may be useful. Additional techniques may range from column chromatography and gel filtration to affinity chromatography and electrophoresis.

Physiological implications of molecular techniques

The various techniques for gene and protein production by genetic manipulation depend, with rare exceptions, on an intimate cooperation between a host cell and an auxiliary entity, such as a bacteriophage, a plasmid, or some entity that combines the properties of both plasmid and phage. Through application of genetic techniques, particularly through the use of plasmids, we may confer altered properties on the host cell, for example, antibiotic resistance or the ability to form a nonhost protein. Conversely, host factors profoundly influence the success with which genetic techniques may be accomplished. Artificial transformation, for example, depends upon the ability to render the host cell permeable to the exogenously supplied, altered genetic material. In addition, the ability of hosts to facilitate production of proteins

whose genes are coded for by nonhost elements may reflect factors such as the ability of host enzymes to recognize nonhost promoter sequences or replication origins for DNA. In addition to these factors, a plasmid may contain regulatory elements that control its own replication in relationship to that of its host cell or factors that mitigate against host cell destruction by virus invasion. The agents of genetic manipulation live in intimate and complex associations with the cells that they invade and it is only by careful study of those associations that we are able to devise, and use, phages, plasmids, and the like for genetic purposes. Together, the agents of genetic engineering offer us opportunities for genetic manipulation and for study of the molecular intricacies of biology.

Applications of molecular biology

We are only beginning to realize the potential applications of molecular biology. We may, initially, distinguish between the use of genetically engineered *products* and genetic *techniques* on one hand, and on the other hand, the use of genetically engineered *organisms*. At the moment, we find more use for products and techniques than we do for the organisms. Genetically engineered products range from hormones and enzymes to therapeutic substances useful in the diagnosis and treatment of disease to agents for the detection and study of as yet uncultivated organisms. Genetic techniques are currently used for processes as diverse as the production of useful products and the solution of crimes. The possibilities are apparently endless.

The use of genetically engineered organisms is, at present, less advanced than is the use of genetically engineered products and techniques. To a large extent this is because the use of organisms is intrinsically more complex than that of products. We can readily envision use of genetically altered microbes, primarily microbes that have acquired genetically engineered plasmids, as agents for enhancement of agricultural productivity, for bioremediation, and for industrial purposes, among other possibilities. Our current, relatively restricted knowledge of the genetic properties of environmentally significant organisms limits our ability to exploit them efficiently for such purposes. Much more information about both the genetics and the physiology of genetically engineered organisms and about the consequences of genetic change is required before we may intelligently use genetically engineered microbes for industrial and environmental purposes. It is likely, if not inevitable, that increasing knowledge of the genetic and physiological properties of as yet unstudied microbes of environmental significance will be as instructional, both for genetics and for microbial physiology, as has thus far been the case for *E. coli* and *S. typhimurium*. The possibilities for use of a variety of genetically altered microbes for the benefit of humankind are tremendous, but so are the challenges.

Ethical questions posed by molecular biology

The potential implications of molecular biology, in the broadest sense of the word, go far beyond the microbial world. This fact was recognized many years ago when a group of distinguished scientists met at Asilomar, California to discuss possible difficulties with molecular techniques and to establish guidelines for their use. Public concern regarding the possible "creation of a monster" was a major impetus for the meeting. Although it was concluded that with proper precautions, molecular techniques are both safe and appropriate, and that the potential benefits of genetic engineering far outweighed its potential dangers, a number of ethical and philosophical questions became apparent. These questions, although not scientific, are complex and continue to be of concern. Initially, we may reflect upon whether or not it is appropriate to deliberately alter genetic systems that have been selected for, by evolution, over the eons. A second different, but equally intriguing, question concerns whether or not we are, by genetic engineering, "creating new life." Additionally, assuming that we have created a novel life form, to whom does it belong? Is a genetically engineered organism the property of its inventors, their employers, their state, their nation, their world? Particularly because of the fact that genetically engineered organisms and their products have substantial eco-

nomic consequences, such questions are not idle. Finally, we may ponder the extent to which we should employ genetic techniques with large life forms, even humankind. In such a situation, who should decide whether a particular technology should be used, by whom, for what purpose, or to what extent? Molecular biology, microbial and otherwise, has many implications.

Summary

Molecular biology, in the broadest sense of the phrase, is a complicated and challenging subject with many implications. At the microbial level, we must understand not only the techniques and mechanisms by which we may genetically alter organisms, but also the physiological consequences of such alterations. This necessity requires that we become molecular physiologists and must understand all critical cell processes at the molecular level.

Careful examination of the nature and interrelationships between genetic and phenotypic changes is useful and essential if we would fully exploit the genetic properties of microbes, both as laboratory entities and in nature. If we would use genetically altered microbes in nonlaboratory environments, we must understand clearly the effects of genetic change on the physiological response of the organism in nature.

At present, protein and nucleic acid production are major ways in which microbes are exploited by the techniques of genetic engineering. Although a variety of techniques are available to accomplish the component processes of genetically engineered protein production, the components of the process are the same: 1) identification and isolation of a gene; 2) cloning of the gene; 3) insertion of the gene into an expression system; 4) formation of the protein; and 5) purification of the product. By whatever means we accomplish the tasks, all of the tasks are required for successful protein formation by genetic methods.

The processes of genetic engineering at the microbial level involve, primarily, exploitation of the properties of bacteriophages and plasmids. Intimate understanding of the nature of these entities, and of the relationship between viruses, plasmids, and host cells is required if we are to accomplish the tasks of genetic engineering effectively and safely. The information is as useful and necessary for understanding basic biological processes as it is for practical purposes.

At the present time, the physiology of microbes is molecularly exploited, primarily by genetically engineered nucleic acid and protein production and by the application of techniques developed for microbial purposes for other uses. Our ability to use genetically altered microbes in nonlaboratory situations is currently limited by the shortcomings in our knowledge, of both the physiology and the genetics of potentially useful organisms, but as our knowledge increases, so will our ability to exploit genetically altered microbes for medical, industrial, agricultural, and ecological purposes.

Questions posed by the advent of molecular biology far transcend the discipline and are a cause for deep reflection regarding both the uses of such techniques and the limits that we may impose upon their use.

Selected References

Sambrook, J., E. F. Fritsch, and T. Maniatis. 1989. *Molecular Cloning—A Laboratory Manual.* Vols. I, II, and III. Cold Spring Harbor Laboratory Press, New York.

Watson, J. D., N. M. Hopkins, J. W. Roberts, J. A. Seitz, and A. M. Weimer. 1987. *Molecular Biology of the Gene.* 4th edition. Vols. I and II. Benjamin Cummings, Menlo Park, California.

Index